Thermal Radiation

Thermal radiation studies have progressed rapidly, not only in theoretical and experimental exploration beyond the conventional use but also in advanced applications. This is a one-stop resource for capturing and discussing these cutting-edge developments, exploring the theory, experiments, and applications of thermal radiation from macro- to nanoscale. Presented in a systematic framework, this book is divided into two parts: the first on macroscopic thermal radiation and the second on micro- and nanoscale thermal radiation. Each part delivers basic theory, numerical methods, advanced experimental techniques, and promising applications, making this an easy-to-follow guide meeting both basic and advanced needs. Supported by more than 180 colorful illustrations, readers can clearly visualize the theory, experiments, and applications in practice.

A book for all, written at a graduate level but undoubtedly a useful tool for researchers, professionals, and even engineers who are interested in this fast-developing area.

Changying Zhao directs a thermal laboratory and has been conducting research on micro/nano thermal radiation, porous media heat transfer, and thermal energy storage for over 30 years. He has published more than 300 high-quality papers in top-ranking journals such as *Nature Materials*, *Physical Review Letters*, *Nano Letters*, *Advanced Materials*, and *International Journal of Heat and Mass Transfer*. Professor Zhao was awarded the William Begell Medal in 2023 for Excellence in Thermal Science and Engineering on the basis of his lifetime achievements.

Thermal Radiation

From Macro to Nano

CHANGYING ZHAO
Shanghai Jiao Tong University

CAMBRIDGE
UNIVERSITY PRESS

Shaftesbury Road, Cambridge CB2 8EA, United Kingdom

One Liberty Plaza, 20th Floor, New York, NY 10006, USA

477 Williamstown Road, Port Melbourne, VIC 3207, Australia

314–321, 3rd Floor, Plot 3, Splendor Forum, Jasola District Centre,
New Delhi – 110025, India

103 Penang Road, #05–06/07, Visioncrest Commercial, Singapore 238467

Cambridge University Press is part of Cambridge University Press & Assessment,
a department of the University of Cambridge.

We share the University's mission to contribute to society through the pursuit of
education, learning and research at the highest international levels of excellence.

www.cambridge.org
Information on this title: www.cambridge.org/9781316516652

DOI: 10.1017/9781009030793

When citing this work, please include a reference to the DOI 10.1017/9781009030793

First published 2024

A catalogue record for this publication is available from the British Library

Library of Congress Cataloging-in-Publication Data
Names: Zhao, Changying, 1969– author.
Title: Thermal radiation : from macro to nano / Changying Zhao.
Description: First published 2024. | New York, NY : Cambridge University
Press, 2024. | Includes bibliographical references and index.
Identifiers: LCCN 2023054390 | ISBN 9781316516652 (hardback) |
ISBN 9781009030793 (ebook)
Subjects: LCSH: Heat – Radiation and absorption. | Materials – Thermal
properties.
Classification: LCC QC331 .Z43 2024 | DDC 536/.3–dc23/eng/20240521
LC record available at https://lccn.loc.gov/2023054390

ISBN 978-1-316-51665-2 Hardback

Contents

Preface

Thermal radiation is a ubiquitous aspect of nature, and any object with a finite temperature is capable of emitting thermal radiation. What is thermal radiation? Different people with different backgrounds may give different answers to this question. For people who design boilers and heaters, thermal radiation refers to light rays from high-temperature bodies. For people who are interested in microscopic phenomena, thermal radiation is an electromagnetic (EM) wave generated by thermal motions of charges in the matter. For many physicists, the quantum origin of thermal radiation is more interesting, and it can be understood as the fluctuations of quantized energy levels inside the matter. It would be instructive to discuss the profound implications of thermal radiation systematically in a single book.

The writing of this book was inspired by the rapid development of study on micro- and nanoscale thermal radiation (MNTR), which has made great progress not only in theoretical and experimental exploration but also in many applications associated with energy harvesting and conversion, radiative engineering for cooling, camouflage, imaging, and so on. In order to make the book accessible to readers who may not have any background in thermal radiation, we present a comprehensive introduction to various aspects of thermal radiation, including theory, numerical methods, and experimental techniques, in a coherent fashion starting from the macroscopic description of thermal radiation heat transfer to microscopic treatment in the form of EM waves.

This book is divided into two parts. The first part includes Chapters 1–5, which deals with macroscopic thermal radiation, mainly based on geometric optics considerations. Chapter 1 introduces the fundamental concepts of macroscopic thermal radiation. The radiative transfer equation (RTE) and numerical methods for solving it are discussed in Chapters 2 and 3. Then Chapter 4 gives an overview of experimental techniques for measuring macroscopic thermal radiative properties. In Chapter 5, several typical applications at macroscopic scale are introduced. The second part is dedicated to MNTR, covering Chapters 6–11. In Chapter 6, an overview of MNTR is presented. Chapter 7 gives a comprehensive introduction of the theoretical foundations of MNTR. Then comes Chapter 8, which aims to describe some commonly used numerical methods at micro- and nanoscales. Chapter 9 contains experimental techniques for MNTR. In Chapter 10, a variety of methods to tailor thermal radiative properties at micro- and nanoscales are introduced. In Chapter 11, important applications of MNTR are discussed.

This book would not have been possible without my graduate students' hard work and dedication. They have provided immense help to me by proofreading and providing illustrations, and some of the materials in this book are based on their research. Special thanks are extended to them, including Boxiang Wang, Mengqi Liu, Wenbin Zhang, Jie Chen, Xujing Liu, Shenghao Jin, Fan Yi, Jiahao Zhou, and Zhen Gong. I would also like to emphasize that the selection of topics in this book is somewhat biased by our views and interests in this field, and thus many important topics are omitted. These include the ray tracing and Distributions of Ratios of Energy Scattered or Reflected methods for solving RTE, far-field emissivity measurement at high temperatures, concentrating solar power plants, the Green function method for near-field thermal radiation, among many others. Since there is a huge amount of literature in this field, we also apologize to the researchers whose work we may not have appropriately cited.

Part I

General Principles of Thermal Radiation and Applications

1 Fundamentals of Thermal Radiation

The terms *radiative heat transfer* and *thermal radiation* are commonly used to describe the science of heat transfer caused by electromagnetic (EM) waves. The main goal of this chapter is to introduce the nature of thermal radiation, the fundamental laws of thermal radiation, and the methods for computing radiative heat exchange between two or more surfaces.

1.1 Basic Characteristics of Thermal Radiation

1.1.1 The Nature of Thermal Radiation

Radiation is one of the fundamental modes of heat transfer, and the research on the mechanism and nature of radiation is still ongoing. Our current understanding of radiation is based on classical EM theory and quantum physics. In 1865, Maxwell published the complete equations of EM waves, believing that light is a form of EM radiation [1]. Once the energy is radiated, it propagates as an EM wave, regardless of whether there is a vacuum or matter along its path.

In 1905, Einstein built on the idea of quantization of radiation proposed by Planck, who believed that light is a stream of energy quanta moving at the speed of light [2]. This energy quantum is called a photon whose energy is proportional to its frequency. Radiation is the energy-transfer process by which an object emits photons to the outside. Later, Einstein further pointed out that photons have wave–particle duality [3]. From the relationship between the frequency of photon energy and the EM wavelength, we can simply glimpse the relationship between wave and particle properties.

In general, the energy properties of radiation are explained by Einstein's light-quantum hypothesis, while the propagation properties are explained by Maxwell's EM field theory. This book mainly focuses on the conversion and transfer of radiative energy, so the basic properties of thermal radiation will be explored under the guidance of EM field theory. The core content of EM field theory is the famous Maxwell's equations. These consist of Gauss's laws for the electric field and the magnetic field, Faraday's law of EM effect, and the Ampere–Maxwell law. No doubt, Maxwell's equations are as sacred in the field of electrodynamics as Newton's second law is in classical physics.

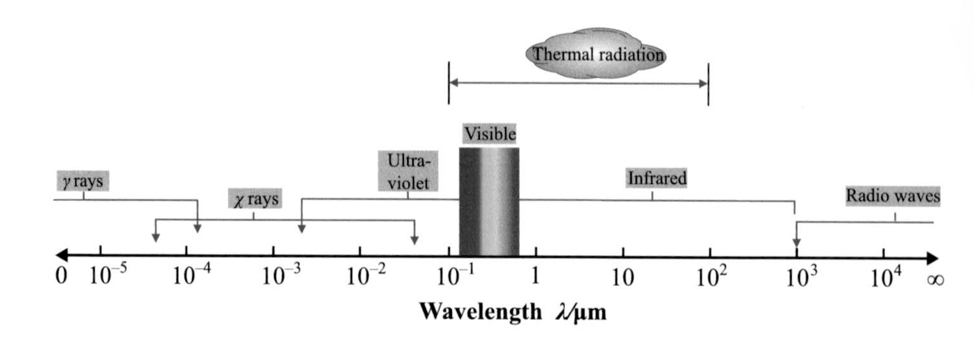

Figure 1.1 The spectrum of EM waves.

According to the EM field theory, the electric field and the magnetic field are interrelated and mutually excited to form a unified EM field. The EM field propagates in vacuum at the speed of light, which is called the EM wave. The wavelength range of EM waves is very wide, covering cosmic rays with wavelengths of less than 10^{-9} m to radio waves with wavelengths of several hundred meters. Figure 1.1 shows the wavelength distribution of various EM waves.

The EM wave generated by thermal motion is called thermal radiation, whose wavelength is in the range of 0.1–100 μm, mainly including the visible region and most of the infrared region. In vacuum, the wavelength of visible light is 0.38–0.76 μm and that of infrared light is 0.76–1000 μm. In the range of temperatures in industry (i.e., below 2000 K), the radiation wavelengths are between 0.8 and 100 μm. The sun is a heat source with a surface temperature of about 5800 K, and the energy of solar radiation is concentrated in the wavelength range of 0.2–2 μm. As long as the temperature of an object is higher than absolute zero (0 K), it always emits continuous thermal radiation outward. At the same time, the object also constantly absorbs the incident thermal radiation on its surface from the surrounding environment and converts the absorbed radiation energy into heat energy. When it is in thermal equilibrium with the surrounding environment, the thermal radiation on its surface is still evolving, but its net radiative heat transfer is equal to zero.

1.1.2 The Effect of Surfaces on Radiation

When the total radiation energy Q from the outside strikes a surface, part of the energy is reflected by the surface (Q_ρ), part is absorbed (Q_α), and part is transmitted (Q_τ) as shown in Fig. 1.2. According to the law of conservation of energy,

$$Q = Q_\rho + Q_\alpha + Q_\tau. \tag{1.1.1}$$

Dividing both sides of this equation by Q, we can obtain

$$1 = \frac{Q_\rho}{Q} + \frac{Q_\alpha}{Q} + \frac{Q_\tau}{Q}. \tag{1.1.2}$$

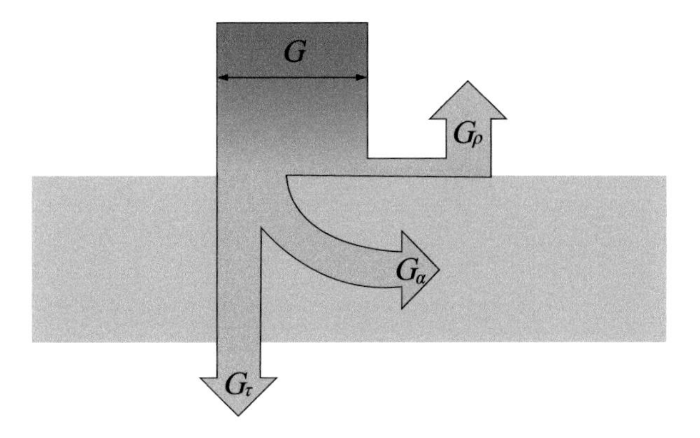

Figure 1.2 Reflection, absorption, and transmission of thermal radiation.

The three parts of energy, Q_ρ/Q, Q_α/Q, and Q_τ/Q, are called reflectivity, absorptivity, and transmissivity of the body, denoted by ρ, α, and τ, respectively. Therefore, Eq. (1.1.2) can be further expressed as

$$\rho + \alpha + \tau = 1. \tag{1.1.3}$$

When the radiation energy is incident on a solid or liquid surface, its absorption occurs only at a very thin layer of the surface owing to tightly arranged molecules. For metal conductors, this thickness is of the order of 1 μm; for most nonconductive materials, this thickness is usually less than 1 mm. Therefore, it can be considered that neither solid nor liquid is allowed to penetrate thermal radiation, that is, $\tau = 0$. Thus, Eq. (1.1.3) can be simplified as

$$\alpha + \rho = 1. \tag{1.1.4}$$

In addition, since thermal radiation cannot penetrate through thick solids and liquids, the absorption of radiation energy takes place only over a very thin surface. In the same way, their radiation should occur at the thin layer of the surface. Therefore, the thermal radiation of solids and liquids is a surface process, which makes the calculation of radiation heat transfer easier. Like visible light, the reflection phenomenon of radiation is also divided into specular reflection and diffuse reflection, which depends on the size of the irregularity of the surface of the object (i.e., the surface roughness) and the magnitude of the wavelength of input radiation. When the wavelength of the input radiation is larger than the irregularity of the object surface, the reflection follows the law of geometric optics and forms specular reflection, as shown in Fig. 1.3. The reflection angle is equal to the incident angle. By contrast, when the wavelength of the input radiation is smaller than the irregularities of the object surface, as shown in Fig. 1.4, diffuse reflection is formed.

Gas has little ability to reflect radiation energy, so the reflectivity can be considered to be 0 ($\rho = 0$). Therefore, Eq. (1.1.3) can be simplified as

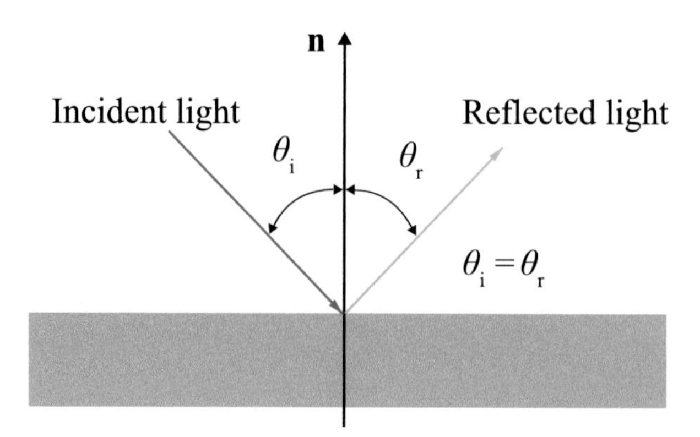

Figure 1.3 Specular reflection.

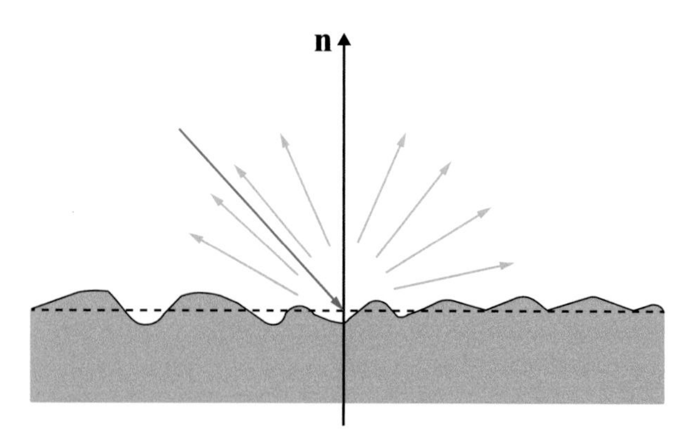

Figure 1.4 Diffuse reflection.

$$\alpha + \tau = 1. \tag{1.1.5}$$

The above discussion shows that the absorption and transmission of thermal radiation by gas characterize a volumetric process.

1.1.3 Blackbody Model

The radiation properties of real objects are usually very complex. Therefore, some ideal physical models are abstracted in the study of thermal radiation. An object with the absorption rate of $\alpha = 1$ is called an absolute blackbody [4], which means that it can absorb radiation of all wavelengths. As shown in Fig. 1.5, it is a typical blackbody model that is composed of an opening surface of an isothermal cavity. After repeated absorption and reflection, the incident radiation entering the cavity can finally leave the hole with very little radiation energy. Therefore, the opening surface of the isothermal cavity can be regarded as a surface that

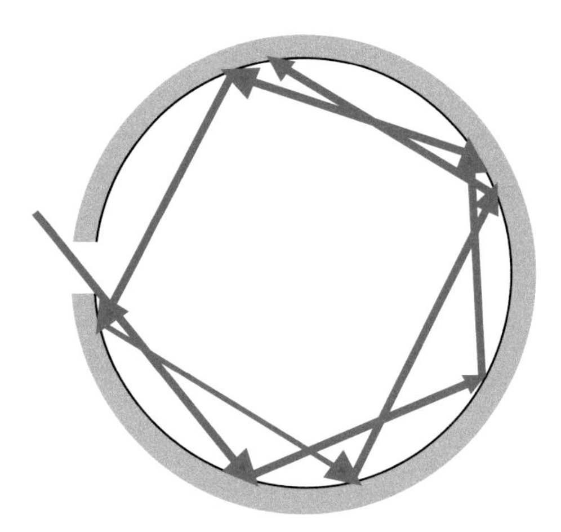

Figure 1.5 Blackbody model.

completely absorbs thermal radiation, namely, an artificial blackbody. And we cannot infer the absorptive capacity of an object to the projection of full-band radiation energy simply by its color. A black object is not necessarily a blackbody.

1.2 Basic Laws of Blackbody Thermal Radiation

Through the introduction of the basic characteristics of thermal radiation, the qualitative understanding of thermal radiation is concluded: thermal radiation is directly related to temperature, and it has spectral characteristics. In addition, the transmission of radiation energy has a certain directivity. Therefore, this section will continue to study the above characteristics of thermal radiation quantitatively, that is, systematically focus on the basic laws of thermal radiation, which respectively reveal the amount of energy radiated from a unit blackbody surface to the outside at a certain temperature from different angles and its distribution law with space direction and wavelength.

1.2.1 Hemispherical Emissive Power and Spectral Emissive Power

In order to quantitatively describe the laws of thermal radiation, the following concepts need to be introduced from the aspects of space geometric properties and energy properties.

Hemispherical Emissive Power

The hemispherical emissive power, $E(\mathrm{W \cdot m^{-2}})$, is defined as the rate at which radiation is emitted per unit area at all possible wavelengths and in all possible directions of the hemispherical space.

Hemispherical Spectral Emissive Power

The hemispherical spectral emissive power, $E_\lambda(\mathrm{W\cdot m^{-2}\cdot \mu m^{-1}})$, is defined as the rate at which radiation of wavelength λ is emitted per unit surface area with per unit wavelength interval $\mathrm{d}\lambda$ and in all possible directions of the hemispheric space.

Obviously, the relationship between the hemispherical emissive power and the hemispherical spectral emissive power is as follows:

$$E=\int_0^\infty E_\lambda \mathrm{d}\lambda. \tag{1.2.1}$$

Solid Angle

The solid angle, Ω (sr), is defined by a small conical region between the rays of a sphere, and it is measured as the ratio of the area $\mathrm{d}A_c$ on the sphere to the square of the sphere's radius. Accordingly,

$$\Omega=\frac{A_\mathrm{c}}{r^2},\ \mathrm{d}\Omega=\frac{\mathrm{d}A_\mathrm{c}}{r^2}. \tag{1.2.2}$$

In the spherical coordinate system of Fig. 1.6, φ is called the azimuthal angle, θ is called the zenith angle [5], and from Fig. 1.6, we can conclude that

$$\mathrm{d}A_\mathrm{c}=r\ \mathrm{d}\theta\cdot r\sin\theta\mathrm{d}\varphi. \tag{1.2.3}$$

Rearranging Eq. (1.2.3), it follows that

$$\mathrm{d}\Omega=\sin\theta\mathrm{d}\theta\mathrm{d}\varphi. \tag{1.2.4}$$

Directional Radiation Intensity

Directional radiation intensity, $I(\ \mathrm{W\cdot m^{-2}\cdot sr^{-1}})$, is defined as the rate at which radiation energy is emitted at all wavelengths in a direction per unit area of the emitting surface normal to this direction and per unit solid angle about this direction.

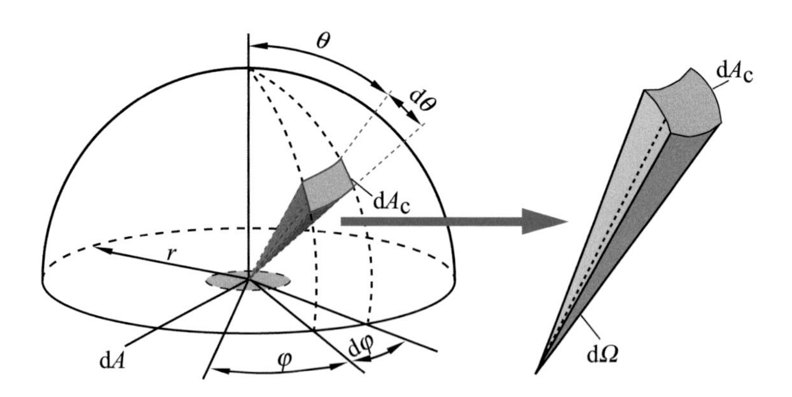

Figure 1.6 The solid angle subtended by $\mathrm{d}A_c$ at a point on $\mathrm{d}A$ in the spherical coordinate system.

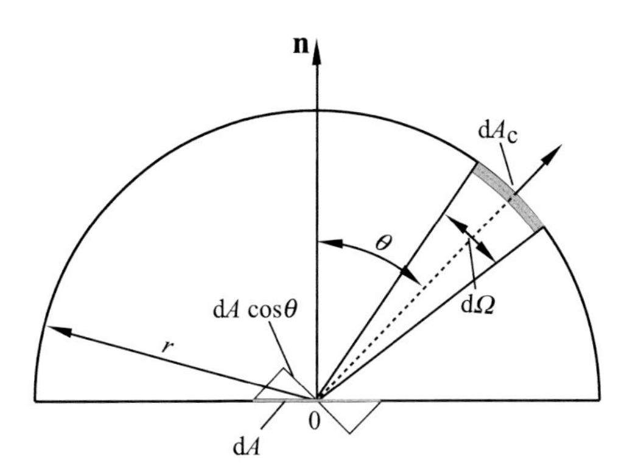

Figure 1.7 The projection of dA normal to the direction of radiation.

And d$\varphi(\theta)$ means the energy emitted from the unit area of the blackbody to the solid angle of the element around the latitude angle of space, and then the experiment shows that

$$\frac{\mathrm{d}\varphi(\theta)}{\mathrm{d}A\ \mathrm{d}\Omega} = I\cos\theta. \tag{1.2.5}$$

That is,

$$I = \frac{\mathrm{d}\varphi(\theta)}{\mathrm{d}A\ \mathrm{d}\Omega\cos\theta}, \tag{1.2.6}$$

where I is a constant, independent of direction, and $\mathrm{d}A\cos\theta$ is the normal area in the direction θ (Fig. 1.7).

1.2.2 Planck's Law

Planck's law reveals how the blackbody spectral emissive power varies with wavelength in thermodynamic equilibrium. It is

$$E_{\mathrm{b}\lambda} = \frac{c_1\lambda^{-5}}{\exp\left[c_2/(\lambda T)\right]-1}, \tag{1.2.7}$$

where $E_{\mathrm{b}\lambda}$ is the blackbody spectral emissive power, $\mathrm{W\cdot m^{-3}}$; λ is the wavelength, m; T is the absolute temperature of the blackbody, K; c_1 is the first radiation constant, $3.7419\times10^{-16}\ \mathrm{W\cdot m^2}$; and c_2 is the second radiation constant, $1.4388\times\ 10^{-2}\ \mathrm{m\cdot K}$

The blackbody spectral emissive power distribution is plotted in Fig. 1.8, from which we see that the blackbody has a maximum spectral emissive power and that the corresponding wavelength λ_{m} depends on temperature:

$$\lambda_{\mathrm{m}}T = 2.8976\times10^{-3}\ \mathrm{m\cdot K}\approx 2.9\times10^{-3}\ \mathrm{m\cdot K}. \tag{1.2.8}$$

Equation (1.2.8) is known as Wien's displacement law, and the blackbody temperature can be calculated according to the spectrum of the blackbody. It

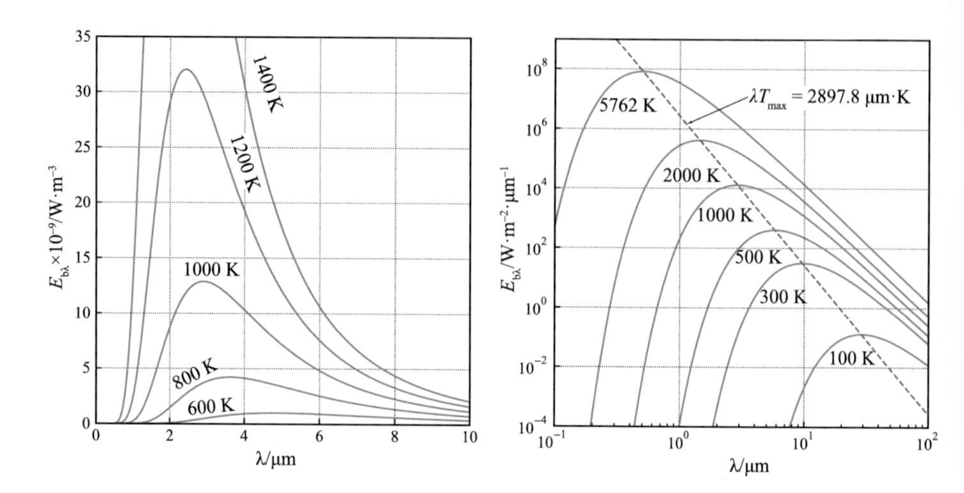

Figure 1.8 Spectral blackbody emissive power.

is concluded completely based on the empirical summary of experimental data, but it can be deduced mathematically from the Planck distribution [6]:

$$\frac{\partial E_{b\lambda}}{\partial \lambda} = \frac{5c_1\lambda^{-6}}{\exp[c_2/(\lambda T)]-1}\left\{\frac{c_2\exp[c_2/(\lambda T)]}{5\lambda T\{\exp[c_2/(\lambda T)]-1\}}-1\right\}=0. \tag{1.2.9}$$

We set $x=c_2/(\lambda_m T)\,x=c_2/(\lambda_m T)$; rearranging Eq. (1.2.9), it follows that

$$\frac{x\exp x}{5(\exp x-1)}-1=0. \tag{1.2.10}$$

Equation (1.2.10) is the transcendental equation of the variable x, and the solution is as follows:

$$x=c_2/(\lambda_m T)=4.9651. \tag{1.2.11}$$

Hence,

$$\lambda_m T=c_2/4.9651=2.8976\times10^{-3}\ \text{m}\cdot\text{K}. \tag{1.2.12}$$

1.2.3 Stefan–Boltzmann Law

The Stefan–Boltzmann law points out that the blackbody emissive power is proportional to the fourth power of the blackbody's temperature,

$$E_b=\sigma T^4, \tag{1.2.13}$$

where σ is the Stefan–Boltzmann constant, $5.67\times10^{-8}\ \text{W}\cdot(\text{m}^2\cdot\text{K}^4)^{-1}$. Equation (1.2.13) can be deduced from Eq. (1.2.7) as follows:

$$E_{\mathrm{b}} = \int_0^\infty E_{\mathrm{b}\lambda} \mathrm{d}\lambda = \int_0^\infty \frac{c_1 \lambda^{-5}}{\exp\left[c_2/(\lambda T)\right] - 1} \mathrm{d}\lambda. \tag{1.2.14}$$

We set $x = c_2/(\lambda T)$, it follows that

$$\mathrm{d}\lambda = \frac{-c_2}{Tx^2} \mathrm{d}x, \tag{1.2.15}$$

$$E_{\mathrm{b}} = \frac{c_1}{c_2{}^4} T^4 \int_0^\infty \frac{x^3}{\exp(x) - 1} \mathrm{d}x, \tag{1.2.16}$$

where

$$\begin{aligned} \int_0^\infty \frac{x^3}{\exp x - 1} \mathrm{d}x &= \int_0^\infty x^3 \left[\sum_{n=1}^{\infty} \exp(-nx) \right] \mathrm{d}x \\ &= \sum_{n=1}^{\infty} \int_0^\infty x^3 \exp(-nx) \mathrm{d}x \\ &= \sum_{n=1}^{\infty} \frac{3!}{n^4} = \frac{\pi^4}{15}. \end{aligned} \tag{1.2.17}$$

Hence,

$$E_{\mathrm{b}} = \frac{\pi^4 c_1}{15 c_2{}^4} T^4 = \sigma T^4. \tag{1.2.18}$$

For the blackbody emissive power in a prescribed wavelength interval from 0 to λ, it can be obtained by integrating as follows:

$$E_{\mathrm{b}(0-\lambda)} = \int_0^\lambda E_{\mathrm{b}\lambda} \mathrm{d}\lambda. \tag{1.2.19}$$

The fraction of the emission in this wavelength range is determined as follows:

$$F_{\mathrm{b}(0-\lambda)} = \frac{\int_0^\lambda E_{\mathrm{b}\lambda} \mathrm{d}\lambda}{\sigma T^4} = \int_0^\lambda \frac{c_1 (\lambda T)^{-5}}{\exp\left[c_2/(\lambda T)\right] - 1} \frac{1}{\sigma} \mathrm{d}(\lambda T) = f(\lambda T). \tag{1.2.20}$$

This function is called the blackbody radiation function; its value can be easily obtained according to the given value of λT. Meanwhile, the fraction of the radiation between any two wavelengths λ_1 and λ_2 may also be easily obtained

$$E_{\mathrm{b}(\lambda_1-\lambda_2)} = F_{\mathrm{b}(\lambda_1-\lambda_2)} E_{\mathrm{b}} = \left(F_{\mathrm{b}(0-\lambda_2)} - F_{\mathrm{b}(0-\lambda_1)}\right) E_{\mathrm{b}}. \tag{1.2.21}$$

1.2.4 Lambert Law

The Lambert law tells us that the directional radiation intensity of the blackbody is a constant, independent of the direction. This law also shows that the energy emitted from the unit area of the blackbody varies according to the law of cosine of the latitude angle of space: it reaches its maximum in the direction perpendicular to the surface and zero in the direction parallel to the surface.

The relationship between the Lambert law and the Stefan–Boltzmann law can be derived as follows. Considering Eq. (1.2.6), the blackbody emissive power can be written as

$$E_{\mathrm{b}} = \int_{\Omega=2\pi} \frac{\mathrm{d}\phi(\theta)}{\mathrm{d}A}\,\mathrm{d}\Omega = I_{\mathrm{b}} \int_{\Omega=2\pi} \cos\theta\,\mathrm{d}\Omega = I_{\mathrm{b}} \int_0^{2\pi} \mathrm{d}\varphi \int_0^{\pi/2} \sin\theta\cos\theta\,\mathrm{d}\theta = I_{\mathrm{b}}\pi. \tag{1.2.22}$$

The above equation shows that the blackbody emissive power is π times the directional radiation intensity of the blackbody.

1.2.5 Kirchhoff's Law

Concepts Used to Describe Radiation by the Real Surface

(1) Emissivity

We define the emissivity, ε, as the ratio of the radiation emitted by the surface to the radiation emitted by a blackbody at the same temperature:

$$\varepsilon_T = \frac{E(T)}{E_{\mathrm{b}}(T)}. \tag{1.2.23}$$

(2) Absorptivity

We define the absorptivity, α, as the fraction of the total radiation absorbed by a surface:

$$\alpha = \frac{G_{\mathrm{abs}}}{G}, \tag{1.2.24}$$

where G_{abs} and G represent the absorbed irradiation and incident irradiation, respectively. If the incident radiation originates from an ideal blackbody, then G can be replaced by E_{b}.

The Relationship between Emissivity and Absorptivity in Thermal Equilibrium

Kirchhoff's law reveals the relationship between the emissivity and absorptivity of a real surface. Consider two parallel plates that are very close to each other (Fig. 1.9); all of the radiation energy emitted from one plate is incident on the other. Assume that plate 1 has a blackbody surface, and its emissive power, absorptivity, and surface temperature are E_{b}, α, and T_1, respectively. Plate 2 is the surface of any object, and its emissive power, absorptivity, and surface temperature are E, α, and T_2, respectively. The energy emitted per unit area per unit time by plate 2 is absorbed entirely when it is incident on the surface of plate 1. Meanwhile, the energy emitted from plate 1 is absorbed only when $\alpha E_{/rmb}$ is incident on plate 2, and the rest of the energy $(1-\alpha)E_b$ is reflected back to plate 1 and absorbed entirely. The energy difference of plate 2 is the heat flux of the radiative heat transfer between the two plates:

$$q = E - \alpha E_{\mathrm{b}}. \tag{1.2.25}$$

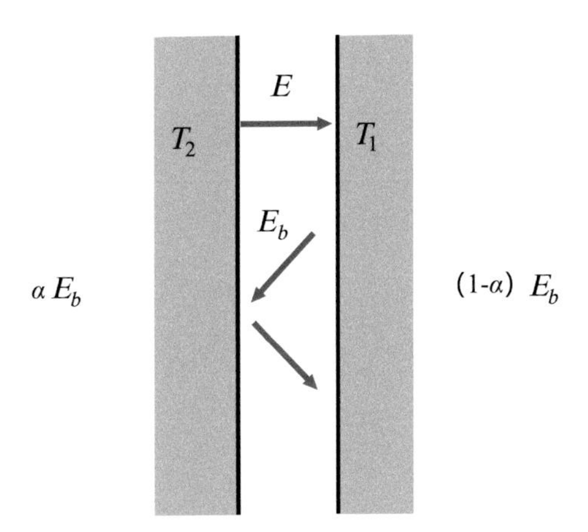

Figure 1.9 Derivation model of Kirchhoff's law.

When the system is at the thermal equilibrium condition, that is, when $q=0$, we obtain

$$\frac{E}{\alpha}=E_{\rm b}. \tag{1.2.26}$$

Extending this relation to any object, the following equations can be written

$$\frac{E_1}{\alpha_1}=\frac{E_2}{\alpha_2}=\cdots=E_{\rm b}, \tag{1.2.27}$$

$$\alpha=\frac{E}{E_{\rm b}}=\varepsilon. \tag{1.2.28}$$

The meaning of Eq. (1.2.27) can be expressed as follows. At the thermal equilibrium condition, the ratio of an object's radiation to its absorption of radiation from a blackbody is the same as the emissive power of the blackbody at the same temperature. Similarly, the meaning of Eq. (1.2.28) can be briefly expressed as: at thermal equilibrium condition, the absorptivity of any object to the blackbody's incident radiation is equal to the emissivity of the object at the same temperature.

The Relationship between the Absorptivity and Emissivity of the Diffuse Gray Surface

In reality, most of the radiation is not emitted from the blackbody; to broaden the application scope of Kirchhoff's law, an assumption called the diffuse gray body, of which the emissivity does not change with direction and absorptivity does not change with wavelength, is introduced.

First of all, assume that a diffuse gray body and a blackbody are at a thermal equilibrium condition, and then the blackbody is removed to allow another nonblackbody to radiate different temperatures that are incident on its surface, whereas the diffuse gray body still keeps its temperature unchanged. Since the

Table 1.1 Three expressions of Kirchhoff's law.

Level	Expression	Constraint condition
Spectral, direction	$\varepsilon(\lambda,\varphi,\theta,T)=\alpha(\lambda,\varphi,\theta,T)$	Unconditional, θ is the latitudinal angle
Spectral, hemispherical	$\varepsilon(\lambda,T)=\alpha(\lambda,T)$	Diffuse surface
Total, hemispherical	$\varepsilon(T)=\alpha(T)$	Should be at a thermal equilibrium condition with blackbody radiation or a diffuse gray surface

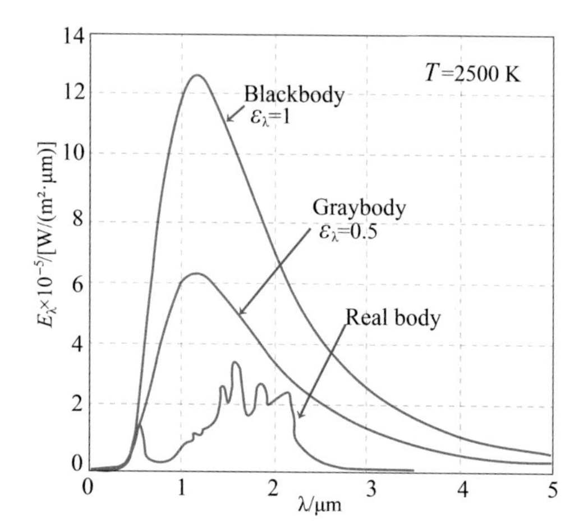

Figure 1.10 Schematic diagram of the spectral emissive power of the gray body and the real surface.

emissivity and absorptivity of the diffuse gray body do not change, the energy emitted by the diffuse gray body at the same temperature should be equal to the energy absorbed, that is, the absorptivity at the same temperature is equal to the emissivity. In conclusion, the total hemispherical absorptivity for a diffuse gray body is equal to its total hemispherical emissivity, regardless of whether the diffuse gray body is at a thermal equilibrium condition with other substances or the environment and regardless of whether other substances are blackbodies. This the conclusion simplifies the calculation of radiative heat transfer and establishes the relationship between the absorptivity and emissivity of the real surface.

According to Kirchhoff's law, the larger the emissive power of an object is, the greater its absorption capacity will be, so a blackbody has the largest emissive power at the same temperature. In addition, Kirchhoff's law can be divided into three levels [7] according to the application conditions, as shown in Table 1.1, and each level corresponds to different constraint conditions. Figure 1.10 qualitatively shows the variation of spectral emissive power with the wavelength of a gray body and the real surface. (For a diffuse gray body at a specific temperature, its spectral emissivity $\varepsilon(\lambda)$ is a constant.)

1.3 Gas Radiation Characteristics and Solar Radiation

1.3.1 Gas Radiation Characteristics

Being different from solid and liquid radiation, gas radiation has the following two characteristics. First, gas radiation is strongly wavelength selective, and gas is not a gray body. The radiating gas can be composed of molecules, atoms, ions, and free electrons with various energy levels. The energy associated with the motion of vibration and the rotation of a molecule has specific quantized values, and hence gas emits and absorbs radiation in discrete energy intervals dictated by the allowed states within the molecule. Gas molecules tend to have radiative and absorptive abilities only within a specific wavelength range. For example, ozone absorbs almost all UV wavelengths of less than 0.3 μm [8], so the ozone layer in the atmosphere protects life on the Earth from UV damage. As a greenhouse gas, carbon dioxide has three main absorption bands: 2.65–2.8, 4.15–4.45, and 13.0–17.0 μm [9]. This makes it difficult for radiation from the ground to penetrate the atmosphere into the universe. In addition, water vapor also has three main bands: 2.55–2.84, 5.6–7.6, and 12–30 μm [10]. Figure 1.11 schematically shows the main bands of carbon dioxide and water vapor.

Another property of gas radiation is that the radiation and absorption of gas occur throughout the volume. In a container filled with gas, radiation and absorption occur along its path, regardless of the direction along which the radiation propagates. To study the absorption of a certain part of gas in a container, it is necessary to consider the influence of the whole container, including the size, shape, and wall characteristics of the container. Besides, emission, absorption, and scattering occur all the time in the radiation path, which involves the surrounding gas in the study of radiation. To comprehensively study the gas radiation in a container, a more complicated model is needed to describe it, which will be introduced in Chapter 2.

1.3.2 Emissivity and Absorptivity of Water Vapor and Carbon Dioxide

Many factors affect gas emissivity and absorptivity. This section will introduce some of the key influencing factors and a theoretical system describing gas emissivity and absorptivity. In most engineering applications, we only care about the total radiation ability. Therefore, we can temporarily ignore the spectral properties and only concentrate on the total gas emissivity at a certain temperature. In Section 1.3.1, we also pointed out the volumetric properties of gas radiation, that is, the shape and size of the volume also have a certain influence on gas radiation.

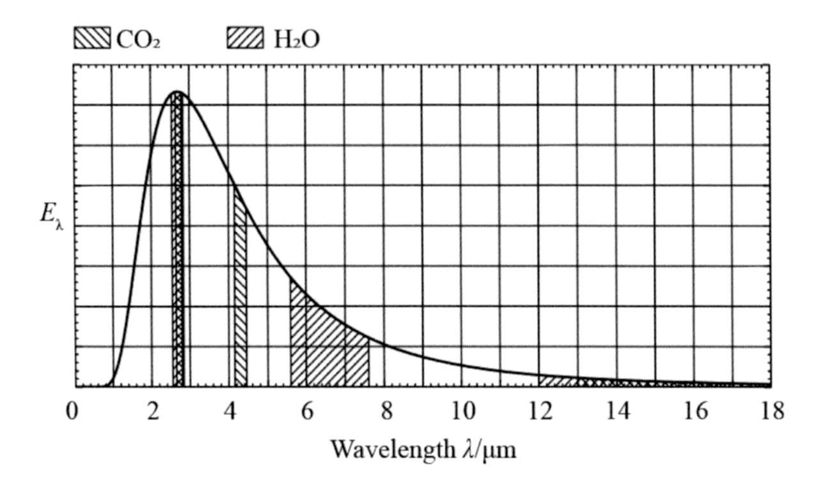

Figure 1.11 Schematic diagram of main optical bands of CO_2 and H_2O.

Mean Beam Length

The radiation ability of gas is related to the shape of the gas volume and the location of the research object. Parameters describing the radiation path need to be developed based on the shape of the volume and the location of the search object. Assume the radiation from a hemispherical gas volume to a differential area element located in the center. In this case, all the paths between the hemisphere and the area have the same length as the radius of the hemisphere, R, as shown in Fig. 1.12. And the mean beam length is R itself [11]. For gas volumes with other shapes, the equivalent hemisphere method can be applied to obtain the mean beam length. The so-called equivalent hemisphere is a hemisphere filled with the same gas in the same state as in the volume, and the radiation power on the center of the hemisphere from the equivalent hemisphere is equivalent to that on the studied area from the gas volume. The radius of such an equivalent hemisphere is the mean beam length of the gas. The equivalent hemisphere method is only a simple approximation, and there are more accurate formulas for other typical volume gases [12, 13]. In simple processing, the mean beam length of gas with any geometry can be calculated as follows, where V is the volume of gas (m^3) and A is the area of cladding (m^2):

$$s = 3.6\frac{V}{A}. \tag{1.3.1}$$

Emissivity

The mean beam length takes the volumetric properties of gas radiation into account, while the emissive power of the gas on the wall or on a specified point on the wall is also affected by the temperature, composition of the gas, and the number of absorbent gas molecules along the path. The number of gas molecules

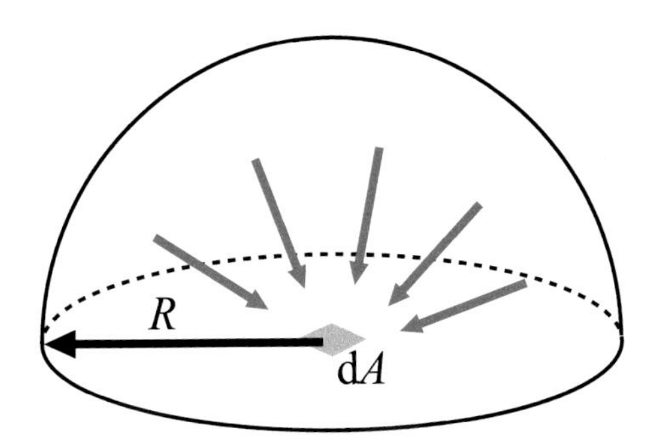

Figure 1.12 Schematic diagram of gas radiation to the center in hemispheres.

along the path can be expressed by the product of the partial pressure (p) of gas and the mean beam length (s):

$$\varepsilon_{\mathrm{g}} = f\left(T_{\mathrm{g}}, ps\right). \tag{1.3.2}$$

For water vapor, in addition to the synthetic parameter ($p_{\mathrm{H_2O}}s$) that affects the gas emissivity, there is also a separate effect of $p_{\mathrm{H_2O}}$. After extrapolating the single effect of $p_{\mathrm{H_2O}}$ to the limit case, where $p_{\mathrm{H_2O}}$ is zero under certain conditions, as the basis for drawing the graph line $\varepsilon^*_{\mathrm{H_2O}} = f\left(T_{\mathrm{g}}, p_{\mathrm{H_2O}}s\right), p = 10^5$ Pa, the separate effects of the total pressure $p \neq 10^5$ Pa, and $p_{\mathrm{H_2O}}$ is then corrected by introducing a coefficient $C_{\mathrm{H_2O}}$. Thus, the emissivity of water vapor is

$$\varepsilon_{\mathrm{H_2O}} = C_{\mathrm{H_2O}} \varepsilon^*_{\mathrm{H_2O}}. \tag{1.3.3}$$

Similarly, the emissivity of carbon dioxide is confirmed by Eq. (1.3.4):

$$\varepsilon_{\mathrm{CO_2}} = C_{\mathrm{CO_2}} \varepsilon^*_{\mathrm{CO_2}}. \tag{1.3.4}$$

When both water vapor and carbon dioxide exit in the mixture, a correction quantity needs to be introduced for the overlapping part of the wavebands of two gases. The gas emissivity is calculated by the following formula [13]:

$$\varepsilon_{\mathrm{g}} = C_{\mathrm{H_2O}} \varepsilon^*_{\mathrm{H_2O}} + C_{\mathrm{CO_2}} \varepsilon^*_{\mathrm{CO_2}} - \Delta\varepsilon. \tag{1.3.5}$$

Absorptivity

When the gas emits radiation energy, it also absorbs radiation from the wall and/or other gas. Kirchhoff's law is no longer applicable to obtain gas absorptivity mainly for two reasons. Gas radiation is strong wavelength selective, so gas cannot be regarded as a gray body. Besides, gas diffuses in the whole container, and there is heat transfer between the gas and the wall. Hence, the internal temperature is not necessarily balanced, that is, the thermal equilibrium state is not necessarily satisfied. Similar to the emittance calculation, we can write the absorptivity of a mixture of water vapor and carbon dioxide to the radiation from the blackbody shell:

$$\alpha_{\mathrm{g}} = C_{\mathrm{H_2O}} \alpha^*_{\mathrm{H_2O}} + C_{\mathrm{CO_2}} \alpha^*_{\mathrm{CO_2}} - \Delta\alpha, \tag{1.3.6}$$

where C_{H_2O} and C_{CO_2} are the same as that in Eqs. (1.3.3) and (1.3.4), respectively, and $\alpha^*_{H_2O}$, $\alpha^*_{CO_2}$, $\Delta\alpha$ can be calculated by the following empirical formulas in which T_w is the wall temperature [13]:

$$\alpha^*_{H_2O} = \left[\varepsilon^*_{H_2O}\right]_{T_W, p_{H_2O}s\left(\frac{T_W}{T_g}\right)} \left(\frac{T_g}{T_W}\right)^{0.45}, \tag{1.3.7}$$

$$\alpha^*_{CO_2} = \left[\varepsilon^*_{CO_2}\right]_{T_W, p_{CO_2}s\left(\frac{T_W}{T_g}\right)} \left(\frac{T_g}{T_W}\right)^{0.65}, \tag{1.3.8}$$

$$\Delta\alpha = [\delta\varepsilon]_{T_W}. \tag{1.3.9}$$

1.3.3 Solar Radiation

The Sun is a nearly spherical body that has a diameter of 1.39×10^9 m and is located at a distance of 1.50×10^{11} m from the Earth. The Sun, where a thermonuclear reaction occurs continually, radiates to the Earth at a rate of 1.7×10^{17} W, of which 30% is reflected and 23% is absorbed by the atmosphere, and the rest reaches the Earth's surface. Figure 1.13 shows the blackbody radiation spectrum of 5 770 K and the solar spectrum at the outer edge of the atmosphere and that on the ground. The radiation reaching the outer surface of the Earth's atmosphere has spectral properties (shown in Fig. 1.13) close to that of a blackbody at 5 770 K. But through the atmosphere, the energy spectrum reaching the ground would appear to fluctuate because of the strong selective absorption of the gases. At the average distance between the Sun and the Earth, the solar radiation energy received by the unit surface area perpendicular to the solar rays at the outer edge of the atmosphere is $(1\,370 \pm 6)\,\mathrm{Wm^{-2}}$. This value is called the solar constant [14], and it is independent of geographical location or time of day (see Fig. 1.14). In fact, the amount of solar input per unit area received at the horizontal surface of the outer edge of the atmosphere is

$$G_{s,o} = S_c f \cos\theta. \tag{1.3.10}$$

1.4 View Factor of Radiative Heat Transfer

Radiative heat transfer between surfaces is closely related to geometrical factors, such as surface geometry and orientations, which are usually considered as the view factor. The concept of view factor was put forward in the 1920s with the appearance and development of the radiation heat transfer calculation method on solid surfaces.

1.4.1 Definition and Calculation of View Factor

To separate geometric relations between surfaces from radiation intensity and make the view factor only contains the geometric relations, the enclosures are summed to be opaque, diffuse, and gray [15]. For surfaces that do not meet the

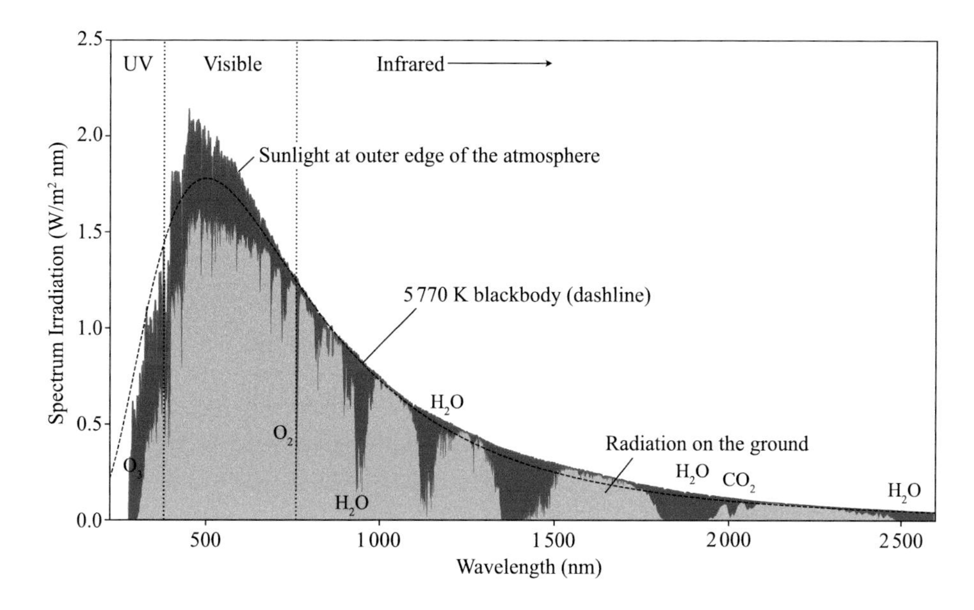

Figure 1.13 Spectral distribution of solar radiation.

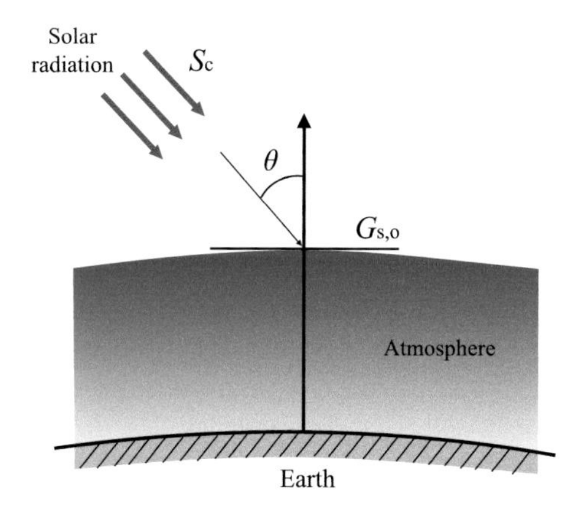

Figure 1.14 Solar radiation at the outer edge of the atmosphere.

first condition, that is, nondiffuse surfaces, the influence of geometric factors on radiative heat transfer is related to the direction, so the concept of view factor cannot be generally used, but its calculation method and principle are basically similar to that of the view factor. For the convenience of discussion, the object is treated as a blackbody in the study of the view factor, but the conclusions obtained are suitable for a diffuse gray surface.

Let there be two surfaces 1 and 2, both of which meet the above conditions, such as there exist two diffuse surfaces between transparent media. Then, the view factor of surface 1 to surface 2, $X_{1,2}$, is the fraction of uniform diffusive radiation leaving surface 1 that is intercepted by surface 2. The definition of the view factor can thus be written as

$$X_{1,2}=\frac{\text{The incident radiation from surface 1 to surface 2}}{\text{The effective radiation from surface 1}}. \tag{1.4.1}$$

The relative spatial position of two surfaces directly affects the radiative heat transfer between them and hence affects the view factor. For example, when two opposite-placed surfaces are infinitely close to each other, the incident radiation from one to the other is equivalent to the effective radiation, and the view factor is 1. However, for two surfaces in the same plane, since neither surface can receive the incident radiation from the other, the radiative heat transfer and view factor are zero. Besides, the shape of the surface will also affect the value of the view factor. This section will specifically study the influence of the shape and relative position of the surfaces on the view factor and how to compute view factors between surfaces.

1.4.2 Properties of the View Factor

According to the definition of the view factor and the spatial geometric relationship, when the assumptions are satisfied, four basic algebraic properties of view factors can be obtained. For two surfaces with other special relative positions and geometric relations, there may be other properties of the view factor. Here we will only introduce the basic three properties and their derivations.

Reciprocity Rule

For the view factor from a differential area element, $\mathrm{d}A_1$, to another element, $\mathrm{d}A_2$, denoted as $X_{\mathrm{d}1,\ \mathrm{d}2}$, as shown in Fig. 1.15, where the subscripts d1 and d2 represent $\mathrm{d}A_1$ and $\mathrm{d}A_2$ respectively. According to the definition,

$$X_{\mathrm{d}1,\ \mathrm{d}2}=\frac{\text{Irradiation from d1 to d2}}{\text{Effective radiation of d1}}=\frac{I_{\mathrm{b}1}\cos\theta_1\ \mathrm{d}A_1\ \mathrm{d}\Omega_1}{E_{\mathrm{b}1}\ \mathrm{d}A_1}=\frac{\mathrm{d}A_2\cos\theta_1\cos\theta_2}{\pi r^2}. \tag{1.4.2}$$

Similarly,

$$X_{\mathrm{d}2,\ \mathrm{d}1}=\frac{\mathrm{d}A_1\cos\theta_1\cos\theta_2}{\pi r^2}. \tag{1.4.3}$$

So

$$\mathrm{d}A_1X_{\mathrm{d}1,\ \mathrm{d}2}=\mathrm{d}A_2X_{\mathrm{d}2,\ \mathrm{d}1}. \tag{1.4.4}$$

The relativity of the view factor between two finite surfaces can be obtained by analyzing the radiative heat transfer between two isothermal blackbody surfaces. Thus, the relativity expression of the view factor between two finite surfaces is

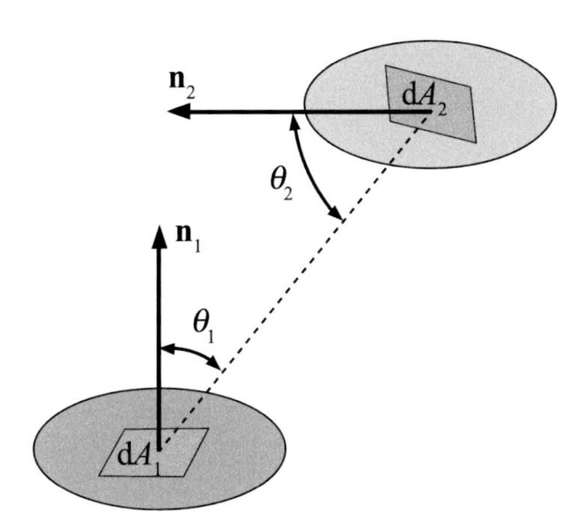

Figure 1.15 Reciprocity rule proof of the infinitesimal surface.

$$A_1 X_{1,2} = A_2 X_{2,1}. \tag{1.4.5}$$

This property is called the reciprocity rule.

Summation Rule

Assuming that the surface A_k forms an enclosure with other surrounding surfaces, all the radiation leaving the surface A_k is intercepted by the enclosure surfaces. Therefore, the effective radiation of A_k is equal to the radiation intercepted by all surfaces of the enclosure, that is,

$$Q_k = \sum_{i=1}^{n} Q_{k,i}. \tag{1.4.6}$$

Wherein, n is the number of closed body surfaces, as shown in Fig. 1.16. Using the definition of the view factor and Eq. (1.4.6), we can obtain

$$Q_k = \sum_{i=1}^{n} Q_k X_{k,i}. \tag{1.4.7}$$

Therefore,

$$\sum_{i=1}^{n} X_{k,i} = 1. \tag{1.4.8}$$

This property is called the summation rule.

Consider a set of surfaces $s = \{2, 2', 2'', \ldots\}$ shown in Fig. 1.17; any surface in the set is covered by surface 1. According to the summation rule, we have $X_{1,j} + X_{1,1} = 1$, for $j \in s$; thus, it can be obtained that

$$1 - X_{1,1} = X_{1,2} = X_{1,2'} = X_{1,2''}. \tag{1.4.9}$$

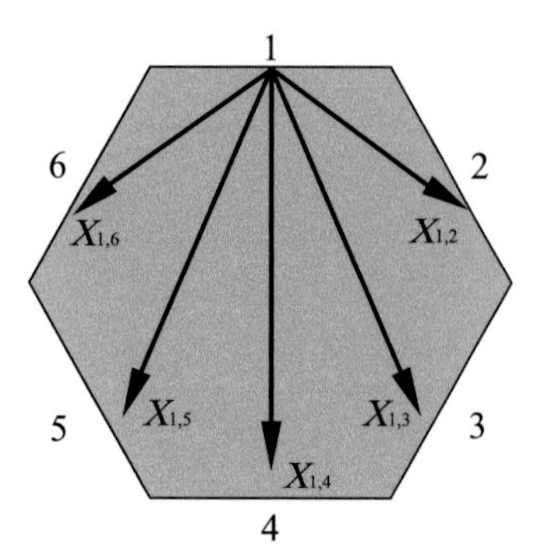

Figure 1.16 Proof of summation rule.

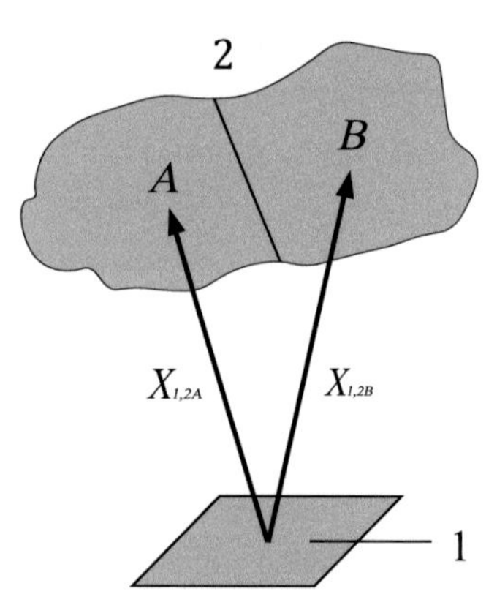

Figure 1.17 Proof of equivalence rule.

Superposition Rule

Consider the view factor of surface 1 against surface 2 as shown in Fig. 1.18. Since the total energy falling on surface 2 from surface 1 is equal to the sum of the radiation energy falling on the parts of surface 2,

$$A_1 E_{b1} X_{1,2} = A_1 E_{b1} X_{1,2A} + A_1 E_{b1} X_{1,2B}. \tag{1.4.10}$$

So

$$X_{1,2} = X_{1,2A} + X_{1,2B}. \tag{1.4.11}$$

If surface 2 is further divided into several small pieces, then

$$X_{1,2} = \sum_{i=1}^{n} X_{1,2i}. \tag{1.4.12}$$

When the superposition rule of view factor is used, only the second term in the subscript symbol is additive, while the first one does not have a relation similar to eq. (1.4.12), that is, $X_{1,2} \neq X_{2A,1} + X_{2B,1}$. This property is called the superposition rule.

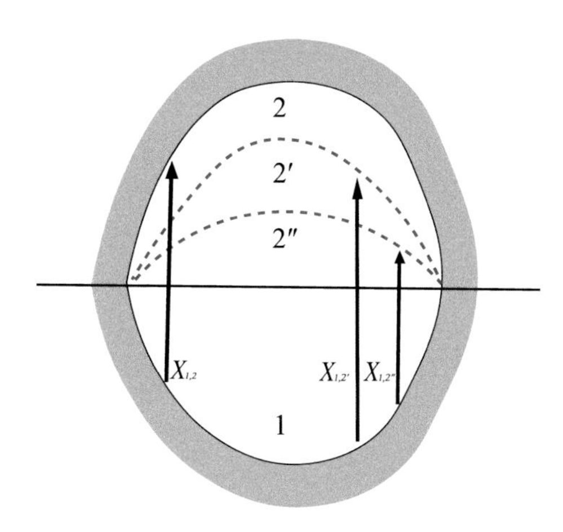

Figure 1.18 Proof of superposition rule.

1.4.3 Calculation Methods of the View Factor

There are many methods [16] for calculating the view factor. The most basic method is the integral method, whereas the most used one in engineering is the algebraic analysis. The concept of view factor was proposed very early, and many research findings were achieved in the 1950s and 1960s. For most of the systems with typical geometries, view factors have been calculated and compiled into manuals [17, 18].

Method of Direct Integration

The integral expression of the view factor between any two diffusive gray surfaces can be derived from the view factor of the differential area elements 1 and 2 in Eq. (1.4.2):

$$X_{1,2} = \frac{1}{A_1} \int \left(\int \frac{\cos\theta_1 \cos\theta_2 \, \mathrm{d}A_2}{\pi r^2} \right) \mathrm{d}A_1. \tag{1.4.13}$$

This integral formula is a quadruple integral and is rather complicated to obtain an analytical result. For complex cases, the numerical method may be applied to calculate the view factor. Literature [18] gives some formulas of the view factor between two-dimensional geometric structures, three typical three-dimensional geometric structures, and plots for engineering use. To expand the scope of calculation, these lines are often plotted in logarithmic coordinates, and attention should be paid to the logarithmic coordinates and the surface indicated by the subscripts 1 and 2.

Method of Algebraic Analysis

For the surface satisfying the condition of the view factor: (1) The surface should be a diffuse surface and (2) there should be uniform effective radiation on all

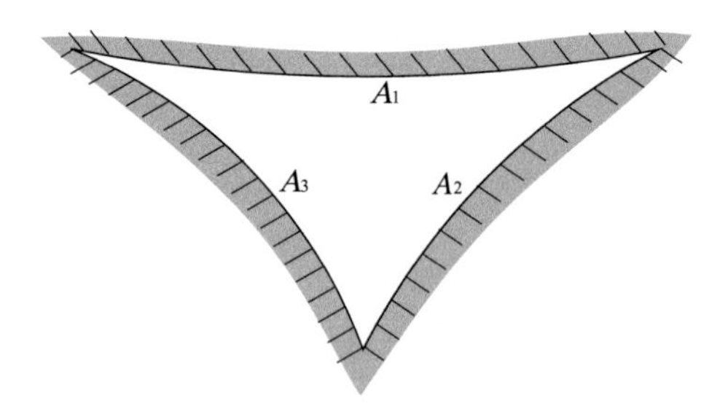

Figure 1.19 Enclosure with three surfaces.

surfaces. Applying the properties of the view factor, the method of obtaining the view factor by solving algebraic equations is called the method of algebraic analysis.

Figure 1.19 shows a composed of three convex surfaces extending infinitely along the direction perpendicular to the paper surface. The radiative energy spilling from both ends of the system can be ignored, and the system can be considered a closed system. Assume that the areas of the three surfaces are A_1, A_2, and A_3 respectively. According to the reciprocity rule and summation rule of the view factor, we have

$$\begin{aligned} X_{1,2}+X_{1,3}&=1, \quad A_1X_{1,2}=A_2X_{2,1},\\ X_{2,1}+X_{2,3}&=1, \quad A_1X_{1,3}=A_3X_{3,1},\\ X_{3,1}+X_{3,2}&=1, \quad A_2X_{1,3}=A_3X_{3,2}. \end{aligned}$$

By solving the above equation group, the view factors can thus be obtained as follows:

$$X_{i,j}=\frac{A_i+A_j-A_k}{2A_i}, \qquad \text{for } i\neq j\neq j\in\{1,2,3\}. \tag{1.4.14}$$

The other five view factors are also found. Since the three surfaces are of equal length in the direction perpendicular to the paper surface, it is simplified as

$$X_{i,j}=\frac{l_i+l_j-l_k}{2l_i}, \qquad \text{for } i\neq j\neq j\in\{1,2,3\}. \tag{1.4.15}$$

For a system containing nonadjacent surfaces, as shown in Fig. 1.20, the crossline method can be used to determine the view factor between A_1 and A_2. The auxiliary lines *ad* and *bc* were added between A_1 and A_2 to form an enclosure *abcd*. It is easy to obtain the view factor from the conclusion in formula (1.4.14) of the summation rule of the view factor

$$X_{ab,cd}=\frac{(bc+ad)-(ac+bd)}{2ab}. \tag{1.4.16}$$

Thus, for a system consisting of multiple surfaces extending infinitely in length in one direction, the view factor between any two surfaces can be summarized as

$$X_{1,2}=\frac{\text{Sum of crossed lines} \; - \; \text{Sum of uncrossed lines}}{\text{Twice the cross-sectional length of the surface 1}}. \tag{1.4.17}$$

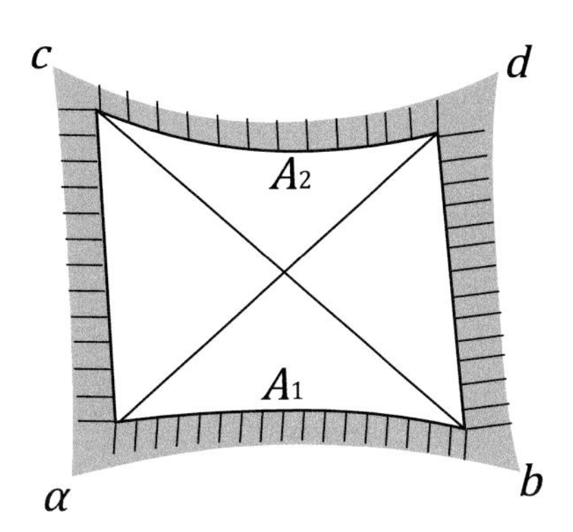

Figure 1.20 Diagram of the crossline method.

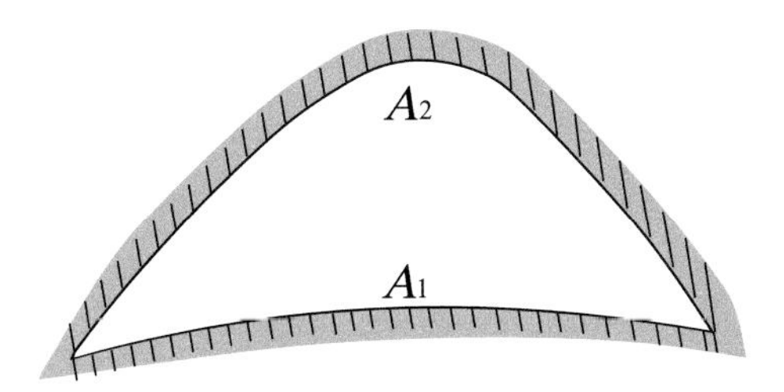

Figure 1.21 Enclosure composed of two black surfaces.

1.5 Calculation of Radiation Exchange in Multisurface Enclosures

1.5.1 Radiation Exchange in Enclosures Composed of Two Surfaces

Radiation Exchange in Enclosures Composed of Black Surfaces

When calculating the radiative heat transfer in an enclosure where all surfaces are black, no reflections need to be considered. Figure 1.21 shows an enclosure model [19] formed by two black surfaces. Thus, the net radiation exchange between the two surfaces is

$$\Phi_{1,2} = A_1 E_{\mathrm{b}1} X_{1,2} - A_2 E_{\mathrm{b}_2} X_{2,1} = A_1 X_{1,2} \left(E_{\mathrm{b}1} - E_{\mathrm{b}2}\right) = A_2 X_{2,1} \left(E_{\mathrm{b}1} - E_{\mathrm{b}2}\right). \tag{1.5.1}$$

Radiation Exchange in Enclosures Composed of Diffuse Gray Surfaces

Different from the enclosures composed of two black surfaces, the surface absorption of a gray body is less than 1, and it can only be absorbed after multiple reflections. Moreover, the energy emitted by the gray body includes both its own radiation energy and the reflected radiation energy. Therefore, the radiative heat transfer in enclosures composed of diffuse gray surfaces is more complicated. The term q, which is the net radiation leaving the surface, represents the net effect of radiative interactions occurring at the surface. It is equal to the difference between the surface radiosity and irradiation and can be expressed as Eq. (1.5.2). The outflow energy consists of the sum of the surface's own radiation εE_b and the energy reflected by the solid surface ρG. It is called radiosity, denoted as J:

$$q = J - G. \tag{1.5.2}$$

The net radiation exchange can also be expressed as

$$q = \varepsilon E_b - \rho G. \tag{1.5.3}$$

The relationship between the radiosity J and the net radiation exchange q is obtained by establishing the Eqs. (1.5.1) and (1.5.2):

$$J = E_{\mathrm{b}} - \left(\frac{1}{\varepsilon} - 1\right) q. \tag{1.5.4}$$

Radiation exchange in enclosures consisting of two gray surfaces is analyzed using the concept of radiosity.

In a two-dimensional enclosure composed of two isothermal opaque gray surfaces (areas A_1 and A_2), the radiative heat transfer between the two surfaces is

$$q_{1,2} = A_1 J_1 X_{1,2} - A_2 J_2 X_{2,1}, \tag{1.5.5}$$

and $q_{1,2} = -q_{2,1}$ is obtained in conjunction with Eqs. (1.5.4) and Eq. (1.5.5).

$$q_{1,2} = \frac{E_{\mathrm{b}1} - E_{\mathrm{b}2}}{\frac{1-\varepsilon_1}{\varepsilon_1 A_1} + \frac{1}{A_1 X_{1,2}} + \frac{1-\varepsilon_2}{\varepsilon_2 A_2}}. \tag{1.5.6}$$

(1) When surface 1 is a nonconcave surface, $X_{1,2} = 1$, then

$$q_{1,2} = \frac{A_1 \left(E_{\mathrm{b}1} - E_{\mathrm{b}2}\right)}{\frac{1}{\varepsilon_1} + \frac{A_1}{A_2}\left(\frac{1}{\varepsilon_2} - 1\right)}. \tag{1.5.7}$$

(2) When $A_1/A_2 \to 1$ and surface 1 is a nonconcave surface, such as two parallel infinite plates, then

$$q_{1,2} = \frac{A_1 \left(E_{\mathrm{b}1} - E_{\mathrm{b}2}\right)}{\frac{1}{\varepsilon_1} + \frac{1}{\varepsilon_2} - 1}. \tag{1.5.8}$$

(3) When $A_1/A_2 \to 0$ and surface 1 is a nonconcave surface, then

$$q_{1,2} = \varepsilon_1 A_1 \left(E_{\mathrm{b}1} - E_{\mathrm{b}2}\right). \tag{1.5.9}$$

The Two-Surface Enclosure Network

The parameters in Eq. (1.5.6) are similar to the EM parameters in Ohm's law. Heat transfer (q) is analogous to the current intensity; $E_{\mathrm{b}1} - E_{\mathrm{b}2}$ is analogous to the electric potential difference; the surface radiation thermal resistance $\left(\frac{1-\varepsilon}{\varepsilon A}\right)$ and the space radiation thermal resistance $\left(\frac{1}{A_1 X_{1,2}}\right)$ are analogous to the electrical resistance. E_b is analogous to the source electromotive force. J is analogous to

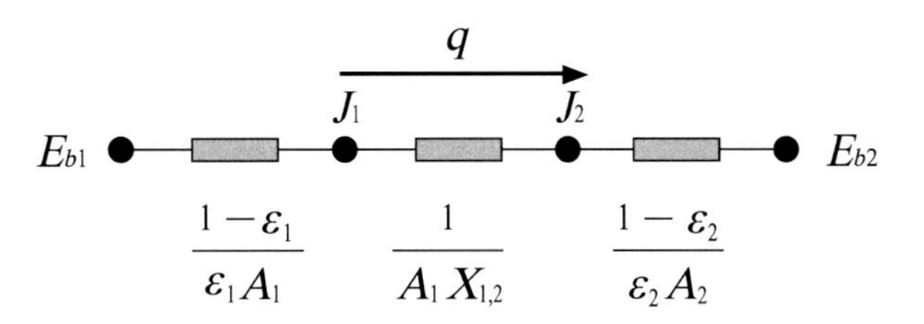

Figure 1.22 Radiation heat transfer equivalent network diagram of the two-surface enclosure.

the node voltage. The two-surface gray enclosure network is shown in Fig. 1.22. This method is called the network method of radiation exchange [20, 21].

1.5.2 Radiation Heat Transfer of Multisurface in an Enclosure

In a multisurface enclosure, the net radiation heat transfer of a surface is the sum of the heat transfer of the other surfaces. The network method can be used to obtain simultaneous equations for calculating the effective radiation of each surface. As for a three-surface enclosure: (1) draw the equivalent network diagram as shown in Fig. 1.23; (2) list the current equation of the nodes according to each node J in the network graph; (3) get J_1, J_2, and J_3 by solving Eq. (1.5.10) and then obtain the net radiation heat transfer:

$$
\begin{aligned}
J_1 &: \frac{E_{b1}-J_1}{\frac{1-\varepsilon_1}{\varepsilon_1 A_1}} + \frac{J_2-J_1}{\frac{1}{A_1 X_{1,2}}} + \frac{J_3-J_1}{\frac{1}{A_1 X_{1,3}}} = 0, \\
J_2 &: \frac{E_{b2}-J_2}{\frac{1-\varepsilon_2}{\varepsilon_2 A_2}} + \frac{J_1-J_2}{\frac{1}{A_1 X_{1,2}}} + \frac{J_3-J_2}{\frac{1}{A_1 X_{1,3}}} = 0, \\
J_3 &: \frac{E_{b3}-J_3}{\frac{1-\varepsilon_3}{\varepsilon_3 A_3}} + \frac{J_1-J_3}{\frac{1}{A_1 X_{1,3}}} + \frac{J_2-J_3}{\frac{1}{A_2 X_{2,3}}} = 0.
\end{aligned}
\tag{1.5.10}
$$

1.6 Strengthening and Weakening of Thermal Radiation

The strengthening and weakening of thermal radiation is an important part of heat transfer. The physical mechanisms of radiation heat transfer, heat conduction, and convection heat transfer are different, so the control methods are also different.

1.6.1 Principles of Strengthening and Weakening Thermal Radiation

According to the network method of radiative heat transfer, the method of strengthening or weakening the radiative heat transfer between two surfaces changes the surface resistance and the space resistance.

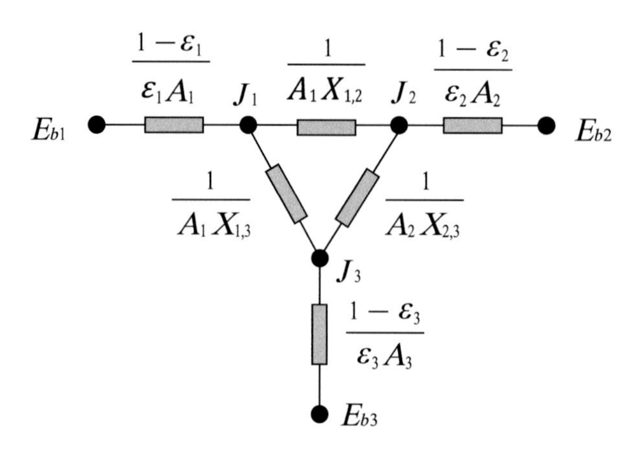

Figure 1.23 The heat transfer equivalent network diagram of a three-surface enclosure.

1. *Changing the surface resistance.* According to the definition of surface resistance ($\frac{1-\varepsilon}{\varepsilon A}$), changing the surface resistance can be achieved by changing the surface area or surface emissivity [22–24]. It is worth noting that when using the method of changing surface reflectivity to control radiation heat transfer, the surface emissivity that has the greatest impact on radiant heat transfer should be changed first.
2. *Changing the space resistance.* According to the definition of space resistance ($\frac{1}{A_i X_{i,j}}$), the area A_i generally depends on the specific heat dissipation or insulation surface [25–27]. Therefore, the view factor between the surfaces is generally adjusted to change the space resistance.

1.6.2 Application of Radiation Heat Transfer

In engineering applications, one of the most effective methods to weaken radiation heat transfer is using radiation shields, 1. The principle of radiation shields When inserting a thin metal plate between two plates, the radiation heat transfer between the two plates will be reduced. The thin metal plate is called a radiation shield [28–31]. When the radiation shield is not added, the radiation thermal resistances between two plates compose of two surface resistances and one space resistance. After adding the radiation shield, two surface resistances, and one space resistance will be added. Therefore, the total radiation resistance increases and the radiation heat transfer between two plates decreases. This is the principle of radiation shields. Take the insertion of a radiation shield between two parallel large plates as an example to illustrate the influence of the radiation shield on radiation exchange. Radiation network diagrams with or without the radiation shield between the parallel large plates are shown in Fig. 1.24.

Since the plate is infinite, the view factor is

$$X_{1,3} = X_{3,1} = X_{1,2} = 1 \tag{1.6.1}$$

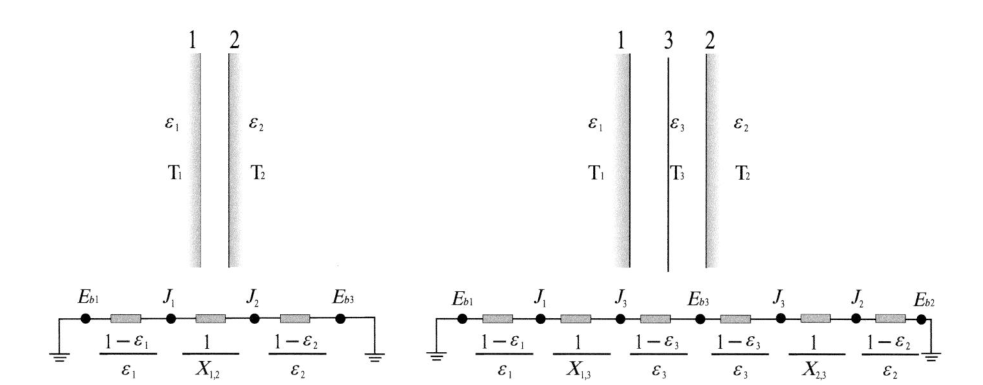

Figure 1.24 Radiation heat transfer with or without a radiation shield between two large plates.

because of

$$A_1 = A_2 = A_3 = A. \tag{1.6.2}$$

Then the heat transfer without and with the radiation shield would be as follows. Without the radiation shield,

$$q_{1,2} = \frac{\sigma\left(T_1^4 - T_2^4\right)}{\frac{1-\varepsilon_1}{\varepsilon_1 A_1} + \frac{1}{A_1 X_{1,2}} + \frac{1-\varepsilon_2}{\varepsilon_2 A_2}} = \frac{\sigma\left(T_1^4 - T_2^4\right)}{\frac{1}{\varepsilon_1} + \frac{1}{\varepsilon_2} - 1}. \tag{1.6.3}$$

With a radiation shield,

$$\begin{aligned} q_{1,3,2} = q_{1,3} = q_{3,2} &= \frac{E_{\mathrm{b}1} - E_{\mathrm{b}2}}{\frac{1-\varepsilon_1}{\varepsilon_1 A_1} + \frac{1}{A_1 X_{1,3}} + \frac{1-\varepsilon_{3,1}}{\varepsilon_{3,1} A_3} + \frac{1-\varepsilon_{3,2}}{\varepsilon_{3,2} A_3} + \frac{1}{A_3 X_{3,2}} + \frac{1-\varepsilon_2}{\varepsilon_2 A_2}} \\ &= \frac{\sigma\left(T_1^4 - T_2^4\right) A}{\frac{1}{\varepsilon_1} + \frac{1}{\varepsilon_{3,1}} - 1 + \frac{1}{\varepsilon_{3,2}} + \frac{1}{\varepsilon_2} - 1}. \end{aligned} \tag{1.6.4}$$

Obviously, $q_{1,3,2} < q_{1,2}$. If $\varepsilon_1 = \varepsilon_2 = \varepsilon_{3,1} = \varepsilon_{3,2} = \varepsilon$, we will have $q_{1,3,2} = q_{1,2}/2$. It can be proved that the radiation heat transfer when inserting a radiation shield (thin metal plate) with the same frequency on the surface of the two parallel large flat walls is $1/(n+1)$ of the radiation heat transfer without radiation shields.

Finally, it is very convenient to use the network method to analyze the radiation shields. When the emissivity of each surface is different, the network method can be used to calculate the radiation heat transfer and the temperature of the radiation shields.

1.7 Summary

This chapter introduces the basic concepts of thermal radiation, and then mainly discusses the calculation method of the radiative heat transfer between objects, focusing on the radiation heat transfer between the surfaces of an enclosure.

Some basic concepts are listed below:

Irradiation: Rate at which radiation is incident on a surface from all directions per unit area of the surface.

Radiosity: Rate at which radiation leaves a surface due to emission and reflection in all directions per unit area of the surface.

Blackbody: The ideal emitter and absorber. Modifier refers to ideal behavior.

Diffuse: Modifier referring to the directional independence of the intensity associated with an emitter, reflected, or incident radiation.

Gray surface: A surface for which the spectral absorptivity and emissivity are independent of wavelength over the spectral regions of surface irradiation and emission.

Planck's law: Spectral distribution of emission from a blackbody.

Stefan–Boltzmann law: Emissive power of a blackbody.

Wien's displacement law: Locus of the wavelength corresponding to peak emission by a blackbody.

Kirchhoff's law: Relation between emission and absorption properties for surfaces.

View factor: The percentage of radiation energy emitted by one surface that falls on another surface

Basic properties of the view factor: Under the assumption of the uniform surface radiant heat flow and the diffuser, the view factor is a pure geometric factor and has nothing to do with surface emissivity and temperature. From the perspective of energy balance, the relativity, completeness, and additivity of the view factor can be derived.

Effective radiation: The total radiation energy emitted from the unit surface includes self-radiation and emitted radiation. The introduction of effective radiation simplifies the calculation of radiation heat transfer among gray-body surfaces and avoids the complexity of analyzing multiple absorption and reflection.

Surface resistance of radiation heat transfer: Determined by the surface area and emissivity, $\frac{1-\varepsilon}{A\varepsilon}$.

Space resistance of radiation heat transfer: Determined by the area and shape of the surface and the relative position of the other surface, $1/(AX_{1,2})$.

2 Radiative Transfer Equation

2.1 Basic Concepts in Radiative Transfer Equation

For radiative transfer between different surfaces, or within a medium that does not emit, absorb, or scatter radiation (known as a "nonparticipating medium"), the radiative intensity in any given direction is constant along its path. Different from these situations, in a "participating medium" (a medium that does absorb, emit, and/or scatter radiation), intensity along any given path could change due to one or more of these phenomena. The intensity field within a participating medium is governed by the radiative transfer equation (RTE) according to the energy balance. The radiative intensity – defined as radiative energy transferred per unit time, solid angle, spectral variable, and area normal to the pencil of rays, instead of the emissive power – is applied to describe the radiative transfer within a medium.

An EM wave passing through a medium containing small particles will be absorbed or scattered. For example, a colorful rainbow, blue sky, and red sunset, are the results of absorption and scattering of sunlight when passing the atmosphere containing various dust and smoke particles. The scattering is due to three separate phenomena: (i) diffraction (waves never pass through the particle, but their direction of propagation is altered by the presence of the particle); (ii) reflection by a particle (waves reflected from the surface of the particle); and (iii) refraction in a particle (waves that penetrate into the particle and, after partial absorption, reemerge traveling in a different direction). The vast majority of photons are scattered elastically, that is, their wavelength (the frequency of the incident beam) remain unchanged, which is called elastic scattering.

In practical applications, a participating medium is composed of different particles; even though the irregular particles have different sizes and shapes, most particles can be simplified as spheres, ellipsoids, or cylinders if they are averaged. Here, we are concerned with how small particles interact with EM waves – how they absorb, emit, and scatter radiative energy. The scattering and absorption of radiation by single spheres was first discussed during the nineteenth century by Lord Rayleigh [32], who obtained a simple solution for spheres with small size parameters, ($x \ll 1$), whose diameters are much smaller than the wavelength of radiation. The theory was further developed by Lorenz in the 1890s [33, 34] and Gustav Mie in 1908 [35] and was known as the "Lorenz–Mie theory" and the "Mie theory," respectively. Detailed derivations can be found in the books by van de Hulst [36], Kerker [37], and Bohren and Huffman [38].

Absorption takes place when light interacts with a particle, converting EM wave energy into the internal thermal energy of the particle. And the absorption may cause a change in the particle properties, such as temperature in general. Scattering, like absorption, takes place when light interacts with a particle while the incident EM wave energy is scattered in all directions. It is a kind of redirection of incident energy and has no influence on the internal energy of the particle.

The amount of scattering and absorption by a particle is usually expressed in terms of the scattering cross section, C_{sca}, and the absorption cross section, C_{abs}. The total amount of absorption and scattering, or extinction, is expressed in terms of the extinction cross section:

$$C_{\text{ext}} = C_{\text{abs}} + C_{\text{sca}}. \tag{2.1.1}$$

In addition, efficiency factors Q are also frequently used, which is nondimensionalized with the surface area of the sphere. The absorption efficiency factor, the scattering efficiency factor, and the extinction efficiency factor can be respectively indicated as

$$Q_{\text{abs}} = \frac{C_{\text{abs}}}{\pi a^2}, \quad Q_{\text{sca}} = \frac{C_{\text{sca}}}{\pi a^2}, \text{ and } \quad Q_{\text{ext}} = \frac{C_{\text{ext}}}{\pi a^2}. \tag{2.1.2}$$

For a single particle, different sizes and shapes usually contribute to different abilities of absorption and scattering. Figure 2.1 shows the typical behavior of efficiency factors, which are some representative results of Lorenz–Mie calculations. The figure demonstrated the variation of the extinction efficiency ($k = 0$) for different refractive indexes m of dielectric media. It can be observed that there is a primary oscillation in the variation of Q_{ext} with the size parameter. Note also that the oscillations become smaller for larger size parameters and $Q_{\text{ext}} \to 2$ as $x \to \infty$ (for dielectrics as well as metals).

When absorption and scattering take place, the intensity of the incident light may be attenuated or augmented. The attenuation is called extinction that can be attributed to the absorption of the participating medium and scattering away from the direction of the incident light (or out-scattering). The augmentation takes place because of the emission and scattering from other directions into the incident direction (or "in-scattering"). When considering the augmentation contributed by "in-scattering," scattering from all directions should be summed up. The phase function is defined to describe the scatting process more specifically, especially in the scattering direction. It indicates the probability that radiation energy incident on a particle in the direction of (θ', Φ') will be scattered in the direction of (θ, Φ) with a solid angle Ω. Furthermore, the scattering angle θ between the incident and scattered directions satisfies

$$\cos\Theta = \cos\theta\cos\theta' + \sin\theta\sin\theta'\cos(\Phi - \Phi'). \tag{2.1.3}$$

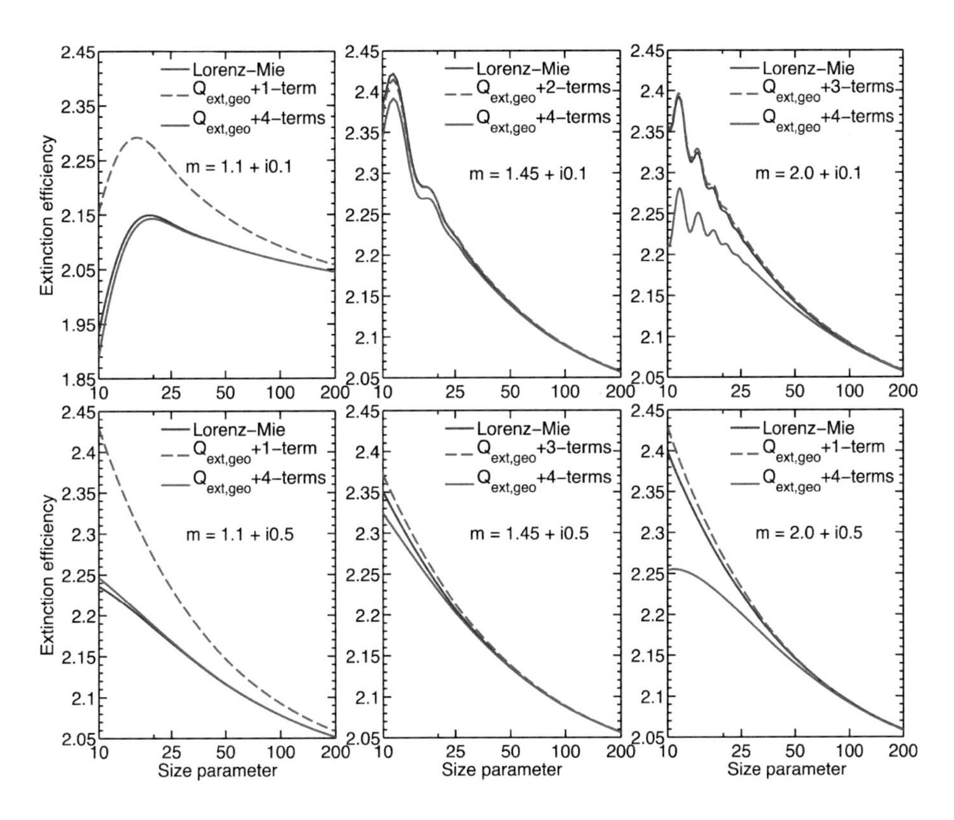

Figure 2.1 Extinction efficiency factors for dielectric spheres for several refractive indexes. Reproduced from [39]. OSA Open Access Publishing Agreement.

The fraction of the energy that is scattered in any given direction is denoted by the scattering phase function $\Phi(\theta)$, which is normalized such that

$$\frac{1}{4\pi}\int_{4\pi}\Phi\left(\hat{s}_i,\hat{s}\right)\mathrm{d}\Omega=1. \tag{2.1.4}$$

For a simpler analysis, the directional scattering behavior may be described by the average cosine of the scattering angle, known as the asymmetry factor, and related to the phase function by

$$g=\frac{1}{4\pi}\int_{4\pi}\phi_\lambda\left(\theta\right)\cos\theta\,\mathrm{d}\Omega. \tag{2.1.5}$$

In the case of isotropic scattering (i.e., equal amounts are scattered in all directions, and $\Phi\equiv 1$), the asymmetry factor vanishes; g also vanishes if scattering is symmetrical about the plane perpendicular to the beam propagation.

If the particle scatters more radiation into the forward directions ($\theta<\pi/2$), then g is positive; if more radiation is scattered in the backward direction ($\theta>\pi/2$), then g is negative. For very small particles (known as the Rayleigh scattering), scattering is symmetric to the plane perpendicular to the incident

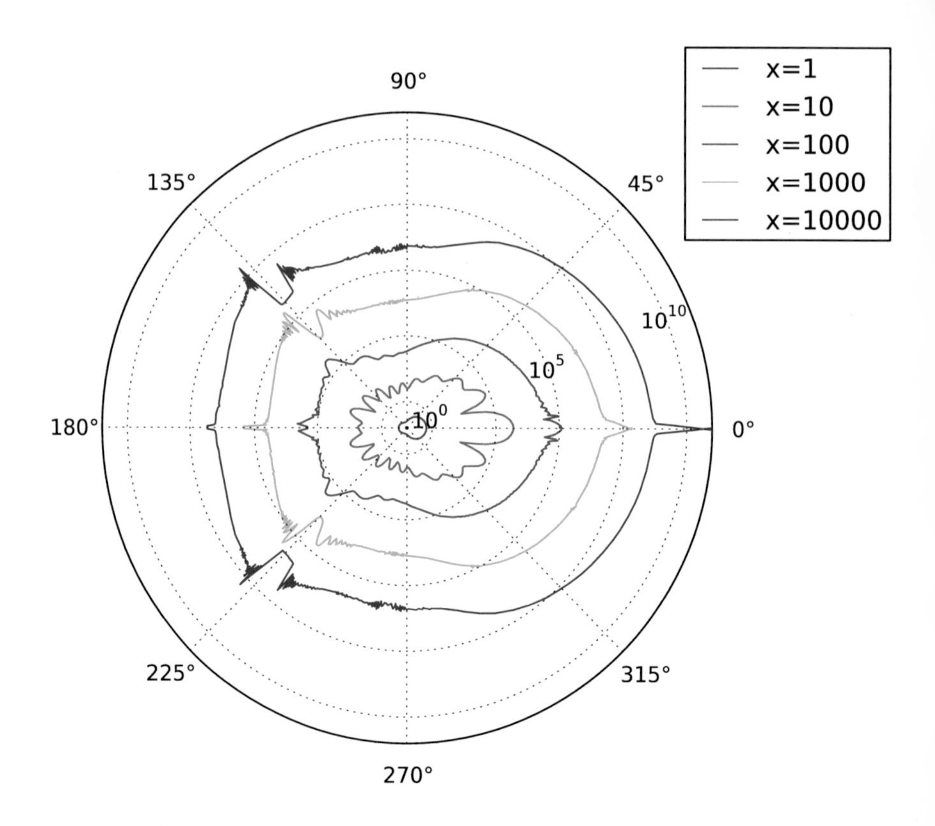

Figure 2.2 Scattering phase functions for various size parameters. It should be noted that the scattered intensity increases with the size of the scattering particles, denoted by the size parameter. Reproduced from [40]. Approved for Public Release, unlimited distribution.

beam and is nearly isotropic with slight forward- and backward-scattering peaks. Figure 2.2 shows some cases of scattering phase functions for various size parameters. For particles with refractive indexes close to unity (known as the Rayleigh–Gans scattering), nearly all of the scattered energy is scattered in forward directions with some scattering in a few preferred directions. This behavior becomes more extreme as the size parameter increases. For $x = 10$, the scattering behavior demonstrates rapid maxima and minima at varying scattering angles with a strong forward component.

For a group of particles, the number of particles along the path of an incident beam and the distances between neighboring particles have an essential influence on the amount of absorption energy. If the scattering is independent, then the effects of all the particles can be simply linearly cumulated. For simplicity, it is often assumed that particle clouds consist of spheres that are all equally large. In order to quantify the absorption ability, the absorption coefficient κ_λ is defined as shown in Eq. (2.1.6); a that relation reflects the reducing proportion of the incident radiative intensity, I_λ, along the path dS due to absorption:

$$\frac{dI_\lambda}{dS} = -\kappa_\lambda I_\lambda. \tag{2.1.6}$$

Naturally, the absorption coefficients have the units of m^{-1} and satisfy the linear relation, which means that the properties of a group can be a summation of individual contribution

$$\kappa_\lambda = \sum_i \kappa_{\lambda,i}, \tag{2.1.7}$$

where i indicates different particles or other absorptive components.

The scattering coefficients, σ_λ, can be used to qualify the scattering process. It is defined as Eq. (2.1.8) that describes the proportion of radiative intensity whose direction was changed along the path in the incident radiative intensity. In addition, the linearity property can be applied to the scattering coefficients, where $N_T C$ is the number of particles per unit volume:

$$\frac{dI_\lambda}{dS} = -\sigma_\lambda I_\lambda, \tag{2.1.8}$$

$$\sigma_\lambda = \sum_i \sigma_{\lambda,i} = N_T C_{\text{sca}}. \tag{2.1.9}$$

Every substance will spontaneously emit EM radiation only if its temperature is higher than absolute 0°. Therefore, an incident beam can gain energy from the emission of the participating medium. With the help of the emission coefficient j_λ, the emission energy can be described as

$$(dI_\lambda)_e = j_\lambda dS, \tag{2.1.10}$$

which indicates that the emitted intensity $(dI_\lambda)_e$ along the path dS should be proportional to the length of the path. Therefore, the extinction coefficient can be defined as

$$\beta_\lambda = \kappa_\lambda + \sigma_\lambda. \tag{2.1.11}$$

For clouds of particles with nonuniform size, it is customary to describe the number of particles as a function of radius in the form of a particle distribution function. Several different forms for the distribution function have been used by various researchers.

2.2 The Scalar Radiative Transfer Equation

2.2.1 The Radiative Transfer Equation

The RTE is built based on the variation of the intensity of an incident beam along the direction $\hat{s}$ (see Fig. 2.3). The energy along the direction can be divided into two parts. The first part is the attenuation caused by absorption and scattering away from the direction $\hat{s}$ (out-scattering). Absorbed energy is converted into internal energy, while scattered energy is redirected to other directions in space.

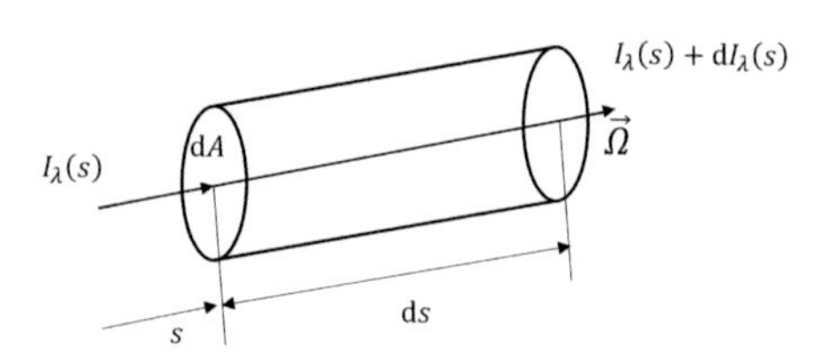

Figure 2.3 The variation of the intensity of an incident beam along the direction $\hat{s}$.

The second part is the augmentation by emission and scattering into the direction $\hat{s}$ (in-scattering) from the other direction. According to the energy balance, the radiative energy propagating along the direction of $\hat{s}$ within a small pencil of rays can be built.

$$\mathrm{d}I_\lambda(s) = W_\lambda(s)\,\mathrm{d}s = [-W_{a\lambda}(s) - W_{s\lambda}(s) + W_{em\lambda}(s) + W_{is\lambda}(s)]\,\mathrm{d}s, \tag{2.2.1}$$

$$\frac{\mathrm{d}I_\lambda(s)}{\mathrm{d}s} = -W_{a\lambda}(s) - W_{s\lambda}(s) + W_{em\lambda}(s) + W_{is\lambda}(s), \tag{2.2.2}$$

where $W_{a\lambda}(s), W_{em\lambda}(s), W_{s\lambda}(s)$, and $W_{is\lambda}(s)$ show the absorption, emission, out-scattering, and in-scattering energy, respectively. The absorption, attenuation by scattering, and emitted intensity are proportional to the magnitude of the incident energy as well as the distance the beam travels through the medium, which can be written as

$$W_{a\lambda}(s) = k_{a\lambda} I_\lambda(s)\,\mathrm{d}s, \tag{2.2.3}$$

$$W_{em\lambda}(s) = k_{em\lambda} I_{b\lambda}(s) = k_{a\lambda} I_{b\lambda}(s), \tag{2.2.4}$$

$$W_{s\lambda}(s) = k_{s\lambda} I_\lambda(s). \tag{2.2.5}$$

The emission coefficient is equal to the absorption coefficient for thermodynamic equilibrium. Scattering from all directions into direction $\hat{s}$ will increase the energy, as shown in Fig. 2.4. The total amount of energy scattered away from direction $\hat{s}$ is $k_{s\lambda} I_\lambda\left(s, \vec{\Omega}'\right)$. The fraction $\phi_\lambda\left(\vec{\Omega}, \vec{\Omega}'\right)/4\pi$ is scattered into the cone $\mathrm{d}\Omega$ around the direction $\hat{s}$. The function $\phi_\lambda\left(\vec{\Omega}, \vec{\Omega}'\right)$ is called the scattering phase function and describes the probability that a ray from one direction, $\hat{s}_i$ will be scattered in a certain other direction, $\hat{s}$. The constant 4π is arbitrary and is included for convenience. Therefore, the in-scattering energy from the other direction can be calculated by integration over all solid angles

$$W_{is\lambda}(s) = \int_{4\pi} \frac{1}{4\pi} \cdot k_{s\lambda} I_\lambda\left(s, \vec{\Omega}'\right) \cdot \phi_\lambda\left(\vec{\Omega}, \vec{\Omega}'\right) \mathrm{d}\Omega'. \tag{2.2.6}$$

An energy balance on the radiative energy propagating along the direction of $\hat{s}$ within a small pencil of rays can be built. The change in intensity is the summation of the contribution from emission, absorption, scattering away from

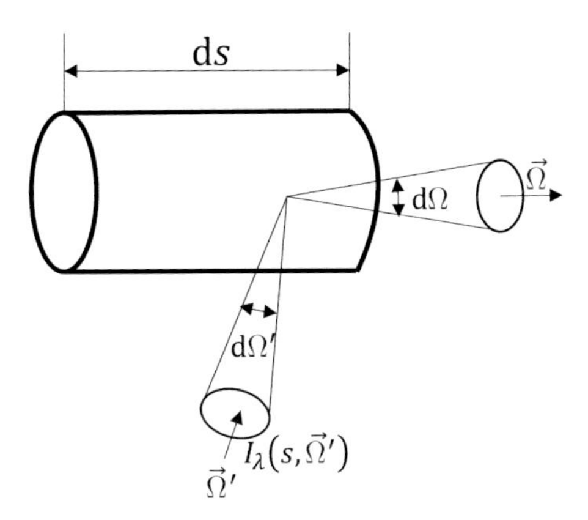

Figure 2.4 Redirection of radiative intensity by scattering.

the direction $\hat{s}$, and scattering in the direction $\hat{s}$, which can be described by Eq. (2.2.7):

$$\begin{aligned} &\mathrm{I}_\eta(s+\mathrm{d}s,\hat{\mathbf{s}},t+\mathrm{d}t)-\mathrm{I}_\eta(s,\hat{\mathbf{s}},t) \\ &\quad = j_\eta(s,t)\,\mathrm{d}s-\kappa_\eta I_\eta(s,\hat{\mathbf{s}},t)\,\mathrm{d}s-\sigma_{s\eta}I_\eta(s,\hat{\mathbf{s}},t)\,\mathrm{d}s \\ &\qquad +\frac{\sigma_{s\eta}}{4\pi}\int_{4\pi} I_\eta(\hat{\mathbf{s}}_i)\,\Phi_\eta(\hat{\mathbf{s}}_i,\hat{\mathbf{s}})\,\mathrm{d}\Omega_i\,\mathrm{d}s. \end{aligned} \tag{2.2.7}$$

In this equation, photons travel along a ray from s to $s+\mathrm{d}s$, which indicates that the equation is Lagrangian in nature. The ray propagates at the speed of light c, which relates $\mathrm{d}s$ and $\mathrm{d}t$ through $\mathrm{d}s=c\mathrm{d}t$. According to the Taylor series, the outgoing intensity can be written as

$$\mathrm{I}_\eta(s+\mathrm{d}s,\hat{\mathbf{s}},t+\mathrm{d}t)=\mathrm{I}_\eta(s,\hat{\mathbf{s}},t)+\mathrm{d}t\frac{\partial I_\eta}{\partial t}+\mathrm{d}s\frac{\partial I_\eta}{\partial s}. \tag{2.2.8}$$

And Eq. (2.2.7) becomes

$$\frac{1}{c}\frac{\partial I_\eta}{\partial t}+\frac{\partial I_\eta}{\partial s}=j_\eta(s,t)-\kappa_\eta I_\eta(s,\hat{\mathbf{s}},t)-\sigma_{s\eta}I_\eta(s,\hat{\mathbf{s}},t)+\frac{\sigma_{s\eta}}{4\pi}\int_{4\pi} I_\eta(\hat{\mathbf{s}}_i)\,\Phi_\eta(\hat{\mathbf{s}}_i,\hat{\mathbf{s}})\,\mathrm{d}\Omega_i, \tag{2.2.9}$$

which is called the RTE. In this equation, all quantities may change with the spatial location, time, wavenumber, and intensity; and the phase function also relates to the direction $\hat{s}$.

If we substitute the extinction coefficient defined in Eq. (2.1.11), then Eq. (2.2.9) will convert into its equilibrium, quasi-steady form:

$$\frac{\mathrm{d}I_\eta}{\mathrm{d}\tau_\eta}=\hat{\mathbf{s}}_i\cdot\nabla I_\eta=\kappa_\eta I_{b\eta}-\beta I_\eta+\frac{\omega_\eta}{4\pi}\int_{4\pi} I_\eta(\hat{\mathbf{s}}_i)\,\Phi_\eta(\hat{\mathbf{s}}_i,\hat{\mathbf{s}})\,\mathrm{d}\Omega_i. \tag{2.2.10}$$

If the nondimensional optical coordinates and single-scattering albedo are used, then the RTE can be written as

$$\frac{\mathrm{d}I_\eta}{\mathrm{d}\tau_\eta} = -I_\eta - (1-\omega_\eta) I_{b\eta} + \frac{\omega_\eta}{4\pi}\int_{4\pi} I_\eta(\hat{\mathbf{s}}_i)\,\Phi_\eta(\hat{\mathbf{s}}_i,\hat{\mathbf{s}})\,\mathrm{d}\Omega_i, \tag{2.2.11}$$

where

$$\tau_\eta = \int_0^s (\kappa_\eta + \sigma_{s\eta})\,\mathrm{d}s = \int_0^s \beta_\eta \mathrm{d}s \tag{2.2.12}$$

and

$$\omega_\eta \equiv \frac{\sigma_{s\eta}}{\kappa_\eta + \sigma_{s\eta}} = \frac{\sigma_{s\eta}}{\beta_\eta}. \tag{2.2.13}$$

In Eq. (2.2.2), the last two terms are often combined, which are known as the *source function* for the radiative intensity, and the equation can be rewritten as

$$S_\eta(\tau_\eta,\hat{\mathbf{s}}) = (1-\omega_\eta) I_{b\eta} + \frac{\omega_\eta}{4\pi}\int_{4\pi} I_\eta(\hat{\mathbf{s}}_i)\,\Phi_\eta(\hat{\mathbf{s}}_i,\hat{\mathbf{s}})\,\mathrm{d}\Omega_i, \tag{2.2.14}$$

which can be further deceptively simplified into

$$\frac{\mathrm{d}I_\eta}{\mathrm{d}\tau_\eta} + I_\eta = S_\eta(\tau_\eta,\hat{\mathbf{s}}). \tag{2.2.15}$$

When the source function is known, Eq. (2.2.15) can be further integrated by multiplying with e^{τ_η}, and it thus becomes

$$\frac{\mathrm{d}}{\mathrm{d}\tau_\eta}\left(I_\eta e^{\tau_\eta}\right) = S_\eta\left(\tau'_\eta,\hat{\mathbf{s}}\right) e^{\tau_\eta}. \tag{2.2.16}$$

It can be integrated from $s'=0$ to $s'=S$ along the propagation pathway in the medium, and we thus obtain

$$I_\eta(\tau_\eta) = I_\eta(0)\mathrm{e}^{-\tau_\eta} + \int_0^{\tau_\eta} S_\eta\left(\tau'_\eta,\hat{\mathbf{s}}\right)\mathrm{e}^{-\left(\tau_\eta,-\tau'_\eta\right)}\mathrm{d}\tau'_\eta, \tag{2.2.17}$$

where τ'_η is the optical coordinate at $s=s'$. The physical meaning of the first term on the right-hand side (RHS) of Eq. (2.2.17) is the contribution to the local intensity by the intensity entering the enclosure at $s=0$, and it decays exponentially with the increase of the optical distance τ_η. The second term on the RHS is contributed from the summation of local emission over the optical pathway of emission. The local emission, $S_\eta\left(\tau'_\eta\right)\mathrm{d}\tau'_\eta$, is attenuated exponentially along propagation due to self-extinction over the optical distance. In some special conditions, the equation can be simplified further.

For a nonscattering medium, when considering only the emission and absorption of the medium, the source function contains the only term contributed by the local blackbody intensity. Therefore, Eq. (2.2.17) becomes

$$I_\eta(\tau_\eta) = I_\eta(0)\mathrm{e}^{-\tau_\eta} + \int_0^{\tau_\eta} I_{b\eta}\left(\tau'_\eta\right)\mathrm{e}^{-\left(\tau_\eta,-\tau'_\eta\right)}\mathrm{d}\tau'_\eta. \tag{2.2.18}$$

In this circumstance, the radiation intensity can be calculated from the equation explicitly if the temperature field is known. However, the temperature field usually depends on the conservation of energy in another format except for radiation energy.

When the radiation intensity emitted from the medium is small compared with the incident intensity, it is called the cold medium. In this circumstance, the RTE is irrelevant to other modes of heat transfer. Thus, Eq. (2.2.12) becomes

$$\begin{aligned} I_\eta(\tau_\eta, \hat{\mathbf{s}}) = {} & I_\eta(0)\mathrm{e}^{-\tau_\eta} \\ & + \int_0^{\tau_\eta} \frac{\omega_\eta}{4\pi} \int_{4\pi} I_\eta(\tau'_\eta, \hat{\mathbf{s}}_i)\, \Phi_\eta(\hat{\mathbf{s}}_i, \hat{\mathbf{s}})\, \mathrm{d}\Omega_i \mathrm{e}^{-(\tau_\eta - \tau'_\eta)} \mathrm{d}\tau'_\eta. \end{aligned} \tag{2.2.19}$$

If the isotropic scattering is taken into consideration, or $\Phi \equiv 1$, the directional integration in Thus, Eq. (2.2.10) can be calculated. As a result,

$$I_\eta(\tau_\eta, \hat{\mathbf{s}}) = I_\eta(0)\mathrm{e}^{-\tau_\eta} + \frac{1}{4\pi} \int_0^{\tau_\eta} \omega_\eta G_\eta(\tau'_\eta)\, \mathrm{e}^{-(\tau_\eta - \tau'_\eta)} \mathrm{d}\tau'_\eta, \tag{2.2.20}$$

$$G_\eta(\tau) \equiv \int_{4\pi}^{\infty} I_\eta(\tau'_\eta, \hat{\mathbf{s}}_i)\, \mathrm{d}\Omega_i, \tag{2.2.21}$$

where Eq. (2.2.21) is called the *incident radiation function.* Because of the isotropic scattering, there is no need to determine the direction-dependent intensity, and the RTE is much simplified.

For the purely scattering medium, when there is scattering but not absorption or emission in the medium, radiation transfer is irrelevant to other modes of heat transfer. In this circumstance, $\omega_\eta \equiv 1$, and the RTE is simplified to

$$\begin{aligned} I_\eta(\tau_\eta, \hat{\mathbf{s}}) = {} & I_\eta(0)\mathrm{e}^{-\tau_\eta} \\ & + \frac{1}{4\pi} \int_0^{\tau_\eta} \int_{4\pi} I_\eta(\tau'_\eta, \hat{\mathbf{s}}_i)\, \Phi_\eta(\hat{\mathbf{s}}_i, \hat{\mathbf{s}})\, \mathrm{d}\Omega_i \mathrm{e}^{-(\tau_\eta - \tau'_\eta)} \mathrm{d}\tau'_\eta. \end{aligned} \tag{2.2.22}$$

If the scattering is isotropic, we can introduce the incident radiation function to reduce the equation, and it can be rewritten as

$$I_\eta(\tau_\eta, \hat{\mathbf{s}}) = I_\eta(0)\mathrm{e}^{-\tau_\eta} + \frac{1}{4\pi} \int_0^{\tau_\eta} G_\eta(\tau'_\eta)\, \mathrm{e}^{-(\tau_\eta - \tau'_\eta)} \mathrm{d}\tau'_\eta. \tag{2.2.23}$$

Solving the RTE requires knowledge of radiative intensity at a single point in space as boundary conditions. Generally, the intensity can be specified independently by the surface of an enclosure surrounding the participating medium. Boundary conditions depend on different surfaces. For diffusely emitting and reflecting opaque surfaces, the exiting intensity is independent of the direction. If the reflectance of the surface has a specular as well as a diffuse component, then the outgoing intensity also consists of two components. One part of the outgoing intensity is due to diffuse emission as well as the diffuse fraction of reflected energy. In addition, the outgoing intensity has a spectacular reflected component. For opaque surfaces with arbitrary surface properties, reflection from a surface with nonideal radiative properties is governed by the bidirectional reflection function. If the boundary is a semitransparent wall, external radiation may penetrate into the enclosure and must be added to the intensity. If the boundary is located on the interface between two semitransparent media. The interface between two semitransparent media is of interest only if radiation can penetrate an appreciable distance through either medium – if not, then the optically dense medium may be modeled as an "opaque surface." This implies that the absorptive indexes of both media are very small.

Besides the Cartesian coordinate system, cylindrical and spherical coordinate systems are the other two commonly used orthogonal coordinate systems. Here the traditional RTE in a cylindrical coordinate system is presented. For the cylindrical coordinate system, the stream operator can be expanded as

$$\frac{\mathrm{d}}{\mathrm{d}s}=\frac{\mathrm{d}\rho}{\mathrm{d}s}\frac{\partial}{\partial\rho}+\frac{\mathrm{d}\Psi}{\mathrm{d}s}\frac{\partial}{\partial\Psi}+\frac{\mathrm{d}z}{\mathrm{d}s}\frac{\partial}{\partial z}+\frac{\mathrm{d}\theta}{\mathrm{d}s}\frac{\partial}{\partial\theta}+\frac{\mathrm{d}\varphi}{\mathrm{d}s}\frac{\partial}{\partial\varphi}=\mathbf{\Omega}\cdot\nabla_{\mathrm{I}}-\frac{\eta}{\rho}\frac{\partial}{\partial\varphi}, \tag{2.2.24}$$

where $\mathbf{\Omega}=\mu\mathbf{e}_\rho+\eta\mathbf{e}_\psi+\xi\mathbf{e}_z$ is the local direction vector of the beam, $\mu=\sin\theta\cos\varphi$, $\eta=\sin\theta\cos\varphi, \xi=\cos\theta, \mathbf{e}_\rho, \mathbf{e}_\psi$, and $\mathbf{e}_z$ are the unit coordinate vectors, and $\nabla_{\mathrm{I}}=\mathbf{e}_\rho\frac{\partial}{\partial\rho}+\mathbf{e}_\psi\rho^{-1}\frac{\partial}{\partial\Psi}+\mathbf{e}_z\frac{\partial}{\partial z}$ is the gradient operator in Type I cylindrical coordinate system. Hence, the RTE in the Type I cylindrical coordinate system can be written as

$$\mathbf{\Omega}\cdot\nabla I_\lambda-\frac{\eta}{\rho}\frac{\partial I_\lambda}{\partial\varphi}+\beta_\lambda I_\lambda=\kappa_{a,\lambda}I_{b,\lambda}+\frac{\kappa_{s,\lambda}}{4\pi}\int_{4\pi}^{\infty}I_\lambda\left(\boldsymbol{r},\mathbf{\Omega}'\right)\Phi_\lambda\left(\mathbf{\Omega}',\mathbf{\Omega}\right)\mathrm{d}\Omega'. \tag{2.2.25}$$

In addition, other types of cylindrical coordinate systems $(\rho,\Psi,z,\ \theta,\varphi)$ were proposed for radiative transfer analysis, in which the definition of local angular variables was redefined, that is, the zenith angle θ and the azimuthal angle φ. By definition, the optical plane of reflection or refraction at the cylindrical interfaces coincide with the isosurface of the azimuthal angle φ; hence, it facilitates the treatment of reflection/refraction at the cylindrical interfaces/boundaries.

2.2.2 The Radiative Transfer Equation for a Medium with Gradient Refraction

For the RTEs presented in Section 2.2.1, it is assumed that the direction of light beam goes along a straight path in participating media of a uniform refractive index distribution. If the refractive index of the considered medium is a function of the spatial position, which may be the result of material, component, temperature, or pressure, the thermal radiative behavior of the medium can consequently be modified. In recent decades, due to emerging applications in atmospheric radiation, ocean optics, thermo-optical systems, and optoelectronic devices, there has been a growing interest in studying radiation transport in media with a graded refractive index. Based on the Fermat principle, light in the media travels along a curved ray path. Therefore, the formulation of the RTE in refractive media must take the effect of gradient index refraction into consideration. In this section, we provide a brief outline of the RTE in gradient refractive index media and its formulation under different coordinate systems.

By considering the effect of the gradient of refractive index, the governing equation of radiative transfer in gradient index media can be considered as an extension of the classical RTE. The variation of radiative intensity along the light beam can be attributed to two parts, one is attenuation and augmentation processes caused by absorption, scattering, and emission processes as discussed in uniform refractive index media. The other one accounts for the variation of refractive index. The total variation of radiative intensity can be written as

$$\mathrm{d}I_\lambda(s,n(s),s)=\frac{\partial I_\lambda(s,n(s),s)}{\partial s}\,\mathrm{d}s+\frac{\partial I_\lambda(s,n(s),s)}{\partial n}\,\mathrm{d}n, \tag{2.2.26}$$

where the first term on the RHS represents the variation caused by the processes of absorption, scattering, and emission, which can be expressed based on the RTE in uniform refractive index media as

$$\begin{aligned}&\frac{\partial I_\lambda(s,n(s),\boldsymbol{s})}{\partial s}\\&=-\beta_\lambda I_\lambda(\boldsymbol{s},s)\,ds+\kappa_{a,\lambda}I_{b,\lambda}[T(s)]\\&+\frac{\kappa_{a,\lambda}}{4\pi}\int_{4\pi}I_\lambda\left(s,\boldsymbol{\Omega}'\right)\Phi_\lambda\left(\boldsymbol{\Omega}'\cdot\mathrm{s}\right)\mathrm{d}\Omega'.\end{aligned} \tag{2.2.27}$$

The second term stands for the variation only caused by the gradient of refractive index. For transparent gradient index media, the Clausius invariant relation gives $\mathrm{d}\left[I_\lambda/n^2\right]=0$, from which the second term in Eq. (2.2.26) can be obtained as

$$\frac{\partial I_\lambda}{\partial n}\,\mathrm{d}n=2\frac{I_\lambda}{n}\,\mathrm{d}n. \tag{2.2.28}$$

Substituting Eq. (2.2.28) into Eq. (2.2.26), then the RTE in the gradient index medium (GRTE) in the Lagrange form along the ray coordinate can be expressed as

$$n^2\frac{\mathrm{d}}{\mathrm{d}s}\left[\frac{I_\lambda(s,s)}{n^2}\right]+\beta_\lambda I_\lambda(s,\boldsymbol{s})=\kappa_{a,\lambda}I_{b,\lambda}[T(s)]+\frac{\kappa_{s,\lambda}}{4\pi}\int_{4\pi}I_\lambda\left(s,\boldsymbol{\Omega}'\right)\Phi_\lambda\left(\boldsymbol{\Omega}'\cdot\mathrm{s}\right)\mathrm{d}\Omega'. \tag{2.2.29}$$

By expanding the Lagrangian stream operator in the Eulerian frame, the transient GRTE is obtained:

$$\begin{aligned}&\frac{n_\lambda}{c}\frac{\partial I_\lambda(s,t,\boldsymbol{s})}{\partial t}+n^2\frac{\mathrm{d}}{\mathrm{d}s}\left[\frac{I_\lambda(s,t,\boldsymbol{s})}{n^2}\right]+\beta_\lambda I_\lambda(s,t,\boldsymbol{s})\\&=\kappa_{a,\lambda}I_{b,\lambda}[T(s)]+\frac{\kappa_{s,\lambda}}{4\pi}\int_{\Delta\pi}I_\lambda\left(s,t,\boldsymbol{\Omega}'\right)\Phi_\lambda\left(\boldsymbol{\Omega}'\cdot\mathrm{s}\right)\mathrm{d}\Omega'.\end{aligned} \tag{2.2.30}$$

It should be noted that the GRTE itself does not reflect any information about the curved ray path. To solve the equation, the ray equation shown below must be solved

$$\frac{\mathrm{d}}{\mathrm{d}s}(ns)=\nabla n. \tag{2.2.31}$$

Assuming the radiative intensity is expressed as $I\left(r,\Omega\right)=I(x,y,z,\theta,\varphi)$ in the Cartesian coordinate system, the stream operator can be expanded as

$$\begin{aligned}\frac{\mathrm{d}}{\mathrm{d}s}&=\frac{\mathrm{d}x}{\mathrm{d}s}\frac{\partial}{\partial x}+\frac{\mathrm{d}y}{\mathrm{d}s}\frac{\partial}{\partial y}+\frac{\mathrm{d}z}{\mathrm{d}s}\frac{\partial}{\partial z}+\frac{\mathrm{d}\theta}{\mathrm{d}s}\frac{\partial}{\partial\theta}+\frac{\mathrm{d}\varphi}{\mathrm{d}s}\frac{\partial}{\partial\varphi}\\&=\boldsymbol{\Omega}\cdot\nabla+\frac{\mathrm{d}\theta}{\mathrm{d}s}\frac{\partial}{\partial\theta}+\frac{\mathrm{d}\varphi}{\mathrm{d}s}\frac{\partial}{\partial\varphi}.\end{aligned} \tag{2.2.32}$$

The wavelength subscript has been omitted without loss of generality. The explicit formulation of the two angular derivatives $\frac{d\theta}{ds}$ and $\frac{d\varphi}{ds}$ can be obtained by the ray equation

$$\frac{\mathrm{d}\varphi}{\mathrm{d}s}=\frac{1}{\sin\theta}\left(s_1\cdot\frac{\nabla n}{n}\right),\ \frac{\mathrm{d}\theta}{\mathrm{d}s}=\frac{1}{\sin\theta}\left[(\xi\mathbf{\Omega}-\mathbf{k})\cdot\frac{\nabla n}{n}\right], \tag{2.2.33}$$

where s_1 is an auxiliary vector defined as $s_1 = -\sin\varphi\mathbf{i} + \cos\varphi\mathbf{j}$. Substituting Eq. (2.2.32) and Eq. (2.2.33) into the GRTE in ray coordinate Eq. (2.2.26) and after some manipulations, the final conservative form of the GRTE in the Cartesian coordinate system can be obtained as

$$\mathbf{\Omega}\cdot\nabla I(\mathbf{r},\mathbf{\Omega})+\frac{1}{\sin\theta}\frac{\partial}{\partial\theta}\left[I(\mathbf{r},\mathbf{\Omega})(\xi\mathbf{\Omega}-\mathbf{k})\cdot\frac{\nabla n}{n}\right]+\frac{1}{\sin\theta}\frac{\partial}{\partial\varphi}\left[I(\mathbf{r},\mathbf{\Omega})s_1\right]\cdot\frac{\nabla n}{n}$$
$$+(\kappa_a+\kappa_s)I(\mathbf{r},\mathbf{\Omega})=\kappa_a I_b(r)+\frac{\kappa_s}{4\pi}\int_{4\pi}I_\lambda\left(r,\mathbf{\Omega}'\right)\Phi\left(\mathbf{\Omega},\mathbf{\Omega}'\right)\mathrm{d}\Omega'. \tag{2.2.34}$$

It can be seen that the second and third terms about the gradient refractive index in Eq. (2.2.34) are related to the two derivatives of the angular variables θ and φ, which are caused by the gradient distribution of refractive index and called angular redistribution. When incident light propagation through a gradient medium, the direction of rays continuously changes, following curvilinear paths instead of straight paths. Snell's law, $n\sin\theta=$ const. also indicates the phenomenon, yielding $\sin\theta\mathrm{d}n+n\cos\theta\,\mathrm{d}\theta=0$ and $\mathrm{d}\theta=-(\tan\theta/n)\,\mathrm{d}n$. It indicates that a small change in n will lead to a variation of θ, causing bending rays in a gradient medium. For example, laser beams used in optical diagnostics may deviate from the original alignment due to the variation in the refractive index of the medium caused by local heating or cooling. Besides, the mirage phenomenon in deserts is also the result of spatial variations of the refractive index.

2.3 The Vector Radiative Transfer Equation

Generally, polarization effects are not taken into consideration in most heat transfer applications, as radiation is randomly polarized or nonpolarized. In some applications with laser sources, partially or fully polarized light is employed. Therefore, the influence of polarized beams on the emitted light needs to be investigated. We shall give here only a very brief introduction to polarization. More detailed accounts on the subject are referred to in the books by van de Hulst [36], Chandrasekhar [41], and others.

2.3.1 Stokes Vector

Polarization is a fundamental property of EM waves. The Stokes vectors are widely used to characterize the polarization states of light, which can be used to derive the ellipsometric characteristics of the plane EM wave, presented by Eqs. (2.3.1) and (2.3.2)

$$\varepsilon_x = a_x \mathrm{e}^{-\frac{\mathrm{i}\delta(t)}{2}}, \tag{2.3.1}$$

$$\varepsilon_y = a_y \mathrm{e}^{-\frac{\mathrm{i}\delta(t)}{2}}, \tag{2.3.2}$$

where $a_x^2 = \langle A_x^2(t)\rangle$ and $a_y^2 = \langle A_y^2(t)\rangle$ represent the square of the amplitudes $A_x(t)$ and $A_y(t)$ respectively of the EM wave over the measurement time t. δ is the phase shift $(0 \le \delta \le 2\pi)$. The Stokes parameters of the explicit form can be obtained as

$$\begin{aligned}
s_0 &= \langle \varepsilon_x(t)\varepsilon_x^*(t)\rangle + \langle \varepsilon_y(t)\varepsilon_y^*(t)\rangle \\
s_1 &= \langle \varepsilon_x(t)\varepsilon_x^*(t)\rangle - \langle \varepsilon_y(t)\varepsilon_y^*(t)\rangle \\
s_2 &= \langle \varepsilon_x(t)\varepsilon_y^*(t)\rangle + \langle \varepsilon_y(t)\varepsilon_x^*(t)\rangle \\
s_3 &= \mathrm{i}\left(\langle \varepsilon_x(t)\varepsilon_y^*(t)\rangle + \langle \varepsilon_y(t)\varepsilon_x^*(t)\rangle\right).
\end{aligned} \tag{2.3.3}$$

The quantities s_0, s_1, s_2, and s_3 are the so-called Stokes parameters and constitute a complete set of measurable parameters, which allow for expressing coherency matrix Φ in the following manner:

$$\phi = \frac{1}{2}\begin{pmatrix} s_0 + s_1 & s_2 - \mathrm{i}s_3 \\ s_2 + \mathrm{i}s_3 & s_0 - s_1 \end{pmatrix}. \tag{2.3.4}$$

The coherency matrix is a Hermitian matrix, in which a pair of necessary and sufficient conditions need to be satisfied. The nonnegativity of Φ entails the following constraints:

$$\mathrm{tr}\,\phi = s_0 \ge 0, 4\det\phi = s_0^2 - s_1^2 - s_2^2 - s_3^2 \ge 0. \tag{2.3.5}$$

Hence, any set of four parameters s_0, s_1, s_2, and s_3 satisfying conditions Eq. (2.3.5) can be considered a physically realizable set of Stokes parameters, which are usually arranged as a 4×1 Stokes vectors. It should be noted that the Stokes parameters are also frequently noted as I, Q, U, V:

$$\vec{\boldsymbol{I}} = \begin{pmatrix} s_0 \\ s_1 \\ s_2 \\ s_3 \end{pmatrix}. \tag{2.3.6}$$

Thus, the relation between ϕ and $\vec{I}$ can also be expressed as

$$\vec{I} = L\phi, \tag{2.3.7}$$

where

$$L = \begin{bmatrix} 1 & 0 & 0 & 1 \\ 1 & 0 & 0 & -1 \\ 0 & 1 & 1 & 0 \\ 0 & \mathrm{i} & -\mathrm{i} & 0 \end{bmatrix}. \tag{2.3.8}$$

The conversion of Stokes vectors in a coordinate frame from XY to $X'Y'$ can be performed using an orthogonal transformation of the form

$$s' = M_G(\theta) s M_G(\theta) = \begin{bmatrix} 1 & 0 & 0 & 0 \\ 0 & \cos 2\theta & \sin 2\theta & 0 \\ 0 & -\sin 2\theta & \cos 2\theta & 0 \\ 0 & 0 & 0 & 1 \end{bmatrix}, \quad (2.3.9)$$

where the orthogonal matrix $M_G(\theta)$ corresponds to a proper counterclockwise rotation by the angle θ about the axis Z, from the original X reference axis to X'.

A Stokes vector constitutes two parts that are usually expressed as

$$\mathbf{s} = \left(\sqrt{s_1^2 + s_2^2 + s_3^2}, s_1, s_2, s_3\right)^T + \left(s_0 - \sqrt{s_1^2 + s_2^2 + s_3^2}, 0, 0, 0\right)^T, \quad (2.3.10)$$

where $\mathbf{s}$ can be interpreted as an incoherent item of a pure state (called the *characteristic component*) and an unpolarized state (called *the 2D unpolarized component*). The first item on the RHS defines the *characteristic polarization ellipse* of the whole polarization states. Furthermore, $\mathbf{s}$ can be parameterized as

$$\mathbf{s} = \begin{bmatrix} 1 \\ P\cos 2\varphi \cos 2\chi \\ P\sin 2\varphi \cos 2\chi \\ P\sin 2\chi \end{bmatrix} = PI \begin{bmatrix} 1 \\ \cos 2\varphi \cos 2\chi \\ \sin 2\varphi \cos 2\chi \\ \sin 2\chi \end{bmatrix} + (1-P)I \begin{bmatrix} 1 \\ 0 \\ 0 \\ 0 \end{bmatrix}, \quad (2.3.11)$$

where $I = s_0$ is the intensity of the reference plane containing the polarization ellipse. The azimuth angle φ, with $0 \leq \varphi < \pi$, is the direction of the major semiaxis of the characteristic polarization ellipse with respect to the given reference axis X. The ellipticity angle χ, with $-\pi/4 \leq \chi \leq \pi/4$, is the characteristic polarization ellipse. $P = \sqrt{s_1^2 + s_2^2 + s_3^2}$ represents the degree of polarization of the 2D state. P is a dimensionless quantity whose values are restricted to $0 \leq P \leq 1$. $P = 1$ represents totally polarized states. $P = 0$ corresponds to unpolarized states; in other words, it corresponds to the states with a completely random temporal distribution of the polarization ellipse, where the field components have no correlations with each other. The intermediate values $0 < P < 1$ represent the partially polarized states; the higher the value of the degree of polarization P, the stronger the correlation between the field components.

The above parameters provide an interpretation of 2D states of polarization. The Stokes parameters can also be interpreted in terms of meaningful physical quantities. s_0 is the intensity that can be considered as the sum of the intensities associated with the components of the electric field with respect to any orthonormal generalized basis: $s_0 = I_x + I_y = I_{+45°} + I_{-45°} = I_r + I_l \cdot s_1$ denotes the difference between the respective intensities corresponding to the components of the electric field with respect to the canonical generalized basis $(e_x, e_y), s_1 = I_x - I_y$. $s_1 = 1(\varphi = 0, \chi = 0)$ for linearly x-polarized states, whereas $s_1 = -1(\varphi = \pi/2, \chi = 0)$ for linearly y-polarized states. The Stokes vector can be measured by simple procedures. The parameter s_1 of a plane wave can be realized by intensity measurements with two linear polarizers (usually called an analyzer) at $0°$ and $90°$. The two consecutive intensities are measured by placing the two linear polarizers before the detector. s_2 denotes the difference between the respective intensities corresponding to the components of the electric field with respect to the basis $\left(e_{+\pi/4}, e_{-\pi/4}\right)$, $s_2 = I_{+\pi/4} - I_{-\pi/4}$. $s_2 = 1(\varphi = \pi/4, \chi = 0)$ for linearly $+45°$-polarized states, whereas $s_2 = -1(\varphi = 3\pi/4, \chi = 0)$ for linearly $-45°$-polarized states. The parameter s_2 of a plane wave can be realized by intensity measurements by respectively placing a linear analyzer at $+45°$ and $-45°$ before the detector. s_3 denotes the difference between the respective intensities corresponding to the components of the electric field with respect to the basis (e_r, e_l), $s_3 = I_r - I_r$.$s_3 = 1(\chi = \pi/4)$ represents the right-handed circular polarized states, whereas $s_3 = -1(\chi = -\pi/4)$ represents the left-handed circularly polarized states. The parameter s_3 of a plane wave can be realized by intensity measurements by respectively placing a right-circular analyzer and a left-circular analyzer before the detector. Note that a circular polarizer (analyzer) can be achieved by the serial combination of a linear total polarizer and a quarter-wave plate whose respective eigenvalues make an angle of $45°$.

In summary, 2D unpolarized states entail the equality of the intensities associated with the respective pair of the orthogonal components of an electric field with respect to the bases (e_x, e_y), $s_1 = I_y - I_x = 0$; $\left(e_{+\pi/4}, e_{-\pi/4}\right)$, $s_2 = I_{+\pi/4} - I_{-\pi/4} = 0$; and (e_r, e_l), $s_3 = I_r - I_l = 0$. In fact, the fulfillment of these three equalities implies necessarily the equality $I_1 = I_2$ of the intensities I_1 and I_2 associated with the respective pair of the orthogonal components of an electric field with respect to any generalized orthonormal bases e_1, e_2. The indicated invariance is an essential and characteristic property of unpolarized states due to various random distributions of unpolarized waves. Different types of unpolarized states can be distinguished by the correlations of the Stokes parameters.

Any 2D polarization states can be considered the result of the incoherent superposition of the characteristic component $s_p = \left(\sqrt{s_1^2 + s_2^2 + s_3^2}, s_1, s_2, s_3\right)^T$ and the 2D unpolarized component $s_u = \left(s_0 - \sqrt{s_1^2 + s_2^2 + s_3^2}, 0, 0, 0\right)^T$.

A 2D state represented by a given Stokes vector $\boldsymbol{s} = (I, s_1, s_2, s_3)^T$ is polarimetrically indistinguishable from an incoherent combination of two states propagating in the same direction, namely, a pure state s_p with intensity $I_p = \sqrt{s_1^2 + s_2^2 + s_3^2} = PI$ and a 2D unpolarized state with intensity $I_u = I - \sqrt{s_1^2 + s_2^2 + s_3^2} = (1 - P)I$. For pure states that are characterized by the equality $P = 1$, the total intensity is associated with the pure contribution (the characteristic component), and the shape of the polarization ellipse is constant for time intervals larger than the measurement time. For 2D unpolarized states that propagate in a well-defined direction and satisfy the equality $P = 0$, the total intensity is associated with the unpolarized contribution where the shape of the polariza-

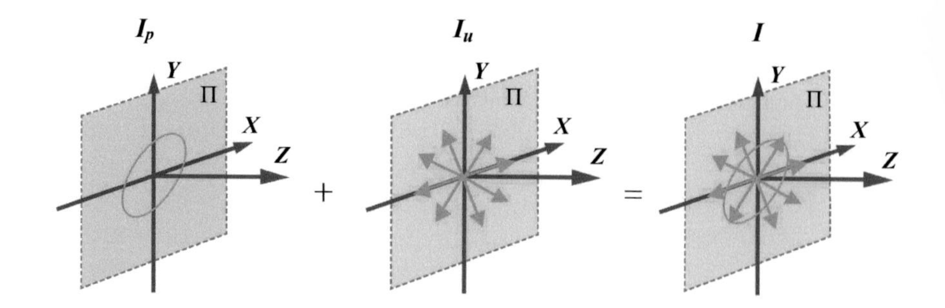

Figure 2.5 The degree of polarization P of a 2D state is defined as the ratio of the intensity I_p of the totally polarized component to the intensity $I = I_p + I_u$ of the whole state.

tion ellipse fluctuates in a completely random manner during the measurement time.

The degree of polarization P is just the ratio of the intensity $I_p = \sqrt{s_1^2 + s_2^2 + s_3^2} = PI$ of the characteristic component to the intensity $\mathrm{I} = I_p + I_u$ of the entire state (Fig. 2.5). Thus, P is a dimensionless and nonnegative quantity limited by the double inequality $0 \leq P \leq 1$. Moreover, P is invariant with respect to any rotation of the underlying reference frame XY about the direction of propagation Z. Furthermore, from a more general point of view, P is invariant with respect to any change in the generalized underlying reference bases (e_1, e_2), that is, with respect to any unitary transformation of the basis vectors (e_1, e_2).

2.3.2 The Expression of the Vector Radiative Transfer Equation

For microscopically isotropic and symmetric plane-parallel scattering media, the vector radiative transfer equation (VRTE) can be substantially simplified as follows [42]:

$$\frac{\mathrm{d}\vec{\boldsymbol{I}}(\bar{r},\theta,\varphi)}{\mathrm{d}\tau(\bar{r})} = -\vec{\boldsymbol{I}}(\bar{r},\theta,\varphi) + \frac{\Lambda(\bar{r})}{4\pi}\int_{-1}^{+1}\mathrm{d}(\cos\theta')\int_{0}^{2\pi}\mathrm{d}\varphi'\overline{\mathrm{Z}}(\bar{r},\theta,\theta',\varphi-\varphi')\,\vec{\boldsymbol{I}}(\bar{r},\theta',\varphi'), \tag{2.3.12}$$

where $\vec{I}$ is the Stokes vector, $\bar{r}$ is the position vector, θ and φ are the angles characterizing the incident direction, that is, the zenith and the azimuth angles,

$$\mathrm{d}\tau(\bar{r}) = \rho(\bar{r})\,\langle\sigma_{ext}(\bar{r})\rangle\,\mathrm{d}s \tag{2.3.13}$$

is the optical path-length element, ρ is the local particle number density, $\langle\sigma_{ext}(\bar{r})\rangle$ is the local ensemble-averaged extinction coefficient measured along the unit vector of the direction of light propagation, Λ is the single-scattering albedo, θ'

and φ' are the angles that characterize the scattering direction of the zenith and azimuth angles, respectively, and $\bar{Z}$ is the normalized phase matrix given by

$$\bar{Z}(\bar{r},\theta,\theta',\varphi-\varphi') = R(\phi)M(\theta)R(\Psi), \tag{2.3.14}$$

where $M(\theta)$ is the single-scattering Mueller matrix, θ is the scattering angle, and $R(\phi)$ is the Stokes rotation matrix for angle ϕ given by

$$R(\phi) = \begin{bmatrix} 1 & 0 & 0 & 0 \\ 0 & \cos 2\phi & -\sin 2\phi & 0 \\ 0 & \sin 2\phi & \cos 2\phi & 0 \\ 0 & 0 & 0 & 1 \end{bmatrix}. \tag{2.3.15}$$

The phase matrix, Eq. (2.3.4), links the Stokes vectors of the incident and scattered beams, specified relative to their respective meridional planes. To compute the Stokes vector of a scattered beam with respect to its meridional plane, one must calculate the Stokes vector of the incident beam with respect to the scattering plane, multiply it by the scattering matrix (to obtain the Stokes vector of the scattered beam with respect to the scattering plane), and then compute the Stokes vector of the scattered beam with respect to its meridional plane: $\Phi = -\varphi; \Psi = \pi - \varphi$ and $\Phi = \pi + \varphi; \Psi = \varphi$. The scattering angle θ and the angles Φ and Ψ are expressed via the polar and azimuth incident and scattering angles

$$\begin{aligned} \cos\theta &= \cos\theta'\cos\theta + \sin\theta'\sin\theta\cos(\varphi'-\varphi), \\ \cos\phi &= \frac{\cos\theta - \cos\theta'\cos\theta}{\sin\theta'\sin\theta}, \\ \cos\psi &= \frac{\cos\theta' - \cos\theta'\cos\theta}{\sin\theta\sin\theta}. \end{aligned} \tag{2.3.16}$$

The first term on the RHS of Eq. (2.3.12) describes the change in the specific intensity vector over the distance ds that is caused by extinction and dichroism; the second term describes the contribution of light illuminating a small volume element centered at $\bar{r}$ from all incident directions and scattered in the chosen direction. For real systems, the form of the VRTE tends to be rather complex and often intractable. Therefore, a wide range of analytical and numerical techniques have been developed to solve the VRTE. Because of an important property of the normalized phase matrix, Eq. (2.3.4) is dependent on the difference of the azimuthal angles of the scattering and incident directions rather than on their specific values [43]. An efficient analytical treatment of the azimuthal dependence of the multiply scattered light, using a Fourier decomposition of the VRTE, is possible. The following techniques and their combinations can be used to solve VRTE: the transfer matrix method, the singular eigenfunction method, the perturbation method, the small-angle approximation, the adding–doubling method, the matrix operator method, the invariant embedding method, and the Monte Carlo method (MCM).

2.4 Approximate Methods for the Solution of the RTE

Exact analytical solutions to the radiative transfer are exceedingly difficult, and explicit solutions are impossible for all but the very simplest situations. Therefore, research on radiative heat transfer in participating media has generally proceeded in three directions: (i) exact (analytical) solutions of highly idealized situations, (ii) approximate solution methods for more involved scenarios, and (iii) numerical solutions. The RTEs are an integral differential equation for the radiance as a function of six variables: x, y, z, θ, φ, and t respectively related to three spatial, two-directional coordinates and time. It is worth mentioning that the time-dependent term in the RTE is always neglected because the speed of light is much larger than the moving velocity of particles in media. However, for some cases, that time-dependent term must be taken into consideration, and the free additional variable for time (or frequency) cannot be neglected. Hence, even though the exact analytical solutions of the RTE can be easily deducted for simple geometries with uniform optical parameter distributions, it is hard to solve it considering all the variables. Therefore, for more general light propagation problems, such as media with complex geometries and nonuniform optical parameter distributions, many approximated methods to solve the RTE have been proposed. For a self-absorbing medium, approximate methods for calculating radiative transfer in multidimensional geometries can be classified into two groups. The first group is directional averaging approximations, including 2-flux, 4-flux, multiflux, and so on. The second group is differential approximations, such as moment, modified moment, spherical harmonics, and so on. Moment methods are widely used to solve the differential form of the RTEs, which are based on the expansion of the intensity along the transport direction of light. High-order approximations, such as the discrete ordinates (S_N) or spherical harmonics (P_N), are used to solve the RTE, which transforms the RTE into a large system of differential equations (S_N or P_N equations). The S_N- and P_N-approximations yield exact solutions to the RTE as the order N approaches ∞. The S_N- and P_N-methods have a wide application range in heat transfer. The number of P_N equations grows as $(N+1)^2$ for a 3D medium. The S_N method solves a set of $N(N+2)$-coupled equations. Consequently, by truncating the RTEs to a small number of terms, the S_N and P_N equations form a large system of equations, which requires expensive computational techniques. The P_N-approximation with its $(N+1)^2$ equations have been simplified by approximating the mixed spatial derivatives with simple Laplacian operators and a system of only $(N+1)/2$-simplified spherical harmonics (SP_N) equations for N being odd can be obtained. The S_N-approximation consists of coupled diffusion equations (DEs) and covers most of the radiative transfer properties in media [44]. The analytical solutions of the RTEs for half-space problems can be solved by using Green's functions approach. Besides, for an optically thin or optically thick medium, the RTEs can further be mathematically simplified. For example, the DE is a low-order approximation to the RTE and does not cover all transport properties of light in media. It is easy to solve the DE, which does not require a large number of computational solution techniques. In this part, we discuss some general approximation methods to simplify the RTE to determine the local radiative flux and the local spectral intensity.

For a weak scattering medium, the scatterers are rarefied and independent of each other, and the scattering volume is limited; iteration may be used to solve the RTE. In the first approximation, the iterative solution drives the known first-order approximation of the radiative transfer theory. In this approximation, it is assumed that the total intensity incident on the scatters in a medium is approximately equivalent to the attenuated incident intensity. Hence, the solution within the first-order approximation can be presented as

$$I(\bar{r},\bar{s}) = I_{ri}(\bar{r},\bar{s}) + I_d(\bar{r},\bar{s}), \tag{2.4.1}$$

$$I_d(\bar{r},\bar{s}) = \int_0^s \exp\left[-(\tau-\tau_1)\right]\left[\frac{\mu_s}{4\pi}\int_{4\pi} I_{ri}(\bar{r}_1,\bar{s}')\,p(\bar{s},\bar{s}')\,\mathrm{d}\Omega'\right]\mathrm{d}\bar{s}', \tag{2.4.2}$$

where I_{ri} is the attenuated incident intensity, I_d is the diffuse intensity, and $\tau = \int_0^s \rho\sigma_e \mathrm{d}s, \tau_1 = \int_0^{s_1} \rho\sigma_e \mathrm{d}s$ are the optical path lengths.

The so-called first-order solution applies to optically thin and weak scattering media ($\tau < 1$ and single-scattering albedo $\Lambda = \sigma_s/\sigma_e < 0.5$). Assume a narrow beam like a laser; this approximation may be applied to optically denser tissues ($\tau > 1, \Lambda < 0.9$) [45].

For flux theory, the radiation intensity I itself is not often of interest, but its integrals yielding the energy characteristics of the radiation field are used. If the incident light is diffuse and the medium is sufficiently turbid, providing the light diffusion scattering, experimental results are well described by the two-flux Kubelka–Munk theory [46]. This theory relies on a model of two light fluxes propagating in the forward and backward directions. The extension of the two-flux Kubelka–Munk theory to a four-flux theory makes it possible to describe a collimated beam incidence onto the medium. The four-flux model is actually two diffuse fluxes traveling to meet each other (the Kubelka–Munk model) and two collimated laser beams, the incident beam, and the beam reflected from the rear boundary of the sample.

A seven-flux model is the simplest 3D representation of scattered radiation and an incident laser beam in a semi-infinite medium [47]. Of course, the simplicity and the possibility of expeditious calculations of the radiation dose or rapid determination of tissue optical parameters (the solution of the inverse scattering problem) are achieved at the expense of accuracy.

There exist various numerical procedures to solve the RTE. A more exact solution of the RTE is possible using the discrete ordinates method (multiflux theory) in which Eq. (2.2.9) is converted into a matrix differential equation for illumination along many discrete directions. The solution approximates an exact one as the number of angles increases. Many computer programs are available to solve the RTE under different conditions.

In an optically thick medium, a light beam travels only a short distance before being scattered or absorbed. The local intensity at a point depends on the radiation from the immediate vicinity of the position, including the emission and scattering process. Radiation of one position is greatly attenuated in an optically thick medium before reaching another position being considered, resulting in obviously different conditions. Energy transfer is obtained according to the gradient of the conditions at that position, which is only determined by the conditions of the surroundings being considered. With a longer optical path, the transport direction of the scattered light is broadened, and at last, it becomes almost isotropic: The scattered light "forgets" the initial direction of wave propagation in the interior of the medium – the depth regime. In this circumstance, the intensity is nearly to be isotropic inside the medium, the radiative energy can be considered as a heat stream similar to heat conduction; thus the expression of the RTE can be transformed into a diffusion relation. Owing to this almost isotropic radiation, the diffusion approximation makes the classic differential RTE simplified to the solution of differential equations. The DE has been widely used as an effective approach to investigate multiple scattering processes due to its brevity. To derive the DE, the RTE should be applied to an infinite system so that equations for the coefficients of beam intensity could be approximated as expansions of spherical harmonic functions. The reduction of this system brings about the so-called P_L-approximation, the simplest of which is a series of equations for four functions that are equal to the diffusion approximation. Other approaches also exist to conduct the DEs with the diffusion coefficient. However, all differences vanish when passing to the limit of weak absorption, which is valid for many applications. It is easier to understand that from the physics of the diffusion approximation. Actually, the description of scattering as a certain diffusion process only becomes adequate if the strong scattering process is dominant compared with the absorption process.

As a simplification of the RTE, the diffusion approximation can be solved with an analytical solution. The differential DE equation can also be solved by standard techniques such as finite-difference method or discrete dipole approximation (DDA) schemes. The diffusion approximation requires that the medium is optically thick with small temperature gradients; in most cases, the optical thickness of the medium should be larger than about 10. This value depends on the wavelength conditions, the optical characteristic, and the geometry of the media. In some cases, this value may need to be larger. However, sometimes good results for some cases with much smaller optical thicknesses are predicted, even though the mean optical thickness does not meet this criterion. The diffusion approximation is not valid near boundaries for certain types of conditions. It must be careful when strong absorption and small geometries of media are considered. If the boundary of a cold medium is adjacent to a much higher temperature environment, there will be a great difference in the radiation entering and leaving the medium. In other words, the radiation in boundaries "has not yet forgotten" the initial conditions. Due to the large anisotropic stream, the diffusion approximation is not accurate near the boundary. If the process of heat conduction is dominant near a boundary, the inaccuracy of the radiative diffusion can be neglected; otherwise, a temperature jump boundary condition should be used to improve accuracy.

The diffusion theory provides a good approximation for a small scattering anisotropy factor $g \leq 0.1$ and a large albedo $\Lambda \to 1$. For many biological tissues, $g \approx 0.6$–0.9 and can be as large as $0.990-0.999$ for blood. This significantly restricts the applicability of the diffusion approximation. It is argued that this approximation can be used at $g < 0.9$ when the optical thickness τ of an object is of the order 10–20. However, the diffusion approximation is inapplicable for an input beam near the object's surface where single- or low-step scattering prevails.

The first-order approximation is valid only when the volume density, equal to the relation between a volume occupied by the particle and the entire volume of the medium, is substantially less than 1%. The diffusion approximation is very credible when the volume density is much greater than 1%. For volume density of the order of 1%, neither the near-order approximation nor the diffusion approximation can be valid, and we need to solve the RTE.

To model light propagation in highly scattering media, it is useful to derive the DE from the RTE, since the DE is often an excellent approximation and is a much simpler equation to solve. For this purpose, we expand the angular photon density $u(\boldsymbol{r}, \widehat{\Omega}, t)$, the source term $q(\boldsymbol{r}, \widehat{\Omega}, t)$, and the phase function $f\left(\widehat{\Omega}', \widehat{\Omega}\right)$ into the spherical harmonics $Y_l^m(\widehat{\Omega})$. As a result of the completeness property of the spherical harmonics, any function $h(\theta, \varphi)$ (where the angles θ and φ identify a direction $\widehat{\Omega}$ in the space) can be expanded in the Laplace series

$$h(\theta, \varphi) = \sum_{l=0}^{\infty} \sum_{m=-l}^{l} h_{lm} Y_l^m(\widehat{\Omega}), \tag{2.4.3}$$

where h_m are the coefficients independent of u and w, and $\widehat{\Omega}$ is the unit directional vector defined by $\widehat{\Omega} = \sin\theta \cos\varphi \hat{\mathbf{x}} + \sin\theta \cos\varphi \hat{\mathbf{y}} + \cos\theta \hat{\mathbf{z}}$. Accordingly, we expand $u(\boldsymbol{r}, \widehat{\Omega}, t), q(\boldsymbol{r}, \widehat{\Omega}, t)$, and $f\left(\widehat{\Omega}', \widehat{\Omega}\right)$ into spherical harmonics as follows:

$$u(\boldsymbol{r}, \widehat{\Omega}, t) = \sum_{\substack{l=0 \\ \infty}}^{\infty} \sum_{m=-l}^{l} u_{lm}(\boldsymbol{r}, t) Y_l^m(\widehat{\Omega}), \tag{2.4.4}$$

$$q(\boldsymbol{r}, \widehat{\Omega}, t) = \sum_{l=0}^{\infty} \sum_{m=-l}^{l} q_{lm}(\boldsymbol{r}, t) Y_l^m(\widehat{\Omega}), \tag{2.4.5}$$

$$f\left(\widehat{\Omega'}, \widehat{\Omega}\right) = \sum_{l=0}^{\infty} \frac{2l+1}{4\pi} f_l P_l\left(\widehat{\Omega}', \widehat{\Omega}\right) = \sum_{l=0}^{\infty} \sum_{m=-l}^{l} f_l Y_l^{m*}(\widehat{\Omega}) Y_l^m(\widehat{\Omega}). \tag{2.4.6}$$

In Eq. (2.4.6), we have assumed that the phase function $f\left(\widehat{\Omega}',\widehat{\Omega}\right)$ depends only on the dot product $\widehat{\Omega}'\cdot\widehat{\Omega}$ (on the cosine of the scattering angle γ). Equation (2.4.6) is based on a Legendre series representation and on the addition theorem for Legendre polynomials. Combining the orthogonality property of the spherical harmonics, an infinite set of coupled partial differential equations (PDEs) for the unknowns u_{lm} can be obtained. For the fixed integer values $l=L$ and $m=M$, we find that the PDE for u_{LM} contains u_{lm}, with indexes l ranging only from $L-1$ to $L+1$, and m ranging from $M-1$ to $M+1$, while the only coefficients of the source and phase function expansion are q_{LM} and f_L, respectively. Therefore, the expansion of the RTE into spherical harmonics leads to an infinite set of equations with indexes L (ranging from 0 to ∞) and M (ranging from $-L$ to L). Truncation of the Laplace series at $L=N$ leads to the so-called P_N-approximation. The Laplace series with $L \lesseqgtr 1$. The four equations for u_{LM} in the P_1-approximation are those for indexes $(L,M)=(0,0),(1,-1),(1,0),(1,1)$. It can be shown that these four equations can be consolidated into the following two equations:

$$\frac{\partial}{\partial t}U(\boldsymbol{r},t)+v\mu_a U(\boldsymbol{r},t)+\nabla\cdot J(\boldsymbol{r},t)=S_{\mathbf{0}}(\boldsymbol{r},t), \tag{2.4.7}$$

$$\frac{1}{v}\frac{\partial}{\partial t}J(\boldsymbol{r},t)+(\mu_s\langle 1-\cos\gamma\rangle+\mu_s)\,J(\boldsymbol{r},t)+\frac{1}{3}v\nabla U(\boldsymbol{r},t)=S_{\mathbf{1}}(\boldsymbol{r},t). \tag{2.4.8}$$

The two equations are the coupled PDEs for the photon density $U(\boldsymbol{r},t)$ and the photon flux $J(\boldsymbol{r},t)$. The source terms in Eqs. (2.4.7) and (2.4.8), $S_{\mathbf{0}}(\boldsymbol{r},t)$ and $S_{\mathbf{1}}(\boldsymbol{r},t)$, respectively, are related to the source coefficients $q_{0,0}, q_{1,-1}, q_{1,0}$, and $q_{1,1}$. In particular, we observe that $S_{\mathbf{1}}(\boldsymbol{r},t)=0$ for an isotropic source term. One can define the quantity $\mu_s{}'=\mu_s\langle 1-\cos\gamma\rangle$, which is termed the reduced scattering coefficient according to the scattering coefficient μ_s and the average cosine of the scattering angle $\langle\cos\gamma\rangle$, which depends on the phase function. The coupled Eqs. (2.4.7) and (2.4.8) of the P_1-approximation can be written as a single equation, which for homogeneous media (i.e., for spatially independent μ_s', μ_a) simplifies to the following equation:

$$\nabla^2 U(\boldsymbol{r},t)=\frac{3}{v^2}\frac{\partial^2 U(\boldsymbol{r},t)}{\partial t^2}+\frac{1}{D}\left(1+\frac{3D}{v}\mu a\right)\frac{\partial U(\boldsymbol{r},t)}{\partial t}+\frac{v\mu_a}{D}U(\boldsymbol{r},t)-\frac{3}{v^2}\frac{\partial S_0(\boldsymbol{r},t)}{\partial t}$$
$$-\frac{1}{D}S_0(\boldsymbol{r},t)+\frac{3}{v}\nabla S_1(\boldsymbol{r},t), \tag{2.4.9}$$

where the diffusion coefficient is defined as $D=v/\left[3\left(\mu_s'+\mu_a\right)\right]$. Over the years, it has also become clear that one method may only be valid under certain conditions. For example, diffusion and low-order spherical harmonics methods do well for optically thick media, in which the discrete ordinate methods perform poorly, while the opposite is true in optically thin conditions. This naturally leads to hybrid methods, employing separate RTE solvers for separate subdomains (spatially or spectrally). Research in the development of such hybrid methods is still in its infancy and is expected to become an important topic in the future.

3 Numerical Methods for Solving the Radiative Transfer Equation

3.1 Overview of Computational Methods in Radiative Transfer

The RTE can be used to simulate the propagation of light in a medium. Due to the mathematical complexity of the RTE, analytical solutions only exist for simple and certain cases. For more complex problems in practical applications, numerical simulation is important to analyze the process of radiative transfer. The development of computers based on a parallel computing environment facilitates the solution of very complex problems. However, considering the complexity and high dimensionality of the radiative transfer process in participating media, which contains a three-dimensional space and a frequency dimension, the numerical simulation of the radiative process is always time-consuming. Thus, developing efficient and precise numerical methods is crucial for most practical cases. Until recently, many numerical methods have been devised for the RTE solution. In this chapter, numerical methods to deal with general aspects of radiative transfer in multidimensional geometries are introduced, which can be classified in terms of different principles. Generally, the methods can be classified into two broad categories [48]. The first group is based upon stochastic simulation, which includes various implementations of MCMs [49–51], the Distributions of Ratios of Energy Scattered Or Reflected (DRESOR) method [52], and the second group is the deterministic methods, such as the spherical harmonics method (or the P_N-approximation) [53–55], the discrete ordinates method (DOM) [56–58], the finite volume method (FVM), the finite element method (FEM), the radiation element method, the spectral element method, the spectral method, and the meshless method [48]. There are also other classified categories for solving radiative transfer in multidimensional geometries, as shown in Fig. 3.1. Except for the optically thin and optically thick medium, we have discussed in Chapter 2, there are four main categories of the self-absorbing medium:

(1) Directional averaging approximations

(2) Differential approximations

(3) Energy balance methods

(4) Hybrid methods

The FVM and the DOM based on approximations of the RTE facilitate the solution of transmission problems in combustion systems. The energy balance method is appropriate for calculating overall heat transfer rates in the furnace or

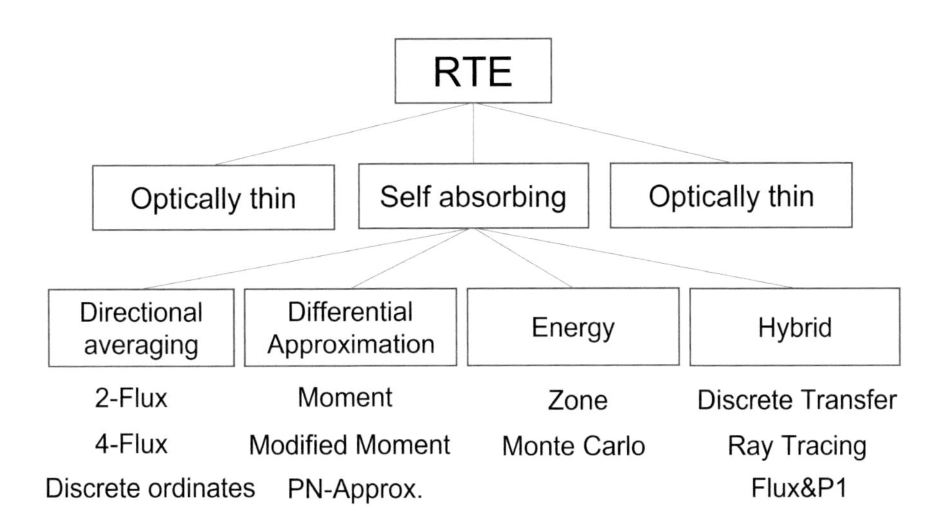

Figure 3.1 Approaches for calculating radiative transfer.

heat fluxes when radiating species concentrations and temperatures are known. The discrete transfer method (DTM) is assumed as a hybrid method that is based on the statistical MCM and flux methods for choosing the finite number of directions of the RTE [59]. The differential method based on the P_1-approximation, the DTM, the DOM, and the FVM applies to complex multidimensional geometries and band calculations, which can be carried out with moderate computing resources. Besides, the line-by-line (LBL) approach is one of the most accurate models for solving the local radiative flux vector and local volumetric radiant energy of a heat source, which shows promising perspectives on industrial combustion systems [48]. However, this method requires excessive computational resources and needs to be simplified for practical applications. It should also be noted that all approximate methods suffer from numerical errors more or less. The MCM suffers from statistical errors. The FVM, DOM, and FEM are affected by angular and space discretization errors. For example, the numerical errors of the DOM method are mainly caused by two general kinds of numerical errors, ray effects, and false scattering. Several strategies have been proposed to eliminate these errors correspondingly [55, 59, 60]. Many methods, like FVM, can be viewed as extensions of the discrete ordinates equations with a certain angular quadrature scheme. Therefore, the errors can be avoided by reducing ray effects and false scattering. Besides, the impact of numerical errors depends on practical applications. Different kinds of errors bring in some nonphysical characteristics to the results, making the solution difficult to be explained and sometimes destroying the solution. So, it is very important to clarify the sources and features of numerical errors, which will benefit to interpret the results of numerical simulation and propose reasonable methods to correct or eliminate the errors. Unlike the spherical harmonics and DOMs, the energy balance methods, such as the zone method (ZM) and its variant method, like the boundary element method (BEM), approximate spatial behavior by breaking up an enclosure into finite and isothermal subvolumes. The disadvantage of the BEM and ZM is that they consume excessive data storage of the local switching factors to calculate the local quantities. Besides, it is difficult to evaluate integrals with singular integrands and generate appropriate meshes in areas with large temperature and species concentration gradients [55, 59, 60]. The MCM is a statistical method, in which the history of bundles of photons is traced as they travel through the enclosure. While the statistical nature of the MCM makes it difficult to match it with other calculations, it is the only method that can satisfactorily deal with effects of irregular radiative properties (nonideal directional and/or nongray behavior).

3.2 Monte Carlo Method

The MCM was first invented to study neutron transport during the atomic bomb program of World War II. After a long time of development, MCMs have become established, powerful, and reliable tools for the study of a variety of fields, from planetary atmospheres to astrophysical applications [61]. The common feature of these fields is that they all need to deal with complex geometries (i.e., a nonspherical medium) and a sophisticated process of scattering. Demands for accurate radiation modeling have accompanied recent advances in finite-scale atmospheric modeling to resolve clouds and satellite measurements by sensors with improved spatial and spectral resolution. A large number of studies have revealed that 3D radiative transfer is particularly important when examining radiation processes at a cloud-resolving scale [62]. The MC model can be employed to calculate path-length statistics when the amounts of gaseous species are traced by differential optical absorption spectroscopy techniques. After a long time of efforts on a physical basis and corrections, the MC model has been regarded as a gold standard of modeling to study complex processes and has regularly been used to validate other models. The effectiveness of the diffusion theory is usually validated by comparing its prediction with the results of MC simulations. The MCM also plays an important role when the diffusion theory or other approximation methods are invalid to deal with some situations. MCMs have been employed in many scientific fields to simulate complex processes, including light propagation in random media. Key examples include radiative transfer in highly scattering materials, such as biological tissues, clouds, and pharmaceuticals [63]. The method has also been used to understand the generation of fluorescence and Raman signals in random media. Research has shown that the MC inversion technique can be used to determine the near-infrared optical properties of human skin ex vivo and derive optical properties from spatially resolved spectroscopic data used for characterizing the intraoperative human brain. Absorption and reduced scattering coefficients can also be derived from point measurements of diffuse reflection based on inverse MC. Furthermore, MC-based data evaluation can be used to extract the optical properties of tissue heterogeneities. Investigating time-dependent photon migration applies to characterize in vivo optical and physiological characteristic samples. The main disadvantage of the MCM is time-consuming and large amounts of calculation. In addition to the development of computer processors that make MC available for photon migration modeling, other developments of variance-reducing techniques are widely used. In a general way, the absorption process is simulated by reducing photon weight instead of photon termination. Some approximations of complex functions can also be employed to increase the calculation speed of photon migration. Another efficient method, but conceptually different, to increase the computational speed is to rescale single MC processes to avoid multiple simulations with different optical properties. Erik Alerstam proposes a technical approach that can dramatically improve the speed of the migration of the MC photon in MC simulation. Besides, parallel computing based on programmable graphics processing unit (GPU) is an efficient approach to remarkably improve the speed of photon migration simulations [64–71].

In MC methods, the radiative transfer process is simulated by introducing a large number of "MC photons" (or "packets"). MC photons behave like their physical counterparts. They experience movement, scattering, or absorption during photon migration like real photons. In the simulations, decisions about the propagation behavior of a particular MC photon are randomly determined. The population of MC photons provides an accurate description of the radiation field and the evolution of photon migration.

3.2.1 Monte Carlo Basics

In the MCM for radiative transfer, the absorption coefficient is defined as the probability of a photon to be absorbed per unit length, and the scattering coefficient is defined as the probability of a photon to be scattered per unit length. Using these probabilities, a random sampling of photon trajectories is generated. The probabilistic methods are used to simulate the transport of individual MC photons through a medium. In the "random walk" process, we only need to describe all the radiation sources, trace the path for each photon describing all interactions, and tabulate parameters of interest, such as intensity, flux, wavelength, angle, and position of exit. When a large amount of photons are processed, these parameters should converge to a steady state and become statistically significant. Here, we will take the steady-state MCM as an example to introduce the detailed process of the MCM for the RTE.

At its core, the main processing loop works in the whole packet population, propagating each packet separately. In this loop, each packet is moved through the domain until a termination event occurs. For example, an absorption interaction could lead to escape through one of the domain boundaries or reach the end of a predefined time interval. The instruction "update estimators" refers to

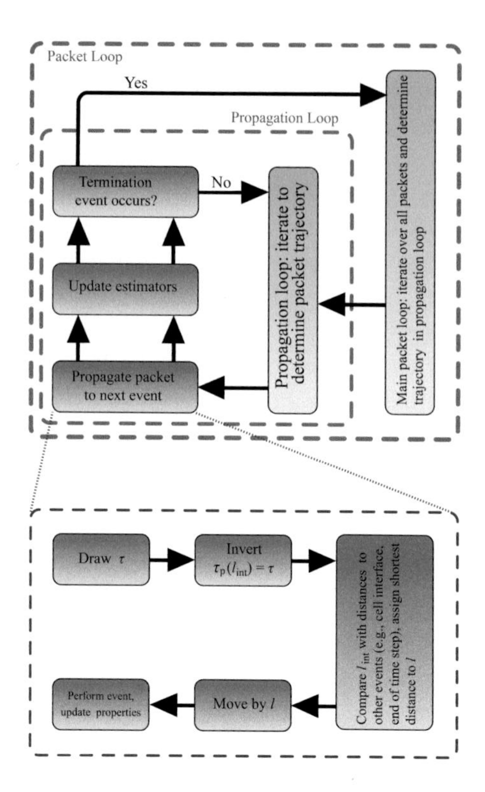

Figure 3.2 Illustration of the packet propagation process.

all activities related to recording packet properties, which are used to reconstruct physical information about the radiation field from the packet ensemble.

Photon launching. The first step of the MC simulation is photon launching. Initially, photons are emitted with an initial weight (w) set to 1.0. And the trajectory of each photon can be specified by the trajectory cosines in Cartesian coordinates as

$$\begin{aligned} ux &= \sin\theta\cos\varphi, \\ uy &= \sin\theta\sin\varphi, \\ uz &= \cos\theta, \end{aligned} \tag{3.2.1}$$

where θ is the angle that the trajectory makes with respect to the $\boldsymbol{z}$ axis and φ is the angle that the trajectory makes with respect to the $\boldsymbol{x}$ axis. Under different circumstances, different photon-launching methods should be adopted.

Under the assumption of local thermodynamic equilibrium, if the emission of a participatory object in the medium is isotropic, the object emits photons in all possible directions with equal opportunities We can discretize it into a number of isotropic point sources. For each isotropic point source, one can take random numbers in the region of $[0, 1]$ to determine the direction of the photon emitted as follows:

$$Rnd_1 = \frac{\int_0^{2\pi}\int_0^{\theta}\sin\theta \mathrm{d}\theta \mathrm{d}\varphi}{\int_0^{2\pi}\int_0^{\pi}\sin\theta \mathrm{d}\theta \mathrm{d}\varphi} = \frac{1}{2}(1-\cos\theta). \tag{3.2.2}$$

Then the direction of the photon emitted by an ordinary object under the local thermodynamic equilibrium (θ, φ) can be expressed as

$$\begin{aligned} \theta &= \arccos\left(1 - 2 \cdot Rnd_1\right), \\ \varphi &= 2\pi \cdot Rnd_2, \end{aligned} \tag{3.2.3}$$

where Rnd_1 and Rnd_2 are the random numbers in the region of [0,1].

So that the trajectory of photons emitted from the isotropic point source can be expressed as

$$\begin{aligned} ux &= \sin\left(\arccos\left(1 - 2 \cdot Rnd_1\right)\right)\cos\left(2\pi \cdot Rnd_2\right), \\ uy &= \sin\left(\arccos\left(1 - 2 \cdot Rnd_1\right)\right)\sin\left(2\pi \cdot Rnd_2\right), \\ uz &= \cos\left(\arccos\left(1 - 2 \cdot Rnd_1\right)\right). \end{aligned} \tag{3.2.4}$$

Photon step size. As mentioned at the beginning of Section 3.2.1, in the MCM, the paths of photons in the medium are based on the probability functions for the involved physical quantities, including the absorption coefficient κ_a, the scattering coefficient κ_s, and the phase function $P(\Omega, \Omega')$. The probability functions are established according to these basic physical quantities of radiative objects. Once the step size of each photon is determined, the transmission process inside the medium, such as whether it reaches the boundaries, can be judged by the random number in the region of $[0, 1]$. Actually, the absorption and scattering coefficients are defined as the probability for absorption and scattering for a photon propagating over an infinitesimal distance ds, respectively. In order to calculate the absorption in the medium, the absorption process of photons can be connected with the extinction coefficient κ_e. Considering the actual situation that the photon step size in the medium is exponentially distributed, an appropriately normalized probability density function can be established to describe the probability of a specific photon step size as

$$p(s) = \frac{\mathrm{e}^{-\kappa_e \cdot s}}{\kappa_e}, \tag{3.2.5}$$

where the value $1/\kappa_e$ represents the average free path length of photons before either absorption or scattering occurs and the $\mathrm{e}^{-\kappa_e \cdot s}$ denotes the exponential distributed length of a specific photon with step size s. And this probability density function has a property as follows:

$$\int_0^{\infty} p(s)\mathrm{d}s = \int_0^{\infty} \frac{\mathrm{e}^{-\kappa_e \cdot s}}{\kappa_e}\mathrm{d}s = 1. \tag{3.2.6}$$

According to the above relationship, the step size s (transport distance) of one particular photon in the medium can be calculated based on a random number $Rnd_3 \in [0, 1]$ in the MCM:

$$Rnd_3 = F(s) = 1 - \mathrm{e}^{-\kappa_e \cdot s}. \tag{3.2.7}$$

Then the step size s of the photon can be derived as

$$s = \frac{-\ln(1 - Rnd_1)}{\kappa_e}. \tag{3.2.8}$$

In the sense of probability, $1 - Rnd_3$ is equal to Rnd_3 due to the uniform distribution of the random number between $[0, 1]$. As a result, Eq. (3.2.8) can be simplified as

$$s = \frac{-\ln(Rnd_3)}{\kappa_e}. \tag{3.2.9}$$

A series of random numbers Rnd_3 will generate a series of step sizes s that follow the probability density function $p(s)$.

After specifying the step size s of each photon, the position that the photon can reach can be expressed as

$$\begin{aligned} x &= x_0 + ux \cdot s, \\ y &= y_0 + uy \cdot s, \\ z &= z_0 + uz \cdot s, \end{aligned} \tag{3.2.10}$$

where x_0, y_0, and z_0 are the origin position of the photon, and ux, uy, and uz are the trajectory cosines.

Drop. If the photons cannot reach the boundaries of the medium, the photons must interact with the medium inside. The possible interactions are scattering or absorption. Similarly, a fraction and a random number can be compared in the region of [0,1] to identify the interactive process. The fraction of absorption and scattering can be expressed as

$$\omega_\lambda = \frac{\kappa_s}{\kappa_e}, Rnd_5 \in [0, 1]. \tag{3.2.11}$$

If $Rnd_5 < \omega_\lambda$, the photon will be scattered by the medium.

Else, the photon will be absorbed by the medium and will be recorded as absorbance.

Spin. If the photon is scattered by the medium, it is necessary to specify the new trajectory according to some scattering functions. The two angles of scatter are θ and φ, the deflection and azimuthal scattering angles, respectively. The most common function to determine the deflection scattering angle θ is the Henyey–Greenstein (HG) phase function, which was presented to describe the scattering of light from distant galaxies by galactic dust. According to the Eqs. (3.2.1)–(3.2.4), the deflection scattering angle θ and the azimuthal scattering angle φ from the incoming direction (ux, uy, uz) and the new scattering direction (ux', uy', uz') can be specified; then the new photon trajectory can be derived as

$$\begin{aligned} ux' &= \sin\theta \frac{ux \cdot uz \cdot \cos\varphi - uy \cdot \sin\varphi}{\sqrt{1 - uz^2}} + ux\cos\theta, \\ uy' &= \sin\theta \frac{uy \cdot uz \cdot \cos\varphi - ux \cdot \sin\varphi}{\sqrt{1 - uz^2}} + uy\cos\theta, \\ uz' &= -\sin\theta\cos\varphi\sqrt{1 - uz^2} + uz\cos\theta. \end{aligned} \tag{3.2.12}$$

If the trajectory is extremely close to alignment with the z axis, that is, nearly $(uz, uy, uz) = (0, 0, \pm 1)$, the new photon trajectory can be derived as

$$\begin{aligned} ux' &= \sin\theta\cos\varphi, \\ uy' &= \sin\theta\sin\varphi, \\ uz' &= \cos\theta(uz \geq 0). \end{aligned} \tag{3.2.13}$$

Finally, the new photon trajectory can be updated, and the new photons can be transmitted to the corresponding direction. The photon step sizes will be determined by new random numbers and the cycle continues.

Terminate. After each transport process, check whether the total number of photons has already reached the maximum number N_{photons} requested by the input file. Otherwise, a new photon will be emitted and the cycle continues. If so, the simulation is completed, and the results of the MCM will be ready to be output.

MC simulations are a relatively simple and traditional method to calculate the light transport process in micro/nanoscale discrete disordered media (DDM) or other macromedia with average absorption and scattering properties. Simulations are similar to the experiments involving a number of photons (N_{photons}). When the diffusion theory or other analytical expressions of light transport fail to solve the RTE equation, MC simulations can be adopted to deal with the situation. However, similarly, compared with the above method, this method does not apply to all work.

3.2.2 MC for Polarized Light

The investigation of polarized light propagation in participating media is an important topic in different research areas, such as biomedical optics and optical imaging [72]. Maxwell's theory can be used to accurately calculate the propagation of polarized light in scattering media. However, the accurate solution requires an extremely large amount of calculation to study the analog quantity, which makes it inappropriate for many applications. As an approximation of Maxwell's theory, radiative transfer theory allows us to study light propagation also in a large volume of scattering media. Different from absorbing scatterers with the complex refractive index, the wave character of polarization cannot be neglected in the radiative transfer process. Most researches of tracking the polarization status of scattering photons describe polarization in terms of the Stokes vector.

MC simulation has always been an effective method to investigate polarized light transport in media. Kattawar and Plass [73] were the first to employ the MCM to calculate the polarization propagation of the light after multiple scattering. They defined the degree of polarization of different optical thicknesses and solar zenith angles of clouds and haze [74]. MCMs are also applied to analyze the variation of polarization properties of optical pulses transmitted through random media and dusty spiral galaxies. Ambirajan and Look [75] developed an inverse MC model to analyze the transport of circularly polarized incident light with a plane slab geometry. Martinez and Maynard [76] proposed an MC model to study the Faraday effect in an optically active medium with the slab geometry. The MC model using a local coordinate system applies to track the reference frame of polarization [77, 78]. Based on the system, the diffusely backscattered Mueller matrix for the suspension of polystyrene (PS) spheres can respectively be characterized numerically and experimentally. Three-dimensional MC simulation is developed to study optically active molecules in turbid media [79] and light propagation in birefringent media [74]. Several researches have focused on improving the efficiency of polarized light transport simulations. Predicting the two-dimensional Mueller matrix accurately facilitates to investigate the total reflectance and transmission through layered media [80, 81]. Kaplan et al. [82] measured experimentally the Mueller matrix back-scattered from mono-disperse solutions of microspheres at different concentrations and scattering angles and developed an MC program whose speed was optimized by using the next-event point-estimator method [83]. Min Xu [84] proposed an MC implementation considering the coherent phenomenon, which traced the electric field through a multiply scattering media based on Stokes vectors. There are three main different ways of implementing MC programs, which track the polarization status of scattered light when it propagates in solutions of Mie and Rayleigh scatterers. The first is called the "meridian plane MC," which is known for the relative position between the reference plane and the meridian plane. The second MC program is "Euler MC," known for Euler angles used in spherical rotation systems. The third MC program is "quaternion MC," in which quaternions are used to propagate the local coordinate system.

3.2.3 Reciprocity Monte Carlo

Any radiation exchange process must follow the principle of reciprocity. The reciprocity MCM is attractive in high accuracy but still suffers from time-consuming calculation. In this method, the number of exchange factors is independently evaluated, $\frac{(N_V + N_S)^2}{2}$, where N_V and N_S are the number of volume and surface elements in the computational domain [85]. Each evaluation requires several hundreds of samples to accurately calculate the triple-integral function [86], causing the simulation computationally complex even based on MC standards. In practical situations, each exchange factor is not calculated independently, large amounts of photon packets may be emitted from a given volume element at a time. Thus, their contributions to the calculation of the integrals may be counted simultaneously. On the other hand, rays may return back to a given volume element from all other volume or surface elements. The former is similar to the forward MCM, while the latter is analogous to the backward MCM. In principle, it is worth noting that the ray-tracing procedure is only used to calculate the multidimensional integrals here. Instead of connecting a pair of preselected points, a ray is emitted from one point, and the points along the ray's path receive contributions toward the integral calculation. On the basis of this broad idea, Tessé et al. [87] developed two different methods to calculate the direct exchange factors. The emission reciprocity method (ERM) enables rays to be traced forward from an emitting volume, and the absorption reciprocity method (ARM) enables rays to be traced backward from all other volume and surface elements to a given volume element. Both these methods follow the principle of reciprocity, which can be combined in a single radiative heat transfer process. The sampling tracing rays lead to the different computed exchange factors in the two methods. Dupoirieux et al. [88] used the optimized reciprocity method (ORM) based on a combination of the ARM and the ERM in a 1D slab. The exchange factor calculated by the ERM or ARM is determined by checking the statistical error with estimating the statistical error, σ_m, of either method a priori. Dupoirieux et al. [89–95] estimated the ratio of statistical errors as $\frac{\sigma_{m,\mathrm{ERM}}}{\sigma_{m,\mathrm{ARM}}} = I_{b,i}I_{b,j}$. In other words, the ERM is preferable to the ARM if V_j is hotter than V_i and vice versa. In recent work, instead of choosing between the ERM and the ARM to compute a given exchange factor, the two methods are combined to produce the so-called bidirectional reciprocity method (BRM). The weight used to combine the two methods comes from the estimation of minimizing the statistical error in the computed results. The method is suitable for a 1D flat plate and shows higher efficiency than the ORM without any loss of accuracy.

3.3 Discretization of the Differential Form of the RTE

3.3.1 Spherical Harmonics Method

The spherical harmonics method is one of the basic methods to solve the RTE, also known as the P_N-approximation. This method provides a vehicle to obtain an approximate solution of arbitrarily high order (i.e., accuracy), by transforming the equation of transfer into a set of simultaneous PDEs. This approach was first proposed by Jeans [96] in his work on radiative transfer in stars. The advantage of this method is that it is less affected by the rays, which can significantly reduce the accuracy of the DOM. The drawback of the method is that low-order approximations are usually only accurate in media with near-isotropic radiative intensity, and accuracy improves only slowly for higher order approximations, while mathematical complexity increases extremely rapidly. This has prompted several researchers in the neutron transport community [97] to develop an approximate spherical harmonics method, known as the simplified P_N-approximation, or SP_N. While more readily taken to a higher order, this method does not approach the exact solution in the limit.

The spherical harmonics method transforms the RTE into a set of relatively simple PDEs and eliminates the integral term. As a function of three spatial and two angular coordinates, the radiance is expanded into a series of spherical harmonics Y_{nm} by deleting its angular variable Ω. Different levels of accuracy in the solution of the P_N method can be obtained according to the order N of truncation. The low-order approximations ($N<3$) are usually only effective in optically thick media with large transport albedos, and the accuracy of optically thin media only increases slowly with an increased order of truncation.

The derivation of the P_N-approximation for tissue with 3D geometry and luminescent sources starts with the expansion of $\psi(\boldsymbol{r},\boldsymbol{\Omega},\mathrm{t})$ and $Q(\boldsymbol{r},\boldsymbol{\Omega},\mathrm{t})$ into spherical harmonics as follows:

$$\psi(\boldsymbol{r},\boldsymbol{\Omega},\mathrm{t})=\sum_{n=0}^{\infty}\sum_{m=-n}^{n}\left(\frac{2n+1}{4\pi}\right)^{1/2}\psi_{nm}(\boldsymbol{r},\mathrm{t})Y_{nm}(\boldsymbol{\Omega}) \tag{3.3.1}$$

and

$$Q(\boldsymbol{r},\boldsymbol{\Omega},\mathrm{t})=\sum_{n=0}^{\infty}\sum_{m=-n}^{n}\left(\frac{2n+1}{4\pi}\right)^{1/2}Q_{nm}(\boldsymbol{r},\mathrm{t})Y_{nm}(\boldsymbol{\Omega}). \tag{3.3.2}$$

The differential scattering coefficient $\mu_S\left(\boldsymbol{r},\boldsymbol{\Omega},\boldsymbol{\Omega}'\right)$ is expressed in Legendre polynomials:

$$\mu_s\left(\boldsymbol{r},\boldsymbol{\Omega},\boldsymbol{\Omega}'\right)=\mu_s(\boldsymbol{r})\sum_{n=0}^{\infty}\left(\frac{2n+1}{4\pi}\right)g^n P_n(\boldsymbol{\Omega})P_n\left(\boldsymbol{\Omega}'\right). \tag{3.3.3}$$

Then, the spherical harmonics expansions of $\psi(\boldsymbol{r},\boldsymbol{\Omega},\mathrm{t})$ and $Q(\boldsymbol{r},\boldsymbol{\Omega},\mathrm{t})$ and the Legendre polynomial expansion $\mu_s\left(\boldsymbol{r},\boldsymbol{\Omega},\boldsymbol{\Omega}'\right)$ are substituted into Eq. (3.3.3). The RTE is multiplied by $Y^*_{n'm'}(\boldsymbol{\Omega})$ and integrated over 4π. As a result, coupled PDEs with the unknown moments $\psi_{nm}(\boldsymbol{r},\mathrm{t})$ are generated. Coefficients whose exponent n' exceeds a defined order N are set to zero. Therefore, the number of equations is truncated to $(N+1)^2$, leading to the P_N-approximation.

Despite the P_N-method's elegant mathematical formulation, its applications are limited in the radiative transfer of luminescence light. The mathematical complexity increases significantly for P_N-approximations of order $N>3$. In addition, partial-reflective boundary conditions based on Eq. (3.3.3) are very difficult to implement for the P_N-method for general 3D tissue geometries, so there is no successful solution so far. However, two other important approximations, namely the SP_N and the diffusion equations, have been derived from spherical harmonics expansions and will be presented later. Both approximations can easily be derived from the planar-geometry approximation.

So far, the lower order approximations of this method, such as the P_1 and P_3 approximation, have been widely used.

In the spherical harmonics method, radiative intensity $I(\boldsymbol{r},\boldsymbol{\Omega})$ related to the angle, which is also the solution of the RTE, is expanded into a series of spherical harmonics methods and approximated as

$$I(\boldsymbol{r},\boldsymbol{\Omega})-\sum_{l=0}^{\infty}I_l^m(\boldsymbol{r})Y_l^m(\boldsymbol{\Omega}), \tag{3.3.4}$$

where $Y_l^m(\mathbf{\Omega})$ are the spherical harmonics that are orthogonal functions in a solid angular space and $I_l^m(\boldsymbol{r})$ are the corresponding expansion coefficients.

Then substitute Eq. (3.3.1) into the RTE to obtain the governing equations of $I_l^m(r)$. Finally, the expansion coefficients are formulated into a set of PDEs to be solved, which can be solved by performing weighted angular integration of spherical harmonics with different orders. Here, this book takes the P_1-approximation as an example to illustrate this method: For the P_1-approximation, the expansion in Eq. (3.3.1) is truncated for $l>1$; then the angular-dependent radiative intensity $I(\boldsymbol{r},\mathbf{\Omega})$ can be expressed as

$$I(\boldsymbol{r},\mathbf{\Omega})=a(\boldsymbol{r})+\boldsymbol{b}(\boldsymbol{r})\cdot\mathbf{\Omega}. \tag{3.3.5}$$

Substituting Eq. (3.3.2) into the RTE the following governing equations of $a(\boldsymbol{r})$ and $b(\boldsymbol{r})$ can be obtained as

$$\nabla\cdot\boldsymbol{b}(\boldsymbol{r})=3\kappa_a\left[I_b-a(\boldsymbol{r})\right], \tag{3.3.6}$$

$$\boldsymbol{b}(\boldsymbol{r})=-\frac{1}{\beta-\kappa_S g}\nabla a(\boldsymbol{r}), \tag{3.3.7}$$

where g, the asymmetry factor of the scattering phase function, is defined as

$$g=\frac{1}{4\pi\mathbf{\Omega}}\int_{4\pi}\mathbf{\Omega}'\Phi\left(\mathbf{\Omega}'\cdot\mathbf{\Omega}\right)\mathrm{d}\Omega'. \tag{3.3.8}$$

Finally, by substituting Eq. (3.3.4) into Eq. (3.3.3), and substituting Eq. (3.3.2) into the definition formula of the radiative heat flux and incident radiation, the P_1-approximation equations can be summarized as follows:

$$\begin{aligned} -\nabla\cdot\left[\tfrac{1}{3(\beta-\kappa_s g)}\nabla G\right]&=\kappa_a\left(4\pi I_b-G\right),\\ \boldsymbol{q}&=-\tfrac{1}{3(\beta-\kappa_s g)}\nabla G,\\ I(\boldsymbol{r},\mathbf{\Omega})&=\tfrac{1}{4\pi}[G(\boldsymbol{r})+3\boldsymbol{q}(\boldsymbol{r})\cdot\mathbf{\Omega}], \end{aligned} \tag{3.3.9}$$

where $\boldsymbol{q}(\boldsymbol{r})=\int_{4\pi}I(\boldsymbol{r},\mathbf{\Omega})\mathbf{\Omega}\mathrm{d}\Omega$ is the radiative heat flux vector, and $G(\boldsymbol{r})=\int I(\boldsymbol{r},\mathbf{\Omega})\mathrm{d}\Omega$ is the incident radiation.

Using Eq. (3.3.8), the boundary condition for diffuse emission and reflection boundary can be determined as

$$\frac{2-\varepsilon}{\varepsilon}\frac{2}{3\left(\beta-\kappa_s g\right)}\boldsymbol{n}_w\cdot\nabla G+G=4\pi I_{bw}. \tag{3.3.10}$$

At radiative equilibrium, where $\nabla\cdot q=0$ and $G=4\pi I_b$, Eq. (3.3.7) can be expressed as

$$\boldsymbol{q}=-\frac{4\pi}{3\left(\beta-\kappa_s g\right)}\nabla I_b. \tag{3.3.11}$$

It can be seen from the above equation that only one equation needs to be solved (Eq. (3.3.5)) for radiative heat transfer, which makes it highly efficient for analyzing engineering radiative transfer problems. Furthermore, the accuracy of the spherical harmonics method can be improved by using higher order spherical harmonics, such as the P_3-approximation and the simplified P_N-approximation [98]. The general P_N-approximation for one-dimensional absorbing/emitting, and anisotropically scattering cylindrical media, and the P_3-approximation for 1D slabs, concentric cylinders, and concentric spheres have been developed in terms of different situations. Higher order solutions, up to P_{11}, for a gray, anisotropically scattering medium between concentric spheres have been considered for uniform heat generation and for an isothermal medium.

The P_1 or differential approximation enjoys great popularity because of its relative simplicity and compatibility with standard methods for the solution of the (overall) energy equation. The fact that the P_1-approximation may become very inaccurate in optically thin media has prompted a number of investigators to seek enhancements or modifications to the differential approximation to make it reasonably accurate for all conditions. The directional intensity at any given point inside the medium is due to two sources: radiation originating from a surface (due to emission and reflection) and radiation originating from within the medium (due to emission and in-scattering). The contribution due to radiation emanating from walls may display very irregular directional behavior, especially in optically thin situations (due to surface radiosities varying across the enclosure surface, causing irradiation to change rapidly over incoming directions). Intensity emanating from inside the medium generally varies very slowly with direction because emission and isotropic scattering result in an isotropic radiation source. Irregular directional behavior only occurs for highly anisotropic scattering radiation.

The modified differential approximation (MDA) [99] separated wall emission from medium emission in simple black and gray-walled enclosures with gray, nonscattering media, evaluating radiation due to wall emission with exact methods and radiation from medium emission with the differential (or P_1) approximation. While very accurate, the model was limited to nonscattering media in simple, mostly 1D enclosures. Wu et al. [100] demonstrated, for 1D plane-parallel media, that the MDA may be extended to scattering media with reflecting boundaries. Finally, Modest [101] showed that the method can be applied to 3D linear-anisotropically scattering media with reflecting boundaries. The P_1-approximation, higher order P_N-, and SP_N-methods can also benefit from this approach.

The *improved differential approximation* (IDA) has been extended to linear anisotropically scattering three-dimensional media (at radiative equilibrium or not [102]. In this method, wall and medium contributions are broken up similar to the MDA, except that the emission from within the medium is also partially assigned to the wall term; the evaluation of flux and incident radiation at points within the medium require an additional surface integral.

3.3.2 Discrete Ordinates Method

Similar to the P_N-method, the DOM can transform the RTE for a gray medium into a combination of PDEs, which was first proposed by Chandrasekhar in 1960 [103]. The DOM is based on a discrete representation of the directional variation of the radiative intensity [104]. The equation of transfer is solved by preselecting a set of discrete directions spanning the total solid angle range of 4π. Thus, the DOM can be considered as a finite differencing of the directional dependence of the RTE. Integrals over solid angles, like the source term, and the heat flux of radiation are evaluated by numerical quadrature. Recently, many numerical heat transfer models use finite volumes instead of finite differences. And the finite solid angles may also be employed for directional (or angular) discretization. The variation of the discrete ordinates for the RTE enjoys increasing popularity, known as the FVM or the finite angle method (FAM). It should be noted that the FAM uses solid angles for discretization and the FVM uses spatial discretization here. As a relatively straightforward formulation of high-order implementations, the DOM and its finite angle cousin, the FAM, get increasing attention, which is in high favor among RTE solvers. Some versions of them have been incorporated in commercial computational fluid dynamics (CFD) codes. Related reviews of the superiority and disadvantages of the DOM and FAM have been summarized by Charest et al. [105] and by Coelho [106]. These reviews provide a complete description and more details of the method for general cases.

The basic principle of this method is presented as follows. The DOM transforms the RTE into a set of K-coupled differential equations by using numerical quadratures instead of integration

$$\int_{4\pi} \psi(\boldsymbol{r},\boldsymbol{\Omega})\mathrm{d}\Omega \approx \sum_{k=1}^{K} w_k \psi_k(\boldsymbol{r}). \tag{3.3.12}$$

The total number K of the discrete ordinates, $\boldsymbol{\Omega}_k = (\Omega_x, \Omega_y, \Omega_z)_k$, is equal to $N(N+2)$, where N is the number of distinct direction cosines of the S_N-approximation, as shown in Fig. 3.3. The quadrature weights w_k associated with the discrete ordinates $\boldsymbol{\Omega}_\mathrm{k}$ are determined by rotational symmetry constraints.

The S_N-approximation to the steady-state RTE for the radiance $\psi(\boldsymbol{r},\boldsymbol{\Omega}) = \psi_k(r)$ can now be approximately written as

$$(\boldsymbol{\Omega}_k \cdot \nabla) + \mu_t(\boldsymbol{r}))\, \psi_k(\boldsymbol{r}) = \mu_s(\boldsymbol{r}) \sum_{k'=1}^{K} \omega_{k'} p_{kk'} \psi_{k'}(r) + \frac{Q(\boldsymbol{r})}{4\pi}. \tag{3.3.13}$$

The HG phase function can easily be included with

$$p_{kk'} = \frac{1-g^2}{4\pi\left(1+g^2-2g\boldsymbol{\Omega}_k \cdot \boldsymbol{\Omega}_{k'}\right)^{3/2}}. \tag{3.3.14}$$

The boundary condition of partial reflection is transformed into equations for the directions $\boldsymbol{\Omega}_k$ of all boundary sources S_k and partially reflected light ψ_k at the tissue–air interface. Each incident ordinate $\boldsymbol{\Omega}_{k'}$ at its inner medium boundary

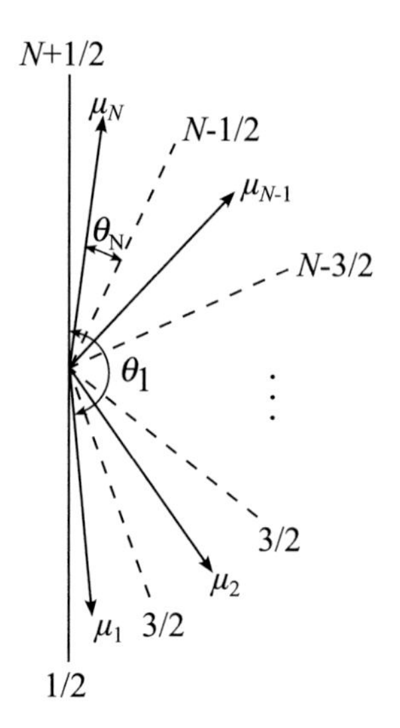

Figure 3.3 Directional discretization and discrete ordinates values for one-dimensional problems.

has its associated ordinate $\mathbf{\Omega}_k$ for partially reflected light due to the imposed symmetry considerations, and we obtain Eq. (3.3.15) for all $\mathbf{\Omega}_k \cdot \boldsymbol{n} < 0$:

$$\psi_k = S_k + R(\mathbf{\Omega}_{k'} \cdot \boldsymbol{n})\, \psi_{k'}. \tag{3.3.15}$$

The reflectivity $R(\mathbf{\Omega}_k \cdot \boldsymbol{n})$ and the angle of refraction are only a function of the angle between the incident ordinate and the normal $\boldsymbol{n}$. Therefore, the S_N method provides a relatively simple means to include partial-reflective boundary conditions. This makes the S_N method more attractive than the P_N-approximation, because the complex boundary conditions of the spherical harmonics expansion are difficult to implement in the latter. Equations (3.3.14) with their boundary conditions (3.3.15) constitute a system of coupled first-order PDEs, and their solutions can be found using iterative techniques. Finally, the fluence $\Phi(\boldsymbol{r})$ and the partial boundary current $J^+(\boldsymbol{r})$ are obtained by using the following formulas:

$$\Phi(\boldsymbol{r}) = \sum_{k=1}^{K} \omega_k \psi_k(r) \tag{3.3.16}$$

and

$$J^+(\boldsymbol{r}) = \sum_{k=1}^{K} \omega_k \left[1 - R(\mathbf{\Omega}_k \cdot \boldsymbol{n})\right] \psi_k(r), \mathbf{\Omega}_k \cdot \boldsymbol{n} > 0. \tag{3.3.17}$$

The potential problem of using the S_N-approximation arises when it is applied to strongly absorbing media either found at short wavelengths extending from 600 nm down to 400 nm. In such media, the RTE is accurately solved along the discrete ordinates, but the ray effect is encountered, resulting in inaccurate solutions near the source. The S_N-approximation has extensively been applied in tissue optics.

Strongly anisotropic light scattering with an anisotropy factor g larger than 0.7 typically requires a high-order S_N-approximation with many discrete ordinates to meet conditions. Instead of using a large number of discrete ordinates (such as the S_{16} method with 288 discrete ordinates), which significantly increases the computational workload, it is intelligent to employ the delta-Eddington (DE) approximation to reduce the number of discrete ordinates. For example, the S_6-approximation with 48 ordinates needs to be used.

Usually, the solution of the RTE requires discretization of both angular and spatial domains. On the contrary, the idea of DOM is to represent the angular space through a set of discrete directions as $\boldsymbol{\Omega}_m = \mu_m \boldsymbol{i} + \eta_m \boldsymbol{j} + \xi_m \boldsymbol{k}$ $(\mu_m^2 + \eta_m^2 + \xi_m^2 = \mathbf{1})$, at which the radiative intensity is solved only. During the calculation of this method, each direction is associated with a quadrature weight. Both the directions and the weight should be carefully selected to ensure the accuracy of angular integration, which is very important for discretizing the internal scattering term and calculating the radiative heat flux. Then, after the angular discretization, the original RTE becomes a set of coupled PDEs, which can then be discretized and solved by traditional techniques for solving PDEs.

As mentioned above, the DOM includes two prime parts: angular discretization and spatial discretization:

$$\begin{gathered}\boldsymbol{\Omega}_m \cdot \nabla I_m(\boldsymbol{r}) + \beta I_m(\boldsymbol{r}) \\ = \kappa_a I_b(\boldsymbol{r}) + \frac{\kappa_s}{4\pi} \sum_{m'=1}^{M} I_{m'}(\boldsymbol{r}) \Phi\left(\boldsymbol{\Omega}_{m'} \cdot \boldsymbol{\Omega}_m\right) w_{m'}, \\ m = 1, \ldots, M,\end{gathered} \tag{3.3.18}$$

where w_m is the weight of direction $\boldsymbol{\Omega}_m$ for the angular quadrature and M is the total number of discrete discretions. For the opaque and diffuse boundary, the boundary condition of each discrete ordinates equation can be expressed as

$$\begin{gathered}I_w\left(\boldsymbol{\Omega}_m\right) = \varepsilon_w I_{bw} + \frac{1-\varepsilon_w}{\pi} \sum_{\boldsymbol{n}_w \cdot \boldsymbol{\Omega}_{m'} > 0} I_w\left(\boldsymbol{\Omega}_{m'}\right) \left|\boldsymbol{n}_w \cdot \boldsymbol{\Omega}_{m'}\right| w_{m'}, \\ \boldsymbol{\Omega}_m \cdot \boldsymbol{n}_w < 0.\end{gathered} \tag{3.3.19}$$

According to the forward definition, the radiative heat flux and incident radiation can be expressed as

$$\begin{gathered}q(\boldsymbol{r}) = \sum_{m=1}^{M} I_m(\boldsymbol{r}) \boldsymbol{\Omega}_m w_m, \\ G(\boldsymbol{r}) = \sum_{m=1}^{M} I_m(\boldsymbol{r}) w_m.\end{gathered} \tag{3.3.20}$$

In fact, the selection of discrete directions (ordinates) and the design of the related angular quadrature are very critical for the accuracy of the method. Several criteria can be proposed for the selection of angular discretization and the weights, as shown in the following [107]:

1. *The symmetry criterion.* The discrete set of directions and weights should be the same after the rotation of $\pi/2$ about each axis.

2. *The whole space moment preserving criterion.* The angular quadrature defined based on the selected directions should satisfy the zeroth, first, and second moments integrated over 4π. This criterion can be expressed by the following equations:

$$\begin{aligned} \int_{4\pi} \mathrm{d}\Omega = 4\pi = \sum_{m=1}^{M} w_m, \\ \int_{4\pi} \mathbf{\Omega}\,\mathrm{d}\Omega = \mathbf{0} = \sum_{m=1}^{M} \Omega_m w_m, \\ \int_{4\pi} \mathbf{\Omega\Omega}\,\mathrm{d}\Omega = \frac{4\pi}{3}\boldsymbol{\delta} = \sum_{m=1}^{M} \Omega_m \Omega_m w_m, \end{aligned} \tag{3.3.21}$$

where $\mathbf{0}$ is the zero vector and $\boldsymbol{\delta}$ is the unit tensor.

3. *The half-space moment preserving criterion*: The defined angular quadrature should preserve the first moment integration over 2π.

This standard can be expressed by the following equations:

$$\begin{aligned} \int_{\mu>0} \mu\,\mathrm{d}\Omega = \pi = \sum_{\mu_m} \mu_m w_m, \\ \int_{\eta>0} \eta\,\mathrm{d}\Omega = \pi = \sum_{\eta_m} \eta_m w_m, \\ \int_{\xi>0} \xi\,\mathrm{d}\Omega = \pi = \sum_{\xi_m} \xi_m w_m. \end{aligned} \tag{3.3.22}$$

At present, the most well-known discrete ordinates set is the so-called S_N set that was originally utilized to simulate neutron transmission. The corresponding angular discretization is called the S_N-approximation by utilizing the S_N discrete ordinates set. Here, N represents the number of discrete direction cosines used for each principal direction, and the total number of directions for the S_N-approximation is $M = N(N+2)$. Except for the S_N sets, some other discrete ordinates sets have been proposed, such as T_N proposed by Thurgood et al. [108].

After the angular discretization, the resulting discrete ordinates equations (Eq. (3.3.12)) in each direction $\mathbf{\Omega}_m$ can be discretized by common methods for the PDE solution, including the finite difference method and the FVM, which is based on spatial discretization.

3.3.3 Finite Volume Method

The DOM, in its standard form, suffers from a number of serious drawbacks, such as false scattering and ray effects [109]. While false scattering is a manifestation of spatial discretization, ray effects are caused by angular discretization. Furthermore, the fact that half-range moments must be satisfied for the accurate evaluation of surface fluxes makes it very difficult to apply the method to irregular geometries [21]. Perhaps, the most serious drawback of the method is that it does not ensure the conservation of radiative energy. This is a result of the fact that the standard DOM uses a simple quadrature for angular discretization, even though generally a finite volume approach is used for spatial discretization. The so-called FVM (for radiation) uses exact integration over finite solid angles, which is analogous to integrals over finite volumes that are performed in a spatial finite volume formulation. It is of this analogous nature of the two formulations that the method came to be termed the FVM, when, in fact, what is meant by

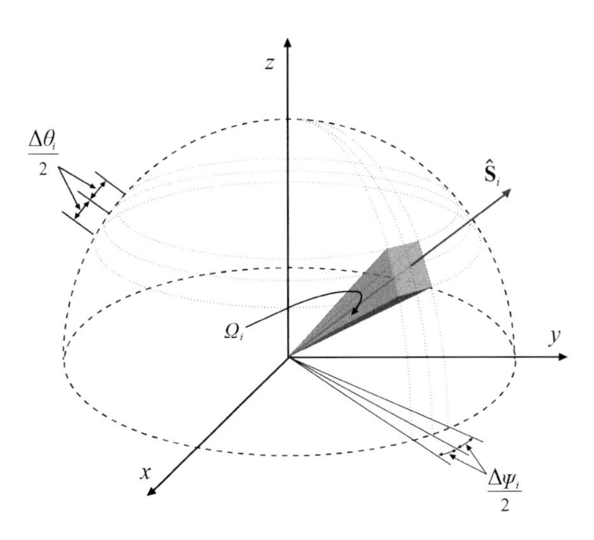

Figure 3.4 The angular domain discretization in the FVM.

this term is a method that is analogous to the FVM, but in angular space. To remove this confusion, especially since the FVM (in space) can also be used in conjunction with the original DOM – henceforth, we will refer to this variation of the DOM as the FAM. The FAM is fully conservative in angular space: Exact satisfaction of all full- and half-moments can be achieved for arbitrary geometries, and there is no loss of radiative energy. The angular grid can be adapted to each special situation, such as collimated irradiation [110]. As mentioned in Section 3.3.3, the FVM is a common method for solving the PDE that can also be used for solving the RTE, which was first extended to the RTE in the 1990s. In the FVM, both the angular domain and the spatial domain are discretized by utilizing control volume integration, which makes the angular discretization in the FVM different from the DOM [111]. Similar to the DOM, the FVM also includes two main parts: angular discretization and spatial discretization.

Angular Discretization

In the process of angular discretization, the angular domain discretization in the FVM uses a structured grid, as shown in Fig. 3.4.

The angular domain discretization is directly based on the directions of θ and φ without any criteria. By integrating the RTE at a small control angle of Ω_{ml}, where the superscript m and l represent the index of θ and φ discretization, respectively, the RTE can be converted into the equation shown below based on the assumption that the radiative intensity is constant over Ω_{ml}:

$$\int_{\Omega_{ml}} \boldsymbol{\Omega}\,\mathrm{d}\Omega \cdot \nabla I_{ml}(\boldsymbol{r}) + \beta \Omega_{ml} I_{ml}(\boldsymbol{r}) = \Omega_{ml} S_{ml}(\boldsymbol{r}), \tag{3.3.23}$$

where $S_{ml}(\boldsymbol{r})$ is given as

$$S_{ml}(\boldsymbol{r}) = \kappa_a I_b(\boldsymbol{r}) + \frac{\kappa_s}{4\pi} \sum_{l'=1}^{N_\varphi} \sum_{m'=1}^{N_\theta} I_{m'l'}(\boldsymbol{r}) \bar{\Phi}(\boldsymbol{\Omega}_{m'l'} \cdot \boldsymbol{\Omega}_{ml}) \Omega_{m'l'}, \tag{3.3.24}$$

where $\bar{\Phi}(\boldsymbol{\Omega}_{m'l'} \cdot \boldsymbol{\Omega}_{ml})$ can be expressed as

$$\bar{\Phi}(\boldsymbol{\Omega}_{m'l'} \cdot \boldsymbol{\Omega}_{ml}) = \frac{1}{\Omega_{m'l'} \Omega_{ml}} \int_{\Omega_{ml}} \int_{\Omega_{m'l'}} \Phi(\boldsymbol{\Omega}_{m'l'} \cdot \boldsymbol{\Omega}_{ml}) \, \mathrm{d}\Omega' \, \mathrm{d}\Omega. \tag{3.3.25}$$

Then, dividing both sides of Eq. (3.3.23) by Ω_{ml}, the discrete ordinates equation can be obtained and written as follows:

$$\overline{\boldsymbol{\Omega}}_{ml} \cdot \nabla I_{ml}(\boldsymbol{r}) + \beta \Omega_{ml} I_{ml}(\boldsymbol{r}) = S_{ml}(\boldsymbol{r}), \tag{3.3.26}$$

where $\overline{\boldsymbol{\Omega}}_{ml}$ is the averaged direction vector expressed as

$$\overline{\boldsymbol{\Omega}}_{ml} = \frac{1}{\Omega_{ml}} \int_{\Omega_{ml}} \boldsymbol{\Omega} \, \mathrm{d}\Omega. \tag{3.3.27}$$

Spatial Discretization

Although the angular discretization methods of the DOM and the FVM used for the RTE solution are different from each other, the spatial discretization approach is the same. As mentioned in Section 3.3.2, the spatial discretization of the DOM can be completed by the FVM.

Here we will give the detailed process of the spatial discretization of the DOM based on the FVM, and then the spatial discretization of the FVM for the RTE solving will also be given subsequently.

Spatial discretization of DOM. In this part, the 2D grid was used to define the FVM discretization scheme, as shown in Fig. 3.5.

For each grid cell with a length of Δx and a width of Δy, the unknown radiative intensities are stored at its center. Take the central grid cell as an example:

For the spatial discretization of the DOM, by integrating Eq. (3.3.12) over the selective control volume, and then using the Gaussian divergence theorem for the first term, we can obtain the discrete equation as follows:

$$\begin{aligned} \Delta x \left(\mu_m I_m(\boldsymbol{r_e}) - \mu_m I_m(\boldsymbol{r_w})\right) + \Delta y \left(\eta_m I_m(\boldsymbol{r}_n) - \eta_m I_m(\boldsymbol{r}_s)\right) \\ + \beta_P I_m(\boldsymbol{r}_P) \Delta x \Delta y = S_m(\boldsymbol{r}_P) \Delta x \Delta y, \end{aligned} \tag{3.3.28}$$

where the terms with $\boldsymbol{r}_P$ denote the value at the cell center of the grid cell P, and the terms with r_e, $\boldsymbol{r_w}, \boldsymbol{r_n}$, and $\boldsymbol{r_s}$ represent the value at the center of the grid cell faces, as shown clearly in Fig. 3.2. $S_m(\boldsymbol{r_p})$ is the source term expressed as

$$S_m(\boldsymbol{r}_P) = \kappa_a I_b(\boldsymbol{r}_P) + \frac{\kappa_s}{4\pi} \sum_{m'=1}^{M} I_{m'}(\boldsymbol{r}_P) \Phi(\Omega_{m'} \cdot \Omega_m) . w_{m'}. \tag{3.3.29}$$

Since the radiative intensities are only stored at the center of cell, the final algebraic equations can be obtained by interpolating the radiative intensity from the cell faces to that at the cell center. Based on the assumption of a uniform

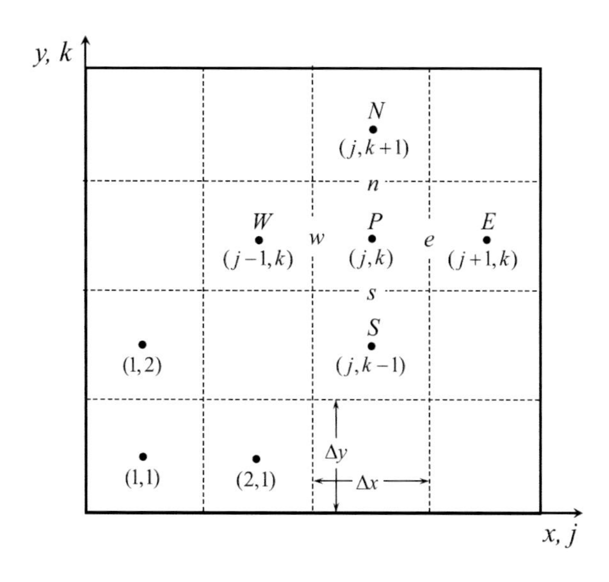

Figure 3.5 The 2D grid of the FVM discretization scheme. Arrangement of cells, faces, and indexing patterns for a 2D orthogonal structured grid.

grid and the utilization of interpolation, the face values can be interpolated using the neighboring cell values as

$$I_m(\boldsymbol{r_n}) = \alpha_y I_m(\boldsymbol{r}_P) + (1-\alpha_y) I_m(\boldsymbol{r}_N), \tag{3.3.30a}$$

$$I_m(\boldsymbol{r_e}) = \alpha_x I_m(\boldsymbol{r}_P) + (1-\alpha_x) I_m(\boldsymbol{r}_E), \tag{3.3.30b}$$

$$I_m(\boldsymbol{r}_s) = \alpha_y I_m(\boldsymbol{r}_S) + (1-\alpha_y) I_m(\boldsymbol{r_P}), \tag{3.3.30c}$$

$$I_m(\boldsymbol{r_w}) = \alpha_x I_m(\boldsymbol{r}_W) + (1-\alpha_x) I_m(\boldsymbol{r}_P), \tag{3.3.30d}$$

where α_x and α_y are, respectively, the interpolation parameters of the x and y directions.

Then substituting Eq. (3.3.30a) into Eq. (3.3.28), we can derive the final discretization expressed as

$$a_P I_m(\boldsymbol{r}_P) = a_N I_m(\boldsymbol{r}_N) + a_S I_m(\boldsymbol{r}_S) + a_W I_m(\boldsymbol{r}_W) + a_E I_m(\boldsymbol{r}_E) + b, \tag{3.3.31}$$

where the discrete coefficients are defined as follows:

$$a_N = \eta_m \Delta y (\alpha_y - 1), \tag{3.3.32a}$$

$$a_E = \mu_m \Delta x (\alpha_x - 1), \tag{3.3.32b}$$

$$a_S = \eta_m \Delta y \alpha_y, \tag{3.3.32c}$$

$$a_W = \mu_m \Delta x \alpha_x, \tag{3.3.32d}$$

$$a_P = a_N + a_E + a_S + a_W + \beta_P \Delta x \Delta y, \tag{3.3.32e}$$

$$b = S_m(\boldsymbol{r}_P) \Delta x \Delta y. \tag{3.3.32f}$$

Actually, different differencing schemes can be obtained by defining different interpolation parameters as shown below:

1. The first-order upwind scheme (the step scheme):

$$\alpha_x = \max\left[\mu_m / \left|\mu_m\right|, 0\right], \tag{3.3.33}$$

$$\alpha_y = \max\left[\eta_m / \left|\eta_m\right|, 0\right]. \tag{3.3.34}$$

2. The central difference scheme (the diamond scheme)

$$\alpha_x = \alpha_y = 1/2. \tag{3.3.35}$$

The resulting linear systems (Eq. (3.3.31)) can be solved element by element or sparse solvers until convergence. Furthermore, because the source term contains the radiative intensity in other directions, it is necessary to carry out global iteration to continuously update the source term with the scattering medium and reflection boundary conditions. Moreover, the discretization shown in Eqs. (3.3.28)–(3.3.35) can be directly extended to 3D problems.

Spatial discretization of the FVM. Because Eq. (3.3.26) can be solved in the same way as that of Eq. (3.3.12), because they both have the same mathematical form. Integrating Eq. (3.3.26) into the spatial control volume and then applying the Gaussian divergence theorem to the first term, the following equation can be derived:

$$\sum_{i=1}^{N_f} A_{f_i} I_{f_i}^{ml} \overline{\boldsymbol{\Omega}}^{ml} \cdot \boldsymbol{n}_{f_i} + \beta_P I_P^{ml} \Delta V_P = S_P^{ml} \Delta V_P, \tag{3.3.36}$$

where the subscript f_i represents the value at the ith face of control volume centered at P, whereas the subscript P denotes the value at the point P, $\boldsymbol{n}$ is the surface normal, A is the surface area, and ΔV_P is the volume of the control volume which is centered at P.

Then, in order to complete discretization, the intensity defined at the surface should be interpolated to nodal values (volume center), which can generally be written as

$$I_{f_i}^{ml} = \alpha_i I_P^{ml} + (1 - \alpha_i) I_{P_i}^{ml}, \tag{3.3.37}$$

where P_i is the center of the neighboring cell of the ith face of the cell P.

Finally, the FVM discretization can be written as

$$a_P I_P^{ml} = \sum_{i=1}^{N_f} a_{P_i} I_{P_i}^{ml} + b^{ml}, \tag{3.3.38}$$

where the discrete coefficients can be expressed as

$$\begin{aligned} a_P &= \textstyle\sum_{i=1}^{N_f} A_{f_i} \alpha_i \overline{\boldsymbol{\Omega}}^{ml} \cdot \boldsymbol{n}_{f_i} + \beta_P \Delta V_P, \\ a_{P_i} &= A_{f_i} (\alpha_i - 1) \overline{\boldsymbol{\Omega}}^{ml} \cdot \boldsymbol{n}_{f_i}, \\ b^{ml} &= S_P^{ml} \Delta V_P. \end{aligned} \tag{3.3.39}$$

Similar to Eq. (3.3.31), Eq. (3.3.37) can be solved element by element and iterated until the convergence of each control angle. If the first-order upwind is used, the interpolation parameters can be expressed as

$$\alpha_i = \max\left[\overline{\boldsymbol{\Omega}}^{ml} \cdot \boldsymbol{n}_{f_i} / \left|\overline{\boldsymbol{\Omega}}^{ml} \cdot \boldsymbol{n}_{f_i}\right|, 0\right]. \tag{3.3.40}$$

Other Related Methods

The nature of directional radiation leads to much research on developing various approximate methods to discretize directions or averages over solid angle ranges. Here, the most frequently related models are briefly discussed.

Flux methods. The 2-flux method is appropriate to deal with 1D problems, especially for cases where the requirement of the accuracy of radiative fluxes is not so high [112–117]. It should be noted that the accuracy of this method is limited to 1D problems in isotropic scattering media. The 6-flux method is developed to deal with strong anisotropic scattering in the presence of collimated irradiation [118, 119]. In this approach, the intensity is divided into six parts, including one forward, one backward, and four sideways components. The way of dividing or averaging over the total solid angle components of 4π is arbitrary. The method has been widely used by various researchers for several 2D and 3D problems [120–122].

Discrete transfer method. The DTM was developed several years before the DOM. The DTM is similar to DOM under the condition that discrete directions are selected. The rays of intensity are traced from surface to surface, related to the MCM. In principle, nodal points are built on the boundary of the enclosure, from which tracing rays go in chosen directions. When the rays propagate through internal finite volumes, the intensity decreases as the ray reduces energy in the volume and increases by the emission and in-scattering process. The disadvantage of the method is also analogous to the standard DOM. The ray-tracing process is nonconservative and susceptible, which needs iterations for nonblack walls and excessive computation of storage and time (much more cumbersome than beam tracing). A solid angle is built at the emission point, causing inaccuracies as the rays transverse through the internal volumes like the MCM. Different from the statistical scatter of the MCM, the efficiency of the DTM is relatively low as each ray carries a single piece of statistics, including a given location, direction, and wavelength, whereas a statistically chosen bundle in an MC simulation carries multiple statistics (particularly important in nongray, reflecting, and/or scattering environments). Nevertheless, the method benefits from its early arrival and early ability to deal with irregular geometries, which has been integrated into several commercial software. Two-dimensional and 3D calculations have been performed mostly for furnaces and other combustion applications [123–125]. Several comparisons between the DTM and other methods have also been made. Coelho et al. [124] study the radiation transfer in 2D enclosures with obstacles, using a combination of the DTM, DOM, FAM, MCM, and the zonal method. The results show the DOM and the FAM to be the most economical. Finally, Keramida et al. [125] employed the DTM and 6-flux methods to model natural gas-fired furnaces, presenting the 6-flux method to be superior. In addition, the DTM has also been recently applied to solve the radiative transfer in multidimensional gradient media.

YIX method. In this method, first developed by Tan and Howell [127], the RTE is expressed in an integral form in which the radiative flux, q, and the incident radiation, G can be evaluated at any point (inside the medium or on the boundaries) as a triple integral (in a distance away from the point and in a solid angle). Integration from each point (along discrete ordinates) involves certain geometric functions that can be predetermined. While this method has the potential to be more efficient than the DOM and the FAM, the setup work for each problem is significant and generalization is difficult. The method appears to have been used only by the group that developed it [59, 128–131].

DRESOR method. The determination of the angular distribution of the intensity with high angular resolution is desirable in many applications. For example, in many measurement techniques, the intensity is measured only along certain lines of sight, and this information is used to determine radiative properties. While the DOM and the FAM both compute intensities along various directions, the chosen directions are generally fairly limited. The MCM is capable of computing the intensity in any arbitrary direction. However, this requires tremendous computational time and resources, and the results are often fraught with statistical errors. The distributions of ratios of energy scattered or reflected (DRESOR) method was developed by Zhou, Cheng et al. [107, 132, 133] with the goal of computing intensities with high angular resolution. In the DRESOR method, the source term of the RTE, given by the RHS of Eq. (3.3.29), is first split into two parts: a term representing the emission from an isothermal unit volume without self-absorption and a term representing the energy scattered by the medium in a unit volume around a distant point and into a unit solid angle around the direction $\hat{s}$. Likewise, the RHS of the boundary condition of the RTE is split into two terms: a term representing the emission from an isothermal unit area without self-absorption and a term representing the energy irradiated onto a point on the boundary and then reflected into a unit solid angle around the direction $\hat{s}$. In the DRESOR method, the aforementioned scattered and reflected terms are written in terms of four ratios. These ratios are computed using a ray-tracing procedure similar to the MCM. It is this calculation that makes the DRESOR method computationally very expensive. In recent years, in an effort to reduce the computational expense of this method, the so-called equation solving the DRESOR method has been proposed and demonstrated for 3D problems. Since its inception, the DRESOR method has been extended to the solutions of the 1D transient RTE [52, 134], to 1D (both plane-parallel and cylindrical) media with the graded refractive index, and to plane-parallel nongray media [135]. In addition to the specific methods discussed in Sections 3.3.2 and 3.3.3, other variations and improvements to the DOM and the FAM have been proposed. Recently, a method known as the generalized source finite volume method (GS

FVM) has been developed. This method is fundamentally based on the FAM and also enables the calculation of the intensity with high angular resolution. Intensities generated by this method have been found to agree well with results generated using the backward MCM. To mitigate ray effects, a method based on the solution of the so-called incident energy-transfer equation has been proposed. In this method, rather than solving the standard RTE (for intensity), a new equation is developed for the incident radiation. The method has only been demonstrated for a plane-parallel medium.

3.4 Machine Learning for the Solution of the RTE

Another promising technique for solving the RTE is the machine learning method. Machine learning is a field of computer science that enables computer systems to find relationships between inputs and outputs, even if they cannot be expressed by explicit algorithms. Their algorithms enable computers to learn from experiences without actually modeling the physical and chemical laws of control of the system. The main focus of machine learning is to automatically extract information from data through computational and statistical methods, which may provide the global solution models for nonlinear inverse problems when relationships between dependent variables and independent variables are not clear. Due to its forecasting and ability to predict, machine learning has found many applications in energy systems. Inspired by biological neural network information processes, artificial neural networks are a group of algorithms used for machine learning, which model data processing of artificial neurons. By training with a data set consisting of a given set of inputs and corresponding outputs, a model can be generated, which can be used to predict proper results from inputs that have the same characteristics as the training set. The multilayer perceptron (MLP) neural network is one of the most popular types of artificial neural networks in machine learning. The MLP consists of an input layer, one or more hidden layers, and an output layer. Each layer contains several nodes called neurons. Figure 3.6 shows a representative MLP neural network architecture for temperatures and species concentration retrieval from infrared spectral emission measurements. The leftmost layer in the figure, called the input layer, consists of a group of neurons representing the input characteristics (infrared spectral intensities). Each neuron in the hidden layers transforms the values from the previous layer with a weighted linear summation, followed by a nonlinear activation function. The output layer receives the values from the last hidden layer and converts them into output values (temperatures/concentrations). The number of neurons in the input and output layers is determined by the input and output dimensions, respectively. For different problems, there is no specific method to determine the number of hidden layers and their neurons, which is usually selected by trial and error [136–139]. The training of neural networks is completed by adjusting appropriate weights between neurons to minimize the error of the objective

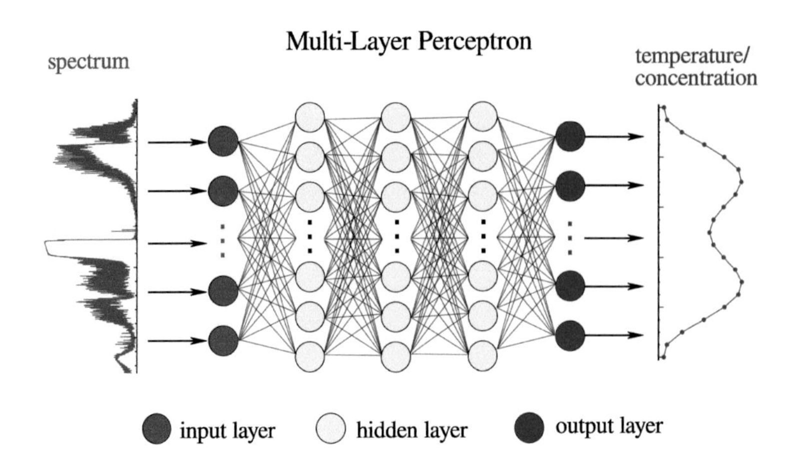

Figure 3.6 Schematic of a representative MLP neural network architecture for temperatures and species concentration retrieval from infrared spectral emission measurements of combustion gases. Reproduced with permission from [140].

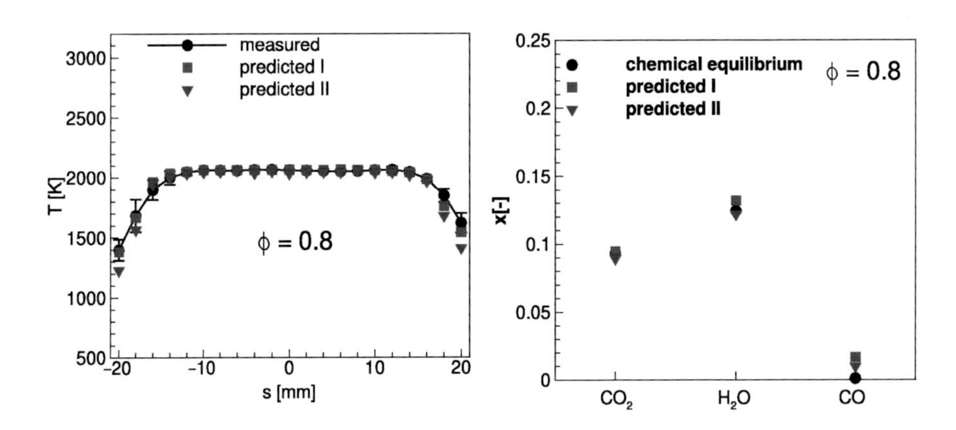

Figure 3.7 Predicted temperatures and species concentrations of the NPL standard flame. Reproduced with permission from [140].

function, where the output values generated by the network are compared to the corresponding actual values.

Learning is an iterative process, starting with a set of random weights, and using a relatively large number of samples, which should contain information evenly distributed over the whole range of the system, so as to obtain a sufficiently low error of the objective function. After training, the model can be directly used to predict new outputs by feeding new inputs. Another optimization method based on machine learning is Bayesian optimization that attempts to find the global optimal value with a minimum number of steps. Bayesian optimization is carried out by maintaining a probabilistic belief about F and designing a so-called acquisition function to directly sample in areas that may be improved over the current best observation.

The temperature and mean species concentrations from spectral radiative intensities obtained from the well-characterized NPL standard flame on a Hencken burner can be inferred and predicted [140], as shown in Fig. 3.7. Spectral measurements were carried out at the heights of 10 and 20 mm above the burner center with a resolution of 4 and 8 cm^{-1} for combustion with three different equivalence ratios, resulting in highly stable data with good long-term reproducibility. The temperature fields of the standard flame were previously measured also in NPL using Rayleigh scattering thermometry. The material concentrations in the postflame region are quite uniform. In order to make accurate predictions based on the MLP neural network, a large number of possible temperature and concentration distributions must be used to train the network, using the same spectral resolution as the measured data [141, 142].

4 Experimental Techniques for Macroscale Thermal Radiative Properties

4.1 Bidirectional Reflectance Distribution Function Instrument

4.1.1 Basic Principles

From the 1960s to the 1970s, many published papers believed that the scientific community should standardize different definitions and terminology of reflectance. Nicodemus [143] was the first to officially introduce the concept of the bidirectional reflectance distribution function (BRDF) and to point out that other standard reflectance can be derived from it. The BRDF can be used to evaluate the reflection, absorption, and radiation characteristics of various material surfaces. It organically unifies the reflection and scattering characteristics of the material surface into the same concept. In the field of research on the optical properties of materials, the BRDF is already a widely recognized comprehensive indicator. It is used extensively in remote sensing, geographic information, marine development, natural disaster monitoring, climate research, the military industry information system, and other fields. With the development of digitization and informatization, the research of BRDF also starts by mathematical modeling and digital simulation. However, another area that needs to be developed urgently is the integration of BRDF measurement methods. In maintaining the correctness and authority of the data, the correct absolute measurement and the transmission and unification of the value appear to be particularly important; otherwise, all the modeling simulation and experimental data will lack persuasiveness. The Chinese Academy of Metrology was the first in China to start the above-mentioned research work in the 1980s and established the "Standard Spectral Variable Angle Reflectometer" that can measure the reflectance of materials in the hemispherical space. At the time this book was being written, the Institute of Metrology is further studying the absolute measurement of BRDF and has the ability to provide the absolute value of BRDF. This is a basic measurement study, because all problems are, in the final analysis, the calibration of absolute quantities.

The BRDF is a basic quantity used to describe the reflection characteristics of a material. It completely describes the reflected light distribution of an opaque surface on a small facet, denoted by f_r [144]. It is defined as the ratio of the radiance of the surface scattering in a specified direction to the irradiance of a single incident on the surface:

$$f_{\mathrm{r}}(\theta_{\mathrm{i}},\varphi_{\mathrm{i}};\theta_{\mathrm{r}},\varphi_{\mathrm{r}})=\frac{\mathrm{d}L_{\mathrm{r}}(\theta_{\mathrm{i}},\varphi_{\mathrm{i}};\theta_{\mathrm{r}},\varphi_{\mathrm{r}})}{\mathrm{d}E_{\mathrm{i}}(\theta_{\mathrm{i}},\varphi_{\mathrm{i}})}=\frac{\mathrm{d}L_{\mathrm{r}}(\theta_{\mathrm{i}},\varphi_{\mathrm{i}};\theta_{\mathrm{r}},\varphi_{\mathrm{r}})}{\mathrm{d}L_{\mathrm{i}}(\theta_{\mathrm{i}},\varphi_{\mathrm{i}})\cos\theta_{\mathrm{i}}\,\mathrm{d}\omega_{\mathrm{i}}}, \tag{4.1.1}$$

where θ and φ are, respectively, the zenith angle and the azimuth angle in standard spherical coordinates, and the subscripts i and r, respectively, represent incidence and reflection. $\mathrm{d}L_r$ is the radiance of the reflected surface scattering in a specified direction (θ,φ). $\mathrm{d}E_i$ is the irradiance of a single incident in the direction (θ,φ). $\mathrm{d}L_i$ is the radiance of a single incident in the direction (θ,φ). $\mathrm{d}\omega$ is the solid angle of radiation. f_{r} varies from 0 to ∞, and its dimension is sr^{-1}, which represents the reflection characteristics of the material surface at any observation angle under different incident angles. Therefore, the BRDF is a function of angle, and it is also affected by the surface temperature and roughness of the material.

The measurement of BRDF can be divided into absolute measurement and relative measurement. The absolute measurement is a measurement performed without using any reference standard, whereas a relative measurement is a measurement that uses a reference standard with a known reflectance to compare with the sample.

In 1980, Bartell et al. [145] introduced the concept of BRDF in detail, discussed the BRDF measurement principle from the two angles of the definition method and the reference sample method, and gave the measurement mathematical expression. Point out the problems that should be paid attention to in the measurement of the size of actual samples, the size of the measured beam, and the viewing field of the receiving detector.

Absolute Measurement

According to the definition of BRDF, the method of directly measuring the BRDF is to measure the incident spectral irradiance and the reflected spectral radiance with an illuminance meter and a luminance meter respectively, and the ratio of the two is BRDF. Although this measurement method is simple, it is also easy to introduce large system errors and random errors, it is difficult to realize, and the accuracy is low.

The bidirectional reflection coefficient is introduced into the absolute measurement, which is defined as the ratio of the reflection energy of the sample surface to the reflection energy of the ideal Lambert surface under the same incident and reflection conditions:

$$\beta(\theta_{\mathrm{i}},\varphi_{\mathrm{i}};\theta_{\mathrm{r}},\varphi_{\mathrm{r}})=\frac{\mathrm{d}\phi_{\mathrm{r}}}{\mathrm{d}\phi_{\mathrm{r,\ ideal}}}. \tag{4.1.2}$$

The relationship between it and the bidirectional reflection distribution function is

$$\beta(\theta_{\mathrm{i}},\varphi_{\mathrm{i}};\theta_{\mathrm{r}},\varphi_{\mathrm{r}})=\pi f_{\mathrm{r}}(\theta_{\mathrm{i}},\varphi_{\mathrm{i}};\theta_{\mathrm{r}},\varphi_{\mathrm{r}}). \tag{4.1.3}$$

It is clear that we can directly obtain the BRDF from bidirectional reflection coefficient, and it has the following form:

$$\beta(\theta_\mathrm{i},\varphi_\mathrm{i};\theta_\mathrm{r},\varphi_\mathrm{r}) = \frac{\pi\rho_\mathrm{d}(\theta_\mathrm{i},\varphi_\mathrm{i})V(\theta_\mathrm{r},\varphi_\mathrm{r})}{\cos\theta_\mathrm{r}\int\left[\frac{V(\theta_\mathrm{r},\varphi_\mathrm{r})}{\cos\theta_\mathrm{r}}\right]\mathrm{d}\Omega_\mathrm{r}}, \tag{4.1.4}$$

where ρ_d is the directional hemispherical reflection coefficient, which is a function of the incident direction and can be measured on the spectrophotometer, and V is the output voltage of the phase-locked amplifier.

Relative Measurement

The most commonly used method is to measure the BRDF with a reference sample, which can be divided into the comparison test method and the single reference standard test method. The generally accepted relative measurement method can not only reduce the system error but also suppress stray light. In 1986, Lee et al. [146] discussed the analysis and suppression of stray light in BRDF measurement from the aspects of measuring the optical path, mirror reflection laser beam, and small angle limitation.

Comparison Test Method

According to Eq. (4.1.3), the $\beta(\theta_\mathrm{i},\varphi_\mathrm{i};\theta_\mathrm{r},\varphi_\mathrm{r})$ is usually written as $\beta(\theta_\mathrm{i},\theta_\mathrm{r})$. The measurement of the directional reflection coefficient is realized by comparison, that is to say, the incident and outgoing modes of the beam are the same on the same variable angle test device for the sample to be tested and the standard sample. It has

$$\beta_\mathrm{S}(\theta_\mathrm{i},\theta_\mathrm{r}) = \frac{V_\mathrm{S}(\theta_\mathrm{i},\theta_\mathrm{r})}{V_\mathrm{B}(\theta_\mathrm{i},\theta_\mathrm{r})}\beta_\mathrm{B}(\theta_\mathrm{i},\theta_\mathrm{r}), \tag{4.1.5}$$

where V_S and V_B are the output voltage of the test sample and the standard sample obtained by the detector respectively. If the reflection coefficient $\beta_\mathrm{B}(\theta_\mathrm{i},\theta_\mathrm{r})$ of the standard sample is known, the reflection coefficient $\beta_\mathrm{S}(\theta_\mathrm{i},\theta_\mathrm{r})$ of the sample to be tested can be calculated. The reflection coefficient $\beta_\mathrm{B}(\theta_\mathrm{i},\theta_\mathrm{r})$ of the standard sample can be derived from the known $\rho(0/d)$ or $\rho(6/d)$. The basic requirement is that the standard sample has good Lambert diffusion properties. It is proved that the error of reflection coefficient obtained by this method is less than 0.5% for materials with good Lambert diffusion characteristics.

Single Reference Standard Test Method

In theory, the relative measurement is to measure the reference sample and the sample to be tested twice in each direction and at different incident angles. However, the point-by-point measurement of the reference sample is rarely carried out, that is, the single reference test method is used. When $A_\mathrm{d}\cos\theta_\mathrm{d}/d^2, EA$, and $\tau\Re$ of the tested sample and the standard reference sample are selected as the same parameters, we have

$$f_\mathrm{S} = f_\mathrm{B}\cdot\frac{V_\mathrm{S}}{V_\mathrm{B}}\cdot\frac{\cos\theta_\mathrm{B}}{\cos\theta_\mathrm{S}}, \tag{4.1.6}$$

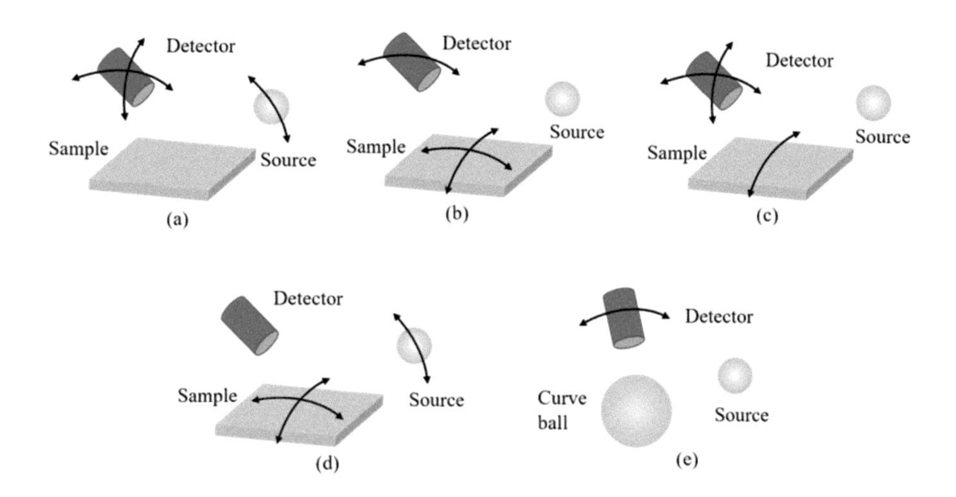

Figure 4.1 Five schemes of BRDF measurement.

where V_S and V_B are the output voltage of the sample to be tested and the standard board obtained by the detector, respectively, θ_S and θ_B are the reflected zenith angle of the sample to be tested and the standard board, and f_B is the BRDF value of the standard board.

Equation (4.1.6) requires less known information for reference samples. There is no need for a sufficient BRDF value, but the voltage reading of the instrument needs to be calibrated in advance.

4.1.2 Experiments and Devices

The instrument used to measure BRDF is usually called the bidirectional reflectometer. Its basic structure comprises a light source (a laser or spectrometer), a corner device and a detector, and a data receiving system. Because the wavelength is either too long or too short to capture, the measurement range of most instruments is in the visible and near-IR range.

The angle device can be used to control the relative position of the detector, sample, and light source [147, 148]. The common measurement scheme is shown in Fig. 4.1. The scheme in Fig. 4.1(a) is suitable for measuring isotropic objects. Here, the sample is fixed, the light source for illumination has 1D freedom of movement, and the detector moves along the entire hemisphere to measure the complete reflected light distribution. In order to measure the complete hemispheric BRDF value, the above process needs to be repeated many times, and the light source must be moved each time to measure the reflection characteristics of the material surface at different incident angles.

In the scheme shown in Fig. 4.1(b), the sample is rotated with 2D degrees of freedom, while the detector only moves with 1D degrees of freedom, the position of the light source is fixed, and the total number of degrees of freedom remains

3D. The advantage of fixing the light source is that there are no restrictions on the size and weight of the light source in practice.

The scheme shown in Fig. 4.1(c) is a combination of the characteristics of the first two device schemes. The sample rotates with 1D degree of freedom, the detector moves with 2D degree of freedom along the hemispherical space, and the light source position is fixed.

In the measurement scheme shown in Fig. 4.1(d), the position of the detector is fixed, the light source rotates in one dimension, and the sample moves in two dimensions.

In the measurement scheme shown in Fig. 4.1(e), a curve ball is used instead of the sample. Since each part of the curve ball has different orientation, data can be obtained from many sampling positions at the same time. Here, the two degrees of freedom of the camera replace the two degrees of freedom of the sample rotation. If the curvature is large enough, it can complete the same measurement as the previous device, which can greatly reduce the measurement time while increasing the sampling density. Given the three-dimensional model of the sample, camera, and light source, the zenith angle, azimuth angle, irradiance of the light source, and the measured values of the reflected brightness can be obtained, and the BRDF can be calculated from the data obtained above.

In 2005, Li et al. [149] developed a three-axis goniometer. The goniometer can quickly measure the BRDF of isotropic samples. The available spectrum measurement range is all visible light and part of the invisible light. The instrument uses a broadband light source, and the detector has a diffraction grating and a linear diode arrangement. The measuring device consists of a stable light source with a wide band and high output; a three-axis rotating mechanical device controlled by a motor; a fixed-position spectroradiometer sensor; and a computer that controls the operation of the instrument, data acquisition, and processing.

Yang et al. [150] measured the BRDF of the dual-layer structure at the wavelength of 635 nm. Measurements of the BRDF and the bidirectional transmittance distribution function (BTDF) of free-standing PTFE sheets are also performed. And they also developed an MC ray-tracing method that incorporates both surface scattering and volumetric scattering. Some results are shown in Fig. 4.2. There is a lot more research on BRDF [151, 152].

4.2 Fourier Transform Infrared Spectrometer

4.2.1 Basic Principles

Fourier transform infrared (FTIR) spectrometer is an instrument that is used to measure IR radiative properties, especially absorptivity and reflectivity [153–155]. Traditionally, people shine monochromatic IR light on the sample's surface to obtain the selective responsibility, which is called dispersive spectrum. Figure 4.3 shows the schematic of an IR spectrum measurement. The light emitter from

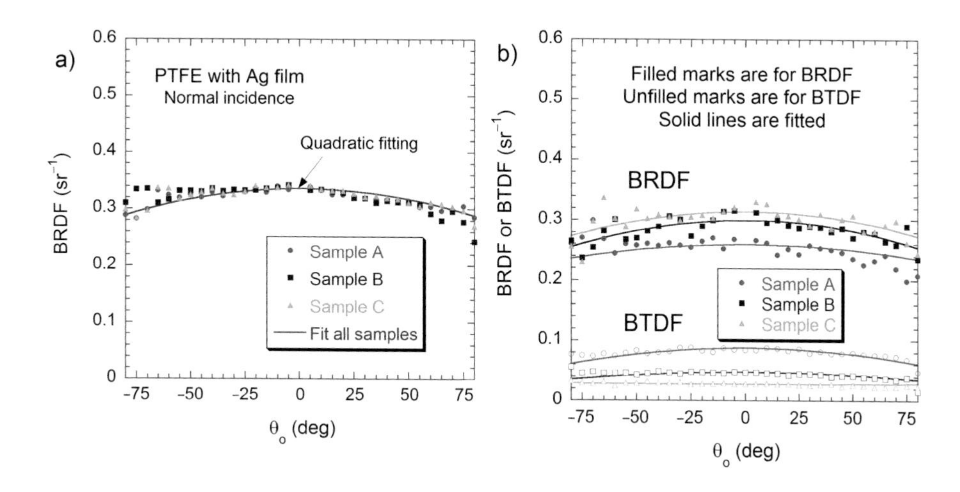

Figure 4.2 (a) The measured BRDF at normal incidence for the three PTFE sheets on the Ag film. (b) The measured BRDF and BTDF at normal incidence for the three PTFE samples. Reprinted from [150], Copyright 2019, with permission from Elsevier.

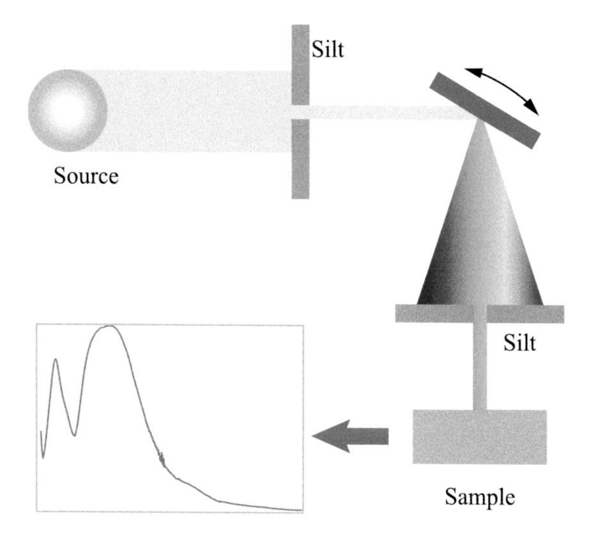

Figure 4.3 The schematic of an infrared spectrum measurement.

the IR source hit samples, and part of them reflects back or transmits through. The photodetectors are placed to receive reflected or transmitting lights and to obtain corresponding reflectivity or transmissivity.

The key of this method is how to control the wavelength to make sure the incident light is monochromatic. Usually, a diffraction order is used to extract the light of a specific wavelength. Diffraction grating is a periodic 1D microstructure where strips or other patterns are placed one by one with a constant period as is

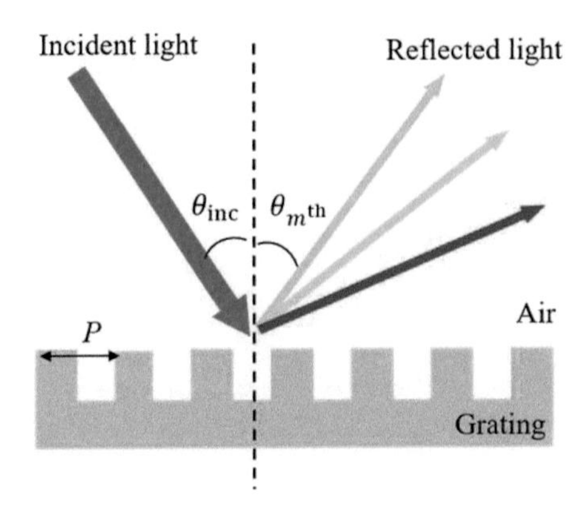

Figure 4.4 Schematic of a diffraction grating.

shown in Fig. 4.4. These fine structures can manipulate the reflection and separate lights of different wavelengths according to their emission angles. In detail, we can quantitatively calculate the reflected lights from gratings. According to the phase-matching condition, the wavevector components of the mth order m in the x direction in the reflection region should satisfy

$$k(m) = k_{\text{inc}} + mK, \tag{4.2.1}$$

where K is the reciprocal vector of the periodic array and k_{inc} is the component wavevector of the incident light. If the reflection region is filled with air (the refractive index is 1.0), this equation can be simplified as

$$\sin\theta_m = \sin\theta_{\text{inc}} + \frac{m\lambda_0}{P}, \tag{4.2.2}$$

where θ_m is the angle of the mth-order diffraction, θ_{inc} is the angle of incident light, λ_0 is the vacuum incident light wavelength, and P is the magnitude of period. Apart from the 0th order, λ_0 has a close one-to-one relationship with θ_m, which means that the monochromatic light can be obtained from a specific direction [156]. And if we can also adjust the detection's orientation, we can obtain continuous monochromatic light.

However, this old method has a high requirement for the IR source, so as to increase the manufacturing cost of the instrument. Therefore, a more convenient method combined with Fourier transform is designed to address this problem and further speed up the detection. Fourier transform enables people to shine lights of various wavelengths on the sample's surface at once and then obtain reflected/transmitted lights to analyze the sample's radiative properties. By transforming the time-domain signal to a frequency-domain signal, the responsibility at different wavelengths can be separated from huge data collection. And the whole process of FTIR is much faster than traditional dispersive spectroscopy.

The center of an FTIR spectrometer is the interferometer, such as the Michelson interferometer [157] shown in Fig. 4.5. The Michelson interferometer consists

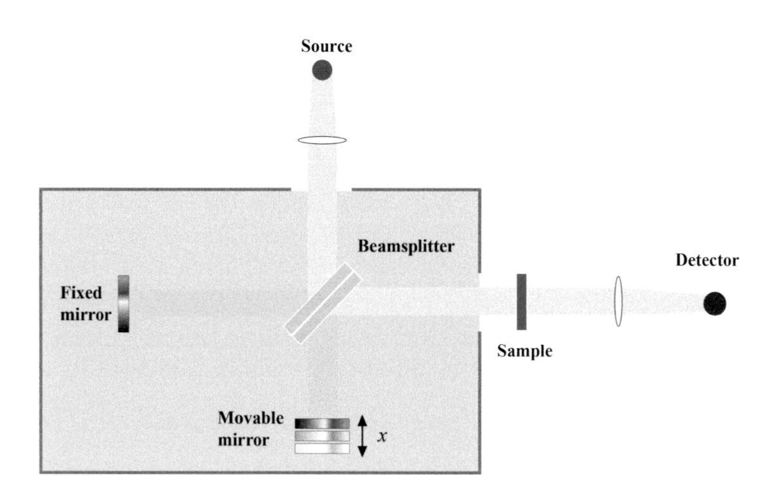

Figure 4.5 Schematic of the Michelson interferometer.

of five components: an IR source, a beam splitter, a fixed mirror, a movable mirror, and detectors. During the measurement, the IR lights emitted from the source first reach the beam splitter. A beam splitter is a type of coated glass. One or more layers of thin films are coated on the surface of optical glass, and when a beam of light is projected onto the coated glass, the beam is divided into two or more beams through reflection and refraction. The beam splitter is usually used at an angle; it can easily separate the incident light into two parts of equal intensity: the reflected light and the transmitted light. Then the reflected light reaches the fixed mirror and reflects again. The transmitted light reaches the movable mirror and also reflects from its surface. These two beams recombine in the beam splitter and cause an interference phenomenon. The interference signal continues to go through the sample and is captured by photodetectors. Considering this measurement process, the light from the source can be a multiwavelength signal, and the spectral information of the sample can also be acquired at once, which saves a lot of time compared with previous dispersive spectroscopy.

Furthermore, it is also important to know how to apply Fourier transform to IR spectroscopy. The movable mirror can change the optical range of the light returning to the beam splitter, that is, by adjusting the moving mirror in different positions, the combined interference light can be obtained in different cases. By changing the difference of optical paths (or called retardation) of lights reflecting from fixed mirrors and movable mirrors, we can have interfered lights of different degrees. First, we assume that the beam splitter and the detector are both ideal, which means the efficiencies of reflection and transmission are 50%, and the detector keeps uniformity in tracking all wavelengths. The intensity of the beam reaching the detector in terms of the retardation δ can be described as

$$I(\delta) = 0.5 I_0(\nu)(1 + \cos 2\pi\nu\delta), \tag{4.2.3}$$

where $I_0(\nu)$ is the intensity of the source at the wavenumber ν. If we allow the movable mirror to move at a certain speed, the received signal by detector is a sinusoidal function consisting of half of the source's intensity and the adjusting part due to the interference. Usually, it is hard to find the ideal beam splitters and detectors. The efficiency and responsibility of the beam splitter and detector are also frequency dependent, separately. Therefore, it is necessary to multiply a factor $F(\nu)$ to eliminate this error during the practical application of the FTIR instrument. This factor $F(\nu)$ includes all parameters influencing the amplitude of the signal. Then the sinusoidal part can be described by

$$I(\delta) = F(\nu) \cos 2\pi\nu\delta. \tag{4.2.4}$$

Let the speed of the movable mirror is V and the retardation $\delta = 2Vt$ after t seconds away from the zero-retardation point. Then

$$I(t) = F(\nu) \cos 2\pi\nu \cdot 2Vt, \tag{4.2.5}$$

where we construct the relationship between the signal and time. Now, we integrate Eqs. (4.2.4) and (4.2.5) as

$$I(t) = \int_{-\infty}^{+\infty} F(\nu) \cos 2\pi\nu \cdot 2Vt \, d\nu. \tag{4.2.6}$$

We further obtain the other pair of Fourier transforms of this equation as

$$F(\nu) = \int_{-\infty}^{+\infty} I(t) \cos 2\pi\nu \cdot 2Vt \, dt. \tag{4.2.7}$$

Considering $I(t)$ is an even function, we can obtain a simpler equation as

$$F(\nu) = 2\int_{0}^{+\infty} I(t) \cos 2\pi\nu \cdot 2Vt \, dt. \tag{4.2.8}$$

Finally, this equation can be used to obtain the relative amplitude of the signal at different wavelengths. However, in practical application, we cannot scan the retardation from 0 to positive infinity. We have to conduct the measurement with a finite resolution. If interested in more details about the effect of finite resolution, please refer to [158].

4.2.2 Experiments and Devices

As mentioned in Section 4.2.1, FTIR obtains measured radiative signals by comparing the incident lights and reflected/transmitted lights. For an object in the environment, it is inevitable that other irrelevant background radiation will interfere with FTIR's detection. Therefore, removing the background radiation is of great importance to obtain accurate and reliable FTIR measurements. Usually, one uses an opaque gold sheet to correct the background radiation. Gold, an inert metal with good chemical stability, can be applied for a long time. Before

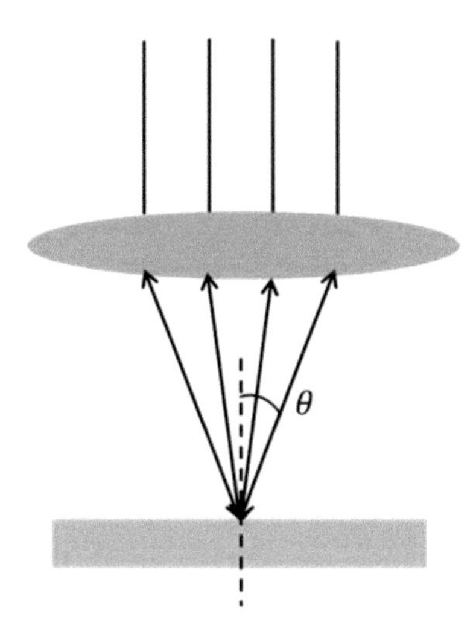

Figure 4.6 The schematic of the lens.

each measurement sample, it is necessary to first put in the standard gold sheet to record the light signal at this time, as a standard to obtain the emissivity and other properties of the later test.

FTIR uses lenses to collect reflected/transmitted light, but any lens has a specific angular collection range. Therefore, FTIR's detection is often an average of a range of angles rather than strictly directional radiative properties. The angular collection range is generally expressed by the numerical aperture (NA), (Fig. 4.6), which satisfies the following equation:

$$\mathrm{NA} = n \cdot \sin\theta, \tag{4.2.9}$$

where n is the refractive index of surrounding media and θ is the half-cone angle. For example, considering a lens with $\mathrm{NA} = 0.5$ in the air, the half-cone angle θ equals 30° and the cone angle equals 60°. This means that the incident lights will hit the sample from θ to 30°, and the measured reflectivity is actually the integrated one in this range as well.

The detection range of FTIR depends on the spectral range of the IR source, including near-IR (0.75–1.40 μm), short-wavelength IR (1.4–3.0 μm), mid-IR (3–8 μm), long-wavelength (8–15 μm), and far-IR region (above 15 μm). After utilizing a proper IR source, we can probe the radiative properties of samples of corresponding spectral spectrums. Figure 4.7 shows the measured absorptivity and emissivity characteristics of a micro nanostructure sample at different temperatures [159]. Both absorptivity and emissivity range from 0 to 1, where 0 means essentially no emission and 1 means equivalent to blackbody emission. Normally, FTIR only obtains absorptivity by detecting reflectivity and transmissivity, while according to Kirchhoff's law, we can also consider absorptivity equals emissivity for an object in thermal equilibrium.

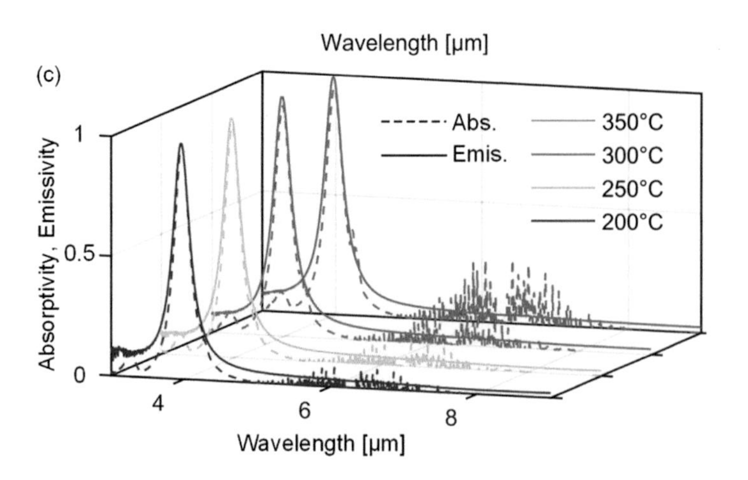

Figure 4.7 The measured absorptivity and emissivity by FTIR. Reproduced from [159]. Copyright © 2017 American Chemical Society.

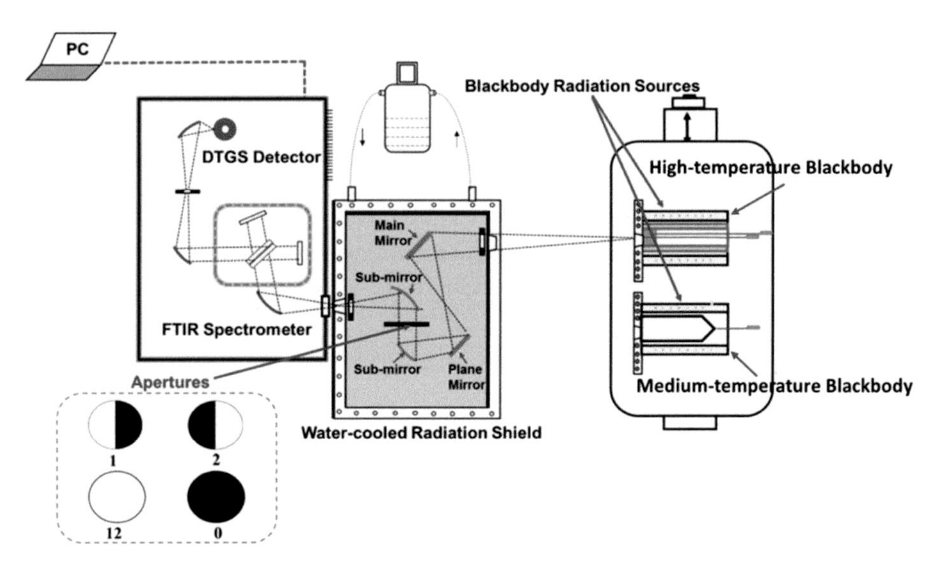

Figure 4.8 Schematic diagram of the linearity measurement system. Reprinted from [160], Copyright 2020, with permission from Elsevier.

Besides, if the temperature of the sample is high enough, the thermal radiation emitted by the sample can be used as a signal source. Song et al. [160] set up an IR spectral emissivity measurement facility by the FTIR spectrum measurement system as shown in Fig. 4.8. On the basis of this system, they obtained the linearity of the spectral responsivity of the system with the wavelength range of 3.9–10.6 μm for different temperatures and a temperature range of 473–1273 K.

4.3 Ellipsometry

4.3.1 Basic Principles

Polarization of Light

As we know, lights in nature can be thought of as a wave of electric and magnetic fields, which oscillates in perpendicular directions. Therefore, if the electric or magnetic field is in a specific direction, we will get polarized light. And by defining two orthogonal directions, such as the x and y axes in a Cartesian coordinate, we can get two orthogonal polarized lights, called transverse magnetic (TM) and transverse electric (TE) waves. The vector E for the TM wave is normal to the plane, and the vector H for TM wave is in the plane; therefore, we call it a TM wave. Conversely, the vector E for TE waves is in the plane, and the vector H is normal to the plane. In our daily life, it is rare to "see" polarized lights because most of the lights are unpolarized, which means that the electric field of light can oscillate in any direction. However, since we already define two orthogonal directions of polarized lights, it is possible for us to study the lights whose electric field oscillates in any direction, by an appropriate linear combination of TM and TE waves.

The reason why we emphasize TM and TE waves is that such two polarizations greatly simplify the derivation and calculation in optics. For TM waves, the directions of electrical and magnetic fields are determined, and the wave equations can be expressed as

$$E = E_x e^{i(\mathbf{k}\cdot\mathbf{r}-\omega t)}, \tag{4.3.1}$$

$$H = H_{yz} e^{i(\mathbf{k}\cdot\mathbf{r}-\omega t)}, \tag{4.3.2}$$

where $\mathbf{k}$ is the wavevector, $\mathbf{r}$ is the location vector, ω is the frequency, t is time, E_x is the magnitude of x component of the electrical field, and H_{yz} is the magnitude of the magnetic field in y, z plane. Similarly, the wave equations of TE waves can be expressed as

$$E = E_{yz} e^{i(\mathbf{k}\cdot\mathbf{r}-\omega t)}, \tag{4.3.3}$$

$$H = H_x e^{i(\mathbf{k}\cdot\mathbf{r}-\omega t)}. \tag{4.3.4}$$

As we mentioned above, most of the lights in nature are unpolarized, like the lights from thermal radiation. However, unpolarized lights can be considered a combination of TM and TE waves, and we can characterize them by analyzing TM and TE waves, separately.

Fresnel Equation

After splitting natural lights into polarized lights, it is much easier to explore the interaction between lights and the surface of matter, which is depicted by Fresnel equations. We start with a 2D plane to simply elaborate the derivative of the Fresnel equation. If the incident light is in TM polarization, the electric field will be in this plane and the magnetic field will be out of the plane, correspondingly. According to the boundary conditions, we have the following equations [161]:

$$E_i \cos\theta_i + E_r \cos\theta_r = E_t \cos\theta_t, \tag{4.3.5}$$

$$-B_i + B_r = -B_t. \tag{4.3.6}$$

Because the electric-field amplitude follows $E = cB/n$, where c is in vacuum and n is the refractive index of the material, Eqs. (4.3.5) and (4.3.6) become

$$E_i \cos\theta_i + E_r \cos\theta_r = E_t \cos\theta_t, \tag{4.3.7}$$

$$-n_i E_i + n_r E_r = -n_t E_t. \tag{4.3.8}$$

Because the incident and reflected lights are normally in the same media, $n_i = n_r$. Besides, the reflection law requests $\theta_i = \theta_r$, and Snell's law requires $n_i \sin\theta_i = n_t \sin\theta_t$.

Ellipsometry is an optical method to study the radiative properties of phenomena occurring at the interface between two media [162]. This instrument can be used to obtain the refractive index of each layer and their own thickness. A simple structure of ellipsometry is depicted in Fig. 4.9 where an oblique light source emits lights through the polarizer and reaches the surface of the sample. Then the reflected light also goes through an analyzer and finally is captured by a detector. Obviously, the principle of this instrument is to use the interaction between the sample and polarized lights to probe the properties of the sample. Usually, two key parameters, the amplitude ratio φ and phase difference Δ, determine the whole detection process. People change the wavelength of the incident light and obtain the spectral responsibility (φ, Δ) in the near-IR region mostly. Then, by fitting with the database of various materials, the refractive index and thickness of each layer can be obtained. This method can handle nanometer-level film and analyze its optical properties, which greatly enhances the control on manufacturing of micro nanoscale samples.

4.3.2 Experiments and Devices

With the development of spectral ellipsometers and the extension to the mid- to far-IR regions, ellipsometry has become a complementary and alternative technique for the study of radiative properties of materials [163–166]. Schoche et al. [163] investigate the anisotropic dielectric response of rutile TiO_2 by applying the spectroscopic ellipsometry in the mid- and far-IR spectral range and the generalized ellipsometry in the mid-IR spectral range. The ordinary and extraordinary dielectric function tensor components and all IR active phonon mode parameters of single-crystalline rutile TiO_2 are determined with high accuracy for wavelengths from 3 to 83 μm. Figure 4.10 shows the experimental (symbols) and the best-matching model-calculated data (solid lines) of the ellipsometric parameters Ψ and $\cos\Delta$ at an angle of incidence (AOI) of 72° for the c-plane rutile TiO_2 surface over the whole measured spectral range from the far- to mid-IR.

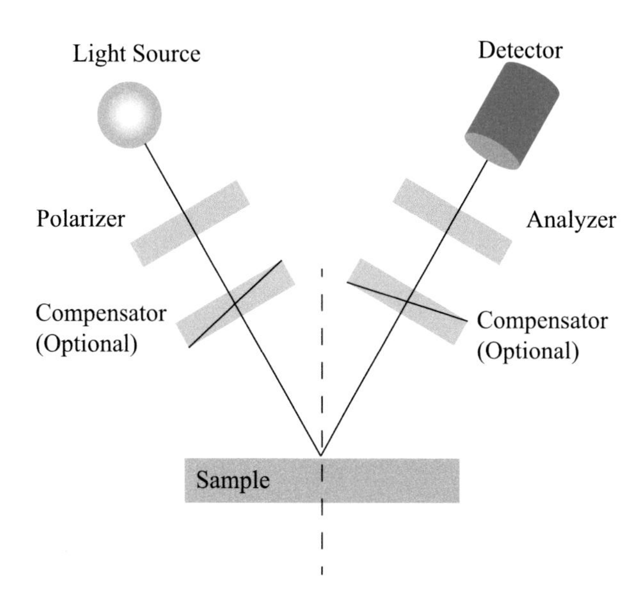

Figure 4.9 The structure of an ellipsometry.

In Fig. 4.11, the results show the calculated p- and s-reflection coefficient spectra for the highly symmetric orientations of the c-plane surface and the a-plane surface of the rutile TiO_2 at $\Phi_a = 72°$. The data were calculated by using the dielectric functions $\varepsilon_{\parallel}$ and $\varepsilon_{\perp}$ obtained in the experiment.

4.4 Measurement of Gas Radiation

4.4.1 Basic Principles

Radiative transfer characteristics of an opaque wall can often be described with good accuracy by a very simple model of gray and diffuse emission, absorption, and reflection. However, the radiative properties of a molecular gas vary so strongly and rapidly across the spectrum that the assumption of "gray" gas is almost never a good one. Much progress in the understanding of molecular gas radiation has been made in the last few decades, in particular the radiation from H_2O and CO_2, for its application in combustion and on the Earth's atmosphere.

Gas radiation is spectrally highly selective and very nongray, which is one of its fundamental characteristics. Meanwhile, different species have very different emission and absorption spectra: monoatomic and symmetric diatomic gases (N_2, O_2, and H_2) are transparent in the IR; polyatomic and asymmetric diatomic (CO_2, H_2O, CO, and CH_4) gases have strong emission and absorption power in the IR. There are three types of radiative transitions as shown in Fig. 4.12: bound–bound transition, bound–free transition, and free–free transition. In bound–bound transition, the molecular energy level is changed by the

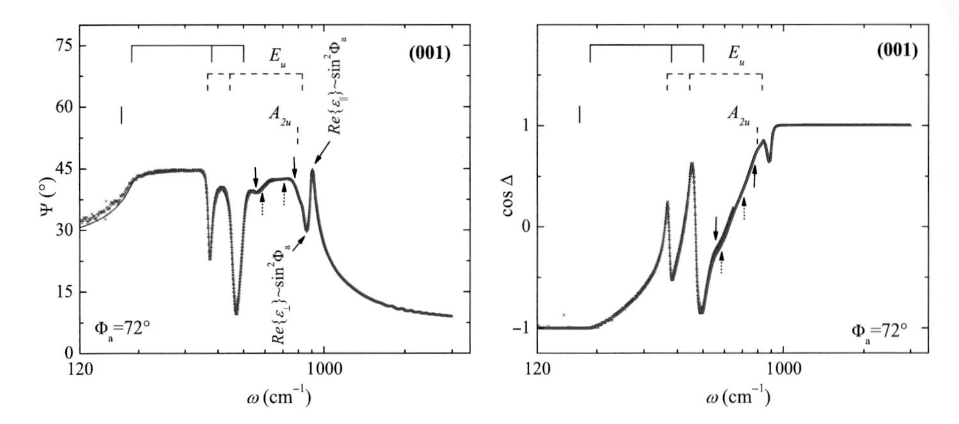

Figure 4.10 Experimental (symbols) and best-match model-calculated data (solid lines) of the ellipsometric parameters Ψ and $\cos\Delta$ for the c-plane rutile TiO_2. Reprinted from [163], with the permission of AIP Publishing.

emission or absorption of a photon between nondissociated atomic or molecular states, and photons must have a certain frequency in order to be captured or released, resulting in discrete spectral lines for absorption and emission. In bound–free transition, photon energy exceeds the required ionization or dissociation energy, and the absorption of a photon may cause the breaking away of an electron or the breakup of the entire molecule. Bound–free transitions result in continuous absorption. In free–free transition, the photon is captured to accelerate a free electron or released to lower the kinetic energy of a free electron. Since kinetic energy levels of electrons are essentially not quantized, these photons may have any frequency or wavelength.

In the following, the atomic and molecular spectral lines will be briefly introduced. Absorbing or releasing of photons in a gas molecule may change the orbit of an electron, and change the molecular vibrational energy level and the rotational energy level. The rotational degrees of freedom and vibrational degrees of freedom are shown in Fig. 4.13. Changing the orbit of an electron requires a relatively large amount of energy, resulting in absorption–emission lines at short wavelengths between the ultraviolet and the near-IR. The vibrational energy level changes require somewhat less energy, so their spectral lines are found in the IR. Changes in rotational energy levels call for the least amount of energy, and thus rotational lines are found in the far-IR.

In the past few decades, a variety of methods to calculate the spectral radiation characteristics of gases have been developed and applied in many fields, such as passive-ranging technology. It is of great significance in engineering application and technical research. It is necessary to obtain the IR radiation characteristics of gas molecules quickly in engineering applications, which puts forward higher requirements for the calculation speed and accuracy of IR radiation characteristics of gas.

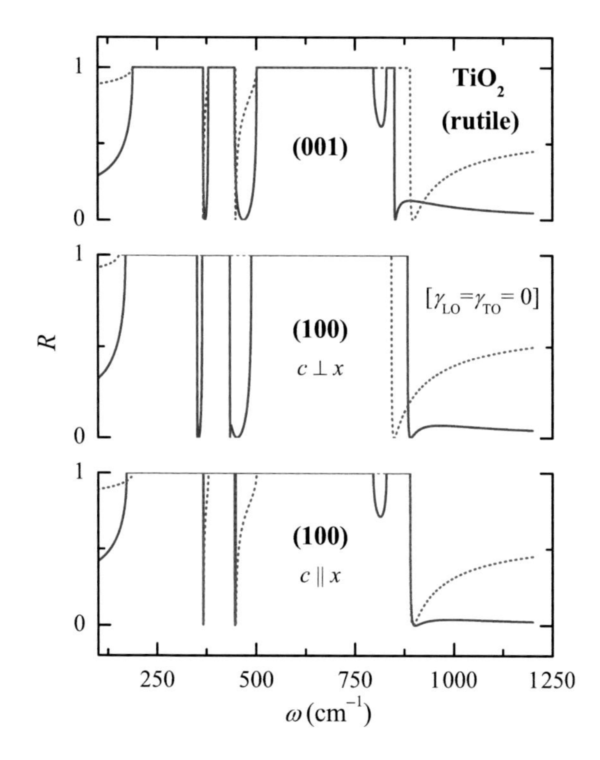

Figure 4.11 Calculated p-(solid) and s-polarized light reflection coefficients (dotted lines) at $\Phi_a = 72°$ for highly symmetric orientations of the rutile TiO_2. Reprinted from [163], with the permission of AIP Publishing.

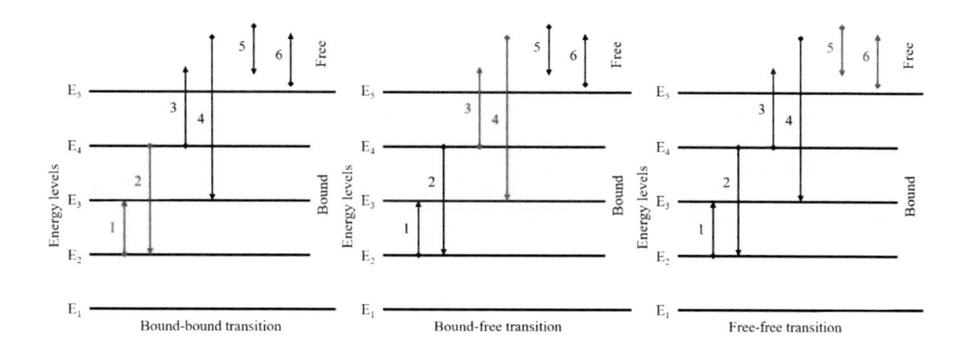

Figure 4.12 Three types of radiative transitions.

At present, the most widely used methods for gas radiation characteristic parameters are the LBL method and the narrowband model. These models need the support of the high-resolution spectral database, such as HITRAN of the United States [167], HITEMP of high-temperature spectral database [168], and CDSD database of Russia [169]. These molecular spectral line parameter databases generally provide the line strength at low temperatures. If we calculate the radiation

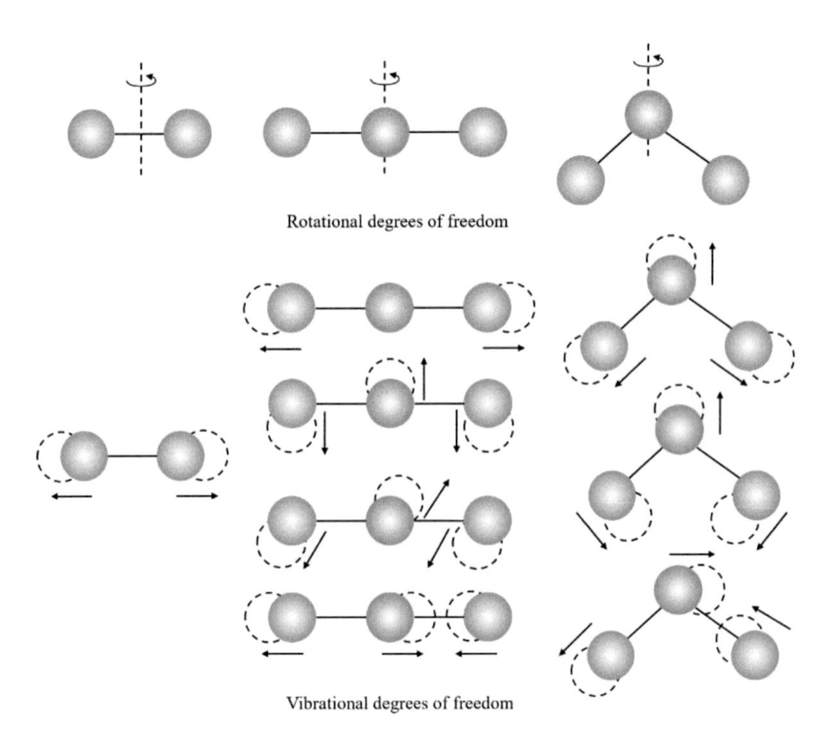

Figure 4.13 Schematic of rotational degrees of freedom and vibrational degrees of freedom.

characteristic parameters of high-temperature gas, we need to pay attention to two points: one is to use a certain method to extend the line strength to high-temperature conditions; the other is that the molecular spectral line parameter database has "hotline" information. With these two points, we can obtain more accurate results at high temperatures.

The LBL method is used to calculate each spectral line of gas molecules. Multiple spectral lines overlap, and the wavenumber resolution is selected. The absorption coefficient at the central wavenumber position is obtained by the superposition of the spectral lines passing through the position. Therefore, the LBL method requires a high-resolution spectral library to provide detailed information on each spectral line. However, it involves a lot of calculation costs, which brings difficulties to practical application. The LBL model is usually used to verify the benchmark solution of approximate methods [170].

Narrowband model usually uses a certain radiation characteristic parameter to express the radiation characteristics of a narrow band. The statistical narrowband (SNB) model and narrowband K-distribution (NBK) model are two typical narrowband models. The basic idea of the SNB model is to assume that the distribution of the intensity and position of all spectral lines in a certain wavenumber interval (usually 5–25 cm^{-1}) conform to certain statistical laws, which can be expressed in the form of certain mathematical functions. Based on

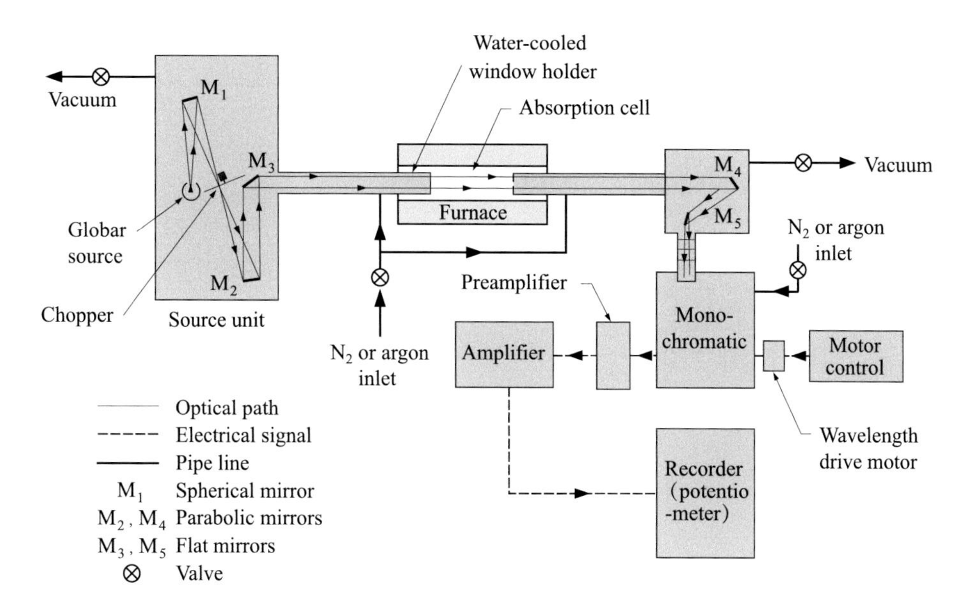

Figure 4.14 Schematic of gas radiation measurement. Reprinted from [174], Copyright 2003, with permission from Elsevier.

this distribution law, the average transmittance within the wavenumber interval can be deduced, which has good accuracy, such as the Goody model [171] and Malkmus model [172]. The principle of the NBK model is that the gas is uniform, the bandwidth cannot be too high, and the absorption coefficient is reordered as a smooth monotone growth function. Compared with the LBL method, the narrowband model can effectively reduce the amount of calculation.

4.4.2 Experiments and Devices

All transmission measurements are similar to each other to a certain extent. They include a light source, a monochromator or an FTIR spectrometer or a tunable laser, a chopper, a test chamber, a detector, related optical components, and an amplifier–recorder device. The test chamber will be full of gas whose properties are to be measured. Only the transmission of the incident beam is measured by measuring the intensity difference between the open and closed conditions. That is, any emission or stray radiation from the test gas will not be part of the signal. A typical gas radiation measurement apparatus is shown in Fig. 4.14 that depicts the device used by Tien and Giedt [173].

Gas radiation measurement may be characterized by the nature of the test gas containment, including the hot window cell, cold window cell, nozzle seal cell, and free jet devices. These methods can be used to carry out the experiments of narrowband measurements, total band absorptance measurements, or total emissivity/absorptivity measurements.

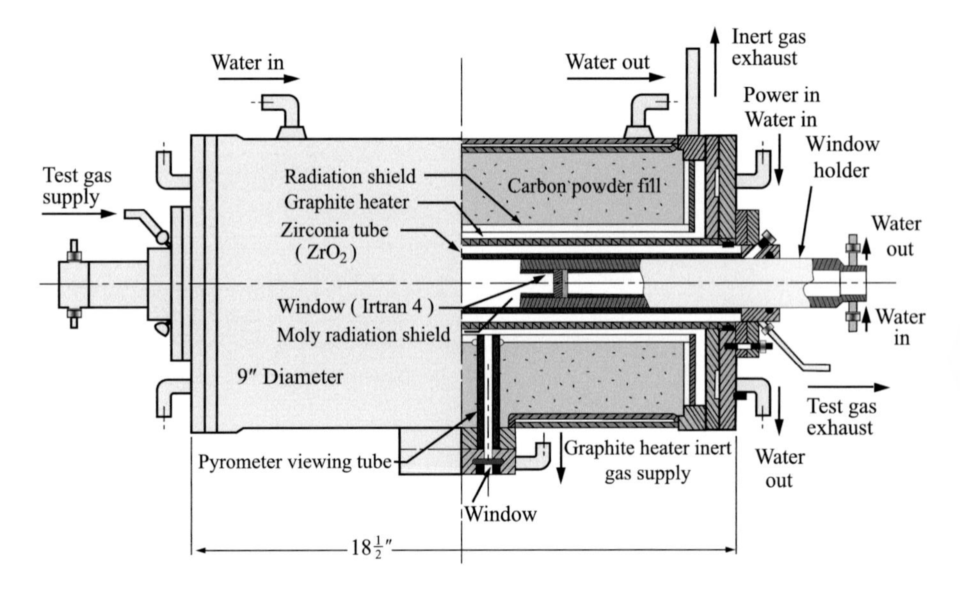

Figure 4.15 Cold window cell. Reprinted from [174], Copyright 2003, with permission from Elsevier.

The hot window cell means that the isothermal gas is sealed within a container and the container has the same temperature as that of the gas. This kind of apparatus is nearly an ideal system for measurements. However, the suitable window material is very hard to find. First, the gas radiation measurement is usually carried out at high temperatures, and it requires the window material to be resistant to high temperatures. Second, the window material should not influence the measurement results. Thus, it should be transparent in the spectral regions where measurements are desired (usually near-IR to mid-IR). Third, the window material should be corrosion-resistant and not succumb to chemical attack from the test gas and other gases. Penner [175] and Goldstein [176] have used similar devices.

The cold window cell lets the probing beam enter and exit the test cell through water-cooled windows. This method has its own advantages compared with the hot window cell. However, this method will fail if the geometric path of the gas is relatively short because serious temperature and density variations will appear along the path. Tien and Giedt [173] designed a high-temperature furnace that has a water-cooled system and allows temperatures up to 2 000 K as shown in Fig. 4.15. Its movable zinc selenide windows are transmissive between 0.5 and 20 μm. Obtaining a truly isothermal gas with this apparatus is challenging when the experimental condition involves high temperature. Thus, the accuracy of the experimental results remains to be discussed.

Nozzle seal cells are open-flow cells. This kind of apparatus may also cause density and temperature gradients near the seal. And another disadvantage is that the mixing flows will introduce some scattering [177]. Hottel and Mangelsdorf

[178] designed a nozzle seal cell to measure the total emissivity of water vapor and carbon dioxide.

Free jet devices are suitable for extremely high temperatures. However, the problem of uncertainty of gas temperature and density distribution along the path is still hard to solve.

4.5 Summary

In this chapter, we focus on the experimental techniques in macroscale thermal radiation. The FTIR spectrometer, UV–VIS–NIR spectrophotometer, bidirectional reflectance distribution function instrument, and gas radiation instruments were introduced from basic principles and experiment measurements. We review some outstanding experiments performed by different research groups for measuring the properties of macroscale thermal radiation. This chapter can serve as a guideline for researchers to design experimental setups.

5 Applications of Macroscale Thermal Radiation

5.1 Application in an Industrial Boiler

Under the vision of carbon peaking and carbon neutrality, on the one hand, the power generation hours of thermal power enterprises will be further compressed to undertake more peak shaving tasks, and thermal power units will operate below 50% low load; on the other hand, thermal power units will burn a higher proportion of economic coal with high moisture and high ash, even sludge, biomass, and other fuels, and the operating conditions of the boiler will seriously deviate from the design conditions. In these cases, problems such as low-load stable combustion, water wall safety, and the clean, economical, and flexible operation of the boiler will be more prominent. There is an urgent need to upgrade the intelligent generation technology of thermal power units [179]. One of the key technologies is to realize the 3D, real-time, and digital monitoring of large furnaces. Whether the 3D spatial distribution of combustion temperature in a large furnace is normal or not, it is related to the safety, economy, and pollutant emission level of the combustion process in the furnace. On the one hand, the overall temperature level is closely related to the load control of the unit, and a certain load output corresponds to a certain temperature level in the furnace; on the other hand, the deviation of the height of the flame center from the design conditions will affect the thermal economy and output of the boiler. The deviation of the flame center on the horizontal plane will cause flame brushing, coking of the water wall, and even the tube explosion of the water wall. The noncontact temperature measurement method is the main development direction of a special ambient temperature measurement means, such as the flame in a boiler of a thermal power unit [180]. Thermal radiation imaging is one such method that can be used to measure the 3D temperature field in a furnace [181].

5.1.1 Theoretical Model

The thermal radiation imaging method for measuring the 3D temperature field in the furnace was first proposed by a famous combustion measurement expert in the 1990s [181]. This method is based on the spontaneous emission information of combustion, and the thermal radiation imaging model is established from the RTE and considers the emission, absorption, and scattering (reflection) effects of

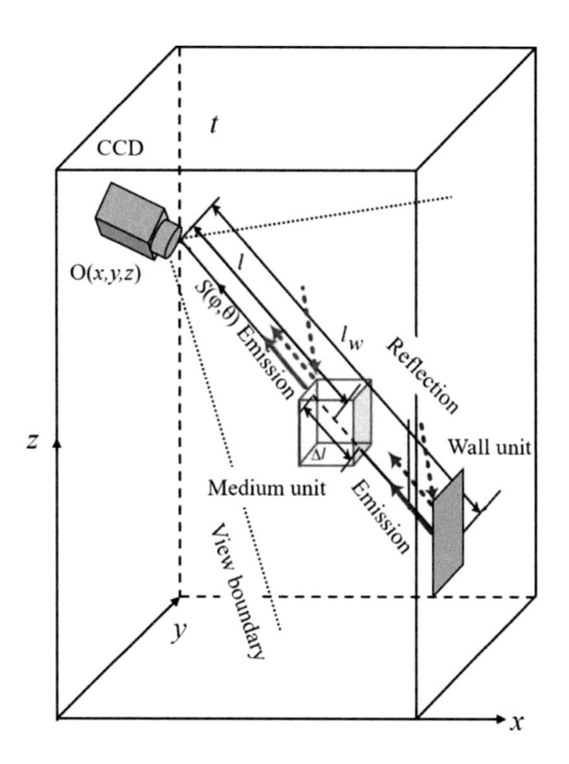

Figure 5.1 Radiative intensity in a line-of-sight direction.

the combustion medium and wall in the furnace. The model connects the combustion temperature and radiation characteristics (component concentration) in the furnace with the thermal radiation image detected by the CCD camera on the furnace boundary and establishes a quantitative relationship. The 3D temperature distribution in the furnace is obtained by solving the inverse problem of radiation transfer combined with the imaging model. The spatial resolution of the thermal radiation imaging method is related to image pixels and has a fast response. Therefore, it has high spatial–temporal resolution and great application potential. At present, it is used in the online monitoring of 3D temperature fields in boilers of more than 10 thermal power units of such as 200, 300, and 660 MW in China [182–184].

The thermal radiation signal from the combustion medium and the wall in the furnace is also an optical signal, and its characterization has many dimensions. As shown in Fig. 5.1, at a certain time t, the CCD camera located at $O(x, y, z)$ on the furnace boundary receives the thermal radiation beam in the direction $S(\varphi, \theta)$, where x, y, and z represent any position in 3D space; φ and θ represent the propagation directions of the thermal radiation beam; λ represents the collected thermal radiation signal; and t represents the time when the thermal radiation signal is captured.

In the furnace system, the combustion medium has emission, absorption, and scattering characteristics, and it is surrounded by walls with emission, absorption, and reflection characteristics. The emission, absorption, and reflection of the solid wall are all carried out on the surface, while the emission, absorption, and scattering of the combustion medium are carried out in the entire furnace volume, which makes thermal radiation imaging different from ordinary optical imaging. In the process of thermal radiation imaging, the CCD camera located on the furnace boundary is regarded as a sensor receiving the radiation energy in the combustion space. The radiation energy projected onto the CCD is gradually reduced by the absorption and scattering of the medium along the radiation path. Therefore, the energy received by the CCD camera can be divided into four parts: wall direct radiation of wall emission energy absorbed by the medium, medium direct radiation of medium emission energy absorbed by the medium, wall indirect radiation of wall and medium emission energy reflected by the wall, and medium indirect radiation of wall and medium emission energy scattered by the medium [174].

By discretizing the field of view angle of the CCD camera into N directions, and the furnace system into M medium units and M' wall units, a linear relationship between the radiation intensity distribution in different directions and the 3D temperature distribution in the furnace can be expressed as follows:

$$\begin{pmatrix} I(1) \\ \vdots \\ I(N) \end{pmatrix} = \begin{pmatrix} A(1,1) & \dots & A(1, M+M') \\ \vdots & \ddots & \vdots \\ A(N,1) & \dots & A(N, M+M') \end{pmatrix} \begin{pmatrix} T^4(1) \\ \vdots \\ T^4(M+M') \end{pmatrix}, \tag{5.1.1}$$

$$\mathbf{I} = \mathbf{A_1 T_g} + \mathbf{A_2 T_w} = \mathbf{AT}, \tag{5.1.2}$$

where $\mathbf{I}$ denotes the radiation intensity distribution received by the CCD camera; $\mathbf{T_g}$ denotes the fourth power of the medium unit temperature; $\mathbf{T_w}$ denotes the fourth power of the wall unit temperature; $\mathbf{A_1}, \mathbf{A_2}$, and $\mathbf{A}$ are the radiation intensity imaging matrix: $\mathbf{A_1}$ represents the radiation contribution of the medium unit to the imaging device; $\mathbf{A_2}$ represents the contribution of the wall element, and $\mathbf{A}$ is determined by the absorption coefficient and scattering coefficient of the medium, wall emissivity, the geometric size of the furnace, and the position of the camera. Formula (5.1.2) establishes the linear relationship between the radiation intensity distribution in different directions and the 3D temperature distribution in the furnace and considers the physical process of thermal radiation absorbed and scattered by the medium and absorbed and reflected by the wall.

5.1.2 Experimental Measurement

With the development of imaging technology, in addition to CCD cameras, new equipment such as light field cameras, and multispectral or hyperspectral cameras can be used to obtain the thermal radiation signal in the furnace. Compared

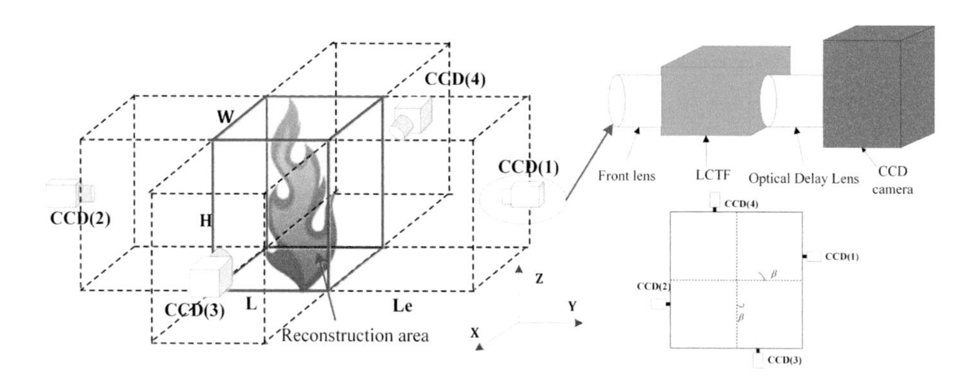

Figure 5.2 Schematic of the reconstruction system. Reprinted from [185], Copyright 2016, with permission from Elsevier.

with traditional cameras, light field cameras can not only record the intensity of light projected on the camera detector but also distinguish the direction of light. The multispectral or hyperspectral camera measures a large number of continuous band radiation intensities in each imaging unit, which can provide flame radiation distribution information, including spatial and spectral information. Ni et al. [185] set up a reconstruction system for the reconstruction of temperature and soot volume fraction distributions as shown in Fig. 5.2. Four CCD cameras are assumed to be identical and were put around the flame marked as CCD (1), CCD (2), CCD (3), and CCD (4). The sensor array size of the camera is $1\,280 \times 1\,024$ pixels with a pixel size of 5.2 μm × 5.2 μm. It is noteworthy that the liquid crystal tunable filters (LCTFs) are fixed in front of the CCD cameras to obtain multispectral images of the reconstructed flame. And a set of optical elements are connected in series with index-matching epoxy. The transparency of each element changes with the wavelength of the incident light, and the transmitted light adds only in the desired bandwidth range. As a result, LCTF is an ideal technology for accurate multispectral imaging, which provides images of an object at multiple wavelengths and generates high-resolution spectra at every pixel.

The size of the reconstruction system is $W \times L \times H = 7$ mm × 7 mm × 36 mm, and the distance from CCD cameras to the edge of the reconstruction area is $L_e = 0.021$ m. The reconstruction region is divided into $7 \times 7 \times 11$ volume elements. The viewing angle of the CCD camera is about 80°. The angle β is determined to be 4.67° degrees. The assumed temperature and soot volume fraction distributions are shown in Fig. 5.3. The profile of soot temperature and the volume fraction distribution on the cross section ($k = 3$) is shown in Fig. 5.4. It can be found that the reconstruction object is an axisymmetric flame and the proposed reconstruction system is capable of both axisymmetric and asymmetric flames.

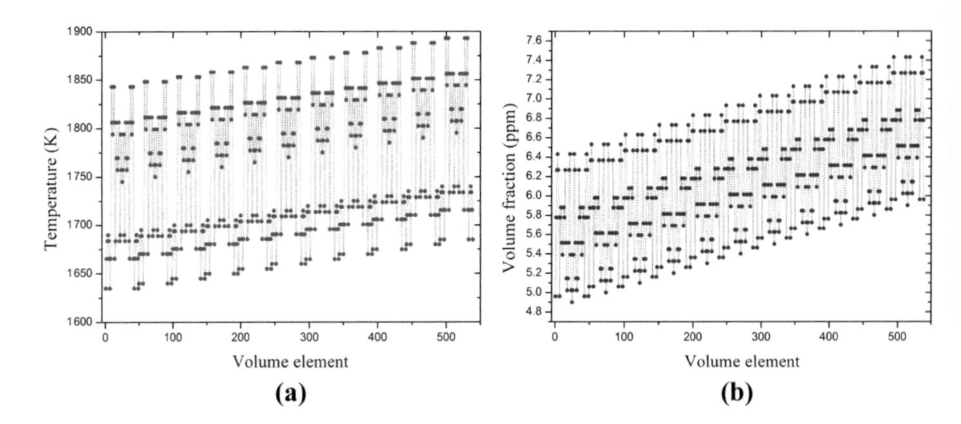

Figure 5.3 Assumed 3D distributions of soot temperature and volume fraction: (a) temperature; (b) soot volume fraction. Reprinted from [185], Copyright 2016, with permission from Elsevier.

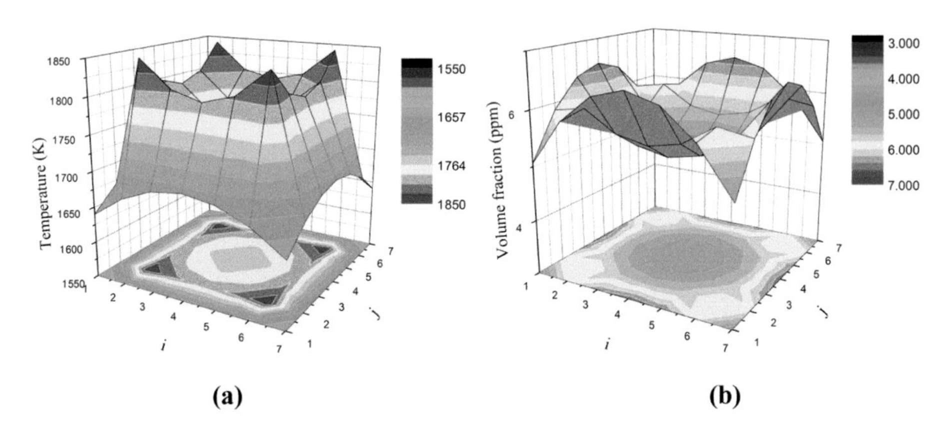

Figure 5.4 Soot temperature and volume fraction distribution profiles on the cross section $k = 3$: (a) temperature; (b) soot volume fraction. Reprinted from [185], Copyright 2016, with permission from Elsevier.

5.2 Application in an Infrared Thermal Imager

Objects with absolute temperatures greater than zero are constantly emitting EM radiation. According to Planck's blackbody radiation law, the EM radiation band range of normal temperature objects is mainly concentrated in the thermal infrared band of 3–14 μm. Because all things have such radiation characteristics, the detection equipment working in the thermal infrared band can achieve uninterrupted imaging of the object day and night. If we can get the spectral information in the thermal infrared band at the same time, then the temperature of the object and gas and the radiation intensity of each spectrum can be inversed. Thus, some objects and gases that cannot be distinguished in visible short

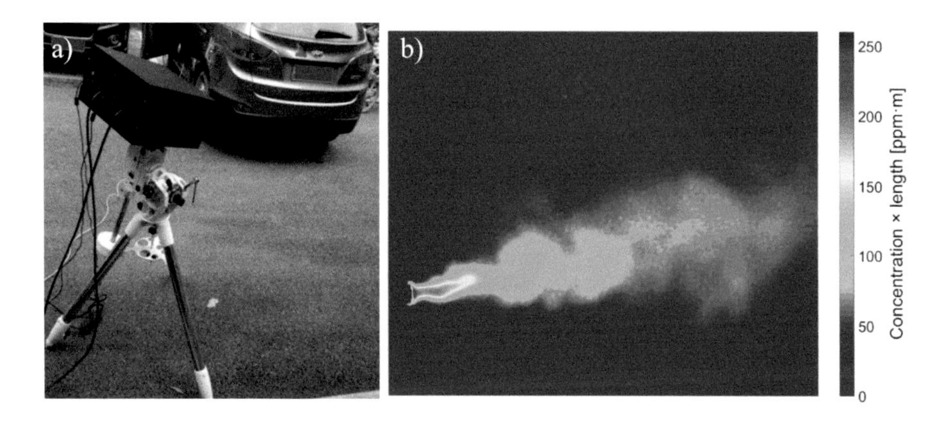

Figure 5.5 (a) Schematic diagram of the detection of vehicle exhausts using a gas correlation spectrometry-based mid-IR imager. (b) Two-dimensional concentration × length of CO in vehicle exhausts. Reproduced from [187]. Copyright 2018 Optical Society of America under the terms of the OSA Open Access Publishing Agreement.

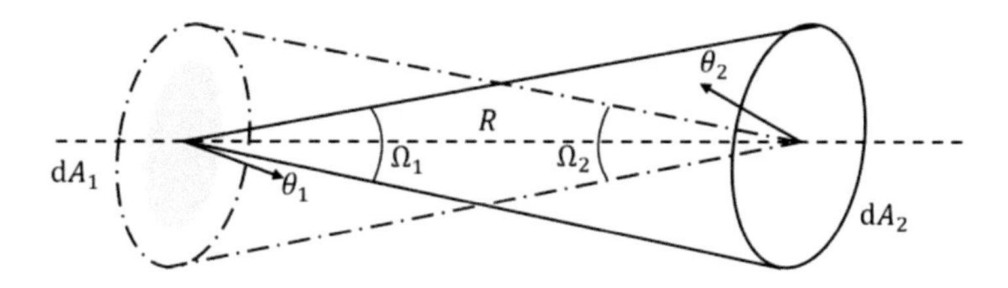

Figure 5.6 Calculation of irradiance generated by a point source.

wavelength bands can be measured and identified, as shown in Fig. 5.5 [186]. It can be seen that thermal infrared spectrum detection plays an important role.

5.2.1 Theoretical Model

Over the past few decades, progress has been made in the infrared thermal imager [188–190]. In this section, we will give a brief introduction to the theoretical model. When the target is a point source, the solid angle of the target to the receiving surface is less than the receiving solid angle Ω_2, as shown in Fig. 5.6. Because there is both the target and the background in the receiving solid angle, the irradiance needs to be calculated by the radiation intensity in the detection direction of the transmitting panel.

The irradiance of the receiving surface element is

$$E = \frac{I_\theta \cos\theta_2}{R^2} = \frac{I_0 \cos\theta_1 \cos\theta_2}{R^2}, \tag{5.2.1}$$

where $I_0 = L\ dA_1$ is the normal radiation intensity of the emitting surface element and θ_1 is the angle between the emission panel discovery and the line of

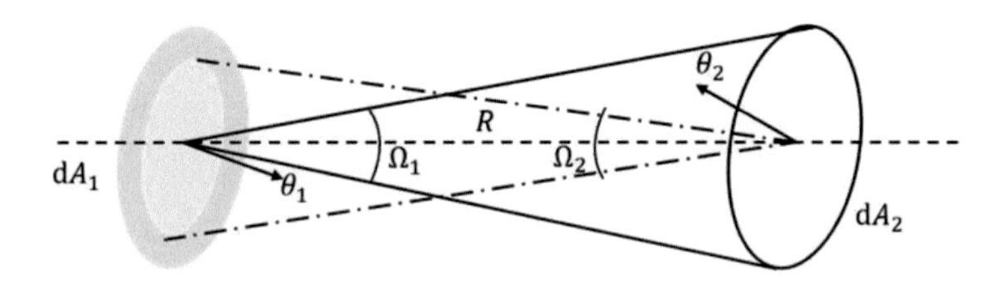

Figure 5.7 Calculation of irradiance generated by the area source.

sight. In fact, the point source is relative to the instantaneous field of view of the observation system, rather than a point in the mathematical sense. An actual target is a point source composed of multiple Lambertian surfaces with different radiation characteristics, and its total radiation intensity in the observation direction is

$$I_\theta = \sum_{i=1}^{n} I_{\theta i} = \sum_{i=1}^{n} L_i A_i \cos\theta_i, \tag{5.2.2}$$

where $I_{\theta i}$ is the radiation intensity in the direction of the surface element i, A_i and L_i are the area and radiance of surface element i, respectively, and $A_i \cos\theta_i$ is the projected area of the surface element i in the viewing direction.

When the target is an area source, the solid angle of the target to the receiving surface element is greater than the receiving solid angle Ω_2, and the receiving surface element can only receive the radiation emitted by part of the surface of the area source, as shown in Fig. 5.7. At this time, the irradiance can be calculated from the irradiance of the area source and the receiving solid angle by using the radiance conservation relationship of the radiation transfer of the area source. If the emission surface element is a diffuse area source with radiance L, the radiation intensity of the emission surface element $\mathrm{d}A_1$ in the θ direction is $L\cos\theta_1\ \mathrm{d}A_1$, and the irradiance of the emitting radiation intensity $\mathrm{d}A_1$ on the illuminated radiation intensity $\mathrm{d}A_2$ can be expressed as

$$E = \frac{I\cos\theta_2}{R^2} = \frac{L\cos\theta_1\ \mathrm{d}A_1 \cos\theta_2}{R^2}. \tag{5.2.3}$$

For the infrared thermal imaging system, the distant target can be considered as a small area source. The areas of the target object area source and the detector image area source are $\mathrm{d}A_1$ and $\mathrm{d}A_2$, respectively, the radiation brightness of the target object with temperature T is $N(T)$, the blackbody radiation emittance is $M(T)$, and the emissivity of the object surface is ε. The object square aperture angle and the image square aperture angle of the beam on the axis are U_1 and U_2, respectively, the optical system efficiency is τ_0, and atmospheric transmittance is τ_a. F is the F number of the optical system.

The radiation flux emitted by the object's surface source is

$$\Phi_1 = \pi N(T)\mathrm{d}A\sin^2 U. \tag{5.2.4}$$

The irradiance reaching the image source is

$$E_2 = \frac{\tau_0 \tau_a \Phi_1}{dA_2} = \frac{\pi \tau_0 \tau_a L(T) dA_1 \sin^2 U_1}{dA_2} = \frac{\pi \tau_0 \tau_a L(T) \sin^2 U_1}{\beta^2}. \tag{5.2.5}$$

The vertical axis magnification of the optical system can be expressed as

$$\beta = \frac{n_1 \sin U_1}{n_2 \sin U_2}, \tag{5.2.6}$$

$$E_2 = \left(\frac{n_2}{n_1}\right)^2 \pi \tau_0 \tau_a N(T) \sin^2 U_2. \tag{5.2.7}$$

The radiation flux received by a pixel on the image plane of the detector is the product of the irradiance received by the pixel surface and the pixel area, expressed as

$$\Phi_2 = E_2 A_d = \frac{A_d}{4F^2} \tau_0 \tau_a M(T). \tag{5.2.8}$$

5.2.2 Experimental Measurement

Thermal parameters are very important pieces of information for many applications. For example, for materials used for heat insulation or heat conduction, the thermal properties of materials are undoubtedly one of the performance evaluation indexes. The type of defect or inclusion under the surface can also be judged by measuring the thermal properties [191], and the thermal properties can be obtained by the infrared thermal imager [192, 193].

Here, we first describe the measurement of the thermal diffusion coefficient of materials by an infrared thermal imager. The surface of the specimen is heated by a laser or flash pulse, and the temperature field changes on the heating surface or back are quickly recorded by an infrared thermal imager. By analyzing the thermal change process, not only the thermal parameters – such as thermal diffusion coefficient, specific heat capacity, thermal conductivity, and heat storage coefficient of the material – but also the information such as material thickness, depth, and internal defects can be obtained.

The experimental principle of measuring the thermal diffusivity of materials by the transmission pulse infrared thermal wave method is shown in Fig. 5.8. The flash lamp and the thermal imager are, respectively, placed on both sides of the tested piece. The flash lamp pulse instantaneously heats one side of the surface of the measured object. After the surface absorbs heat, the temperature rises. At the same time, the heat is transmitted from the surface to the inside and gradually diffuses to the other side of the measured object (the thermal imager side). The surface on this side begins to rise until it reaches the maximum temperature. If the tested part cannot cover the whole field of view, it needs to be shielded to avoid interference caused by the flash pulse directly entering the field of view of the thermal imager. The high-speed infrared thermal imager records the temperature change process of the measured object before and after

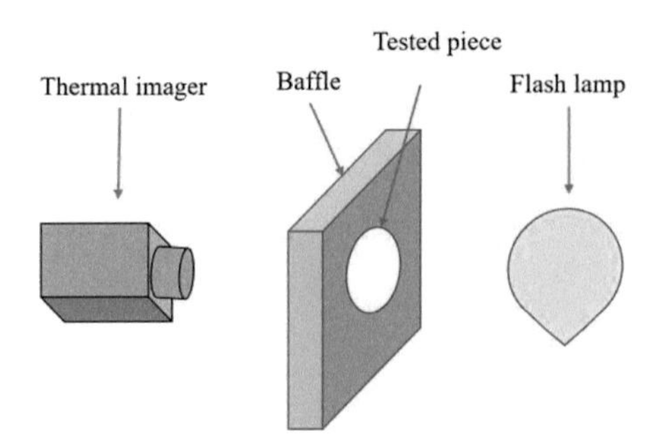

Figure 5.8 Experimental principle of measuring thermal diffusivity of materials by the transmission pulsed infrared thermal wave method.

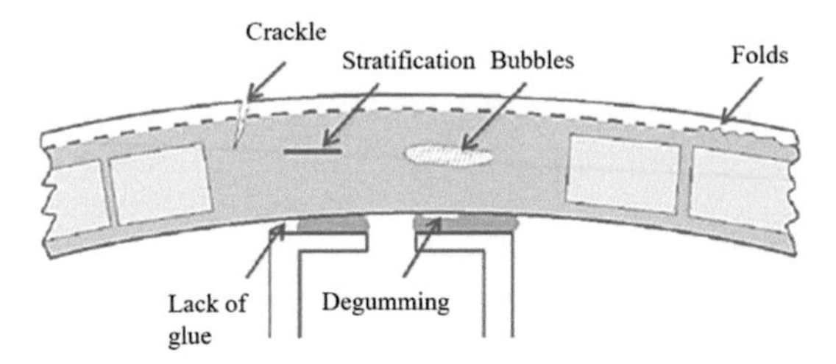

Figure 5.9 Schematic diagram of common defects of fan blades.

heating. Through data processing, analysis, and calculation, the thermal diffusion coefficient of the measured object can be obtained.

In the past 20 years, the research of depth measurement methods has been a hot spot in industry. The existing depth measurement methods of infrared thermal wave technology mostly use the relationship between characteristic time and depth. By analyzing the temperature change process of the defect surface, a peak or valley time is obtained as the characteristic time, and the depth is calculated by using the proportional relationship between the characteristic time and the defect depth [194]. Fan blades are usually bonded by the windward side and leeward side, forming a hollow structure inside. A large number of blades are manufactured by vacuum and negative pressure pouring. The resin is sucked into and infiltrated into the fiber, and then heated and solidified by the mold. The process is very complex, and defects such as bubbles, folds, and lack of glue often appear in the manufacturing process as shown in Fig. 5.9. The quantitative measurement method of infrared thermal waves can be applied to the detection of fan blades.

5.3 Application in Solar Power Plants

Nowadays, our world is experiencing a gradually rising demand for both energy and power, owing to rapid population growth, industrial development, and

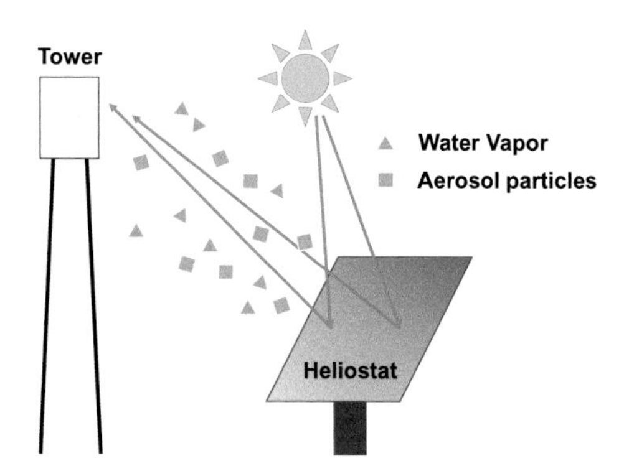

Figure 5.10 Solar radiation between the heliostat field and the receiver.

widespread urbanization. And it is a serious issue all over the world actually. However, a lot of places on Earth have abundant potential for solar energy usage, which is one of the renewable, clean, and freely available sources compared with other fossil energy that is the main reason for global warming or air pollution. There are lots of solar-energy-related technologies, and solar power plants, in particular, have many potential applications. Concentrating solar power (CSP) technology is one of the most important technologies, in which direct solar radiation is reflected from a concentrating system to a receiver. The concentrating system is made up of thousands of heliostats, and the direct solar radiation is reflected to a receiver where it is transformed into process heat.

In terms of the solar power plants considered here, solar radiation can be attenuated between the heliostat field and the receiver because of atmospheric extinction as illustrated in Fig. 5.10. In general, atmospheric extinction is mainly caused by aerosol particles and water vapor, which can scatter and absorb parts of solar energy at certain wavebands. And it is recognized as an important cause of energy loss in the increasingly larger solar power plants. Thus, it is important to obtain atmospheric extinction during the design of these solar power plants. It would be desirable to use extinction maps similar to the solar resource ones and would also be beneficial to have real information on atmospheric extinction during the daily operation of the solar power plants. It is easy to understand that the proportion of solar radiation attenuated by the atmosphere relies strongly on the real conditions of the local atmosphere environment. For example, in two atmospheric conditions as shown in Fig. 5.11, the values of atmospheric extinction will vary significantly [195]. The marked local character of this phenomenon and its temporal variability make it unfeasible to extrapolate atmospheric extinction models to different sites and seasons. Although there are several world maps of solar radiation, they are not detailed enough to be used for the determination of available solar energy in a small area; thus, it has prompted the development of calculation procedures to provide reliable assessment strategies for some areas where the direct measurement cannot be carried out.

Figure 5.11 Attenuation on a clear (left) and hazy (right) day. Reprinted from [195], Copyright 2017, with permission from Elsevier.

In a solar power plant, reflected direct normal irradiance (DNI) by concentrating mirrors is attenuated due to atmospheric extinction as it travels to the receiver. The atmospheric extinction can be considered by a ray-tracing model and plant optimization tools for standard atmospheric conditions. Note that the physical models are universal, which describe the transmission, scattering, and absorption processes of solar radiation in the atmosphere.

As illustrated in Fig. 5.10, a beam of photons can be partly scattered, absorbed, and transmitted when traveling through the heliostat field and the receiving tower. The broadband transmittance per traveled distance in a homogenous medium can be described with the exponential Beer–Lambert–Bouguer law [195].

$$T_x = \frac{\mathrm{DNI_B}}{\mathrm{DNI_A}} \approx \exp\left(-\beta_{\mathrm{ext}} x\right), \tag{5.3.1}$$

where β_{ext} is the atmospheric extinction coefficient and x is the distance between A and B as illustrated in Fig. 5.10. Line AB is termed as the slant range. $\mathrm{DNI_A}$ is the incident DNI at a starting point A, and $\mathrm{DNI_B}$ is the remaining DNI after traveling through an atmospheric layer between A and B. T_x denotes the broadband transmittance for a slant range of x km.

5.3.1 Theoretical Model

In the models of atmospheric extinction in solar radiation, visibility is an important parameter, which is often used as the atmosphere extinction approximately. The definitions of visibility are different from a distinct atmospheric condition [196]. However, visibility is referred to as a definition by a human observer and therefore only a rough estimate. Another way to define visibility is to use the meteorological optical range (MOR). The definition of the MOR is the length of the path in the atmosphere, which is required to reduce the luminous flux in a collimated beam from an incandescent lamp at a color temperature of 2 700 K to 5% of its original value [197]. What's more, the MOR can be defined by using the Koschmieder approximation [198] to connect with the spectral extinction coefficient at 550 nm, which can be expressed as

$$\mathrm{MOR} \approx \frac{-\ln(0.05)}{\beta_{\text{ext, 550 nm}}} \approx \frac{3}{\beta_{\text{ext, 550 nm}}} \approx \frac{-3x}{\ln\left(T_{x,\ \text{550 nm}}\right)}, \tag{5.3.2}$$

where x is the traveled distance through a medium and $T_{x,\ 550\ \mathrm{nm}}$ is the spectral transmittance for the traveled distance x at a wavelength of 550 nm. The standard visual range (SVR) is defined in a similar way but with a threshold of 3% rather 5%.

The data set generated by Vittitoe and Biggs [199] is important to many atmospheric extinction models. The data set has considered two elevations (619 and 1 524 m), two different standard atmosphere environments (mid-latitude summer and winter), and three aerosol conditions (no aerosol, clear conditions (SVR = 23 km) and hazy conditions (SVR = 5 km) at a mean sea level). In addition, the data set was generated for three tower heights (100, 300, and 883 m) and five slant ranges between the heliostat field and the receiving tower. The most of ray-tracing and plant optimization tools for modeling atmospheric extinction are under standard atmospheric conditions through the present research. As for some optimization tools like, for example, Sol TRACE [200], the atmospheric extinction has not been considered, and moderately adjusting the reflectivity of mirrors is merely considered.

The Leary and Hankins Model (L&H Model)

According to the data set generated by Vittitoe and Biggs [199], the Leary and Hankins (L&H) model [201] is defined by fitting to the data set:

$$T_{x,\mathrm{L\&H}} = \begin{cases} 0.679 + 11.76x - 1.97x^2, & (x \le 1\ \mathrm{km}), \\ 100(1 - \exp(-0.1106x)), & (x > 1\ \mathrm{km}), \end{cases} \tag{5.3.3}$$

where x represents the slant range. In order to reduce the computational consumption, the polynomial fitting and the exponential fitting are applied. And the L&H model is setting the slant range of 1 km as the sectioned point.

The Hottel Extinction Model

To estimate the atmospheric attenuation of DNI through clear standard atmospheres, Hottel proposed the Hottel extinction model [202] in 1976. This model is related to the elevation and the climate for two different "visibilities" (23 and 5 km). Many atmospheric extinction models for central power plants utilized this model, and there are three kinds of code as follows: the DELSOL3 code [203], system advisor model (SAM) [204], and HELIOS code [205].

According to the above-mentioned codes, the transmittance can be derived; relevant results are shown in Fig. 5.12. Compared with the clear case and the hazy case, the difference in atmospheric transmittance is obvious, and the former is obviously larger than the latter. We also find that the elevation has a great effect

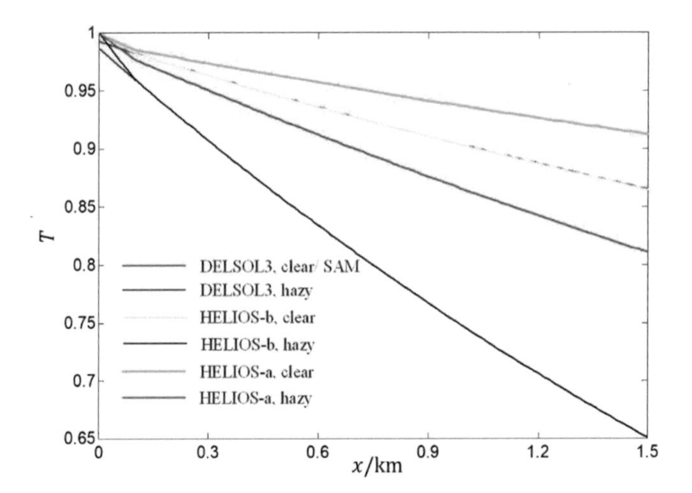

Figure 5.12 Transmittance derived with different codes.

on the atmospheric transmittance since the difference of transmittances between Barstow and Albuquerque is distinct. Due to the low elevation of Dubai, the DELSOL3 is more suitable there.

The Pitman and Vant-Hull Model (P&V)

The Pitman and Vant-Hull model was developed by Pitman and Vant-Hull [206] in 1984, in order to estimate the atmosphere extinction between a heliostat and a receiver. Similar to the L&H model, the P&V model also bases on the database of Vittitoe and Biggs but with more input variables. More specifically, there are five physical variables in this model, including the atmospheric water vapor density ρ_w, the aerosol particle scatter coefficient at a wavelength of 550 nm β, the height of the site above the sea level H_s, the tower focal height H_T, and the slant range R. Besides, this method also contains 10 fitting constants. Notably, in this model, only the scattering of aerosol particles and water vapor is considered, with no consideration of absorption effects. Those variables account for the seasonal variation of the local climate.

As there are several input parameters considered in the P&V model, it is necessary to specify the influence of those variables in atmospheric attenuation. Here, we briefly study their affects, and the related results are shown Figs. 5.13 and 5.14. In the area of Dubai, the height of the site above the sea level H_s is about 150 m, and the height of the tower is 260 m. The concentration of water vapor is chosen to be within 0–25 g/m^3, and the value of visibility is set covering 5–25 km.

Obviously, the changes in attenuation are distinct. Especially for the influence of visibility which is affected by aerosol significantly, the magnitude of atten-

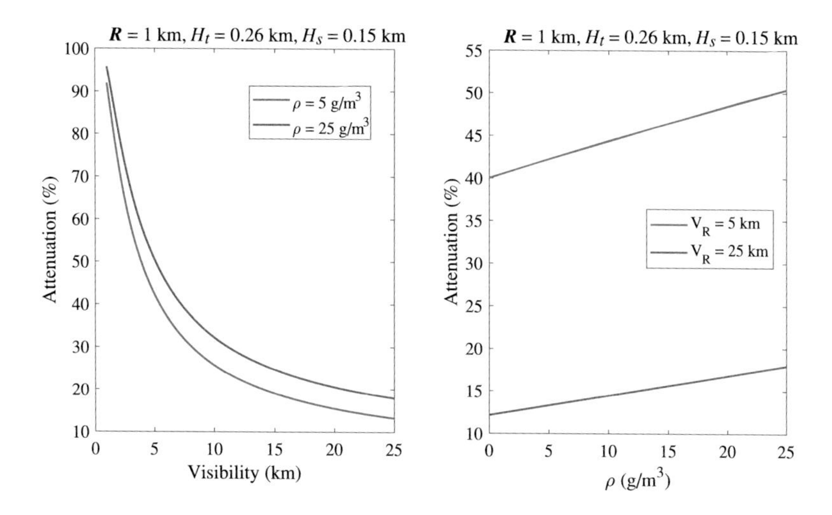

Figure 5.13 The results of attenuation as a function of visibility (left) and the concentration of water vapor (right), respectively. Notably, the $\boldsymbol{R}$ is the slant range in the panels, H_t is the height of the tower, and H_s the height of power plant above.

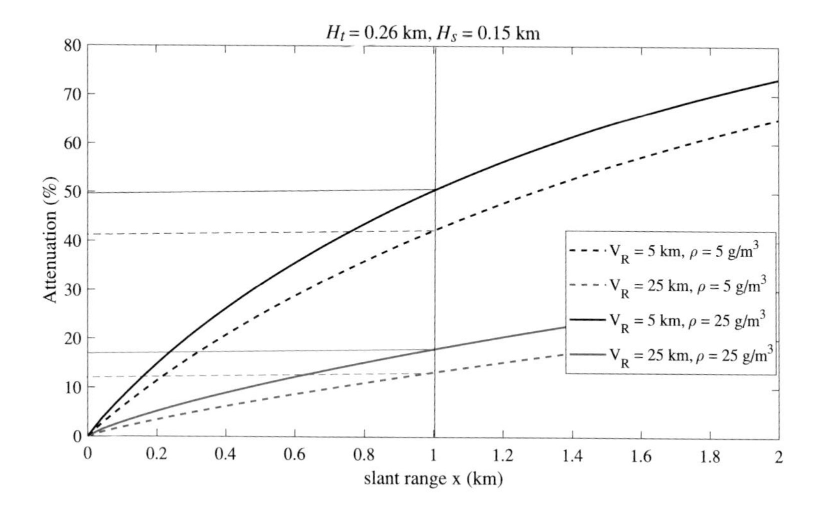

Figure 5.14 The results of attenuation as a function of slant range. The $\mathbf{V_R}$ is visibility and $\boldsymbol{\rho}$ is the concentration of water vapor.

uation increases quickly when the visibility increases. Besides, the increase of attenuation is also obvious when the concentration of water vapor is changed.

5.3.2 Experimental Measurement

Although there are several theoretical models to calculate the atmospheric extinction, the proportion of solar radiation attenuated by the atmosphere re-

Figure 5.15 White and black Lambertian target for extinction measurement Black (IB) and white (IW) zones of interest. Reprinted from [207], Copyright 2018, with permission from Elsevier.

lies much on the real conditions of the local atmosphere environment, and the models are not detailed enough to be used for the determination of the exact value.

In this section, a recent experimental measurement scheme proposed by Ballestrín et al. [207] will be introduced. The aim of this measurement system is to take simultaneous images of the same target at very different distances using two identical optical systems with digital cameras, suitable lenses, and filters. Having Lambertian targets is important for this measurement system since the position of the cameras and the angles relative to the target can never be known for sure at such large distances. In this work, the target was supported on two pillars, approximately 1 m above the ground. It had two side doors that remained closed when it was not used in order to protect it from rain, dust, etc. Two standard steel plates of 2 m×1 m were used for their construction as shown in Fig. 5.15.

The aim was to take simultaneous images of the same target at different distances using two identical cameras with suitable lenses as shown in Fig. 5.16. In order to do this, they work with diffuse solar radiation projected from the target to avoid direct solar radiation and therefore minimize specular influences and directionality. The intensity levels of the digital images would be proportional to the diffuse radiance coming from the target. The intensity difference or light extinction between both images taken was related to the distance between the two cameras, and the extinction for the experimental distance can be obtained in a direct way by

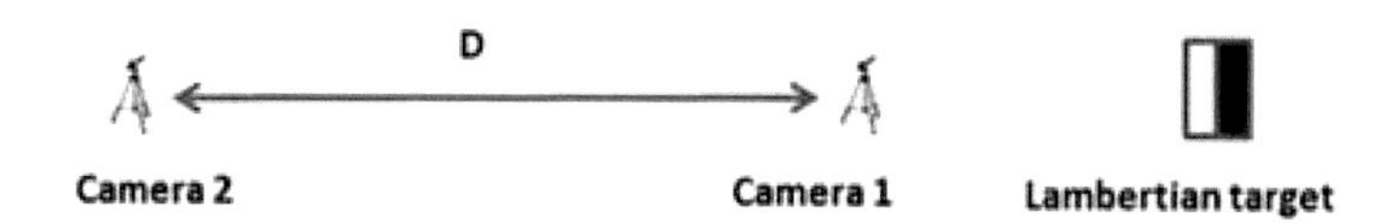

Figure 5.16 Lambertian target and cameras layout. Reprinted from [207], Copyright 2018, with permission from Elsevier.

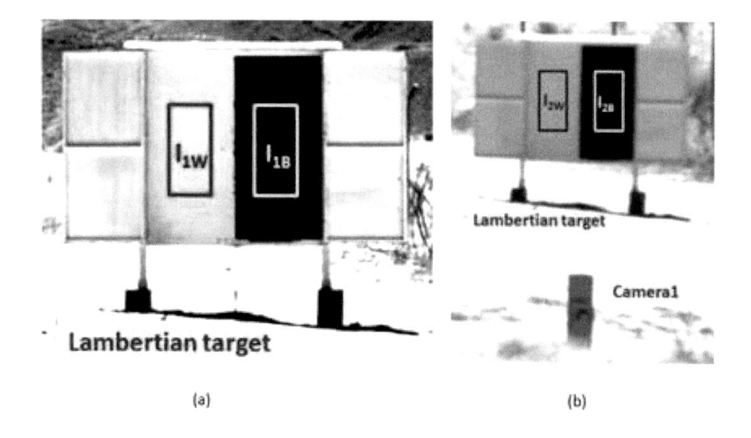

Figure 5.17 Images taken simultaneously with both cameras 03-July-2017 14:00:45. (a) Camera1. (b) Camera2. Reprinted from [207], Copyright 2018, with permission from Elsevier.

$$\mathrm{Ext} = 100\left(1 - \frac{I_2}{I_1}\right), \tag{5.3.4}$$

where I_1 ($I_1 = I_{1W} - I_{1B}$) and I_2 ($I_2 = I_{2W} - I_{2B}$) are the average intensities of the images as shown in Fig. 5.17.

The extinction coefficient can be derived from the Beer–Lambert–Bouguer law by

$$\beta_{\mathrm{ext}} = -\frac{\ln(I_2/I_1)}{D}. \tag{5.3.5}$$

Applying this methodology during different periods will allow us to obtain average extinction coefficients without having to use any simulation tool.

5.4 Summary

Macrothermal radiation theory and analysis methods have been widely used in several real applications. In this chapter, we introduced the applications of macrothermal radiation in industrial boilers, infrared thermal imagers, and solar power plants from theoretical models and experimental measurements. As for the reconstruction of the 3D temperature field in the furnace, intelligent optimization algorithms have attracted extensive attention and have been used to solve the problem with the rapid development of computer technology. These optimization algorithms, including ant colony optimization, genetic algorithm, and artificial neural network, can obtain the global optimal solution without knowing the exact mathematical model of the optimization problem and the gradient of the objective function. We hope this part can offer guidance in engineering applications.

Part II

Micro/Nanoscale Thermal Radiation

6 An Overview of Micro/Nanoscale Thermal Radiation

The treatments for thermal radiation heat transfer discussed in previous chapters are established for macroscopic objects, in which the wave nature of thermal radiation is ignored. This is reasonable because for macroscopic objects with characteristic lengths much larger than the wavelength of thermal radiation (typically, in the range of several hundreds of nanometers to 100 μm), geometric optics can be applied without the need to consider wave interference phenomena. In this sense, we can regard thermal radiation transfer as a process of transport of energy bundles without phase information, or sometimes photons, in the particle picture. Note, in this circumstance, the implication of "photons" is essentially not the same with the quantum of light [208]. However, when the characteristic length of objects is comparable to or smaller than that of thermal radiation, the wave nature of thermal radiation would be remarkable with emerging wave interference phenomena, where radiative heat transfer should be tackled in the framework of Maxwell's equations for EM waves. Moreover, the quantum treatment of the light–matter interaction is needed when the characteristic length scale further reduces to the quantum regime (usually <1 nm) [209] or in the cases with nontrivial couplings (such as ultrastrong coupling (USC) [210] and nonlinear coupling of cavities [211]). The latter regime is out of the scope of this book, and we focus on the treatment of thermal radiation as EM waves.

The scope of micro/nanoscale thermal radiation (MNTR) is to explore, understand, and manipulate thermal radiation heat transfer at microscopic levels. The development of this area largely benefits from the discovery of novel materials with rich microscopic structures and the development of high-precision nanofabrication technologies, which can create a set of artificial materials with desirable micro/nanostructures [6]. In this chapter, we will present a general overview of MNTR. First, we discuss the limitations of macroscopic theories of thermal radiation, in which we attempt to emphasize various important lengths and time scales that affect the treatment of thermal radiation and the validity of different theoretical methods or assumptions. After this, the main research topics in this field will be introduced briefly. The first is the study of microscopic thermal radiation transport mechanisms in novel materials, such as high-temperature superconductors and 2D materials. The second is thermal radiation transport in micro/nanofabrication technologies and relevant micro/nanodevices. Then comes the introduction of scattering and absorption in particulate media. We will also mention the rapid development of nanophotonics and metamaterials and near-field thermal radiation. The majority of the topics presented in this chapter will be introduced in more detail in the latter chapters.

6.1 Limitations of Macroscopic Theory of Thermal Radiation Heat Transfer

Generally speaking, the macroscopic theory of thermal radiation transport depends on the following assumptions [41, 212]: (1) The characteristic length scale of the object is much greater than the wavelength of thermal radiation and other relevant microscopic length scales. In this situation, radiation can be simply treated as light rays, and, therefore, geometric optical principles can be used. (2) The temporal scale of radiation transport under concern is much longer than the characteristic time scale of any microscopic processes, like the relaxation time of carrier transport. In this case, thermal radiation, as energy bundles of EM waves, is converted into thermal energy instantaneously when it is absorbed by the medium, and this conversion is immediately reflected as a rise in the temperature of the medium.

However, with the development of modern micro- and nanofabrication technologies, electronic and optoelectronic devices are miniaturized into even smaller sizes that are usually comparable or even much smaller than the thermal radiation wavelength. Understanding radiation transport in the fabrication and operation of these devices requires the exploration of MNTR transport mechanisms. Novel materials with rich and complicated micro- and nanostructures usually demonstrate exotic thermal radiative properties that cannot be described by macroscopic theories. A typical example is metamaterial with periodic nanoscale structures, which may demonstrate narrowband thermal emission [213], while none of the composite material itself can exhibit such an anomalous behavior. This implies that the totally different thermal radiation transport mechanisms at micro- and nanoscales may lead to a brand new physical panorama. In this sense, it is thus natural to utilize these micro- and nanostructures to control thermal radiation at will, enhance or suppress thermal radiation transfer, and devise new applications [214].

Now let us discuss in more detail the characteristic length scales that are important for MNTR [212, 215]. The first and most important length scale, as mentioned earlier, is the wavelength of radiation λ. This length scale determines whether in the problem under investigation thermal radiation can be regarded as waves or "photons" (energy bundles). When the object is on the scale of λ, the interaction of thermal radiation with the object may involve the propagation, interference, and diffraction of EM waves.

The second length scale is also important, which is the coherence length of thermal radiation L_c. The concept of coherence length describes the maximum optical path difference between two wave trains from the same light source that

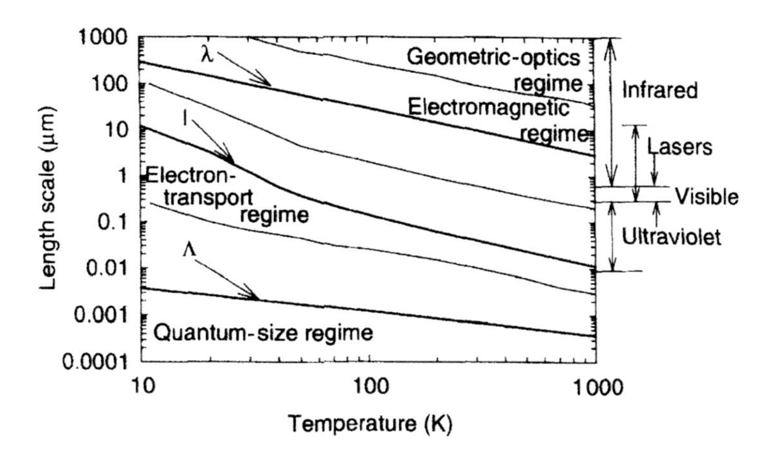

Figure 6.1 Different regimes of thermal radiation transport in the metal of Cu. Reprinted from [41], Copyright 1999, with permission from Elsevier.

can form interference patterns when they are brought together [216]. For a quasi-monochromatic light source with an effective bandwidth of $\Delta\omega$ (expressed in angular frequency), the coherence length is given by

$$L_c = \frac{2\pi c}{\Delta\omega}, \tag{6.1.1}$$

where c is the speed of light in vacuum. For a blackbody of a broad spectrum, the calculation becomes more involved, and its coherence length depends on its temperature as [217–219]

$$L_c T = 0.3\pi\hbar c/k_B \approx 2167.8\ \mu\text{m}\cdot\text{K}, \tag{6.1.2}$$

which means that the coherence length is of the order of or even smaller than the dominant radiation wavelength. (Recall Wien's displacement law Eq. (1.2.8).) Recently, the coherence properties of blackbody radiation have attracted renewed interest with slightly different results of coherence time and length, according to the environment of the blackbody [220]. Moreover, the development of narrowband emitters with engineered micro/nanostructures has been presented to show significantly longer coherence time and lengths, for instance, Klein et al. [221] demonstrated the coherence length of thermal radiation emitted from their heated metallic nanowires at least 20 µm [221]. Chen and Tien [222], as well as Anderson and Bayazitoglu [223], evaluated the effect of partial coherence of thermal radiation on radiative properties of multilayer thin films, and derived practical regime maps for coherent, partially coherent, and incoherent regimes.

The third length scale is the mean free path l of photon-excited carriers in the medium (like free electrons in metals) in the classical Boltzmann theory of particle transport [224]. For free electrons, l is given by

$$l = \frac{m v_F \sigma}{n e^2}, \tag{6.1.3}$$

where m is the electron mass, v_F is the Fermi velocity, n is the electron number density, σ is the electrical conductivity, and e is the elementary charge.

The fourth is the penetration depth of EM waves in the medium $\delta = \lambda/(4\pi\kappa)$, where κ is the imaginary part of the optical constant (or the complex refractive index).

Let us discuss an example given in [41] to quantitatively figure out different radiation transport regimes in the metal Cu, as shown in Fig. 6.1. The first is the geometric-optics regime, where the size L of the object under investigation is much larger than all the characteristic length scales λ, l, δ, and Λ. The second is the EM regime, in which L is of the order of the radiation wavelength λ, while being much larger than the mean free path l and wavelength Λ of free electrons. The penetration depth δ must be either much smaller than l or much larger than l. With these conditions, we can safely assume that the optical constants n and κ used in this regime are the same as those of the bulk material, independent of the size L. This regime is the mostly concerned regime of MNTR, in which EM theory is the main tool to investigate radiation transport, with a variety of interesting phenomena that do not exist in the geometric optics regime. These include the tunneling of evanescent waves, optical resonance modes due to interferences, surface plasmon polaritons (SPP), and so on [225]. In the third regime, L is of the order of the mean free path of electron transport, l, or smaller, but is kept to be larger than the electron wavelength Λ. This regime can also be called the electron-transport regime, in which the bulk properties cannot be used since the scattering of electrons by the boundary of the object will lead to size-dependent optical constants. In this regime, electron-transport theory is needed to study thermal radiation transport in conjunction with Maxwell's equations. If the penetration depth δ is of the order of l, the variation of the electric field in the metal would have an impact on the electron between collisions and should be taken into account, and the effects of all boundaries on the wave have to be considered. If δ is much greater than l, then the variation of the EM field between electron collisions is negligible, but all wave–boundary interactions must be considered. The last regime is the quantum-size regime, where the characteristic length L is of the order of the electron wavelength Λ. Here the quantum-size effects dominate. This regime is not considered in detail in this book.

As a consequence, the MNTR regime is

$$O(L) \approx O(\lambda), O(L) \leq O(l_c), O(L) > O(l) \tag{6.1.4}$$

or

$$O(\delta) < O(l) \text{ or } O(\delta) > O(l); \tag{6.1.5}$$

otherwise, the electron-transport regime is entered since the variation of the electric field acting on the electron between collisions has to be accounted for.[1] Here $O(\cdot)$ is the order of magnitude.

Time scales are also relevant for thermal radiation transport, especially when we consider ultrafast laser propagation and the induced thermal effects (coupled conduction–radiation transport). The first time scale is the pulse duration of the laser, and the second time scale is the coherence time of radiation that indeed determines the wave interference effects. Regarding the coupled conduction–radiation transport problem, there are also two time scales: the relaxation time of carriers and the thermal diffusion time. Nevertheless, the ultrafast process is out of the scope of our book, and a detailed discussion on the time aspects of thermal radiation transport is not given here.

6.2 Thermal Radiation Transport in Novel Materials

Before the 2000s, there were valuable reviews about MNTR in heat transfer journals, for instance, Duncan and Peterson [227], Tien and Chen [217], Longtin and Tien [212], and Kumar and Mitra [41]. After the 2000s, more and more reviews regarding specific topics in this field appeared due to the rapid development of MNTR that has become an active interdisciplinary research area that combines the advances of materials science, optics and photonics, condensed matter physics, micro/nanofabrication technology, etc., disseminated over a wide range of journals, for instance, Xuan [228], Cahill et al. [229, 230], Song et al. [231], Li and Fan [232], Cuevas and García-Vidal [233], Baranov [234], Wang et al. [214], to name a few.

The development of MNTR is largely inspired by the discovery of novel materials. Many efforts were made to study radiative (infrared) properties of superconducting thin films in the 1980s–1990s [235–239], since high-temperature superconducting (HTS) materials, especially La–Ba–Cu–O [240, 241] and Y–Ba–Cu–O [242, 243] that were found in 1986–1987, were receiving much attention at that time [244], in order to determine whether there is a superconducting energy gap in the spectra, and the effects of microscopic carriers, including electrons, phonons, magnons and, most importantly, the quasi-particles called Cooper pairs mediated by the electron–phonon coupling, may lead to intriguing features in the radiative spectra.

In particular, using infrared spectroscopy methods, several fundamental properties of oxide superconductors have been defined, such as the effective mass and density of the free carriers, spectrum, and lifetime of quasi-particle excitations caused by their interaction with the lattice [245]. Studies of reflection spectra in the low-frequency range, $\hbar\omega \leq 2\Delta$, confirmed the appearance of a gap 2Δ in the spectrum of quasi-particles in the superconducting state and led to

[1] However, if for this case, $\delta \ll L$, then the skin effects may be incorporated into the bulk properties, as via the anomalous skin theory [226], the EM theory may be applied.

the estimation of the value of the gap and its temperature dependence $\Delta(T)$. The first experiments carried out on ceramic samples led to contradictory results due to the strong anisotropy of conductivity in copper–oxide compounds and inadequate surface quality. Further experiments with single crystals have removed some discrepancies in the data, although there still remain many unsolved problems [246].

Bonn et al. [247] presented the reflectance spectra for the high-temperature superconductor $YBa_2Cu_3O_7$ and the Kramers–Kronig (K–K)-transformed conductivity from 50 to 900 cm^{-1}. They found that the continuous background is Drude type at high temperatures, but below the superconducting transition, there is a region of suppressed conductivity consistent with a superconducting gap with the weak-coupling Bardeen–Cooper–Schrieffer (BCS) value. They also found the changes to the low-lying phonons and a plasma-type edge at 60 cm^{-1} also associated with superconductivity. Herr et al. [248] showed evidence for strong electron–phonon and electron–electron interactions from the optical spectra of $La_{1.85}Sr_{0.15}CuO_4$. Timusk et al. [249] studied the temperature dependence of the infrared properties of $YBa_2Cu_3O_{7-\delta}$ samples that were a–b oriented.

Rao et al. [235] studied the optical reflectance of thin superconducting $Bi_2Sr_2CaCu_2O_x$ (Bi 2:2:1:2) films in the frequency range 80–48 000 cm^{-1} and the temperature range of $5\ \text{K} < T < 300\ \text{K}$. The dielectric function of the superconductor was calculated by taking into account the effect of the quasi-particle lifetime. In the reflectivity spectra, Bi 2:2:1:2 phonons and $SrTiO_3$ substrate phonons can be identified. Below the superconducting transition temperature $T < T_c$, evidence for a superconducting gap can be observed, which was found to be $2\Delta = 35 \pm 5$ meV $\approx 6k_BT_c$ at $T = 5$ K. Zhang et al. [238] measured the reflectance and transmittance of $YBa_2Cu_3O_7$ (YBCO) films on $LaAlO_3$ substrates that are measured using an FTIR spectrometer at wavelengths from 1 to 100 μm at room temperature and also the reflectance from 2.5 to 25 μm at 10 K using a cryogenic reflectance accessory, in order to determine whether thin-film optics with a constant refractive index can be applied to high-Tc superconducting thin films. The optical constants of YBCO films were obtained by modeling the frequency-dependent complex conductivity in the normal and superconducting states and applying EM-wave theory. It is found that a thickness-independent refractive index can be applied even to a 25 nm film, while for 10 nm film, a deviation of refractive index to those of other films was found due to the boundary scattering effect of electrons. Phelan et al. [236] compared the classical Drude–London theory and the quantum-mechanical Mattis–Bardeen (MB) theory to predict the reflectance spectra of YBCO in the superconducting state. It was shown the quantum-mechanical theory is more successful within 12% of the experimental data. It also recommended using the BCS value for the energy gap. A more recent review on this subject matter, that is, infrared radiative properties of HSC, is given by Timusk and Statt [250], Basov and Timusk [251, 252], as well as by Basov et al. [253], and most recently by Vedeneev [254].

In the 1990s–2010s, MNTR was focusing on understanding the radiative properties of photonic crystals (PCs), plasmonic materials, metamaterials, etc., which can also be regarded as novel materials. This would be discussed separately in Section 6.5. In recent years, 2D materials have also become an extensively studied subject matter of MNTR. The most notable 2D material, graphene, has been widely investigated in the community, focusing the role played by its infrared plasmon polaritons [255, 256], which can also be electrically tuned. Other 2D materials, hexagonal boron nitride (hBN), transition metal dichalcogenides, such as MoS_2 and $MoSe_2$, are also investigated, which show novel features, such as strong phonon polaritons, hyperbolic dispersion (equifrequency surface), and very high dielectric index and excitons, have attracted much interest. The discovery of novel nano- and quantum materials, including 2D materials and their heterostructures [257, 258], and topological insulators and superconductors [259], drastically expands the fundamental limits of our scope into the extreme nanoscale (even the atomistic scale). The tunability of their electronic and optical properties is impressive. Their unconventional and rich optical phenomena can offer a promising platform for realizing exotic thermal radiation phenomena and achieving high thermal radiation emission performance [256, 260].

6.3 Microfabrication and Nanofabrication

There are at least threefold roles of micro- and nanofabrication in the development of MNTR. The first is that, in the many processes of micro- and nanofabrication technologies, we may encounter MNTR problems. The second is that due to the development of micro- and nanofabrication technologies, many micro- and nanoscale devices and technologies like optoelectronic devices (e.g., vertical-cavity surface-emitting laser [VCSELs]), nanosensors, and magnetic recording emerge, in which thermal radiation energy transport is an issue. The third is that, by exploiting a variety of micro- and nanofabrication technologies, we can manipulate radiation transport at micro- and nanoscale. In addition, due to the development of nanofabrication technology, we can measure and probe MNTR which would not be possible without these technologies, for instance, by developing and fabricating high-precision heat flux meter [261–263].

Let us first discuss the first fold in this section. Understanding thermal properties and thermal transport in microstructures, particularly radiation transport, is crucial for thermal control in microfabrication. At high temperatures required by these processes, radiation plays a crucial role in the fabrication of materials [264–266]. Radiation properties of thin films during deposition significantly affect growth rates, and thus microstructure and thermophysical properties [267, 268]. The scattering characteristics of the microgrooved surfaces play a crucial role in the fabrication procedure of large-area spatially coherent gratings using holographic lithography [269]. Periodic surface microstructures are produced when strong lasers interact with metallic and dielectric surfaces, as a consequence of spatially periodic melting and solidification [270]. The production of such structures has not been adequately explained because of a lack of understanding and correct modeling of the thermal processes, such as radiation scattering and interference from the periodic structures during their growth, phase change, and conduction source [41].

Laser fabrication, ablation, and heating inspired the development of MNTR in the 1980s [271]. There is a distinction between ultrafast laser fabrication and traditional machining with continuous wave or large-pulse-width lasers. The latter removes materials by the bulk expulsion of molten material, while in the former method, the molten material is directly vaporized [41]. In this manner, ultrafast lasers can achieve much higher fabrication quality and precision without other parts of the material affected. The reflectivity difference between the (supercooled) liquid and (superheated) solid phases is a source of thermal instability in laser-induced crystal growth, as shown by Grigoropoulos et al. [272] using a simplified model of the fundamental mechanisms governing the laser-induced recrystallization of a thin polysilicon layer on the so-called heat sink structure.

Heat transfer mechanisms during ultrafast laser heating of metals were investigated from a microscopic perspective by Qiu and Tien [273]. Using the Boltzmann transport equation, they modeled the transport of electrons and electron–lattice interactions, with the scattering term of the Boltzmann equation evaluated using quantum mechanical principles. This is because heating involves three processes: the deposition of radiation energy on electrons, the transport of energy by electrons, and the heating of the material lattice through electron–lattice interactions. By solving the Boltzmann equation, they demonstrated that a hyperbolic two-step radiation heating model can be rigorously established that can clearly reveal the hyperbolic nature of the energy flux carried by electrons and the nonequilibrium between electrons and the lattice during fast heating processes when the laser pulse duration is comparable with the relaxation time of electrons, whose energy is dissipated to the lattice. The given model's predictions during subpicosecond laser heating matched well with available experimental data [274]. Later, Heltzel [275] studied the ablation of dielectrics with femtosecond laser pulses using an electron density model.

The use of thermal radiation for the treatment or processing of multilayer thin-film structures is also a microscopic coupled radiation–conduction problem. Wong et al. [276, 277] conducted a parametric study on this topic, and their findings demonstrate that even a small change in layer thickness can cause large changes in the film's reflectivity and temperature. The biggest source of temperature measurement error is the fluctuation of surface emissivity during rapid thermal processing (RTP) of thin-film devices, as shown by Ray [278]. Controversy arose over whether wave or geometric optics should be used in the RTP of thin films after Sorrel et al. [279] modeled thermal radiation transport using optics. Wafer patterning can considerably affect the local radiative characteristics and, by extension, the temperature uniformity during rapid thermal annealing, as shown by Vandenabeele et al. [280].

6.4 Scattering and Absorption in Particulate Media

Knowledge of radiation properties of media containing micro- and nanoparticles is important for various heat transfer applications, many of which have thermal radiation as the dominant mode of energy transfer. Combustion systems, porous matrices in burners and heat tiles, packed bed combustors, deposited soot on furnace walls [281], agglomerated particles in combustion products [282], microsphere insulations, and other applications contain small particles that significantly affect the thermal energy transport. Their scattering, absorption, and emission characteristics play an important role in the overall energy transfer, and an understanding of their radiation characteristics is central to the prediction and evaluation of system performance. For early reviews, see Howell [283] and Tien [284]. These characteristics are also of great importance in the laser diagnostics of emissions.

It is known that when the size of microscopic inhomogeneities in a heterogeneous medium is comparable to the wavelength of EM waves, significant wave interference phenomena can occur [285]. As a consequence, this kind of heterogeneous media can usually interact with EM waves in a much more complicated and stronger manner than conventional bulk and homogeneous materials that typically possess inhomogeneities on a scale much smaller than the wavelength, and thus they have a great potential in controlling the propagation of EM waves. In particular, micro/nanoscale disordered media, which have inhomogeneities with characteristic sizes ranging from a few tens of nanometers to several hundred micrometers, can significantly affect the propagation of thermal radiation, whose wavelength usually lies in the range of 100 nm–100 μm for objects in heat transfer applications.

Therefore, micro/nanoscale disordered media, such as porous dielectric media with micropores and voids [286], particulate media containing micro- and nanoparticles [287], colloidal suspensions of nanoparticles [288], many kinds of coatings [289], foams [290], fibers [291], and soot aggregates [292], have been widely applied in controlling thermal radiation transfer. For example, porous silicon carbide (SiC) material can be utilized as high-efficiency solar absorbers [293], porous zirconia (ZrO_2) coatings are enormously used to provide thermal protection for the metallic components of gas turbines [294], and polymer films containing randomly distributed silica (SiO_2) nanoparticles show an excellent performance for radiative cooling [295].

Since thermal radiation transfer in those micro/nanoscale disordered media is strongly affected by the microscopic structures, in order to tailor their radiative properties and functionalities, a full understanding of underlying physical mechanisms of EM-wave transport as well as the relationship between micro/nanostructures and radiative properties is of critical importance.

However, in such media, radiation is scattered and absorbed in a very complicated way, which brings difficulties to theoretical and experimental investigations. Conventionally, the propagation of radiation is described by the RTE in the mesoscopic scale. The radiative properties entering the RTE, including the scattering coefficient κ_s, absorption coefficient κ_a, and phase function $P(\mathbf{\Omega}', \mathbf{\Omega})$ (where $\mathbf{\Omega}'$ and $\mathbf{\Omega}$ denote incident and scattered directions, respectively), depend on the microstructures, as well as on the permittivity and permeability of the composing materials. In particular, for disordered media consisting of discrete scatterers, that is, DDM, the radiative properties are usually theoretically predicted under the independent scattering approximation (ISA), i.e., in which each discrete inclusion is assumed to scatter EM waves independently as if no other inclusions exist, i.e., without any inter-scatterer interference effects [296–300]. ISA is valid only when the scatterers are far-apart from each other (i.e., the far-field assumption) and no interparticle correlations exist (i.e., independent scatterers) [297–301]. When the two conditions are violated, the scattered waves from different scatterers interfere substantially and consequently, ISA fails [302–305]. In this circumstance, we call the radiation scattering process from a scatterer is "dependent" of the presence of other scatterers. This fact leads many researchers in the thermal radiation community to the considerations of the dependent scattering effect (DSE) in order to correctly predict the radiative properties of DDM [302–314].

For many years, it was generally well known to paint and paper coating technologists that high-concentration packing of pigment particles in a white paint layer can lead to a decrease in its opacity (or "hiding power") [315, 316], due to the interference of light scattered by neighboring particles. In his noted book on light scattering [317], Hendrik C. van Hulst mentioned that the mutual distance between the particles of three times the particle radius may be a sufficient condition for independent scattering. In the 1960s, several experimental works were carried out for optically scattering turbid media composed of dielectric particles, such as TiO_2 and PS particles, and found that at certain particle concentrations, the DSE became prominent, but no clear criterion that could quantitatively determine the departure from ISA was obtained [318–321]. For example, Churchill et al. [318] found a critical value of $\delta/d \sim 1.7$, above which no interference effect was observed, where d is the particle diameter and δ is the center-to-center distance. Harding et al. [319] intended to achieve optimal scattering properties at minimum cost for paint films, and they revealed critical concentrations above which the ability of a particle to scatter light may be precipitously reduced because of optical "overlap" with its neighbors. Blevin et al. [320] experimentally showed that at a high concentration reflectance falls due to interparticle interferences. These works were published in optics and chemistry journals. On the other hand, the general theory of multiple scattering of classical waves, including EM and acoustic waves, was already established even earlier by physicists [322–325], which can provide rigorous treatments for the DSE. However, it was not noticed and applied in these works due to the lack of easy-to-use practical formulas.

To the best of our knowledge, in the thermal radiation heat transfer community, the DSE was first investigated by Hoyt C. Hottel and his colleagues in the 1970s [326, 327]. They experimentally measured the bidirectional reflectance and transmittance spectra of monodisperse PS nanosphere suspensions in water confined between parallel glass slides at different optical thicknesses, where the sphere diameter was in the range of 0.102–0.53 μm and the volume fraction was varied from 1.3×10^{-6} to 0.295. By comparing experimental data with ISA predictions, empirical criteria for the dependent scattering regime were also obtained, where the critical parameter was the clearance-wavelength ratio (c/λ), where the clearance $c = \delta - d$ for spherical particles and λ is the wavelength of incident light [327]. In particular, an experimental correlation between the effective scattering efficiency Q_s and c/λ was established accordingly.

Later, in the 1980s and 1990s, Chang-Lin Tien and his colleagues conducted comprehensive studies on the DSE in particulate media [284, 306, 307, 328–331]. Tien and Drolen reviewed the independent and dependent scattering of thermal radiation transfer in these media [306]. Later many authors carried out investigations into this effect, for example, Wang and Zhao [303], Ma et al. [312], and Singh and Kaviany [332], to name a few. In fact, in the last several decades, many efforts have been made in the study of this mechanism, not only in the field of radiative heat transfer but also in the communities of optics, photonics, biomedical engineering, astrophysics, paint industry (visibility), meteorology (atmospheric sciences), remote sensing, and mesoscopic physics, since the DSE can occur not only for thermal radiation but also for any types of EM waves, including visible light, terahertz waves, and microwaves, only if the packing density is high enough and the distance between adjacent scatterers is comparable with or smaller than the wavelength. Moreover, in the past years, significant progresses have been made in micro- and nanofabrication, nano-optics and photonics, lasers, modulators and detectors, as well as computational capabilities of modern computers, which further reshape the way we study radiative transfer and the DSE.

6.5 Nanophotonics and Metamaterials

The last two decades have witnessed a shift away from the traditional understanding of thermal radiation transfer, thanks to advances in nanophotonics that employ manufactured nanostructures with at least one structural feature at a wavelength or subwavelength scale [6, 333–337]. The thermal radiative characteristics of nanophotonic structures are very different from those of traditional thermal emitters and radiators. The field of nanophotonics holds great promise for the management of thermal radiation. Nanophotonics structures, for instance, may emit heat energy in a coherent, narrow band, polarized, and directed fashion. Enhanced, nonreciprocal, or dynamically adjustable thermal radiation is also possible. The recent breakthroughs in nanophotonic thermal radiation control have opened up a wealth of promising new avenues for harnessing energy.

When the feature sizes of nanophotonic structures are on par with the wavelength of light, wave interference effects give rise to a plethora of opportunities for fine-tuning their spectral responses. It is possible to construct structures with very differing emissivities from those of the underlying materials. By using various photonic resonators, one can significantly enhance a material's emissivity. The absorption spectrum can have a distinct peak in it, thanks to the usage of a lossy resonator. Therefore, a narrowband thermal emitter can be constructed using such a resonator. To achieve this goal, several different types of resonant systems have been used, including arrays of metallic antennas [336, 338], SSP [339], Fabry–Perot cavities [340–342], dielectric microcavities [343], guided resonances in PC slabs [344–347], and metamaterials [348, 349].

In 1986, Hesketh et al. [350] were the first to investigate the size effect of micro/nanostructures due to EM-wave interferences on thermal radiation properties, which significantly departed from the prediction of simple geometric optics. They investigated the thermal emission from 1D deep gratings. Normal spectral (3 μm–14 μm) emissivity measurements were carried out for 45 μm deep, near square-wave gratings made of heavily phosphorus-doped silicon at 400 °C for both *s*- and *p*-polarizations. Different values of the period Γ of the grating were investigated, including 10, 14, 18, and 22 μm, resulting in a Γ/λ ranging from 0.14 to 7.33. They showed that all these deep gratings exhibit significantly higher spectral emissivity than the purely smooth surface of doped silicon in the entire investigated spectral range. For both *s*- and *p*-polarized emission, they found that the resonances in the spectra are due to the excitation of standing waves in the air slots perpendicular to the silicon surface, very similar to those in an organ pipe. They further revealed that the resonant amplitude of the *s*-polarization does not depend significantly on Γ, while that of the *p*-polarization does. Theoretical analysis was later presented in 1988 [351, 352].

In order to produce powerful thermal emission with multiband or broadband characteristics, photonic structures can make use of many resonances [353–356]. Dual-band emissivity can be achieved, for instance, by combining two resonators to create a bipartite checkerboard unit cell [337]. Sawtooth structures [353], trapezoidal structures [357], and even fractal structures [358] can also be used to provide a broadband response by supporting numerous resonances. When numerous resonances with similar resonant frequencies are brought together in a subwavelength region, an analog of superradiance can emerge in thermal radiation [359]. In general, understanding the origin of various resonances and their interactions can serve as a flexible conceptual foundation for engineering the thermal emissivity spectrum.

In contrast to increasing thermal emissivity, a PC structure that allows for a photonic bandgap (PBG) can be used to strongly suppress emissivity throughout a wide range of wavelengths. This bandgap is associated with a frequency band in which the density of states (DOS) in the structure is minimal. The structure near-perfectly reflects light with frequencies within the bandgap. This extreme reflectivity results in a drastically reduced thermal emissivity within the bandgap. Dielectric and metallic PC structures have been presented with strong suppression of thermal emissivity [333, 334]. For instance, Yeng et al. [360] studied a set of air holes in a metal film, in which each air hole can be viewed as a metallic waveguide with a cutoff frequency that grows exponentially with hole size. The array can form a bandgap below the cutoff frequency. The resulting structure has a very low thermal emissivity between almost zero and the cutoff frequency.

In many nanophotonic systems, the emissivity is enhanced in some wavelength ranges while being suppressed in other ranges. As mentioned earlier, PC structures can reduce emissivity in the gap while increasing it near the photonic band edges. To selectively enhance or reduce thermal radiation at different wavelengths, one can also design a metamaterial with certain effective material properties. Examples of metamaterials used to manipulate thermal radiation include those made up of alternating subwavelength layers of metals and dielectrics [348, 349, 361]. This type of metamaterial shows enhanced thermal emission in the ellipsoidal dispersion regime of the effective dielectric function and decreased thermal emission in the hyperbolic dispersion regime [357]. By modulating the layer thickness, the transition wavelength between the two phases can be adjusted.

Absorption/emission spectra of many nanophotonic structures can be designed to depend critically on polarization. This hence allows for highly polarized thermal radiation from them. When compared to a standard blackbody or gray-body thermal emitter, whose thermal emission is normally unpolarized, there is a stark difference. For instance, the nanowire antenna studied by Schuller et al. [362] has resonances that couple only to TE polarization with an electric field parallel to the wire, or TM polarization with an electric field perpendicular to the wire, because of the mirror symmetry of the wire. Thus, the wire's emission can become highly polarized on resonance. The thermal radiation from individual platinum nanoantennas was studied by Ingvarsson et al. [363]. They showed that the antenna's thermal radiation had features very similar to a dipole radiator, with the orientation of the dipole highly associated with the orientation of the nanoantenna. As a result, thermal radiation can be highly polarized. Grating structures [335], PC slabs [364], and cavities [365] can also be used to achieve strongly polarized thermal emission.

Nanophotonic structures can also be engineered to generate circularly polarized thermal emission. For instance, Lee and Chan [366] theoretically proposed that a chiral layer-by-layer PC structure can thermally emit predominantly circularly polarized infrared light (around 12 μm), which arises from the polarization-dependent PBG or the surface plasmons if the PC structure is supported by a metallic substrate. Wu et al. [367] proposed a silicon-based chiral metasurface in which the mirror symmetry is broken for a single meta-atom (like a broken "n" shape). It was shown that highly circularly polarized thermal radiation can be achieved in this structure at a wavelength of about 4.7 μm. Notably, some metasurfaces that can introduce the spin–orbit interaction (SOI) of light, that is, the coupling of the photon spin (or circular polarization) and the extrinsic momentum (wavevector), can be engineered to thermally emit circularly polarized light. For instance, Shitrit et al. [368] demonstrated that, by engineering the space-variant orientation angle for the nanoantennas in the metasurface, a photon-spin-dependent wavevector shift can be induced, which is called the optical Rashba effect as a manifestation of SOI under the broken inversion symmetry. Therefore, the degenerate dispersion relations of different spins are split into two nondegenerate dispersions, leading to spin-dependent thermal emissivities.

Nanophotonic structures can also modify the emissivity's angular or directional profiles. Greffet et al. [335] showed for the first time that the emissivity from a SiC grating structure is highly dependent on the AOI. Since surface phonon polaritons (SPhPs) can exist at the SiC–air interface, strong angular-dependent absorptivity is achieved with the use of grating to enable wavevector-selective resonant stimulation of such SPhPs. At a particular frequency, emissivity is at unity in only one direction, while it is highly suppressed in all other directions. Another variable in regulating a resonant thermal emitter's response is the emissivity cone's angular breadth. The angle-dependent thermal emissivity of a plasmonic metasurface composed of tungsten (W) disks mounted on a silicon nitride (Si_3N_4) layer backed by a platinum (Pt) mirror was studied by Costantini et al. [213]. Gap SPP can exist between the two metal layers in the metasurface structure. At the frequency $\omega = 2\,353$ cm^{-1} (expressed in the unit of wavenumber), the absorption is nearly at unity. The wavelength in free space at this frequency is less than the period. Therefore, there is no greater diffraction order for normal incident light. However, the critical coupling requirement is no longer met at larger incident angles due to the appearance of increased diffraction order, which creates extra routes for radiation from the modes and results in lower absorption/emission at larger incident angles. As a result, by tuning the structure's periodicity, the emitter's angular width can be adjusted. Other photonic structures, such as bull's eye structures [369, 370], a tungsten grating structure [371], and PC [336, 372], have also been suggested for and demonstrated highly directional thermal emission. Furthermore, the combination of a thermal emitter and a photonic structure optimized for a strong angular response, such as a multilayer coating [373] or an angular coupler [374], can result in highly directional thermal emission.

In light of recent developments in metasurface, one can build a metasurface to manipulate the wavefront of thermal emission [368, 375]. This is possible because one can control the wavefront by adjusting the phase and amplitude responses of each element in an ultrathin optical antenna array and induce a phase gradient and geometric phase for wavefront control [376]. It was proposed that a SiC metasurface may be used for thermal focusing [375]. The design is to create nanostructures on the surface of SiC that will scatter surface waves generated thermally. To achieve focus, the surface's scattering elements' size, shape, and spacing must be optimized to achieve the constructive interference of the scattered waves at a target location. Controlling thermal emissions with a wide range of tunable parameters possible by the metasurfaces' degrees of freedom.

We also envisage that the prosperity of topological photonics [377], non-Hermitian optics [378], and quantum nanophotonics [379] has brought new concepts to the micro/nanoscale control of thermal radiation in recent years. For example, the connection between topological photonics and thermal radiation has been made theoretically by a recent work of Silveirinha [380]. Non-Hermitian selective thermal emitters exhibiting passive $\mathcal{PT}$ symmetry at 700 °C were very recently demonstrated by Doiron and Naik [381], in which a $\mathcal{PT}$ phase transition makes the dual emission peaks converges into a single emission peak in the parameter space. Ridolfo et al. [382] theoretically showed that the photon statistics of thermal radiation emitted from a cavity quantum electrodynamics (QED) system in the USC regime vastly differs from the conventional statistics of thermal photons, which implies a possible way to tailor the photon statistics of thermal radiation using quantum nanophotonic tools.

6.6 Near-Field Thermal Radiation

For two macroscopic objects with distances smaller than the wavelength of radiation λ or nanoscale objects much smaller than λ, Planck's law of thermal radiation does not apply. Due to the tunneling of evanescent waves, near-field radiative heat transfer (NFRHT) can be improved by many orders of magnitude when the gap size is below a few hundred nanometers compared to what is expected by typical radiation transfer analysis in the far field.

Total internal reflection (TIR) and the fluctuations of electrons or lattices close to the surface emerge as evanescent waves in the near field. In polar dielectrics like SiC and SiO_2, these fluctuations can be quite prominent due to the high-wavevector SPhP waves. A variety of nanostructures have been proposed like multilayer coatings, or metamaterials can be used to achieve further enhancement and tailoring of NFRHT. Many applications, ranging from sensing to energy harvesting, can make use of the ideas behind near-field radiative transfer [6, 229, 230, 233, 383–389].

The majority of theoretical works on NFRHT is done within the context of fluctuational electrodynamics (FE), which Rytov established in the 1950s [390, 391]. In this semiclassical model, thermal radiation is assumed to be generated by thermally excited, random fluctuating electric currents inside the objects. The fluctuation–dissipation theorem (FDT) [392] gives the correlations of the electric currents, leading to the calculation of thermal radiation energy flows. In this sense, the majority of theoretical works on NFRHT solve the stochastic Maxwell's equations in the presence of random electric currents as radiation sources for a variety of materials and structures. Analytical solutions have been developed for simple geometries such as two parallel plates [393], two spheres [394], or a sphere in front of a plate [395]. For complex geometries, numerical methods should be implemented, which can be regarded as a combination of numerical EM methods and FE [396–400].

Theoretical techniques for computing NFRHT have made great strides in recent years, but there are still fundamental unanswered difficulties. Computational difficulties and delays remain substantial when trying to determine NFRHT between microstructures of unbounded geometry. This is especially challenging and, in many cases, infeasible in the presence of nontrivial temperature profiles and complicated material combinations in a structure. However, the vast majority of FE-based calculations are carried out inside the local approximation, where it is assumed that the dielectric function depends solely on the frequency and not on the wavevector. The function of nonlocal effects in the extreme situation of nanometer-sized gaps is not yet understood [401, 402]. Heat conduction (through electrons or phonons) may also face competition from thermal radiation in the extreme near-field domain [403, 404]. It would be great if one could find new ways to describe the contributions of various types of heat transfer mechanisms on an equal footing [405].

Though the aforementioned NFRHT enhancement was predicted in 1971 [393] and partially verified in numerous preliminary tests in the late 1960s [406–408], it was not unambiguously proven until the late 2000s. Accessing the near-field contribution necessitates precise control of the distance and alignment of macroscopic objects. A significant improvement above the blackbody limit at ambient temperature requires gaps on the nanoscale. Furthermore, it is hypothesized that the greatest improvements will take place between two parallel plates. Because it is so challenging to produce and maintain strong parallelism between macroscopic plates at nanoscale separations, this plate–plate combination is one of the most challenging geometries to implement in practice. Utilizing a sphere–plate arrangement is one way to mitigate these issues; in fact, this was the method employed in the earliest tests that definitively established the near field's contribution to the radiative heat transfer [409–411]. Because of the curvature of the spheres, the NFRHT improvements in sphere–plate arrangements are quite moderate. In recent years, various groups have created innovative approaches to investigate the plate–plate configuration, which promises to further expand this enhancement [386, 407, 412–418]. With the aid of microelectromechanical systems (MEMS) technology, now it is possible to explore gaps as small as 30 nm [386], regardless of very small plates with lateral sizes of several tens of micrometers.

All the experimental investigations have conclusively proven that FE theory can accurately describe the NFRHT for gaps as small as a few tens of nanometers.

The extreme near-field regime, for gap separations $d \leq 10$ nm, has also been investigated experimentally in recent years with the aid of the so-called scanning probe microscopy [384, 419–422]. In this configuration, heat transfer between a scanning probe and a substrate can be measured by integrating the probe with a custom-made thermocouple. For gaps as small as a few nanometers, the atomic force microscope can be used to control the probe–sample distance [384], and for even smaller gaps of a few angstroms, a scanning tunneling microscope (STM) can be applied [419–422]. In 2015, Kim et al. conducted [384] NFRHT experiments for gaps down to 2–3 nm. They found the experimental results of NFRHT between polar dielectrics like SiO_2 and SiN and metals like Au can agree well with the theoretical predictions of FE, without any adjustable parameters [384]. However, in 2017, Kloppstech et al. [422] measured the heat transfer between an Au-coated tip and a surface and found that the heat flux appears to be four orders of magnitude greater than values expected using FE. They showed that the reported signals are inconsistent with typical models for the contributions of phonons and electrons, and hence they suggested that this may be owing to the presence of an extra nonradiative mechanism working in the crossover region between radiation and conduction. Later investigations suggest these large signals inconsistent with theory may be due to contaminants [421, 423, 424]. Nevertheless, it would be valuable to conduct more theoretical and experimental investigations to clarify the heat transfer mechanisms in the extreme near field.

6.7 Summary

To summarize, in this chapter, we have presented a general overview of MTNR that aims to solve problems at microscopic levels that cannot be addressed in a satisfactory manner by macroscopic theories introduced in previous chapters. We start by discussing the limitations of macroscopic theories of thermal radiation, in which we consider a variety of lengths and time scales that may play an important role in the theoretical treatment of thermal radiation, including the wavelength, coherence length, mean free path of carriers, penetration depth, relaxation time, coherence time, and laser duration (if any). Then comes a brief introduction of the main research topics in the field MNTR in the last several decades. The first topic discussed is microscopic thermal radiation transport mechanisms in novel materials like high-temperature superconductors and 2D materials, and then thermal radiation transport in micro/nanofabrication technologies and relevant micro/nanodevices. Scattering and absorption in particulate media is another important theme that has attracted much interest due to its wide range of applications. Afterward, we talk about how the rapid development of nanophotonics and metamaterials in recent years has brought a significant advancement in MNTR research, from which one can manipulate thermal radiation properties at will. Near-field thermal radiation, which also benefits from the development of nanophotonics and nanofabrication, constitutes, to some degree, the central topic of MNTR in recent years.

7 Theoretical Fundamentals of Micro/Nanoscale Thermal Radiation

This chapter aims to introduce the theoretical foundations of MNTR. First, the basics of EM theory, including the macroscopic Maxwell's equations, constitutive relations, boundary conditions, and Fresnel's formulas of reflection and refraction, will be introduced. This is then followed by a general description of optical interactions in solids and related permittivity models. Then, the theory of EM wave scattering will be discussed, with a special emphasis on the theory of multiple wave scattering. In particular, we will discuss the extensively studied DSE. Then come the theoretical foundations of NFRHT, including the fluctuation-dissipation theorem (FDT), EM DOS, surface polaritons, and so on. This is further followed by detailed calculations performed on 1D layered media and two bulk media separated by a vacuum gap.

7.1 Basics of Electromagnetic Wave Theory

In this section, we will introduce the basics of EM wave theory. There are a variety of standard textbooks on this subject; see for example [285, 425, 426].

7.1.1 Maxwell's Equations

In classical and semiclassical viewpoints, all kinds of EM radiation, including gamma rays, visible light, thermal radiation, and microwaves, can be treated as EM waves. The EM wave, or EM field, is represented by the electric field **E** and the magnetic induction (or called the magnetic flux density) **B**, which are both space- and time dependent. The responses of matter over EM waves, described by the electric displacement **D**, electric current density **j**, electric (free) charge density ρ, and magnetic vector **H**, are related with these field vectors by Maxwell's equations as

$$\nabla \cdot \mathbf{D} = \rho, \tag{7.1.1}$$

$$\nabla \times \mathbf{E} = -\frac{\partial \mathbf{B}}{\partial t}, \tag{7.1.2}$$

$$\nabla \cdot \mathbf{B} = 0, \tag{7.1.3}$$

$$\nabla \times \mathbf{H} = \mathbf{j} + \frac{\partial \mathbf{D}}{\partial t}. \tag{7.1.4}$$

These equations are the differential form of Maxwell's equations represented in SI units.[1] Other forms of Maxwell's equations can be found in [425]. By taking the divergence of Eq. (7.1.4), we can obtain the equation of continuity (or charge conservation equation)

$$\frac{\partial \rho}{\partial t} + \nabla \cdot \mathbf{j} = 0, \tag{7.1.5}$$

since the divergence of the curl is zero.

In order to apply Maxwell's equations, the responses of the matter, **D**, **H**, and **j**, with respect to the field vectors, **E** and **B**, should be known, which are the so-called constitutive relations (or material equations [285]). Generally, constitutive relations are very complicated, while for linear (independent of the fields), nondispersive, time-invariant, homogeneous (independent of position), and isotropic (independent of direction) media, they can take the relatively simple form as

$$\mathbf{D} = \varepsilon \mathbf{E}, \tag{7.1.6}$$

$$\mathbf{B} = \mu \mathbf{H}, \tag{7.1.7}$$

$$\mathbf{j} = \sigma \mathbf{E}, \tag{7.1.8}$$

where ε, μ, and σ are the material properties called the electric permittivity, the magnetic permeability, and the electrical conductivity, respectively. Equation (7.1.8) is also known as the microscopic form of Ohm's law. It is valid when there are only conduction currents, without any external electric currents. The vacuum values of permittivity and permeability are $\varepsilon_0 = 8.854 \times 10^{-12}\,\mathrm{F\,m^{-1}}$ and $\mu_0 = 4\pi \times 10^{-7}\,\mathrm{N\,A^{-2}}$, with respect to which the relative electric permittivity and magnetic permeability can be defined as $\varepsilon_r = \varepsilon/\varepsilon_0$ and $\mu_r = \mu/\mu_0$. The relative electric permittivity is also called the dielectric constant.

For anisotropic materials, the material properties should be in tensor forms rather than scalars shown above; for inhomogeneous media, the material properties should be functions of spatial positions. The material properties are usually frequency dependent, and this dependency is called frequency dispersion (or temporal dispersion), which means that the temporal response not only depends on the values of field vectors at the current time t but also on those at all other times t' previous to time t [38, 225, 425]. This can be easily seen from the Fourier transformation of material responses from the frequency domain to the time domain. In correspondence with temporal dispersion, the material properties can also be spatially dispersive, which describes the fact that the macroscopic response of matter is not only a function of the field at a single local position **r** but

[1] Here we follow the definitions of EM fields introduced by Born and Wolf [285], in which the Gaussian (CGS) units are used.

also of the field values at all other positions $\mathbf{r}'$. If Fourier transformed in reciprocal space, the material properties are then a function of the reciprocal vector $\mathbf{k}$. Nevertheless, in this book, we are mainly concerned with conventional isotropic, homogeneous, nonmagnetic materials with $\mu_r = 1$ and nonunity scalar relative permittivity $\varepsilon(\omega)$ with possible frequency dispersions.[2] In addition, we can also note from above that Maxwell's equations treat the medium as a continuum and therefore we usually say the macroscopic Maxwell's equations.

7.1.2 Boundary Conditions

In Section 7.1.1, Maxwell's equations are given for unbounded space domains through which material properties of the medium are continuous. In many cases, one would deal with different space domains with abrupt changes of material properties across the boundaries. In these situations, boundary conditions are usually needed.

For a boundary formed by medium 1 and medium 2 on either side, by applying Gauss' theorem to Eqs. (7.1.1) and (7.1.3), we have the following integral formalism:

$$\int_V \nabla \cdot \mathbf{D} \mathrm{d}V = \int_{\partial V} \mathbf{D} \cdot \mathbf{n} \mathrm{d}S = \int_V \rho \mathrm{d}V \tag{7.1.9}$$

and

$$\int_V \nabla \cdot \mathbf{B} \mathrm{d}V = \int_{\partial V} \mathbf{B} \cdot \mathbf{n} \mathrm{d}S = 0, \tag{7.1.10}$$

where V is a small cylinder volume circumscribing the sharp boundary surface with roof A_1 and floor A_2 surfaces ($A_1 = A_2 = A$) buried in the two different space domains 1 and 2, respectively, as shown in Fig. 7.1, and these two surfaces are assumed to be infinitely close to the boundary surface. In other words, the height h of the cylinder approaches zero. ∂V denotes the entire surface area of this small cylinder, $\mathrm{d}S$ is a differential area on ∂V, and $\mathbf{n}$ is the unit outward normal of ∂V at $\mathrm{d}S$. Since the areas of A_1 and A_2 are very small, $\mathbf{D}$ and $\mathbf{B}$ can be considered to be constant on both surfaces, denoted by $\mathbf{D}_1$, $\mathbf{D}_2$, $\mathbf{B}_1$, and $\mathbf{B}_2$, and in the integrations, the contribution from the cylinder walls can be considered as null, since $h \to 0$. As a result, the following formulas are obtained:

$$\mathbf{n}_{12} \cdot (\mathbf{D}_2 - \mathbf{D}_1) = \rho_S \tag{7.1.11}$$

and

$$\mathbf{n}_{12} \cdot (\mathbf{B}_2 - \mathbf{B}_1) = 0, \tag{7.1.12}$$

with the definition of surface charge density ρ_S as [285]

$$\lim_{h \to 0} \int_V \rho \mathrm{d}V = \rho_S A. \tag{7.1.13}$$

[2] Note metamaterials, which are artificially engineered materials and composed of building blocks made of these conventional materials, can present as magnetic, inhomogeneous, anisotropic, and spatially dispersive in an effective fashion.

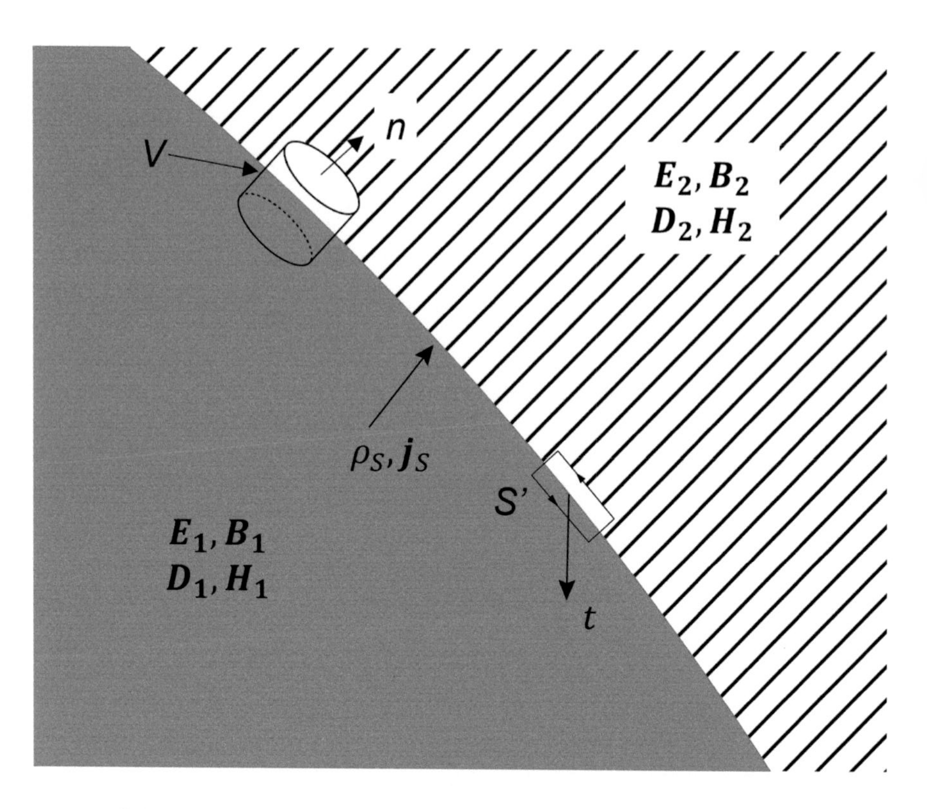

Figure 7.1 Schematic diagram of the boundary surface (heavy line) between different media. The boundary region is assumed to carry an idealized surface charge and the current densities ρ_S and $\mathbf{j}_S$. The small cylinder volume V is half in one medium and half in the other, with the normal **b** being its top, pointing from medium 1 to medium 2. The rectangular contour C is partly in one medium and partly in the other and is oriented with its plane perpendicular to the surface so that its normal **t** is tangent to the surface.

In Eq. (7.1.12) $\mathbf{n}_{12}$ is the unit surface normal pointing from medium 1 to medium 2. The physical significance of these two boundary conditions is evident, that is, the surface charge density leads to an abrupt change of the normal component of the electric displacement across the boundary surface of discontinuity, while the normal component of the magnetic induction is continuous.

In the same spirit, by applying Stokes' theorem to Eqs. (7.1.2) and (7.1.4), the following formalism is derived:

$$\int_{S'} \nabla \times \mathbf{E} \mathrm{d}S = \int_{\partial S'} \mathbf{E} \cdot \mathrm{d}\mathbf{l} = \int_{S'} -\frac{\partial \mathbf{B}}{\partial t} \cdot \mathbf{n}' \mathrm{d}S' \tag{7.1.14}$$

and

$$\int_{S'} \nabla \times \mathbf{H} \mathrm{d}S' = \int_{\partial S'} \mathbf{H} \cdot \mathrm{d}\mathbf{l} = \int_{S'} (\mathbf{j} + \frac{\partial \mathbf{D}}{\partial t}) \cdot \mathbf{n}' \mathrm{d}S', \tag{7.1.15}$$

where S' is a small rectangular area with the surface normal $\mathbf{n}'$ perpendicular to the boundary surface normal $\mathbf{n}_{12}$ (i.e., tangent to the boundary surface), which for this very small S' is well defined. The top and bottom sides s_1 and s_2 (side length $s_1 = s_2 = s$) of the rectangular area are lying in medium 1 and medium 2, respectively, and are parallel to the boundary surface, while its left- and right-hand sides are thus perpendicular to the boundary surface and the side length h is assumed to be infinitely small. ∂S denotes the contour of this small rectangle, and $\mathrm{d}\mathbf{l}$ is the differential vector line element on ∂S. Since s is very small, $\mathbf{E}$ and $\mathbf{H}$ can be considered to be constant on s_1 and s_2, respectively, denoted by $\mathbf{E}_1$, $\mathbf{E}_2$, $\mathbf{H}_1$, and $\mathbf{H}_2$, and in the integrations, the contribution from the perpendicular sides can be considered as null, since $h \to 0$. Considering the time derivatives of $\mathbf{B}$ and $\mathbf{D}$ remain finite, we can obtain the following formulas:

$$\mathbf{t} \cdot (\mathbf{E}_2 - \mathbf{E}_1) = 0 \tag{7.1.16}$$

and

$$\mathbf{t} \cdot (\mathbf{H}_2 - \mathbf{H}_1) = \mathbf{j}_S \cdot \mathbf{n}' \tag{7.1.17}$$

with a similar definition of the surface current density as

$$\lim_{h \to 0} \int_{S'} \mathbf{j} \cdot \mathbf{n}' \mathrm{d}S' = \mathbf{j}_S \cdot \mathbf{n}' s. \tag{7.1.18}$$

Here $\mathbf{t}$ indicates the unit tangent along the boundary surface that abides by a right-hand rule as $\mathbf{t} = \mathbf{n}' \times \mathbf{n}_{12}$. A more familiar form of the Eqs. (7.1.16) and (7.1.17) are

$$\mathbf{n}_{12} \times (\mathbf{E}_2 - \mathbf{E}_1) = 0 \tag{7.1.19}$$

and

$$\mathbf{n}_{12} \times (\mathbf{H}_2 - \mathbf{H}_1) = \mathbf{j}_S. \tag{7.1.20}$$

The physical significance of these two boundary conditions is also evident, that is, the tangential component of the electric field is continuous across the boundary surface of discontinuity, while the surface current density leads to an abrupt change of the tangential component of the magnetic vector.

7.1.3 Energy Law

Next, let us consider the law of energy conservation of the EM field. From Eqs. (7.1.4) and (7.1.2), it follows that

$$\mathbf{H} \cdot (\nabla \times \mathbf{E}) - \mathbf{E} \cdot (\nabla \times \mathbf{H}) = -\mathbf{H} \frac{\partial \mathbf{B}}{\partial t} - \mathbf{E} \cdot \frac{\partial \mathbf{D}}{\partial t} - \mathbf{j} \cdot \mathbf{E}. \tag{7.1.21}$$

By noting the identity for the left-hand side (LHS) of Eq. (7.1.21),

$$\mathbf{H} \cdot (\nabla \times \mathbf{E}) - \mathbf{E} \cdot (\nabla \times \mathbf{H}) = \nabla \cdot (\mathbf{E} \times \mathbf{H}), \tag{7.1.22}$$

we have

$$\nabla\cdot(\mathbf{E}\times\mathbf{H})=-\mathbf{H}\frac{\partial\mathbf{B}}{\partial t}-\mathbf{E}\cdot\frac{\partial\mathbf{D}}{\partial t}-\mathbf{j}\cdot\mathbf{E}. \tag{7.1.23}$$

Integrating Eq. (7.1.23) throughout an arbitrary volume V and applying Gauss' theorem to the LHS yields

$$\int_{\partial V}(\mathbf{E}\times\mathbf{H})\cdot\mathbf{n}\mathrm{d}S=-\int_{V}\left(\mathbf{H}\frac{\partial\mathbf{B}}{\partial t}+\mathbf{E}\cdot\frac{\partial\mathbf{D}}{\partial t}+\mathbf{j}\cdot\mathbf{E}\right)\mathrm{d}V, \tag{7.1.24}$$

with $\mathbf{n}$ being the unit outward normal of ∂V. This is the general law of energy conservation of EM fields. Since it is directly derived from Maxwell's equations, it has the same validity. Of special interest is the situation when the constitutive relations Eqs. (7.1.6) and (7.1.7) are valid, from which we can also understand its physical significance more clearly. In this circumstance, we have

$$\mathbf{E}\frac{\partial\mathbf{D}}{\partial t}=\mathbf{E}\frac{\partial}{\partial t}(\varepsilon\mathbf{E})=\frac{1}{2}\frac{\partial}{\partial t}(\varepsilon\mathbf{E}^2)=\frac{1}{2}\frac{\partial}{\partial t}(\mathbf{E}\cdot\mathbf{D}) \tag{7.1.25}$$

and

$$\mathbf{H}\frac{\partial\mathbf{B}}{\partial t}=\mathbf{H}\frac{\partial}{\partial t}(\mu\mathbf{H})=\frac{1}{2}\frac{\partial}{\partial t}(\mu\mathbf{H}^2)=\frac{1}{2}\frac{\partial}{\partial t}(\mathbf{H}\cdot\mathbf{B}). \tag{7.1.26}$$

By setting the total EM energy density as

$$W=\frac{1}{2}(\mathbf{E}\cdot\mathbf{D}+\mathbf{H}\cdot\mathbf{B}), \tag{7.1.27}$$

with $W_e=(1/2)\mathbf{E}\cdot\mathbf{D}$ and $W_m=(1/2)\mathbf{H}\cdot\mathbf{B}$ being the electric density and magnetic energy density, respectively, we can further define the Poynting vector $\mathbf{S}$ as

$$\mathbf{S}=\mathbf{E}\times\mathbf{H}, \tag{7.1.28}$$

which represents the energy flow (more precisely, the density of the energy flow). Upon these definitions, from Eq. (7.1.24), we obtain

$$-\int_V\frac{\partial W}{\partial t}\mathrm{d}V=\int_{\partial V}\mathbf{S}\cdot\mathbf{n}\mathrm{d}S+\int_V\mathbf{j}\cdot\mathbf{E}\mathrm{d}V. \tag{7.1.29}$$

The physical meaning of this equation is that the negative changing rate of EM energy within a certain volume (which can be considered as the decrease of total EM energy) is equal to the energy flowing out through the boundary surfaces of the volume per unit of time, plus the total work done by the fields on the electric current density within that volume. The latter is represented by $\int_V\mathbf{j}\cdot\mathbf{E}\mathrm{d}V$ and can be understood as the Joule heat generated per unit of time within that volume. This equation is the well-known Poynting theorem, the "work–energy theorem" of electrodynamics. A differential form of it is

$$-\frac{\partial W}{\partial t}=\nabla\cdot\mathbf{S}+\mathbf{j}\cdot\mathbf{E}. \tag{7.1.30}$$

For frequency-dispersive media, the simple formulas Eqs. (7.1.25) and (7.1.26) are not strictly valid, and the expressions of energy densities should be reformulated in a convolution form to take the dispersion into account. A detailed discussion can be found in [225, 427].

7.1.4 Electromagnetic Wave Equation and Harmonic Waves

A basic feature of Maxwell's equations for the EM field is the existence of solutions of propagation waves [425]. Here let us derive the wave propagation equation. For doing so, we consider the case of no charges or currents, that is, $\mathbf{j}=0$ and $\rho=0$. By taking the curl of Eq. (7.1.2) after substituting the material Eq. (7.1.7), we obtain

$$\nabla\left(\frac{1}{\mu}\nabla\times\mathbf{E}\right)+\nabla\times\frac{\partial\mathbf{H}}{\partial t}=0. \tag{7.1.31}$$

Next we differentiate Eq. (7.1.4) with respect to time and apply the material Eq. (7.1.6), yielding

$$\nabla\times\left(\frac{1}{\mu}\nabla\times\mathbf{E}\right)+\varepsilon\frac{\partial^2\mathbf{E}}{\partial t^2}=0. \tag{7.1.32}$$

Noting the identities $\nabla\times(u\mathbf{v})=u\nabla\times\mathbf{v}+\nabla u\times\mathbf{v}$ and $\nabla(\times\nabla\times\mathbf{v})=\nabla\cdot(\nabla\cdot\mathbf{v})-\nabla^2\mathbf{v}$, we then obtain

$$\nabla^2\mathbf{E}-\varepsilon\mu\frac{\partial^2\mathbf{E}}{\partial t^2}+\nabla(\ln\mu)\times\mathbf{E}-\nabla(\nabla\cdot\mathbf{E})=0. \tag{7.1.33}$$

Using the identity $\nabla\cdot(u\mathbf{v})=\nabla u\cdot\mathbf{v}+u\nabla\cdot\mathbf{v}$ and the material Eq. (7.1.6) yields

$$\nabla\cdot(\varepsilon\mathbf{E})=(\nabla\varepsilon)\cdot\mathbf{E}+\varepsilon\nabla\cdot\mathbf{E}, \tag{7.1.34}$$

which is then substituted into Eq. (7.1.33), leading to

$$\nabla^2\mathbf{E}-\varepsilon\mu\frac{\partial^2\mathbf{E}}{\partial t^2}+\nabla(\ln\mu)\times\mathbf{E}+\nabla\left[\mathbf{E}\cdot\nabla(\ln\varepsilon)\right]=0. \tag{7.1.35}$$

Similarly, an equation for $\mathbf{H}$ can be obtained as

$$\nabla^2\mathbf{H}-\varepsilon\mu\frac{\partial^2\mathbf{H}}{\partial t^2}+\nabla(\ln\varepsilon)\times\mathbf{H}+\nabla\left[\mathbf{H}\cdot\nabla(\ln\mu)\right]=0. \tag{7.1.36}$$

If the medium is homogeneous, the gradients regarding permittivity and permeability are zero, giving

$$\nabla^2\mathbf{E}-\varepsilon\mu\frac{\partial^2\mathbf{E}}{\partial t^2}=0 \tag{7.1.37}$$

and

$$\nabla^2\mathbf{H}-\varepsilon\mu\frac{\partial^2\mathbf{H}}{\partial t^2}=0. \tag{7.1.38}$$

These are standard forms of wave equation, implying the existence of EM waves with a propagation velocity (phase velocity) as

$$v=\frac{1}{\sqrt{\varepsilon\mu}}=\frac{c}{n}, \tag{7.1.39}$$

where the constant $c = 1/\sqrt{\varepsilon_0 \mu_0} = 299,792,458\ \mathrm{m\,s^{-1}}$ is the speed of light, and n is the refractive index defined as $n = \sqrt{\varepsilon_r \mu_r}$.

The wave equations have infinite sets of solutions. Of particular interest is the harmonic wave solution at a specific angular frequency ω, since any time evolution of the field can be Fourier decomposed into the linear superposition of harmonic waves at different frequencies. It can be written in the following form:

$$\mathbf{E}(\mathbf{r},t) = \mathrm{Re}\left[\mathbf{E}(\mathbf{r})\exp(-\mathrm{i}\omega t)\right], \tag{7.1.40}$$

since the field $\mathbf{E}(\mathbf{r},t)$ is a real quantity while the spatial amplitude $\mathbf{E}(\mathbf{r})$ can be complex, which is thus called the complex amplitude [285]. Similar expressions can be written for the other fields. Notice that although $\mathbf{E}(\mathbf{r})$ is complex in general, it is convenient to drop the symbol Re in Eq. (7.1.40) and directly deal with the complex function and take the real part of the final expression if the operations on $\mathbf{E}(\mathbf{r},t)$ are linear. However, if we are dealing with expressions involving nonlinear operations such as squaring (e.g., in calculations of the electric or magnetic energy densities), we always have to take the real parts in the first place. This is not necessary when only the time average of a quadratic expression is desired, as will be seen later for the time-averaged Poynting vector for an example (cf. Eq. (7.1.60)) [285].

By substituting the harmonic wave ansatz into Maxwell's equations, Eqs. (7.1.1)–(7.1.4), we obtain

$$\nabla\cdot\mathbf{D} = \rho(\mathbf{r}), \tag{7.1.41}$$

$$\nabla\times\mathbf{E}(\mathbf{r}) = \mathrm{i}\omega\mathbf{B}(\mathbf{r}), \tag{7.1.42}$$

$$\nabla\cdot\mathbf{B}(\mathbf{r}) = 0, \tag{7.1.43}$$

$$\nabla\times\mathbf{H}(\mathbf{r}) = -\mathrm{i}\omega\mathbf{D}(\mathbf{r}) + \mathbf{j}(\mathbf{r}). \tag{7.1.44}$$

It is then obvious that the spatial part $\mathbf{E}(\mathbf{r})$ also depends on ω, while for the brevity of notation, this is usually not included in the argument. The material properties can also be a function of ω, that is, time dispersive as mentioned earlier. Moreover, inserting Eq. (7.1.40) into the wave equation (Eq. (7.1.37)) gives

$$\nabla^2\mathbf{E}(\mathbf{r}) + \varepsilon\mu\omega^2\mathbf{E}(\mathbf{r}) = 0. \tag{7.1.45}$$

This is the standard Helmholtz wave equation, and we find that it leads to the requirement for the wavenumber k as

$$k = \sqrt{\varepsilon\mu}\,\omega = \frac{\omega}{v} = n\frac{\omega}{c}. \tag{7.1.46}$$

A similar equation can be obtained for $\mathbf{H}$ as

$$\nabla^2\mathbf{H}(\mathbf{r}) + \varepsilon\mu\omega^2\mathbf{H}(\mathbf{r}) = 0. \tag{7.1.47}$$

It is straightforward to verify that

$$\mathbf{E}(\mathbf{r}) = \mathbf{E}_0\exp(\mathrm{i}\mathbf{k}\cdot\mathbf{r}) \tag{7.1.48}$$

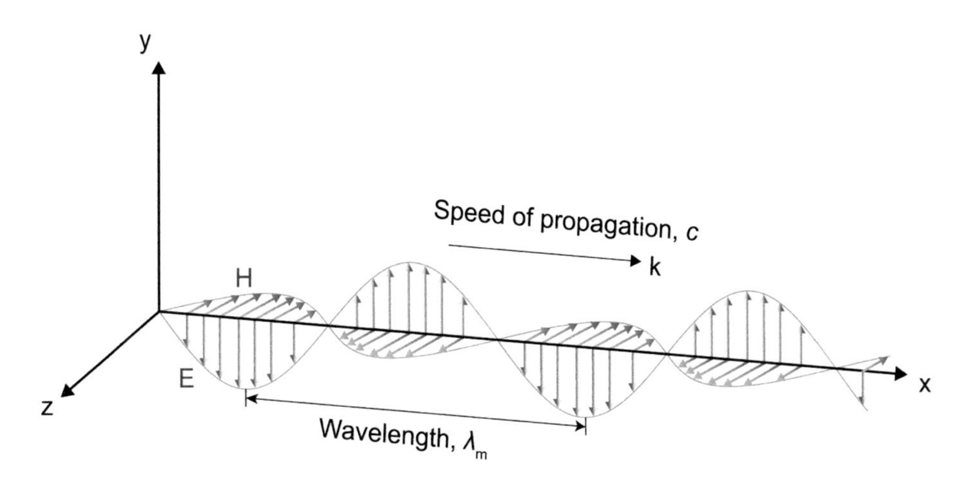

Figure 7.2 A schematic of harmonic plane waves.

and

$$\mathbf{H}(\mathbf{r}) = \mathbf{H}_0 \exp(\mathrm{i}\mathbf{k} \cdot \mathbf{r}), \tag{7.1.49}$$

in which $\mathbf{k} = k\mathbf{s}$, with $\mathbf{s}$ being an arbitrary unit vector and $\mathbf{E}_0$ and $\mathbf{H}_0$ constant, are solutions to Eqs. (7.1.45) and (7.1.47). These solutions represent a plane wave whose propagation direction is $\mathbf{s}$, since the field vectors are constant at

$$\mathbf{k} \cdot \mathbf{r} \;=\; \text{constant}, \tag{7.1.50}$$

which determines the planes that are perpendicular to the vector $\mathbf{k}$ [285]. Also seen is its spatial periodicity when the position vector $\mathbf{r}$ is shifted by integer multiples of $\lambda\mathbf{s}$, where λ is the wavelength in the medium given by $\lambda = 2\pi/k = 2\pi v/\omega$. It is also required that this type of solution should also satisfy Maxwell's equations for harmonic waves in the absence of charges or currents, i.e., Eqs. (7.1.41)–(7.1.44) with $\rho = 0$ and $\mathbf{j} = 0$, in a homogeneous medium, yielding

$$\mathbf{s} \cdot \mathbf{E}_0 = 0, \tag{7.1.51}$$

$$\mathbf{H}_0 = \sqrt{\frac{\varepsilon}{\mu}}\mathbf{s} \times \mathbf{E}_0, \tag{7.1.52}$$

$$\mathbf{s} \cdot \mathbf{H}_0 = 0, \tag{7.1.53}$$

$$\mathbf{E}_0 = -\sqrt{\frac{\mu}{\varepsilon}}\mathbf{s} \times \mathbf{H}_0, \tag{7.1.54}$$

for which we have already applied the material equations. This means that $\mathbf{E}$ and $\mathbf{H}$ are both perpendicular to the propagation direction $\mathbf{s}$, implying such a wave is a transverse wave. More specifically, $\mathbf{E}$, $\mathbf{H}$, and $\mathbf{s}$ form a right-handed orthogonal triad of vectors. A schematic of plane waves is shown in Fig. 7.2.

Now consider the energy flux carried by this harmonic plane wave. Since the optical frequencies are very large (ω is of order 10^{15} s^{-1}), it is generally difficult to observe the instantaneous values of these rapidly oscillating field quantities, but only their time average taken over a relatively large time interval $[-T', T]$, compared to the period of optical oscillations $T = 2\pi/\omega$. In this situation, the time-averaged electric energy density is

$$\langle W_e \rangle = \frac{1}{2T'} \int_{-T'}^{T'} \frac{1}{2} \varepsilon \mathbf{E}^2(\mathbf{r}, t). \tag{7.1.55}$$

Since from Eq. (7.1.40) we have

$$\mathbf{E}(\mathbf{r}, t) = \frac{1}{2} \left[\mathbf{E}(\mathbf{r}) \mathrm{e}^{-\mathrm{i}\omega t} + \mathbf{E}^*(\mathbf{r}) \mathrm{e}^{\mathrm{i}\omega t} \right] \mathrm{d}t, \tag{7.1.56}$$

inserting this equation into Eq. (7.1.55) yields

$$\langle W_e \rangle = \frac{\varepsilon}{16T'} \int_{-T'}^{T'} \left[\mathbf{E}^2(\mathbf{r})) \mathrm{e}^{-2\mathrm{i}\omega t} + 2\mathbf{E}(\mathbf{r}) \cdot \mathbf{E}^*(\mathbf{r}) + \mathbf{E}^{*2}(\mathbf{r})) \mathrm{e}^{2\mathrm{i}\omega t} \right] \mathrm{d}t. \tag{7.1.57}$$

For a fast oscillating wave $T \ll T'$, the first and third terms in the integrand essentially result in zero, giving

$$\langle W_e \rangle = \frac{1}{4} \varepsilon \mathbf{E}(\mathbf{r}) \cdot \mathbf{E}^*(\mathbf{r}). \tag{7.1.58}$$

Similarly, the time-average magnetic energy density is

$$\langle W_m \rangle = \frac{1}{4} \mu \mathbf{H}(\mathbf{r}) \cdot \mathbf{H}^*(\mathbf{r}). \tag{7.1.59}$$

The time-averaged Poynting vector can also be derived similarly as

$$\begin{aligned} \langle \mathbf{S} \rangle &= \frac{1}{2T'} \int_{-T'}^{T'} \mathbf{E}(\mathbf{r}, t) \times \mathbf{H}(\mathbf{r}, t) \mathrm{d}t \\ &= \frac{1}{8T'} \int_{-T'}^{T'} \Big[\mathbf{E}(\mathbf{r}) \times \mathbf{H}(\mathbf{r})) \mathrm{e}^{-2\mathrm{i}\omega t} + \mathbf{E}(\mathbf{r}) \times \mathbf{H}^*(\mathbf{r}) \\ &\quad + \mathbf{E}^*(\mathbf{r}) \times \mathbf{H}(\mathbf{r}) + \mathbf{E}^*(\mathbf{r}) \times \mathbf{H}^*(\mathbf{r}) \mathrm{e}^{2\mathrm{i}\omega t} \Big] \mathrm{d}t \\ &= \frac{1}{4} \left[\mathbf{E}(\mathbf{r}) \times \mathbf{H}^*(\mathbf{r}) + \mathbf{E}^*(\mathbf{r}) \times \mathbf{H}(\mathbf{r}) \right] \\ &= \frac{1}{2} \mathrm{Re} \left[\mathbf{E}(\mathbf{r}) \times \mathbf{H}^*(\mathbf{r}) \right]. \end{aligned} \tag{7.1.60}$$

This formula confirms that it is convenient to drop the symbol Re for complex amplitudes of EM field quantities in the calculation process and take the real part of the final expression for time-averaged quadratic expressions like energy densities and Poynting vector. As a result, the law of energy conservation, Eq. (7.1.29), also takes a simple form as

$$\int_{\partial V} \langle \mathbf{S} \rangle \cdot \mathbf{n} \mathrm{d}S + \int_V \langle \mathbf{j} \cdot \mathbf{E} \rangle \mathrm{d}V = 0. \tag{7.1.61}$$

For a nonconducting medium with $\sigma = 0$, where no mechanical work is done to the currents, this leads to

$$\int_{\partial V} \langle \mathbf{S} \rangle \cdot \mathbf{n} \mathrm{d}S = 0 \tag{7.1.62}$$

or

$$\nabla \cdot \langle \mathbf{S} \rangle = 0, \tag{7.1.63}$$

which means for a nonconducting medium, the time-averaged total flux of energy through any closed surface is zero.

For a plane wave, Eq. (7.1.60) can be further written as

$$\langle \mathbf{S} \rangle = \frac{1}{2}\sqrt{\frac{\varepsilon}{\mu}}|\mathbf{E}_0|^2 \mathbf{s}, \tag{7.1.64}$$

which means that the direction of energy flow is the same as the propagation direction of harmonic plane EM waves in homogeneous isotropic media. Note that it is in general not the case for anisotropic or inhomogeneous media.

Let us return to the harmonic Maxwell's equations with sources, Eqs. (7.1.41)–(7.1.44). By applying Eq. (7.1.7) and dividing μ on both sides into Eq. (7.1.42), and then substituting Eq. (7.1.44), we obtain

$$\nabla \times \left(\frac{1}{\mu} \nabla \times \mathbf{E}(\mathbf{r}) \right) - \omega^2 \left(\varepsilon + \frac{\mathrm{i}\sigma}{\omega} \right) \mathbf{E}(\mathbf{r}) = 0, \tag{7.1.65}$$

from which we can define a complex electric permittivity[3]

$$\tilde{\varepsilon} = \varepsilon + \frac{\mathrm{i}\sigma}{\omega}. \tag{7.1.66}$$

In practice, it is convenient to directly use the complex electric permittivity implicitly, by doing which one does not need to discriminate between the conduction current density and the source current density (if any) [225]. From this expression, the complex refractive index can be defined as

$$m = \sqrt{\tilde{\varepsilon}\mu} = n + \mathrm{i}\kappa, \tag{7.1.67}$$

where n and κ are real and can be derived as

$$n = \mathrm{Re} m \tag{7.1.68}$$

and

$$\kappa = \mathrm{Im} m. \tag{7.1.69}$$

7.1.5 Reflection and Refraction of a Plane Wave

In Section 7.1.2, we have derived formulas for field vectors across the boundary interfacing two different media. The study of the propagation of a plane wave incident on a plane boundary between two homogeneous isotropic media will now be conducted using these formulas. A plane wave splits into two waves when it crosses the boundary between two homogeneous media with differing optical

[3] The original permittivity ε, for a nondispersive medium, is real. However, for a dispersive medium, it is in general complex in the frequency domain, considering that it is Fourier transformed from the time domain. Nevertheless, for convenience, we call the original permittivity the "real permittivity."

constants: a transmitted wave that travels into the second medium and a reflected wave that travels back into the first medium. In fact, the boundary conditions can be used to predict the existence of these two waves because it is clear that they cannot be met without assuming the existence of both the transmitted wave and the reflected wave. We will make the supposition that these waves are also planar to construct formulas for their propagation directions and amplitudes.

When the temporal behavior of a plane wave propagating in a specific direction $\mathbf{s}_i$ at a specific location in space is known, its propagation behavior is then totally determined. In particular, if we use $\mathbf{A}(t)$ to represent the temporal behavior at any given place, $\mathbf{A}[t-(\mathbf{r}\cdot\mathbf{s})/v]$ represents the time behavior at a second point whose position vector with respect to the first point is $\mathbf{r}$. The time evolution of the secondary fields will be the same as that of the incident field at the interface between the two media. Therefore, when equating the arguments of the three wave functions at a point $\mathbf{r}$ on the boundary plane $z=0$, one finds the following result:

$$t-\frac{\mathbf{r}\cdot\mathbf{s}^{(i)}}{v_1}=t-\frac{\mathbf{r}\cdot\mathbf{s}^{(r)}}{v_1}=t-\frac{\mathbf{r}\cdot\mathbf{s}^{(t)}}{v_2}, \tag{7.1.70}$$

where $\mathbf{s}^{(r)}$ and $\mathbf{s}^{(t)}$ signify unit vectors in the direction of propagation of the reflected and transmitted waves, and v_1 and v_2 are the propagation velocities in the two media respectively. As this equation must be valid for all values of $\mathbf{r}=(x,y,0)$ on the boundary, we immediately obtain

$$\frac{s_x^{(i)}}{v_1}=\frac{s_x^{(r)}}{v_1}=\frac{s_x^{(t)}}{v_2} \tag{7.1.71}$$

and

$$\frac{s_y^{(i)}}{v_1}=\frac{s_y^{(r)}}{v_1}=\frac{s_y^{(t)}}{v_2}. \tag{7.1.72}$$

For convenience we can define the plane specified by $\mathbf{s}^{(i)}$ and the boundary normal as the plane of incidence denoted by the xOz plane in Fig. 7.3, from which we can define the incident, reflection, and transmission angles as θ_i, θ_r, and θ_t, respectively, with respect to the z axis. As a result, Eq. (7.1.71) becomes

$$\frac{\sin\theta_i}{v_1}=\frac{\sin\theta_r}{v_1}=\frac{\sin\theta_t}{v_2}. \tag{7.1.73}$$

Hence, $\sin\theta_r=\sin\theta_i$, and

$$\frac{\sin\theta_t}{\sin\theta_i}=\frac{v_2}{v_1}=\sqrt{\frac{\varepsilon_1\mu_1}{\varepsilon_2\mu_2}}=\frac{n_1}{n_2}=n_{12}, \tag{7.1.74}$$

where we have used the expressions of wave velocity and refractive index in Eq. 7.1.39. A further observation from Fig. 7.3 is that

$$\theta_r=\pi-\theta_i. \tag{7.1.75}$$

The above relations are known as the law of reflection or Snell's law. When $n_2>n_1$ or $n_{12}<1$, we say medium 2 is optically denser than medium 1. In this

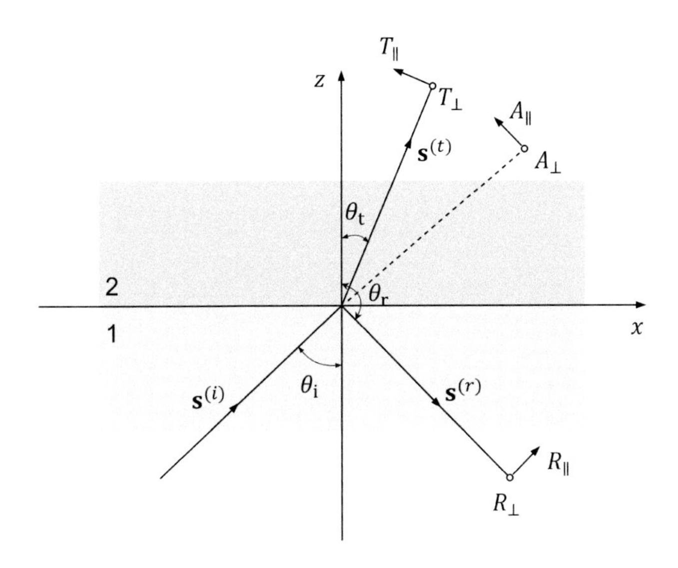

Figure 7.3 Refraction and reflection of a plane wave. Plane of incidence.

circumstance, we can always find a real angle of refraction by Eq. (7.1.74). By contrast, in the case of $n_{12} > 1$, that is, medium 2 is optically less dense than medium 1, we can only find a real θ_t for $\theta_i \leq \arcsin(1/n_{12})$. In this case, when $\theta_i > \arcsin(1/n_{12})$, the so-called TIR phenomenon will occur [225, 285].

Next, we proceed to the evaluation of reflected and transmitted wave amplitudes. As was noted earlier, we consider the most fundamental circumstance where the two media are homogeneous, isotropic, and lossless (real optical constants). We assume that the electric-field vector of the incident wave takes the following form:

$$E_x^{(i)} = -A_{\|} \cos\theta_i \mathrm{e}^{-\mathrm{i}\phi_i}, \tag{7.1.76a}$$

$$E_y^{(i)} = A_{\perp} \mathrm{e}^{-\mathrm{i}\phi_i}, \tag{7.1.76b}$$

$$E_z^{(i)} = A_{\|} \sin\theta_i \mathrm{e}^{-\mathrm{i}\phi_i}, \tag{7.1.76c}$$

where $A_{\|}$ and $A_{\perp}$ are the parallel and perpendicular components of the electric field amplitudes of the incident wave with respect to the plane of incidence, and

$$\phi_i = \omega\left(t - \frac{\mathbf{r}\cdot\mathbf{s}^{(i)}}{v_1}\right) = \omega\left(t - \frac{x\sin\theta_i + z\cos\theta_i}{v_1}\right) \tag{7.1.77}$$

takes the time and space dependence of the propagation phase into account. By Eq. (7.1.52), the components of the magnetic vector are derived as

$$H_x^{(i)} = -A_{\perp} \cos\theta_i \sqrt{\varepsilon_1\mu_1} \mathrm{e}^{-\mathrm{i}\phi_i}, \tag{7.1.78a}$$

$$H_y^{(i)} = -A_{\|} \sqrt{\varepsilon_1/\mu_1} \mathrm{e}^{-\mathrm{i}\phi_i}, \tag{7.1.78b}$$

$$H_z^{(i)} = A_{\perp} \sin\theta_i \sqrt{\varepsilon_1/\mu_1} \mathrm{e}^{-\mathrm{i}\phi_i}. \tag{7.1.78c}$$

Moreover, by assuming the complex amplitudes of transmitted and reflected waves as T and R, we can obtain similar expressions for electric and magnetic vectors. In particular, for transmitted waves,

$$E_x^{(t)} = -T_\parallel \cos\theta_t \mathrm{e}^{-\mathrm{i}\phi_t}, \tag{7.1.79a}$$
$$E_y^{(t)} = T_\perp \mathrm{e}^{-\mathrm{i}\phi_t}, \tag{7.1.79b}$$
$$E_z^{(t)} = T_\parallel \sin\theta_t \mathrm{e}^{-\mathrm{i}\phi_t}, \tag{7.1.79c}$$
$$H_x^{(t)} = -T_\perp \cos\theta_t \sqrt{\varepsilon_2/\mu_2}\mathrm{e}^{-\mathrm{i}\phi_t}, \tag{7.1.79d}$$
$$H_y^{(t)} = -T_\parallel \sqrt{\varepsilon_2/\mu_2}\mathrm{e}^{-\mathrm{i}\phi_t}, \tag{7.1.79e}$$
$$H_z^{(t)} = T_\perp \sin\theta_t \sqrt{\varepsilon_2/\mu_2}\mathrm{e}^{-\mathrm{i}\phi_t}, \tag{7.1.79f}$$

with

$$\phi_t = \omega\left(t - \frac{\mathbf{r}\cdot\mathbf{s}^{(t)}}{v_2}\right) = \omega\left(t - \frac{x\sin\theta_t + z\cos\theta_t}{v_2}\right). \tag{7.1.80}$$

For reflected waves, we obtain

$$E_x^{(r)} = -R_\parallel \cos\theta_r \mathrm{e}^{-\mathrm{i}\phi_r}, \tag{7.1.81a}$$
$$E_y^{(r)} = R_\perp \mathrm{e}^{-\mathrm{i}\phi_r}, \tag{7.1.81b}$$
$$E_z^{(r)} = R_\parallel \sin\theta_r \mathrm{e}^{-\mathrm{i}\phi_r}, \tag{7.1.81c}$$
$$H_x^{(r)} = -R_\perp \cos\theta_r \sqrt{\varepsilon_1/\mu_1}\mathrm{e}^{-\mathrm{i}\phi_r}, \tag{7.1.81d}$$
$$H_y^{(r)} = -R_\parallel \sqrt{\varepsilon_1/\mu_1}\mathrm{e}^{-\mathrm{i}\phi_r}, \tag{7.1.81e}$$
$$H_z^{(r)} = R_\perp \sin\theta_r \sqrt{\varepsilon_1/\mu_1}\mathrm{e}^{-\mathrm{i}\phi_r}, \tag{7.1.81f}$$

with

$$\phi_r = \omega\left(t - \frac{\mathbf{r}\cdot\mathbf{s}^{(r)}}{v_1}\right) = \omega\left(t - \frac{x\sin\theta_r + z\cos\theta_r}{v_1}\right). \tag{7.1.82}$$

Referring to the boundary conditions Eqs. (7.1.19) and (7.1.20) in the absence of surface currents, we can establish the relations among incident, reflected, and transmitted field vectors at the boundary, given by

$$E_x^{(i)} + E_x^{(r)} = E_x^{(t)}, \tag{7.1.83a}$$
$$E_y^{(i)} + E_y^{(r)} = E_y^{(t)}, \tag{7.1.83b}$$
$$H_x^{(i)} + H_x^{(r)} = H_x^{(t)}, \tag{7.1.83c}$$
$$H_y^{(i)} + H_y^{(r)} = H_y^{(t)}. \tag{7.1.83d}$$

We further assume that the two media are nonmagnetic ($\mu_1 = \mu_2 = 1$). Substituting Eqs. (7.1.76) (7.1.81) into Eq. (7.1.83) with some simple manipulations leads to the following relations:

$$\cos\theta_i\left(A_{\parallel}-R_{\parallel}\right)=\cos\theta_t T_{\parallel}, \tag{7.1.84a}$$

$$A_{\perp}+R_{\perp}=T_{\perp}, \tag{7.1.84b}$$

$$\sqrt{\varepsilon_1}\cos\theta_i\left(A_{\perp}-R_{\perp}\right)=\sqrt{\varepsilon_2}\cos\theta_t T_{\perp}, \tag{7.1.84c}$$

$$\sqrt{\varepsilon_1}\left(A_{\parallel}+R_{\parallel}\right)=\sqrt{\varepsilon_2}T_{\parallel}. \tag{7.1.84d}$$

It is immediately noticed that the above equations can be divided into two independent groups: one contains only the parallel components and the other contains only the perpendicular components. This allows us to define the well-known Fresnel reflection and transmission coefficients for parallel and perpendicular polarizations as

$$t_{\parallel}=\frac{T_{\parallel}}{A_{\parallel}}=\frac{2n_1\cos\theta_i}{n_2\cos\theta_i+n_1\cos\theta_t}, \tag{7.1.85a}$$

$$t_{\perp}=\frac{T_{\perp}}{A_{\perp}}=\frac{2n_1\cos\theta_i}{n_1\cos\theta_i+n_2\cos\theta_t}, \tag{7.1.85b}$$

$$r_{\parallel}=\frac{R_{\parallel}}{A_{\parallel}}=\frac{n_2\cos\theta_i-n_1\cos\theta_t}{n_2\cos\theta_i+n_1\cos\theta_t}, \tag{7.1.85c}$$

$$r_{\perp}=\frac{R_{\perp}}{A_{\perp}}=\frac{n_1\cos\theta_i-n_2\cos\theta_t}{n_1\cos\theta_i+n_2\cos\theta_t}. \tag{7.1.85d}$$

From the above equations, we can define the concepts of reflectivity and transmissivity for the energy transport of the parallel polarization across the boundary as $\rho_{\parallel}=|r_{\parallel}|^2$ and $\tau_{\parallel}=|t_{\parallel}|^2$. Similar expressions can be given for the perpendicular polarization. Concerning an arbitrary polarization state whose electric-field vector makes an angle of α_i with the plane of incidence, we have

$$\rho=\rho_{\parallel}\cos^2\alpha_i+\rho_{\perp}\sin^2\alpha_i \tag{7.1.86}$$

and

$$\tau=\tau_{\parallel}\cos^2\alpha_i+\tau_{\perp}\sin^2\alpha_i. \tag{7.1.87}$$

For lossless media, energy conservation requires $\rho+\tau=1$.

7.1.6 Propagating and Evanescent Waves

Then concept of evanescent waves plays a vital role in nanoscale thermal radiation and nano-optics. To some extent, it is the prominent effects of evanescent waves that lead to the pursuit and interest in nanoscale thermal radiation [6, 224, 428]. Look at a harmonic plane wave in the form of $\mathbf{E}\exp\left[\mathrm{i}(\mathbf{k}\cdot\mathbf{r}-\omega t)\right]$. If all components of the wavevector $\mathbf{k}$ are real, this plane wave can propagate freely in space. On the other hand, when at least one of the components of the wavevector is purely imaginary, which is possible when waves encounter inhomogeneities or interfaces [429], the plane wave becomes evanescent. This means that its amplitude decays exponentially in the spatial dimension defined by the imaginary component of the wavevector, which can be easily understood from the plane-wave expression.

The most fundamental situation where evanescent waves emerge is the plane interface between two media with different refractive indices, as shown in Fig. 7.3. We use another form of Fresnel coefficients by using the components of incident and transmitted wavevectors $\mathbf{k}_1$ and $\mathbf{k}_2$ instead of refractive indices in Eq. (7.1.85). By Snell's law, the wavevector components parallel to the boundary are equal on both sides of the boundary (transverse momentum conservation), which can be expressed as k_x and k_y, and the z-component of the wavevector in medium 1 is given by

$$k_{z_1} = \sqrt{k_1^2 - (k_x^2 + k_y^2)}, \tag{7.1.88}$$

$$k_{z_2} = \sqrt{k_2^2 - (k_x^2 + k_y^2)}, \tag{7.1.89}$$

with

$$k_1 = |\mathbf{k}_1| = \frac{\omega}{c}\sqrt{\varepsilon_1 \mu_1} \tag{7.1.90}$$

and

$$k_2 = |\mathbf{k}_2| = \frac{\omega}{c}\sqrt{\varepsilon_2 \mu_2}. \tag{7.1.91}$$

The in-plane wavenumber $k_\parallel = \sqrt{k_x^2 + k_y^2} = k_1 \sin\theta_i = k_2 \sin\theta_t$ is called transverse wavenumber, while the out-of-plane wavevector is usually dubbed longitudinal wavenumber. In this fashion, the Fresnel coefficients in Eq. (7.1.85) are given by

$$t_\parallel(k_x, k_y) = \frac{2\varepsilon_2 k_{z_1}}{\varepsilon_2 k_{z_1} + \varepsilon_1 k_{z_2}} \sqrt{\frac{\mu_2 \varepsilon_1}{\mu_1 \varepsilon_2}}, \tag{7.1.92a}$$

$$t_\perp(k_x, k_y) = \frac{2\mu_2 k_{z_1}}{\mu_2 k_{z_1} + \mu_1 k_{z_2}}, \tag{7.1.92b}$$

$$r_\parallel(k_x, k_y) = \frac{\varepsilon_2 k_{z_1} - \varepsilon_1 k_{z_2}}{\varepsilon_2 k_{z_1} + \varepsilon_1 k_{z_2}}, \tag{7.1.92c}$$

$$r_\perp(k_x, k_y) = \frac{\mu_2 k_{z_1} - \mu_1 k_{z_2}}{\mu_2 k_{z_1} + \mu_1 k_{z_2}}, \tag{7.1.92d}$$

where we have included the possibility of nonunity magnetic permeability (namely, magnetic materials) [225]. As before, we choose the plane of incidence to be the xOz plane, and therefore $k_y = 0$. By this means, the components of the transmitted electric-field vector are expressed as (cf. Eq. (7.1.79))

$$E_x^{(t)} = -t_\parallel(k_x) A_\parallel \frac{k_{z2}}{k_2} \mathrm{e}^{\mathrm{i}k_x x + \mathrm{i}k_{z_2} z}, \tag{7.1.93a}$$

$$E_y^{(t)} = t_\perp(k_x) A_\perp \mathrm{e}^{\mathrm{i}k_x x + \mathrm{i}k_{z_2} z}, \tag{7.1.93b}$$

$$E_z^{(t)} = t_\parallel(k_x) A_\parallel \frac{k_x}{k_2} \mathrm{e}^{\mathrm{i}k_x x + \mathrm{i}k_{z_2} z}. \tag{7.1.93c}$$

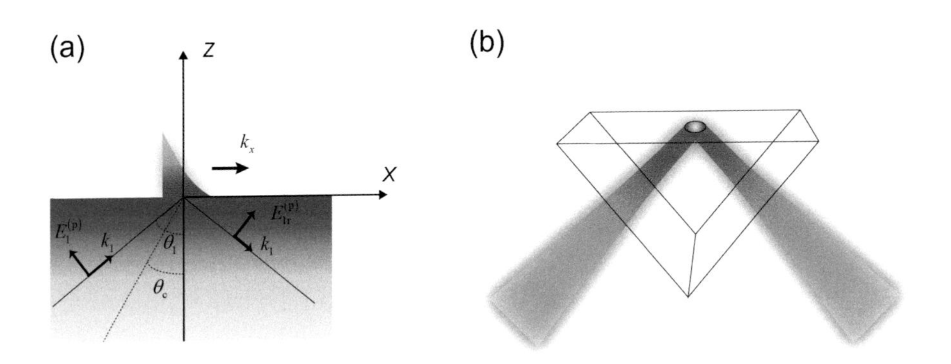

Figure 7.4 Excitation of an evanescent wave by TIR. (a) An evanescent wave is created in a medium if the plane wave is incident at an angle $\theta_1 > \theta_c$. (b) Actual experimental realization using a prism and a weakly focused Gaussian beam.

by suppressing the harmonic time factor $\exp(-\mathrm{i}\omega t)$ in the phase factor given by Eq. (7.1.80).

Let us specify the case in which medium 1 is optically denser than medium 2, that is, the relative index of refraction $n_{12} = \sqrt{\varepsilon_1\mu_1}/\sqrt{\varepsilon_2\mu_2} > 1$. As mentioned earlier, there is a critical angle $\theta_c = \arcsin(1/n_{12})$. We consider the scenario of $\theta_1 > \theta_c$, where k_{z_2} becomes an imaginary quantity that is given by

$$k_{z_2} = k_2\sqrt{1 - n_{12}^2\sin^2\theta_i} = \mathrm{i}\gamma, \tag{7.1.94}$$

in which

$$\gamma = k_2\sqrt{n_{12}^2\sin^2\theta_i - 1} \tag{7.1.95}$$

is a real number. We further express the transmitted field given by Eq. (7.1.93) in terms of above quantities and the incident angle θ_i as

$$E_x^{(t)} = -\mathrm{i}t_{\parallel}(\theta_i)A_{\parallel}\sqrt{n_{12}^2\sin^2\theta_i - 1}\mathrm{e}^{\mathrm{i}k_1\sin\theta_i x}\mathrm{e}^{-\gamma z}, \tag{7.1.96a}$$

$$E_y^{(t)} = t_{\perp}(\theta_i)A_{\perp}\mathrm{e}^{\mathrm{i}k_1\sin\theta_i x}\mathrm{e}^{-\gamma z}, \tag{7.1.96b}$$

$$E_z^{(t)} = t_{\parallel}(\theta_i)A_{\parallel}n_{12}\sin\theta_i\mathrm{e}^{\mathrm{i}k_1\sin\theta_i x}\mathrm{e}^{-\gamma z}. \tag{7.1.96c}$$

It is found that Eq. (7.1.96) describes a surface wave propagating along the interface while decaying exponentially in the transmitted medium, described by the decay constant γ in Eq. (7.1.96), as shown in Fig. 7.4(a). In other words, an evanescent wave can be created in the process of TIR at an AOI larger than the critical angle θ_c. If we take $n_{12} = 1.45$, typical for glass (SiO_2)–air interface in the optical range and $\theta = 60°$, the decay constant becomes $\sim 5.42/\lambda$, which means at a distance of λ, the electric field decays to a level of 0.4% of its value at the interface. The Fresnel formulas Eq. (7.1.92) also work for incident angles larger than the critical angle. However, for the scenarios involving evanescent waves, we can see that the imaginary k_{z2} in the formulas lead to the nontrivial phase shift for the reflected and transmitted waves, which is the origin of the phenomenon called the Goos–Hänchen shift. In practice, evanescent fields can be created by illuminating a beam of light into a prism as sketched in Fig. 7.4(b). Experimental detection of evanescent waves can be achieved by approaching a probe into the subwavelength vicinity of the interface, resulting in a wide-range of near-field optical (NFO) technologies, which will be detailed later in Chapter 9.

7.2 Classical Theory of Permittivity and Other General Concepts

In Section 7.1, constitutive relations are introduced to establish the relationship between material responses and field vectors. Yet the detail form of material properties, especially the relative permittivity (or dielectric function), is not specified. In this section, we will give several important models and relations for describing the complex permittivity. Although simple, they contain profound implications regarding microscopic light–matter interactions.

7.2.1 Drude–Lorentz and Drude Models

The Drude–Lorentz model is one of the simplest and widely used descriptions for a dielectric function [430, 431]. This model is established based on the idea that an electric dipole (ED) is formed in an atom due to the existence of a positively charged nucleus and a negatively charged electron that is bound to the atom nucleus. When driven by the electric field of EM waves, the atom dipoles oscillate, which can emit and absorb EM waves at discrete frequencies. Since the nucleus is much heavier than the electron, it is more or less stationary during oscillation, and thus its motion can be ignored. This constitutes a simple mass-spring oscillator model. We can consider the motion of the bound electron only, described by a displacement $x(t)$, and hence a dipole moment $p(t)=ex(t)$ is formed, where e is the electric charge of the electron. Under the drive of time-varying electric field $E(t)$, the motion is governed by

$$m\ddot{x} = -m\omega_0^2 x - m\gamma\dot{x} + eE(t), \tag{7.2.1}$$

where m is the mass of the electron, ω_0 is the resonance frequency of the bound electron, and γ is the damping constant. The damping constant arises from possible loss mechanisms in realistic situations, for instance, the collisions between electrons and thermally excited phonons. We further assume a time-harmonic electric field:

$$E = E_0 \mathrm{e}^{-\mathrm{i}\omega t}. \tag{7.2.2}$$

After an initial stage, the system will oscillate with the same frequency ω taking the form of $x = X_0\mathrm{e}^{-\mathrm{i}\omega t}$. Then, Eq. (7.2.1) becomes

$$-\omega^2 X = -\omega_0^2 X + \mathrm{i}\gamma\omega X + \frac{eE_0}{m}, \tag{7.2.3}$$

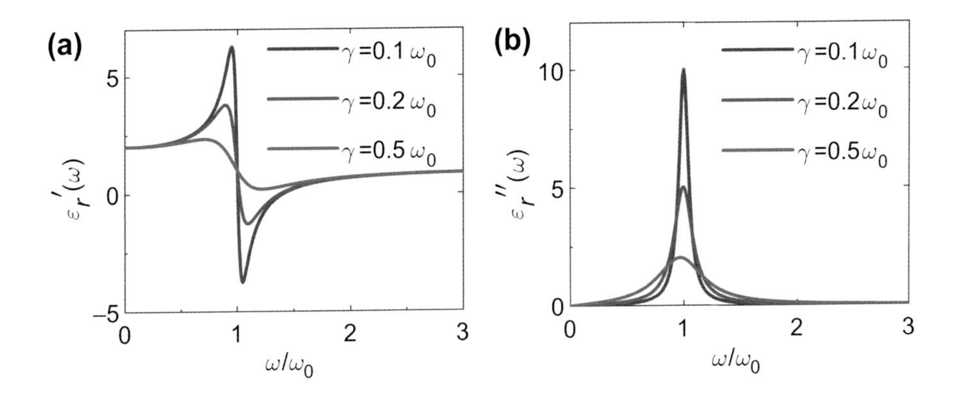

Figure 7.5 Typical behavior of the Drude–Lorentz permittivity for different damping constants γ; see Eq. (7.2.7). (a) Real and (b) imaginary parts of $\varepsilon(\omega)$.

where the time-harmonic factor cancels with each other. The solution of displacement X_0 is given by

$$X_0 = \frac{e}{m}\frac{1}{\omega_0^2 - \omega^2 - i\gamma\omega}E_0. \tag{7.2.4}$$

This equation describes a typical dependence of the displacement amplitude of a harmonic oscillator. It can be seen that the maximum is reached when the angular frequency ω is equal to the resonance frequency ω_0, and its absolute value is dependent on the damping constant γ and the driving amplitude. If we assume that the number density of dipole oscillators is N, the polarization due to these oscillators is given by

$$P = Nex = \frac{Ne^2}{m}\frac{1}{\omega_0^2 - \omega^2 - \mathrm{i}\gamma\omega}E_0 = \varepsilon_0\chi_e E_0, \tag{7.2.5}$$

from which the electric susceptibility χ_e can be determined. Therefore, this equation can be used to determine the complex dielectric constant as

$$\varepsilon(\omega) = \varepsilon_0(1+\chi_e) = \varepsilon_0\left(1 + \frac{Ne^2}{m\varepsilon_0}\frac{1}{\omega_0^2 - \omega^2 - \mathrm{i}\gamma\omega}\right). \tag{7.2.6}$$

If there is a nonresonant background, we should replace the unity in the parentheses by $\epsilon_\infty = 1+\chi_b$, with χ_b being the susceptibility of the background [431]. In this circumstance, the expression for the relative dielectric constant is in the following form:

$$\varepsilon_r(\omega) = 1 + \chi_b + \frac{Ne^2}{m\varepsilon_0}\frac{1}{\omega_0^2 - \omega^2 - \mathrm{i}\gamma\omega}, \tag{7.2.7}$$

where we typically use $\varepsilon_\infty = 1+\chi_b$ to represent the high-frequency limit of the dielectric function, which can be approached by taking $\omega \to \infty$ in Eq. (7.2.7).

In Fig. 7.5, a typical example is presented for the real ε' and imaginary parts ε'' of the Drude–Lorentz dielectric constant described by Eq. (7.2.7), in which for simplicity we take the oscillator strength $S = Ne^2/\left(\varepsilon_0 m\omega_0^2\right)$ to be unity, and χ_b is chosen to be zero. It can be seen from this figure that at the resonance frequency ω_0, the imaginary part reaches its maximum, with the height and width of its peak determined by the damping constant γ. Far from the resonance, the imaginary part becomes negligible. On the other hand, the real part undergoes an anomalous dispersion behavior near resonance, which means that the real part of the refractive index decreases with the increase of frequency. At low frequencies far from the resonance, the oscillator can follow the driving field, resulting in a limit of $\varepsilon_r(0) = 2$ according to Eq. (7.2.6), while at high frequencies far from the resonance, the oscillator is too inert to follow the rapid oscillations of the driving field, leading to a limit of ε_∞ which is unity in Fig. 7.5.

In spite of its simple form, the Drude–Lorentz model can generally work well in describing the frequency-dependent permittivity (dielectric constant), at least close to the resonance. A more general form takes different types of oscillators into account, yielding

$$\varepsilon_r(\omega) = \varepsilon_\infty + \sum_j \frac{N_j e^2}{m_j \varepsilon_0} \frac{1}{\omega_j^2 - \omega^2 - \mathrm{i}\gamma_j\omega} = \varepsilon_\infty + \sum_j \frac{S_j \omega_j^2}{\omega_j^2 - \omega^2 - \mathrm{i}\gamma_j\omega}, \tag{7.2.8}$$

where N_j, m_j, ω_j, and γ_j are the number density, mass, resonance frequency, and damping constant of the jth-type oscillator, and $S_j = N_j e^2/\left(\varepsilon_0 m_j \omega_j^2\right)$ is the corresponding oscillator strength. This model in general takes account of all the transitions in the medium and can be used to describe the full frequency dependence of the permittivity. It is usual to fit the reflectance and transmission spectra of thin films to obtain the oscillator parameters of materials under investigation [6].

For metals, we can assume that the electrons are free, namely, not bound to the ions. Hence, there is no restoring force that is related to the binding energy. In this manner, we can rewrite the Drude–Lorentz model by setting $\omega_0 \to 0$, which results in the well-known Drude model for the permittivity of metals:

$$\varepsilon_r(\omega) = \varepsilon_\infty - \frac{\omega_p^2}{\omega(\omega + \mathrm{i}\gamma)}, \tag{7.2.9}$$

where ε_∞ is the contribution from bound electrons and the plasma frequency is introduced as

$$\omega_p = \sqrt{\frac{Ne^2}{\varepsilon_0 m}}. \tag{7.2.10}$$

At the plasma frequency $\omega = \omega_p$, we obtain $\varepsilon(\omega_p) \approx 0$ if $\varepsilon_\infty = 1$. This is related to longitudinal plasma oscillations of the conduction electrons.

In Fig. 7.6, the real and imaginary parts of the dielectric function described by the Drude model of Ag are shown, which are also compared to experimental data. We can find that the Drude model can reasonably depict the frequency dependence of the permittivity of Ag despite some deviations for the imaginary part due to possible damping mechanisms in experimental measurements.

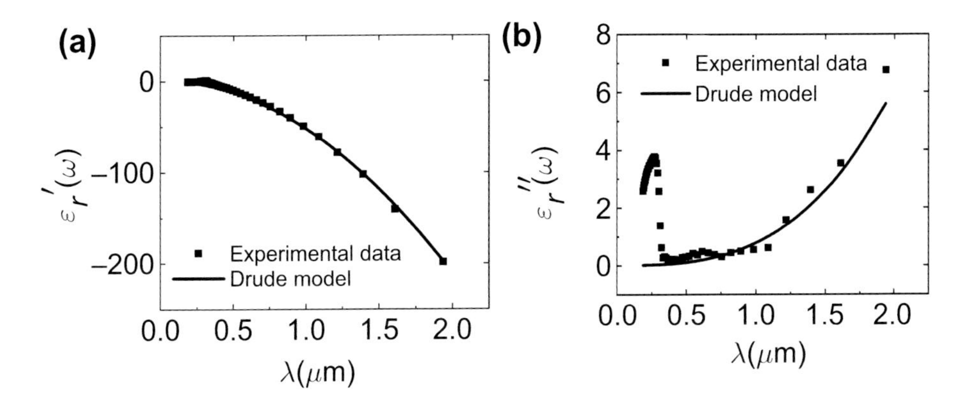

Figure 7.6 (a) Real and (b) imaginary parts of the Drude dielectric function for silver (Ag), and comparison with experimental data. Experimental data are taken from [432]. The Drude model parameters are $\omega_p = 9.013$ eV and $\gamma = 18$ meV [433].

7.2.2 Clausius–Mossotti Relation

In a dilute medium like rarefied gas with a low atom density, the dielectric constant calculation given in Eq. (7.2.8) is valid. However, there is another aspect that we need to take into account in a solid, which is a dense optical medium. The specific atomic dipoles react to the local field they encounter. Since the dipoles themselves produce electric fields that are felt by all the other dipoles, this may or may not be the same as the external field. Therefore, the actual local field that an atom experiences has the following structure:

$$\mathbf{E}_{\text{local}} = \mathbf{E} + \mathbf{E}_{\text{P}}, \tag{7.2.11}$$

where $\mathbf{E}$ and $\mathbf{E}_{\text{P}}$ stand for the external field and the field generated by the other dipoles, respectively. By identifying this difference, it is necessary to use $\mathbf{E}_{\text{local}}$ instead of $\mathbf{E}$ in the derivation of material permittivity presented in Section 7.2.1.

It is actually quite difficult to calculate the correction field caused by the other dipoles in the medium. If we assume that all of the dipoles are parallel to the applied field and are placed on a cubic lattice, we can get an approximate, which was originally presented by Lorentz. As shown in Fig. 7.7, the calculation is carried out by separating the contributions from the nearest dipoles and the remainder of the matter. A hypothetical spherical sphere with a radius large enough to make it reasonable to average the material outside of it is used to create the division. The issue is therefore simplified to computing the impact of an evenly polarized dielectric outside the sphere and adding the fields of the dipoles inside the sphere at the middle dipole. In the end, we obtain

$$\mathbf{E}_{\text{P}} = \frac{\mathbf{P}}{3\varepsilon_0}, \tag{7.2.12}$$

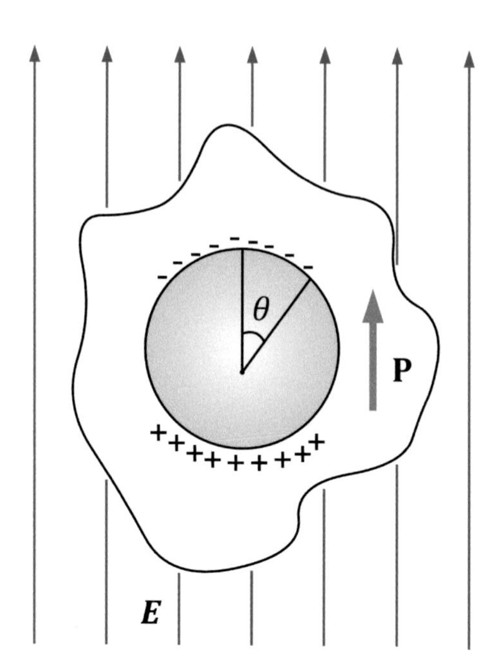

Figure 7.7 Model used to calculate the local field by the Lorentz correction. An imaginary spherical surface drawn around a particular atom divides the medium into nearby dipoles and distant dipoles. The field at the center of the sphere due to the nearby dipoles is summed exactly, while the field due to the distant dipoles is calculated by treating the material outside the sphere as a uniformly polarized dielectric.

where $\mathbf{P}$ represents the polarization of the dielectric outside the hypothetical sphere. A detailed derivation of this expression is given in [425]. Therefore, the local field is calculated as

$$\mathbf{E}_{\mathrm{local}} = \mathbf{E} + \frac{\mathbf{P}}{3\varepsilon_0}. \tag{7.2.13}$$

The macroscopic polarization $\mathbf{P}$ is also related to the local field by

$$\mathbf{P} = N\varepsilon_0\chi_{\mathrm{a}}\mathbf{E}_{\mathrm{local}}, \tag{7.2.14}$$

where χ_{a} is the electric susceptibility of an individual atom that relates its dipole moment $\mathbf{p}$ with the local field as

$$\mathbf{p} = \varepsilon_0\chi_{\mathrm{a}}\mathbf{E}_{\mathrm{local}}. \tag{7.2.15}$$

From Eq. (7.2.4), we are able to obtain the expression of atom susceptibility with only a single resonance as

$$\chi_{\mathrm{a}} = \frac{e^2}{\varepsilon_0 m}\frac{1}{\left(\omega_0^2 - \omega^2 - \mathrm{i}\gamma\omega\right)}. \tag{7.2.16}$$

If there are multiple resonances, an expression similar to Eq. (7.2.8) can also be given. By combining Eqs. (7.2.13) and (7.2.14), we can obtain

$$\mathbf{P} = N\varepsilon_0\chi_\mathrm{a}\left(\mathbf{E} + \frac{\mathbf{P}}{3\varepsilon_0}\right) = (\varepsilon_\mathrm{r} - 1)\,\varepsilon_0\mathbf{E}. \tag{7.2.17}$$

From Eq. (7.2.17), we can solve the relative permittivity as

$$\frac{\varepsilon_\mathrm{r} - 1}{\varepsilon_\mathrm{r} + 2} = \frac{n\chi_\mathrm{a}}{3}. \tag{7.2.18}$$

This equation is the well-known Clausius–Mossotti relation. The relationship works well in gases and liquids. It is also valid for cubic crystals in which the Lorentz correction, namely Eq. (7.2.13), can give an accurate description of the local field effects.

7.2.3 Kramers–Kronig Relation

The Hilbert transform relations connect an analytical function's real and imaginary parts. In particular, the relationship between the real and imaginary components of the dielectric function was first demonstrated by H. A. Kramers (1927) and R. de L. Kronig (1926) independently [425]. These relations are therefore dubbed as the K-K dispersion relations, or K-K relations for abbreviation. The K-K relations are highly helpful in determining optical constants from sparse measurements because they may be thought of as causality in the frequency domain. According to the principle of causality, there cannot be an effect without a cause or an output without an input. A full derivation and proofs can be found in standard textbooks on electrodynamics [38, 285, 425].

Owing to the analyticity of relative permittivity $\varepsilon_r(\omega)$ in the upper half ω-plane, for any point z inside a closed contour C in the upper half ω-plane, we can use Cauchy's theorem and find

$$\varepsilon_r(z) = 1 + \frac{1}{2\pi \mathrm{i}} \oint_C \frac{[\varepsilon_r(\omega') - 1]}{\omega' - z}\,\mathrm{d}\omega'. \tag{7.2.19}$$

The contour C is chosen in such a way that it contains the real ω axis and a large semicircle at infinity in the upper half-plane. From the expression of the Lorentz model, Eq. (7.2.7), without a nonresonant background contribution, it is observed that $\varepsilon_r - 1$ vanishes sufficiently rapidly at infinity. As a consequence, there is no contribution to the above integral from the large semicircle. Thus, the Cauchy integral can be written as

$$\varepsilon_r(z) = 1 + \frac{1}{2\pi \mathrm{i}} \int_{-\infty}^{\infty} \frac{[\varepsilon_r(\omega') - 1]}{\omega' - z}\,\mathrm{d}\omega', \tag{7.2.20}$$

in which z is an arbitrary point in the upper half-plane, while the integral is only carried out along the real axis. Let us take the limit in which the complex frequency approaches the real axis from above, that is, $z = \omega + \mathrm{i}\delta$. Then from Eq. (7.2.20), we can obtain

$$\varepsilon_r(\omega) = 1 + \frac{1}{2\pi \mathrm{i}} \int_{-\infty}^{\infty} \frac{[\varepsilon_r(\omega') - 1]}{\omega' - \omega - \mathrm{i}\delta}\,\mathrm{d}\omega'. \tag{7.2.21}$$

The denominator in the above integrand can be written as

$$\frac{1}{\omega' - \omega - \mathrm{i}\delta} = P\left(\frac{1}{\omega' - \omega}\right) + \pi \mathrm{i}\delta\left(\omega' - \omega\right), \tag{7.2.22}$$

where P means taking the principal value (PV) of the function. By using this formula, Eq. (7.2.21) can be recast into the following form:

$$\varepsilon_r(\omega) = 1 + \frac{1}{\pi \mathrm{i}} P \int_{-\infty}^{\infty} \frac{\left[\varepsilon_r\left(\omega'\right) - 1\right]}{\omega' - \omega} \mathrm{d}\omega'. \tag{7.2.23}$$

By taking the real and imaginary parts of the above equation, we can obtain

$$\mathrm{Re}\,\varepsilon_r(\omega) = 1 + \frac{1}{\pi} P \int_{-\infty}^{\infty} \frac{\mathrm{Im}\,\varepsilon_r\left(\omega'\right)}{\omega' - \omega} \mathrm{d}\omega', \tag{7.2.24a}$$

$$\mathrm{Im}\,\varepsilon_r(\omega) = -\frac{1}{\pi} P \int_{-\infty}^{\infty} \frac{\left[\mathrm{Re}\,\varepsilon\left(\omega'\right) - 1\right]}{\omega' - \omega} \mathrm{d}\omega'. \tag{7.2.24b}$$

These equations are thus the well-known K-K dispersion relation. Since $\varepsilon(-\omega) = \varepsilon^*(\omega^*)$, $\mathrm{Re}\,\varepsilon(\omega)$ is even in ω, while $\mathrm{Im}\,\epsilon(\omega)$ is odd. As a consequence, we can transform the integrals in Eq. (7.2.24) to be conducted only in positive frequencies:

$$\mathrm{Re}\,\varepsilon_r(\omega) = 1 + \frac{2}{\pi} P \int_{0}^{\infty} \frac{\omega' \,\mathrm{Im}\,\varepsilon_r\left(\omega'\right)}{\omega'^2 - \omega^2} \mathrm{d}\omega', \tag{7.2.25a}$$

$$\mathrm{Im}\,\varepsilon_r(\omega) = -\frac{2\omega}{\pi} P \int_{0}^{\infty} \frac{\left[\mathrm{Re}\,\varepsilon_r\left(\omega'\right) - 1\right]}{\omega'^2 - \omega^2} \mathrm{d}\omega'. \tag{7.2.25b}$$

From the derivation procedures, we can see that the K-K relations are quite general, originating from nothing more than the simple presumption that polarization and the electric field are causally related. This means that the knowledge of $\mathrm{Im}\,\varepsilon(\omega)$ permits us to obtain $\mathrm{Re}\,\varepsilon(\omega)$ from Eq. (7.2.25a). The connection between the absorption peak and anomalous dispersion, as already shown in Fig. 7.5, can be established through the K-K relations. More specifically, if we assume that there is a very narrow absorption line at $\omega = \omega_0$ for the medium, in this scenario, the imaginary part of the dielectric constant can be approximated near the resonance as

$$\mathrm{Im}\,\varepsilon\left(\omega'\right) \simeq \frac{\pi K}{2\omega_0} \delta\left(\omega' - \omega_0\right) + \cdots, \tag{7.2.26}$$

where K is a constant and the dots indicate the other (smoothly varying) contributions to $\mathrm{Im}\,\varepsilon$. By applying Eq. (7.2.25a), we can obtain the expression of the real part of the relative permittivity near the resonance frequency as

$$\mathrm{Re}\,\varepsilon(\omega) \simeq \bar{\varepsilon} + \frac{K}{\omega_0^2 - \omega^2}, \tag{7.2.27}$$

where the term $\bar{\varepsilon}$ effectively contains the slowly varying part of $\mathrm{Re}\,\varepsilon$ arising from the integrating contributions of $\mathrm{Im}\,\varepsilon$ far from the resonance. From Eq. (7.2.27), we can find that the real part demonstrates a very rapid variation in the vicinity of the absorption line, as can be observed from Fig. 7.5. Therefore the

K-K relations are extremely useful for connecting the dispersive and absorptive aspects of a process, and if the full spectral data of the imaginary part of the dielectric constant are known, we can obtain a complete description of the real part.

7.3 Scattering of Electromagnetic Waves by Particles

The study of scattering of EM wave by particles is an important and fundamental topic in MNTR, which is critical to the understanding of radiative transfer in disordered and ordered media, the microscopic origin of the RTE, light–matter interaction in micro- and nanoscale objects, and so on. It also constitutes the foundation of a variety of applications in optics and thermal radiation, such as optical forces, energy harvest and conversion, and structural colors and radiative cooling. In this section, we are mainly concerned with isotropic, homogeneous, and nonmagnetic materials. Our main interest is in the propagation and scattering characteristics of a wave in the presence of particles and particle groups. First, we consider a single particle and examine its scattering and absorption characteristics. Second, we investigate the problem of EM wave scattering from a group of particles and introduce the general formulation dealing with this problem. This topic has been exhaustively covered in a number of books [301, 317, 434–436], and therefore only an essential summary is given in this section.

7.3.1 Cross Section and Scattering Amplitude

When an EM wave illuminates a single particle, it interacts with the particle in a rather complicated manner. Some of the incident power is scattered and some is absorbed by the particle. Assume a linearly polarized plane wave propagating in vacuum, with an electric field of unit amplitude, which is expressed as

$$\mathbf{E}_{\mathrm{i}}(\mathbf{r}) = \hat{\mathbf{e}}_{\mathrm{i}} \exp(\mathrm{i}k\hat{\mathbf{i}} \cdot \mathbf{r}), \tag{7.3.1}$$

where $k = \omega\sqrt{\mu_0\varepsilon_0} = 2\pi/\lambda$ is the wavenumber, λ is a wavelength in vacuum, $\hat{\mathbf{i}}$ is a unit vector in the direction of incident wave propagation, and $\hat{\mathbf{e}}_{\mathrm{i}}$ is a unit vector dictating the polarization direction of the electric-field vector.

As shown in Fig. 7.8, the plane wave is incident upon a particle with a relative dielectric constant $\varepsilon_{\mathrm{r}}(\mathbf{r})$, which is a general function of the spatial position and is expressed as $\varepsilon_{\mathrm{r}}(\mathbf{r}) = \varepsilon'_{\mathrm{r}}(\mathbf{r}) + \mathrm{i}\varepsilon''_{\mathrm{r}}(\mathbf{r})$. Let us consider the electric field at a distance R from a reference point within the particle in a direction described by the unit vector $\hat{\mathbf{s}}$. This electric field can be decomposed into the incident field $\mathbf{E}_{\mathrm{i}}$ and the field $\mathbf{E}_{\mathrm{s}}$ scattered by the particle. For a general understanding of the scattering problem, we first investigate the case in which R is sufficiently large, that is, in the far-field region. The criterion can be estimated as $R > D^2/\lambda$, where D is the typical length scale of the particle [435]. If $R < D^2/\lambda$, then the scattered field

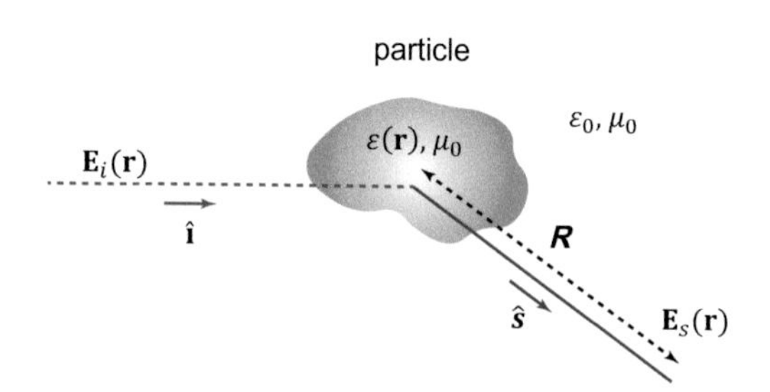

Figure 7.8 A plane wave $\mathbf{E}(\mathbf{r})$ is incident upon a dielectric scatterer, and the scattered field $\mathbf{E}_s(\mathbf{r})$ is observed in the direction $\hat{\mathbf{s}}$ at a distance R.

shows the complicated spatial dependence of amplitude and phase due to the wave interferences from different parts of the particle and is perhaps affected by near-field evanescent waves. In the far-field region, the scattered field $\mathbf{E}_\mathrm{s}$ can be treated as a spherical wave in the following form:

$$\mathbf{E}_s(\mathbf{r}) = \mathbf{f}(\hat{\mathbf{s}}, \hat{\mathbf{i}}) \frac{\mathrm{e}^{\mathrm{i}kR}}{R}, \tag{7.3.2}$$

where $\mathbf{f}(\hat{\mathbf{s}}, \hat{\mathbf{i}})$ is the so-called scattering amplitude in the scattering direction denoted by the unit vector $\hat{\mathbf{s}}$. Note that the scattered wave does not necessarily have the same polarization state as the incident wave.

For an incident power flux density S_i, we further calculate the scattered power flux density S_s at a distance R from the particle in the direction $\hat{\mathbf{s}}$. For doing so, the differential scattering cross section can be defined as

$$\sigma_\mathrm{d}(\hat{\mathbf{s}}, \hat{\mathbf{i}}) = \lim_{R\to\infty} \frac{R^2 S_\mathrm{s}}{S_\mathrm{i}} = |\mathbf{f}(\hat{\mathbf{s}}, \hat{\mathbf{t}})|^2 = \frac{\sigma_\mathrm{i}}{4\pi} p(\hat{\mathbf{s}}, \hat{\mathbf{i}}), \tag{7.3.3}$$

where S_i and S_s represent the magnitudes of the (time-averaged) incident and scattering power flux density vectors that are defined as

$$\mathbf{S}_\mathrm{i} = \frac{1}{2}(\mathbf{E}_\mathrm{i} \times \mathbf{H}_\mathrm{i}^*) = \frac{|E_\mathrm{i}|^2}{2Z_0}\hat{\mathbf{i}}, \tag{7.3.4a}$$

$$\mathbf{S}_\mathrm{s} = \frac{1}{2}(\mathbf{E}_\mathrm{s} \times \mathbf{H}_\mathrm{s}{}^*) = \frac{|E_\mathrm{s}|^2}{2Z_0}\hat{\mathbf{s}}, \tag{7.3.4b}$$

and $Z_0 = \sqrt{\mu_0/\varepsilon_0}$ is the impedance of the vacuum as encountered in Eq. (7.1.64). The differential scattering cross section σ_d defined in Eq. (7.3.3) has the dimensions of area per solid angle, and the dimensionless quantity $p(\hat{\mathbf{0}}, \hat{\mathbf{i}})$ is dubbed as the phase function, as used in the RTE introduced in Chapter 2. σ_t is the total cross section, which will be defined later in Eq. (7.3.7). A difference should be noted that for absorptive media, the integration of phase function over all solid angles is not equal to unity but the albedo, as will be shown later in Eq. (7.3.6).

After obtaining the expression of the differential scattering cross section, we proceed with the consideration of the total scattering cross section, which is related to the total scattered power at all angles surrounding the particle. It is given by

$$\sigma_{\mathrm{s}} = \int_{4\pi} \sigma_{\mathrm{d}}\,\mathrm{d}\Omega = \int_{4\pi} |\mathbf{f}(\hat{\mathbf{s}},\hat{\mathbf{i}})|^2\,\mathrm{d}\Omega = \frac{\sigma_{\mathrm{t}}}{4\pi}\int_{4\pi} p(\hat{\mathbf{s}},\hat{\mathbf{i}})\mathrm{d}\Omega, \tag{7.3.5}$$

where $\mathrm{d}\Omega$ is the differential solid angle at the scattering direction specified by $\hat{\mathbf{s}}$. The ratio ϖ of the scattering cross section to the total cross section is hence the albedo of a single particle, which is calculated as

$$\varpi = \frac{\sigma_{\mathrm{s}}}{\sigma_{\mathrm{t}}} = \frac{1}{\sigma_{\mathrm{t}}}\int_{4\pi} |\mathbf{f}(\hat{\mathbf{s}},\hat{\mathbf{i}})|^2\,\mathrm{d}\Omega = \frac{1}{4\pi}\int_{4\pi} p(\hat{\mathbf{s}},\hat{\mathbf{i}})\,\mathrm{d}\Omega. \tag{7.3.6}$$

Another definition is the absorption cross section called σ_a, that is the cross section of a particle that would absorb the same power as the particle under investigation. The sum of the scattering and the absorption cross sections is therefore the total cross section or the extinction cross section σ_{e}:

$$\sigma_{\mathrm{e}} = \sigma_{\mathrm{s}} + \sigma_{\mathrm{a}}. \tag{7.3.7}$$

It can be helpful to provide a broad overview of how these cross sections relate to the geometric cross section, wavelength, and dielectric constant before we explore precise mathematical representations of the various cross sections. The geometric cross section of a particle, σ_{t}, approaches twice its total cross section, σ_{g}, if the particle's size is substantially larger than its wavelength. Let us use an incident wave with the power flux density S_{i} to illustrate this argument. The particle either scatters out or absorbs the total flux $S_{\mathrm{i}}\sigma_{\mathrm{g}}$ within the geometric cross section σ_{g}. There should be a shadow area behind the particle where there is essentially no wave. The incident wave and the scattered wave from the particle are identical in this shadow region, but they are 180° out of phase, and the scattered flux is equivalent to $S_{\mathrm{i}}\sigma_{\mathrm{g}}$ in magnitude. Therefore, the total scattered and absorbed flux approaches $S_{\mathrm{i}}\sigma_{\mathrm{g}} + S_{\mathrm{i}}\sigma_{\mathrm{g}}$ and the extinction cross section σ_{e} approaches $\sigma_{\mathrm{e}} \to 2S_{\mathrm{i}}\sigma_{\mathrm{g}}/S_{\mathrm{i}} = 2\sigma_{\mathrm{g}}$. In addition, since the total absorbed power for a very large particle cannot exceed $S_{\mathrm{i}}\sigma_{\mathrm{g}}$, the absorption cross section σ_{u} would approach a constant somewhat smaller than the geometric cross section, namely, $\sigma_{\mathrm{u}} \to \sigma_{\mathrm{g}}$.

On the other hand, the scattering cross section σ_{s} is inversely proportional to the wavelength's fourth power and proportional to the square of the particle's volume if the size is substantially lower than a wavelength. To prove this, we observe that the scattered field E_{s} is created by the field in the particle, and so E_{s} at a distance R is proportional to the incident field E_{i}, and the volume V of the scatterer as $|E_{\mathrm{s}}| = |E_{\mathrm{i}}|CV/R$, where C is a constant. By dimension analysis, the constant C in this equation should be of the dimension of the inverse of length squared. Moreover, since this constant is obviously a function of wavelength, it is reasonable to argue that $C \sim \lambda^{-2}$. As a consequence, by combining Eq. (7.3.2), we can obtain

$$|E_\mathrm{s}| \sim |E_\mathrm{i}| \frac{V}{R\lambda^2} \sim |E_\mathrm{i}| \frac{|f(\hat{\mathbf{s}}, \hat{\mathbf{i}})|}{R}. \tag{7.3.8}$$

Therefore, the scattering cross section can be estimated as

$$\sigma_\mathrm{s} \sim |f(\hat{\mathbf{s}}, \mathbf{i})|^2 \sim \frac{V^2}{\lambda^4}. \tag{7.3.9}$$

These characteristics are typical for light scattering by a small particle, which is also named Rayleigh scattering. The absorption cross section σ_a for a small scatterer, on the other hand, is proportional to its volume while inversely proportional to the wavelength. The scattering and absorption efficiencies, defined as the scattering and absorption cross sections divided by the geometric cross section, can thus be estimated as

$$\frac{\sigma_\mathrm{s}}{\sigma_\mathrm{g}} \sim \left(\frac{D}{\lambda}\right)^4 \left[(\varepsilon_\mathrm{r}' - 1)^2 + \varepsilon_\mathrm{r}''^2\right], \tag{7.3.10a}$$

$$\frac{\sigma_\mathrm{a}}{\sigma_\mathrm{g}} \sim \left(\frac{D}{\lambda}\right) \varepsilon_\mathrm{r}''. \tag{7.3.10b}$$

The total cross section σ_t represents the total power loss from the incident wave due to the scattering and absorption of a wave by the particle. This loss is closely related to the behavior of the scattered wave in the forward direction, and this general relationship is embodied in the "forward-scattering theorem," also called the optical theorem.

The extinction cross section σ_e is a measure of the overall power lost by an incident wave as a result of the particle's scattering and absorption. The so-called forward-scattering theorem, which is often known as the optical theorem, captures this general relationship between this kind of loss and the behavior of the scattered wave traveling in the forward direction. In particular, the optical theorem states that the total cross section σ_e is related to the imaginary part of the scattering amplitude in the forward direction $\mathbf{f}(\hat{\mathbf{i}}, \hat{\mathbf{i}})$ in the following manner:

$$\sigma_\mathrm{e} = \frac{4\pi}{k} \mathrm{Im}[\mathbf{f}(\hat{\mathbf{i}}, \hat{\mathbf{i}})] \cdot \hat{\mathbf{e}}_\mathrm{i}, \tag{7.3.11}$$

where $\hat{\mathbf{e}}_\mathrm{i}$ is the unit vector in the direction of polarization of the incident wave, as introduced in Eq. (7.3.1). Proof for the optical theorem can be found in [285]. This theorem is useful when one needs to compute the total cross section when the scattering amplitude is given or to determine the attenuation of the coherent field [436].

There are two approaches to describe the scattering amplitude and cross sections mathematically. It is possible to find precise equations for cross sections and scattering amplitude if a particle has a simple shape, like a sphere or an infinite cylinder. The Mie solution or Mie theory, which is described in Section 7.3.3, is the exact solution for EM scattering by a spherical particle. But in many real-world scenarios, a particle's shape is usually nonspherical or even very irregular. As a result, it is necessary to develop a technique to estimate the cross sections of particles having complex shapes. General integral representations of scattering amplitude can be used to achieve this goal. The method is advantageous for particles with simple shapes as well because it makes computations simple.

Let's have a look at a particle whose relative dielectric constant varies with its internal position as $\varepsilon_{\mathrm{r}}(\mathbf{r})=\varepsilon(\mathbf{r})/\varepsilon_0=\varepsilon_{\mathrm{r}}'(\mathbf{r})+\mathrm{i}\varepsilon_{\mathrm{r}}''(\mathbf{r})$. We assume that this particle has a volume of V and is put in the vacuum without loss of generality. Assuming that the particle is nonmagnetic, we can rewrite the equation for the curl of the magnetic vector, Eq. (7.1.44), into the following form:

$$\nabla\times\mathbf{H}=-\mathrm{i}\omega\varepsilon_0\mathbf{E}+\mathbf{J}_{\mathrm{eq}}, \tag{7.3.12}$$

where

$$\mathbf{J}_{\mathrm{eq}}=\begin{cases}-\mathrm{i}\omega\varepsilon_0\left[\varepsilon_{\mathrm{r}}(\mathbf{r})-1\right]\mathbf{E} & \text{in } V\\ 0 & \text{outside.}\end{cases} \tag{7.3.13}$$

The term $\mathbf{J}_{\mathrm{e}}$ may be regarded as an effective current source that leads to the scattered wave due to the presence of the scatterer. We can further assume that the solution to Maxwell's equations can be decomposed as follows:

$$\mathbf{E}(\mathbf{r})=\mathbf{E}_{\mathrm{i}}(\mathbf{r})+\mathbf{E}_{\mathrm{s}}(\mathbf{r}), \tag{7.3.14a}$$

$$\mathbf{H}(\mathbf{r})=\mathbf{H}_{\mathrm{i}}(\mathbf{r})+\mathbf{H}_{\mathrm{s}}(\mathbf{r}), \tag{7.3.14b}$$

where $\mathbf{E}_{\mathrm{i}}$ and $\mathbf{H}_{\mathrm{i}}$ represent the incident or primary field vectors that exist in the absence of the particle, and $\mathbf{E}_{\mathrm{s}}$ and $\mathbf{H}_{\mathrm{s}}$ denote the scattered wave originating from the presence of the particle. We can introduce the so-called Hertz vector $\mathbf{\Pi}_{\mathrm{s}}$ to solve the above equations [425]. By definition, the electric and magnetic fields can be expressed into the Hertz vector as

$$\mathbf{E}_{\mathrm{s}}(\mathbf{r})=\nabla\times\nabla\times\mathbf{\Pi}_{\mathrm{s}}(\mathbf{r}), \tag{7.3.15}$$

$$\mathbf{H}_{\mathrm{s}}(\mathbf{r})=-\mathrm{i}\omega\varepsilon_0\nabla\times\mathbf{\Pi}_{\mathrm{s}}(\mathbf{r}), \tag{7.3.16}$$

where the Hertz vector can be solved as

$$\begin{aligned}\mathbf{\Pi}_{\mathrm{s}}(\mathbf{r})&=-\frac{1}{\mathrm{i}\omega\varepsilon_0}\int_V G_0(\mathbf{r},\mathbf{r}')\,\mathbf{J}_{\mathrm{eq}}(\mathbf{r}')\,\mathrm{d}V'\\&=\int_V\left[\varepsilon_{\mathrm{r}}(\mathbf{r}')-1\right]\mathbf{E}(\mathbf{r}')\,G_0(\mathbf{r},\mathbf{r}')\,\mathrm{d}V',\end{aligned} \tag{7.3.17}$$

where

$$G_0(\mathbf{r},\mathbf{r}')=\frac{\exp\left(\mathrm{i}k\left|\mathbf{r}-\mathbf{r}'\right|\right)}{\left(4\pi\left|\mathbf{r}-\mathbf{r}'\right|\right)} \tag{7.3.18}$$

is the scalar free-space Green function. A detailed derivation of the above expressions can be found in [285].

Next, we proceed to the scattering amplitude in the far field. Turning back to Fig. 7.8, we consider the far-field position vector $\mathbf{r}=R\hat{\mathbf{s}}$, and in this far-field region, R is much larger than the dimension of the particle; therefore, we can safely approximate $1/\left|\mathbf{r}-\mathbf{r}'\right|$ in the scalar Green function as $1/R$. Moreover, we should be careful that the phase $k|\mathbf{r}-\mathbf{r}'|$ in the scalar Green function cannot be simply approximated as kR, since the small positional difference can be substantial compared to the wavelength. Nevertheless, it is sufficient to keep the leading term of $|\mathbf{r}-\mathbf{r}'|$ as

$$|\mathbf{r}-\mathbf{r}'| = \left(R^2+\mathbf{r}'^2-2R\mathbf{r}'\cdot\hat{\mathbf{s}}\right)^{1/2} \simeq R-\mathbf{r}'\cdot\hat{\mathbf{s}}. \tag{7.3.19}$$

As a result, the scalar Green function in the far-field region (large R) becomes

$$G_0(\mathbf{r},\mathbf{r}') = \frac{\exp(\mathrm{i}kR-\mathrm{i}k\mathbf{r}'\cdot\hat{\mathbf{s}})}{4\pi R}. \tag{7.3.20}$$

It is also noticed that in the far field,

$$\nabla\left(\frac{\mathrm{e}^{\mathrm{i}kR}}{R}\right) \simeq \left(\frac{\mathrm{e}^{\mathrm{i}kR}}{R}\right)(\mathrm{i}k\nabla R) = (\mathrm{i}k\hat{\mathbf{s}})\left(\frac{\mathrm{e}^{\mathrm{i}kR}}{R}\right) \tag{7.3.21}$$

and thus ∇ is equivalent to $\mathrm{i}k\hat{\mathbf{s}}$. Then substituting Eqs. (7.3.20) and (7.3.21) into Eqs. (7.3.15)–(7.3.17) leads to

$$\mathbf{E}_{\mathrm{s}}(\mathbf{r}) = \mathbf{f}(\hat{\mathbf{s}},\hat{\mathbf{i}})\frac{\exp(\mathrm{i}kR)}{R}, \tag{7.3.22a}$$

$$\mathbf{f}(\hat{\mathbf{s}},\hat{\mathbf{i}}) = \frac{k^2}{4\pi}\int_V \{-\hat{\mathbf{s}}\times[\hat{\mathbf{s}}\times\mathbf{E}(\mathbf{r}')]\}\,[\varepsilon_{\mathrm{r}}(\mathbf{r}')-1]\exp(-\mathrm{i}k\mathbf{r}'\cdot\hat{\mathbf{s}})\,\mathrm{d}V', \tag{7.3.22b}$$

which can be regarded as an exact expression for the scattering amplitude $\mathbf{f}(\hat{\mathbf{0}},\hat{\mathbf{i}})$ in terms of the total electric field $\mathbf{E}(\mathbf{r}')$ inside the particle. In general, this field $\mathbf{E}(\mathbf{r}')$ is actually not known *in priori*, and as a consequence, Eq. (7.3.22) is not a closed-form solution of the scattering amplitude. Nevertheless, it is quite convenient in a variety of practical scenarios if we could assume $\mathbf{E}(\mathbf{r}')$ to be some form of function and thus obtain a straightforward approximation for $\mathbf{f}(\hat{\mathbf{s}},\hat{\mathbf{i}})$.

The absorption cross section σ_{a} for the particle can be calculated as the volume integral of the loss inside the volume:

$$\sigma_{\mathrm{a}} = \frac{1}{S_{\mathrm{i}}}\left(\int_V \frac{1}{2}\omega\varepsilon_0\varepsilon_{\mathrm{r}}''|E|^2\,\mathrm{d}V'\right). \tag{7.3.23}$$

When the magnitude of the incident wave is chosen to be unity ($|E_{\mathrm{i}}|=1$), this is given by

$$\sigma_{\mathrm{a}} = \int_V k\varepsilon_{\mathrm{r}}''(\mathbf{r})\,|\mathbf{E}(\mathbf{r})|^2\,\mathrm{d}V. \tag{7.3.24}$$

Therefore, Eqs. (7.3.22) and (7.3.24) are exact integral representations of the scattering and absorption characteristics of the particle, although they are expressed in terms of the unknown total field $\mathbf{E}(\mathbf{r}')$ inside the particle.

Furthermore, it is possible for us to derive similar integral representations for the magnetic field $\mathbf{H}(\mathbf{r})$ as that of $\mathbf{E}(\mathbf{r})$. The vectorial harmonic wave equation (cf. Eq. (7.1.32)) for $\mathbf{E}$ can be recast into the following form:

$$\nabla\times\nabla\times\mathbf{E}(\mathbf{r}) - \frac{\omega^2}{c^2}\mathbf{E}(\mathbf{r}) = \frac{\omega^2}{c^2}[\varepsilon_{\mathrm{r}}(\mathbf{r})-1]\,\mathbf{E}(\mathbf{r}). \tag{7.3.25}$$

However, according to our derivation of Eq. (7.1.36), the wave equation for $\mathbf{H}$ for a homogeneous nonmagnetic medium contains a term involving the gradient of permittivity $\varepsilon_\mathrm{r}(\mathbf{r})$ as

$$\nabla\times\nabla\times\mathbf{H}(\mathbf{r})-\omega^2\mu_0\varepsilon_0\mathbf{H}(\mathbf{r})=\omega^2\mu_0\varepsilon_0\left[\varepsilon_\mathrm{r}(\mathbf{r})-1\right]\mathbf{H}(\mathbf{r})-\mathrm{i}\omega\varepsilon_0\left[\nabla\varepsilon_\mathrm{r}(\mathbf{r})\times\mathbf{E}(\mathbf{r})\right]. \tag{7.3.26}$$

Therefore, an integral equation for $\mathbf{H}$ should contain two terms:

$$\mathbf{H}(\mathbf{r})=\mathbf{H}_\mathrm{i}(\mathbf{r})+\mathbf{H}_\mathrm{s}(\mathbf{r})=\mathbf{H}_\mathrm{i}(\mathbf{r})+\nabla\times\nabla\times\mathbf{\Pi}_\mathrm{ms}(\mathbf{r}), \tag{7.3.27}$$

where the magnetic Hertz vector $\mathbf{\Pi}_\mathrm{ms}$ is given by

$$\begin{aligned}\mathbf{\Pi}_\mathrm{ms}(\mathbf{r})=&\int_V\left[\varepsilon_\mathrm{r}(\mathbf{r}')-1\right]G_0(\mathbf{r},\mathbf{r}')\,\mathbf{H}(\mathbf{r}')\,\mathrm{d}V'\\&-\frac{1}{\omega\mu_0}\int_V G_0(\mathbf{r},\mathbf{r}')\,\nabla'\varepsilon_\mathrm{r}(\mathbf{r}')\times\mathbf{E}(\mathbf{r}')\,\mathrm{d}V'.\end{aligned} \tag{7.3.28}$$

Here $\nabla'\varepsilon_\mathrm{r}(\mathbf{r}')$ represents the gradient with respect to $\mathbf{r}'$, and the second term takes into account not only the effect of depolarization but also the inhomogeneity of the dielectric constant. For a homogeneous sphere, $\nabla'\varepsilon_\mathrm{r}(\mathbf{r})$ becomes a delta function on the surface and therefore the volume integral turns into a surface integral.

7.3.2 Rayleigh Scattering

As indicated in Section 7.3.1, the scattering characteristic of a small particle compared with the wavelength can be deduced from general considerations, which is generally known as the Rayleigh scattering. Now in this section, we provide a thorough analysis for several simple geometries. Take into account a dielectric sphere that is substantially smaller than the wavelength. Due to its small size, the sphere's impinging electric field must behave almost like an electrostatic field. In electrostatics, it is well understood that an electric field E inside a dielectric sphere, when a constant field $\mathbf{E}_\mathrm{i}=E_\mathrm{i}\hat{\mathbf{e}}_\mathrm{i}$ is supplied, is uniform and given by [437]

$$\mathbf{E}=\frac{3}{\varepsilon_\mathrm{r}+2}\mathbf{E}_\mathrm{i}. \tag{7.3.29}$$

Substituting Eq. (7.3.29) into the integral equation (Eq. (7.3.22)), we obtain

$$f(\hat{\mathbf{s}},\hat{\mathbf{i}})=\frac{k^2}{4\pi}\left[\frac{3(\varepsilon_\mathrm{r}-1)}{(\varepsilon_\mathrm{r}+2)}\right]V\left[-\hat{\mathbf{s}}\times(\hat{\mathbf{s}}\times\hat{\mathbf{e}}_\mathrm{i})\right], \tag{7.3.30}$$

where it is noted that $|[-\hat{\mathbf{s}}\times(\hat{\mathbf{s}}\times\hat{\mathbf{e}}_\mathrm{i})]|=\sin\theta_\mathrm{s}$, with θ_s being the angle between the incident direction $\hat{\mathbf{e}}_\mathbf{i}$ and the scattering direction $\hat{\mathbf{s}}$.

This scattered wave's angular dependency is the same when an ED is pointed in the direction of $\mathbf{E}$ in the dielectric. Given that the scattered field is generated by an equivalent current source as mentioned earlier,

$$\mathbf{J}_\mathrm{eq}=-\mathrm{i}\omega\varepsilon_0(\varepsilon_\mathrm{r}-1),\mathbf{E}, \tag{7.3.31}$$

which can be considered as a distribution of ED with electric polarization $\varepsilon_0(\varepsilon_r - 1)\mathbf{E}$. Therefore, the result is to be expected. Moreover, Eq. (7.3.30) is still valid when the particle is lossy, namely, the relative dielectric constant ε_r is complex. From the scattering amplitude, the differential cross section of the particle $\sigma_d(\hat{\mathbf{0}}, \hat{\mathbf{i}})$ can be determined as

$$\sigma_d(\hat{\mathbf{s}}, \hat{\mathbf{i}}) = \frac{k^4}{(4\pi)^2}\left|\frac{3(\varepsilon_r - 1)}{\varepsilon_r + 2}\right|^2 V^2 \sin^2\theta_s, \quad (7.3.32)$$

where $\sin^2\theta_s = 1 - (\hat{\mathbf{s}} \cdot \hat{\mathbf{e}}_i)^2$. This equation dictates that the cross section is proportional to the square of the volume of the scatterer while inversely proportional to the fourth power of the wavelength, consistent with the result of our dimensional analysis in Section 7.3.1. This is known as Rayleigh scattering.

From Eq. (7.3.32), we can explain why the sky is blue because the blue part of the sunlight spectrum was scattered more than other parts of the spectrum due to its dependence on λ^{-4}. It is also obvious that the skylight at a right angle to the sun needs to be linearly polarized. The blue color and polarization, two qualities that were a major scientific conundrum in the nineteenth century, were ultimately explained by Lord Rayleigh. According to Rayleigh, the scatterers need not be water or ice as was previously believed, as air molecules can also contribute to this scattering. Next, let us look at a small dielectric particle's scattering cross section σ_s:

$$\begin{aligned}\sigma_s &= \int_{4\pi} \sigma_d \, d\Omega = \frac{k^2}{(4\pi)^2}\left|\frac{3(\varepsilon_r - 1)}{\varepsilon_r + 2}\right|^2 V^2 \int_0^{\pi} \sin\theta_s d\theta_s \int_0^{2\pi} d\phi \sin^2\theta_s \\ &= \frac{24\pi^3 V^2}{\lambda^4}\left|\frac{\varepsilon_r - 1}{\varepsilon_r + 2}\right|^2 = \frac{128\pi^5 a^6}{3\lambda^4}\left|\frac{\varepsilon_r - 1}{\varepsilon_r + 2}\right|^2.\end{aligned} \quad (7.3.33)$$

The scattering efficiency, defined as the scattering cross section divided by the geometric cross section, is given by

$$\frac{\sigma_s}{\pi a^2} = \frac{8(ka)^4}{3}\left|\frac{\varepsilon_r - 1}{\varepsilon_r + 2}\right|^2. \quad (7.3.34)$$

This is called the Rayleigh equation, valid only for the small values of ka. In practice, the upper limit of the scatterer radius is approximately taken to be $a = 0.05\lambda$, depending on the permittivity. At this radius, the error of the Rayleigh equation (7.3.34) is no more than 4% [434, 435].

The absorption cross section σ_a can be determined by combining Eqs. (7.3.23) and (7.3.29):

$$\sigma_a = k\varepsilon_r''\left|\frac{3}{\varepsilon_r + 2}\right|^2 V, \quad (7.3.35)$$

which leads to an absorption efficiency of

$$\frac{\sigma_a}{\pi a^2} = ka\varepsilon_r''\left|\frac{3}{\varepsilon_r + 2}\right|^2 \frac{4}{3}. \quad (7.3.36)$$

The extinction cross section σ_e is the sum of Eqs. (7.3.34) and (7.3.35). It should be noted that Eqs. (7.3.35) and (7.3.36) do not fulfill the optical theorem, since by applying the optical theorem to Eq. (7.3.30) for a nonabsorbing particle with $\varepsilon_r'' = 0$, we have $\sigma_e = 0$ that is unphysical. Therefore, for a given approximate distribution of an electric field within a particle, it is more reasonable to calculate the scattering cross section by integrating the scattering amplitude $|f|^2$ over all solid angles and determine the absorption cross section by using Eq. (7.3.24) and then calculate the total extinction cross section, rather than directly applying the optical scattering theorem to Eq. (7.3.22).

Let us consider the case of an ellipsoid particle, the surface of which is expressed as

$$\frac{x^2}{a^2} + \frac{y^2}{b^2} + \frac{z^2}{c^2} = 1. \tag{7.3.37}$$

The x-component of the field inside the particle can be calculated as [317, 437]

$$E_x = \frac{E_{ix}}{1 + (abc/2)(\varepsilon_r - 1) A_x}, \tag{7.3.38}$$

where E_{ix} is the x component of the incident field, and

$$A_x = \int_0^\infty \left(s + a^2\right)^{-1} \left[\left(s + a^2\right)\left(s + b^2\right)\left(s + c^2\right)\right]^{-1/2} \mathrm{d}s. \tag{7.3.39}$$

Similar expressions can be obtained for E_y and E_z through an appropriate interchange of a, b, and c. It is thus straightforward to substitute Eq. (7.3.38) into (7.3.22) and (7.3.24) to obtain the scattering and absorption cross sections. Moreover, we can find the expressions $L_1 = abcA_x/2, L_2 = abcA_y/2$, and $L_3 = abcA_z/2$ are the functions of the ratios b/a and c/a only, and they do not depend on the specific values of a, b, and c. We can also verify the relation that $L_1 + L_2 + L_3 = 1$. For a prolate ellipsoid $(a > b = c)$, we can obtain

$$L_1 = \frac{1 - e^2}{e^2} \left(-1 + \frac{1}{2e} \ln \frac{1 + e}{1 - e}\right), \tag{7.3.40}$$

where $e = \sqrt{1 - (b/a)^2}$, and for an oblate ellipsoid $(a < b = c)$, we have

$$L_1 = \frac{1 + f^2}{f^2} \left(1 - \frac{1}{f} \arctan f\right), \tag{7.3.41}$$

where $f = \sqrt{(b/a)^2 - 1}$. Consider a plane wave of unit magnitude $(|E_i| = 1)$ that is traveling in the z direction and is polarized in the x direction. A small dielectric sphere of radius a is illuminated by this wave. The electric field components in the remote region of the sphere are given by

$$E_\theta = \frac{\hat{\boldsymbol{\theta}} \cdot \mathbf{f}(\hat{\mathbf{s}}, \hat{\mathbf{i}})}{R} \exp(ikR), \tag{7.3.42a}$$

$$E_\phi = \frac{\hat{\boldsymbol{\phi}} \cdot \mathbf{f}(\hat{\mathbf{s}}, \hat{\mathbf{i}})}{R} \exp(ikR). \tag{7.3.42b}$$

It is noted that $[-\hat{\mathbf{s}}\times(\hat{\mathbf{s}}\times\hat{\mathbf{e}}_{\mathrm{i}})] = -[\hat{\mathbf{s}}(\hat{\mathbf{s}}\cdot\hat{\mathbf{e}}_{\mathrm{i}}) - \hat{\mathbf{e}}_{\mathrm{i}}]\,\hat{\mathbf{e}}_{\mathrm{i}} = \hat{\mathbf{x}}$, $\hat{\boldsymbol{\theta}}\cdot\hat{\mathbf{s}} = \hat{\boldsymbol{\phi}}\cdot\hat{\mathbf{s}} \doteq 0$, $\hat{\boldsymbol{\theta}}\cdot\hat{\mathbf{x}} = \cos\theta\cos\phi$, $\hat{\boldsymbol{\phi}}\cdot\hat{\mathbf{x}} = -\sin\theta$, and therefore we can straightforwardly obtain

$$E_\theta = E_0\cos\theta\cos\phi\exp(\mathrm{i}kR), \tag{7.3.43a}$$

$$E_\phi = -E_0\sin\phi\exp(\mathrm{i}kR), \tag{7.3.43b}$$

where

$$E_0 = \frac{k^2}{4\pi}\left[\frac{3(\varepsilon_{\mathrm{r}}-1)}{\varepsilon_{\mathrm{r}}+2}\right]\frac{V}{R}. \tag{7.3.44}$$

We also note that $\sin^2\theta_s = 1-\sin^2\theta\cos^2\phi$.

Let us further consider an ellipsoidal dielectric particle whose surface is given by Eq. (7.3.37), with $a<b=c$ being another illustration. Assume that the incident wave has an x polarization and is traveling in the z direction. Then we need to substitute $[1+L_1(\varepsilon_{\mathrm{r}}-1)]^{-1}$ for $3/(\varepsilon_{\mathrm{r}}+2)$ in the example of the spherical example, yielding

$$\sigma_{\mathrm{s}} = \frac{8\pi^3V^2}{3\lambda^4}\left|\frac{\varepsilon_{\mathrm{r}}-1}{1+L_1(\varepsilon_{\mathrm{r}}-1)}\right|^2, \tag{7.3.45a}$$

$$\sigma_{\mathrm{a}} = k\varepsilon_{\mathrm{r}}''\left|\frac{1}{1+L_1(\varepsilon_{\mathrm{r}}-1)}\right|^2 V, \tag{7.3.45b}$$

where the expression of L_1 is presented in Eq. (7.3.41) and $V = (4/3)\pi ab^2$.

7.3.3 The Mie Theory

One of the earliest EM scattering problems, the scattering of EM waves by a spherical particle, was solved rigorously by Gustav Mie over a century ago [34], and Ludvig Lorenz and others independently developed the theory of plane-wave scattering by a dielectric sphere. These particles can be single, homogeneous, or multilayered spherical particles with arbitrary electric and magnetic properties. The anomalous scattering characteristics of single dielectric particles have been extensively studied theoretically and experimentally by many authors in recent years, along with the rapid advancement of nanofabrication and nanophotonics [438, 439], leading to the explosion of Mie theory-based studies on nanoscale light scattering. Mie theory's fundamental goal is to formally resolve the boundary value problem of Maxwell's equations in spherical coordinates. Under these circumstances, a linear combination of vector spherical harmonics (VSHs) or vector spherical wave functions (VSWFs) can be used to formally expand the solution of Maxwell's equations [38, 436]. On the basis of the Hertz potential technique, we provide a brief description of the Mie solution in this section. Many standard textbooks have more in-depth derivations (e.g., [38, 285, 434]).

Consider a sphere with a relative dielectric constant ε_1 and an incident wave propagating in the z direction and polarized in the x direction:

$$\mathbf{E}_{\mathrm{inc}} = \mathrm{e}^{\mathrm{i}kz}\hat{\mathbf{x}}. \tag{7.3.46}$$

Any EM field can be expressed in terms of the two scalar functions Π_1 and Π_2 in the spherical coordinate system, which are the radial components of the electric and magnetic Hertz vectors and can be expressed as

$$\Pi_e = \Pi_1 \hat{\mathbf{r}}, \tag{7.3.47a}$$

$$\Pi_m = \Pi_2 \hat{\mathbf{r}}. \tag{7.3.47b}$$

In spherical coordinates, all TM modes with $H_r = 0$ and all TE modes with $E_r = 0$ can be found to be generated by Π_1 and Π_2, respectively [425, 435]. As noted in Section 7.3.3, both Π_1 and Π_2 satisfy a scalar wave equation:

$$(\nabla^2 + k^2)\,\Pi = 0 \quad \text{outside the sphere,} \tag{7.3.48a}$$

$$(\nabla^2 + k^2 m^2)\,\Pi = 0 \quad \text{inside the sphere,} \tag{7.3.48b}$$

with $m = \sqrt{\varepsilon_r}$ being the complex index of refraction. From Hertz vectors, electric and magnetic fields can be written as [437]

$$\mathbf{E} = \nabla \times \nabla \times (r\Pi_1 \hat{\mathbf{r}}) + \mathrm{i}\omega\mu_0 \nabla \times (r\Pi_2 \hat{\mathbf{r}}), \tag{7.3.49a}$$

$$\mathbf{H} = -\mathrm{i}\omega\varepsilon \nabla \times (r\Pi_1 \hat{\mathbf{r}}) + \nabla \times \nabla \times (r\Pi_2 \hat{\mathbf{r}}), \tag{7.3.49b}$$

where the permittivity is chosen to be $\varepsilon = \varepsilon_0$ outside and $\varepsilon = \varepsilon_r \varepsilon_0$ inside the sphere.

The incident field Eq. (7.3.46) can also be deduced from the scalar Hertz potentials Π_1^i and Π_2^i and can further be expanded into a linear combination of spherical harmonics as

$$r\Pi_1^i = \frac{1}{k^2} \sum_{n=1}^{\infty} \frac{\mathrm{i}^{n-1}(2n+1)}{n(n+1)} \psi_n(kr) P_n^1(\cos\theta)\cos\phi, \tag{7.3.50a}$$

$$r\Pi_2^i = \frac{1}{Z_0 k^2} \sum_{n=1}^{\infty} \frac{\mathrm{i}^{n-1}(2n+1)}{n(n+1)} \psi_n(kr) P_n^1(\cos\theta)\sin\phi, \tag{7.3.50b}$$

where $\psi_n(x) = x j_n(x)$, with $j_n(x)$ being the spherical Bessel function, $P_n^m(\cos\theta)$ represents the associated Legendre polynomials with P_n^1 being used in the above equation, and $Z_0 = \sqrt{\mu_0/\varepsilon_0}$. Equation (7.3.50) is deduced by comparing the spherical harmonic expansions of $\mathbf{E}_{\text{inc}}$ and $\mathbf{E}$ with Eq. (7.3.49) for the radial components [437].

Then the scattered fields can be expanded into spherical harmonic expressions for $r > a$:

$$r\Pi_1^s = \frac{(-1)}{k^2} \sum_{n=1}^{\infty} \frac{\mathrm{i}^{n-1}(2n+1)}{n(n+1)} a_n \zeta_n(kr) P_n^1(\cos\theta)\cos\phi, \tag{7.3.51a}$$

$$r\Pi_2^s = \frac{(-1)}{\eta k^2} \sum_{n=1}^{\infty} \frac{\mathrm{i}^{n-1}(2n+1)}{n(n+1)} b_n \zeta_n(kr) P_n^1(\cos\theta)\sin\phi, \tag{7.3.51b}$$

where two expansion coefficients a_n and b_n will be determined later in Eq. (7.3.53), and $\zeta_n(x) = x h_n^{(1)}(x)$, with $h_n^{(1)}(x)$ being the spherical Hankel function of the first kind [440]. Similarly, the Hertz potential inside the sphere $r<a$ can be expanded as

$$r\Pi_1^{\mathrm{r}} = \frac{1}{(kn)^2}\sum_{n=1}^{\infty}\frac{\mathrm{i}^{n-1}(2n+1)}{n(n+1)}c_n\psi_n(kmr)P_n^1(\cos\theta)\cos\phi, \quad (7.3.52a)$$

$$r\Pi_2^{\mathrm{r}} = \frac{1}{\eta k^2 n}\sum_{n=1}^{\infty}\frac{\mathrm{i}^{n-1}(2n+1)}{n(n+1)}d_n\psi_n(kmr)P_n^1(\cos\theta)\sin\phi, \quad (7.3.52b)$$

with the expansion coefficients c_n and d_n to be determined later.

Now let us determine the expansion coefficients by considering the boundary conditions, which dictate the continuation of $E_\theta, E_\phi, H_\theta$, and H_ϕ across the boundary at $r=a$. Since these boundary conditions will consist of a combination of Π_1 and Π_2, it would be more convenient to separate Π_1 and Π_2 by employing an appropriate linear combination of the field. For instance, we can use $\sin\theta(\partial/\partial\theta)E_\theta + (\partial/\partial\phi)E_\phi$ to give a continuity of $(\partial/\partial r)(r\Pi_1)$. On the basis of considerations, we can obtain the formulas for the continuity of $n^2\Pi_1, (\partial/\partial r)(r\Pi_1), \Pi_2$, and $(\partial/\partial r)(r\Pi_2)$ as the boundary conditions. By applying these conditions, the expansion coefficients can be solved as

$$a_n = \frac{\psi_n(\alpha)\psi_n'(\beta) - m\psi_n(\beta)\psi_n'(\alpha)}{\zeta_n(\alpha)\psi_n'(\beta) - m\psi_n(\beta)\zeta_n'(\alpha)}, \quad (7.3.53a)$$

$$b_n = \frac{m\psi_n(\alpha)\psi_n'(\beta) - \psi_n(\beta)\psi_n'(\alpha)}{m\zeta_n(\alpha)\psi_n'(\beta) - \psi_n(\beta)\zeta_n'(\alpha)}, \quad (7.3.53b)$$

where $\alpha = ka$ and $\beta = kma$. By taking the far-field asymptotic expression of spherical wave functions, the scattered fields E_ϕ and E_θ in the far field can be calculated as

$$E_\phi = -\frac{\mathrm{i}e^{\mathrm{i}kr}}{kr}S_1(\theta)\sin\phi, \quad (7.3.54a)$$

$$E_\theta = \frac{\mathrm{i}e^{\mathrm{i}kr}}{kr}S_2(\theta)\cos\phi, \quad (7.3.54b)$$

with

$$S_1(\theta) = \sum_{n=1}^{\infty}\frac{(2n+1)}{n(n+1)}\left[a_n\pi_n(\cos\theta) + b_n\tau_n(\cos\theta)\right], \quad (7.3.55a)$$

$$S_2(\theta) = \sum_{n=1}^{\infty}\frac{(2n+1)}{n(n+1)}\left[a_n\tau_n(\cos\theta) + b_n\pi_n(\cos\theta)\right], \quad (7.3.55b)$$

where

$$\pi_n(\cos\theta) = \frac{P_n^1(\cos\theta)}{\sin\theta}, \quad (7.3.56a)$$

$$\tau_n(\cos\theta) = \frac{\mathrm{d}}{\mathrm{d}\theta}P_n^1(\cos\theta). \quad (7.3.56b)$$

By applying the optical theorem, the extinction cross section and efficiency σ_{e} can be obtained:

$$Q_{\mathrm{e}} = \frac{\sigma_{\mathrm{e}}}{\pi a^2} = \frac{2}{(ka)^2} \sum_{n=1}^{\infty} (2n+1) \{\mathrm{Re}(a_n + b_n)\}. \tag{7.3.57}$$

The scattering cross section and efficiency can be deduced as

$$Q_{\mathrm{s}} = \frac{\sigma_{\mathrm{s}}}{\pi a^2} = \frac{2}{(ka)^2} \sum_{n=1}^{\infty} (2n+1) \left(|a_n|^2 + |b_n|^2\right). \tag{7.3.58}$$

As a remark, for multilayered spherical particles in which each layer has different dielectric constants, the Mie theory can be derived along the same line. The 2D version of the Mie theory for infinitely long cylinders can also be done similarly [441]. More generally, analytical solutions for arbitrary spheroids with different aspect ratios can also be derived based on spheroidal wave functions using the separation of variables method (SVM) since Mie theory is a member of the family of SVMs to solve the EM properties of regular geometries [442]. For spheroidal and randomly formed objects, for instance, see the works by Yeh [443, 444].

7.3.4 T-Matrix Method

The T-matrix method is an extension of Mie theory for the calculation of the scattering and absorption properties of a single nonspherical particle based on the extended boundary condition method (EBCM) and the VSWF expansion technique. It was originally proposed by Waterman [445] and further developed by many researchers [436, 442, 446]. The basic idea of this approach is to expand the incident and scattered waves into VSWFs and relate the expansion coefficients using the T-matrix. For a particle centered at the origin, the incident and scattered electric field can be expanded into VSWFs as

$$\mathbf{E}_{\mathrm{inc}}(\mathbf{r}) = \sum_{mnp} a_{mnp}^{\mathrm{inc}} \mathbf{N}_{mnp}^{(1)}(\mathbf{r}), \tag{7.3.59}$$

$$\mathbf{E}_{\mathrm{s}}(\mathbf{r}) = \sum_{mnp} a_{mnp}^{\mathrm{s}} \mathbf{N}_{mnp}^{(3)}(\mathbf{r}), \tag{7.3.60}$$

where a_{mnp}^{s} and $a_{m'n'p'}^{\mathrm{inc}}$ are the expansion coefficients. $\mathbf{N}_{mnp}^{(1)}(\mathbf{r})$ and $\mathbf{N}_{mnp}^{(3)}(\mathbf{r})$ are the type-1 and type-3 VSWFs, respectively [38, 317, 436, 442, 447, 448]. n and m are the integers denoting the order and degree of VSWFs with $n \geq 1$ and $|m| \leq n$. The subscript p can only be 1 or 2, which denotes magnetic (TM) or electric (TE) modes, respectively.

For a given incident field, the expansion coefficients a_{mnp}^{inc} can be solved by invoking the mutual orthogonality of VSWFs. In particular, for a plane wave propagating in the z direction $\mathbf{E}_{\mathrm{inc}}(\mathbf{r}) = \mathbf{E}_{\mathrm{inc},0} \exp(\mathrm{i}\mathbf{k} \cdot \mathbf{r}) = \mathbf{E}_{\mathrm{inc},0} \exp(\mathrm{i}kz)$, we can expand it into VSWFs and obtain the expansion coefficients as [436, 442]

$$a_{mn1}^{\mathrm{inc}} = -\mathrm{i}^n \sqrt{\frac{4\pi(2n+1)(n+m)!}{n(n+1)(n-m)!}} \mathbf{E}_{\mathrm{inc},0} \cdot \mathbf{B}_{-m,n}(0,0) \tag{7.3.61}$$

and

$$a_{mn2}^{\text{inc}} = -\text{i}^n \sqrt{\frac{4\pi(2n+1)(n+m)!}{n(n+1)(n-m)!}} \mathbf{E}_{\text{inc},0} \cdot \mathbf{C}_{-m,n}(0,0) \tag{7.3.62}$$

for $m = \pm 1$ (for other m's the coefficients are zero in this circumstance), where $\mathbf{B}_{mn}(\theta,\phi)$ and $\mathbf{C}_{mn}(\theta,\phi)$ are VSHs defined in spherical coordinates. Considering the linearity of Maxwell's equations, the relationship between the expansion coefficients of scattered and incident fields should also be linear, and a corresponding transition matrix (i.e., T-matrix) can be defined through this linear relation [442]:

$$a_{mnp}^{\text{s}} = \sum_{m'n'p'} T_{mnpm'n'p'} a_{m'n'p'}^{\text{inc}}. \tag{7.3.63}$$

As a result, if the T-matrix of an object is known, the expansion coefficients of the scattered field a_{mnp}^{s} can be solved from Eq. (7.3.63), and as a result, the single scattering properties, such as the scattering and extinction cross sections, phase function, and so forth, can be calculated immediately by using the far-field asymptotic forms of VSWFs. So solving the T-matrix of irregularly shaped particles is the main goal of the T-matrix approach.

For the special case of a homogeneous, isotropic spherical particle, the T-matrix is equivalent to the Mie coefficients as follows:

$$T_{mnpm'n'p'} = \begin{cases} b_n \delta_{mm'} \delta_{nn'} \delta_{pp'}, & p=1, \\ a_n \delta_{mm'} \delta_{nn'} \delta_{pp'}, & p=2, \end{cases} \tag{7.3.64}$$

where δ is the Kronecker delta. This relation is invalid for a general nonspherical particle, though. In this scenario, the T-matrix is computed using the EBCM. To employ VSWF expansion at the boundaries, EBCM artificially employs the hypothetical spherical boundaries that encircle (with a radius of $R_>$) and inscribe (with a radius of $R_<$) the nonspherical particle. The internal fields inside the inscribing sphere are related to the incident field by the expansion coefficients a_{mnp}^{int} as

$$a_{mnp}^{\text{inc}} = \sum_{m'n'p'} Q_{mnpm'n'p'} a_{m'n'p'}^{\text{int}}. \tag{7.3.65}$$

Moreover, the scattered field outside the circumscribing sphere can be related to the internal field in the particle thorough a similar formula:

$$a_{mnp}^{\text{s}} = -\sum_{m'n'p'} Q'_{mnpm'n'p'} a_{m'n'p'}^{\text{int}}. \tag{7.3.66}$$

Here the elements $Q_{mnpm'n'p'}$ and $Q'_{mnpm'n'p'}$ can be numerically evaluated by simple surface integrals over the particle surface area S. Thus, the T-matrix can be obtained through a matrix inversion and multiplication procedure according to Eq. (7.3.63). Detailed derivation and application of the T-matrix method can be found in [436] and [442].

The DDA [449] can also be used to tackle the problem of light scattering by various particles. The basic idea behind this numerical approach is to discretize each scatterer into a periodic grid of fictitious dipoles, typically in an arrangement of a cubic lattice. Then, by calculating the EM field for this set of dipoles, and then adding the EM fields generated by each dipole, the scatterer's scattering characteristics can be determined. This method was first put forth by Purcell and Pennypacker in 1973 [450] and further developed by Draine et al. [451–453]. When considering the formalism of this method, it is somewhat equivalent to the coupled-dipole model (CDM), which will be discussed later for the investigation of multiple scattering of EM waves. Nevertheless, there are some nontrivial differences between DDA and CDM, which have recently been reviewed in [454]. Interested readers can refer to [442, 449, 455] for additional theoretical and numerical considerations regarding the practical implementation for realistic scatterers to improve the accuracy and computation speed of DDA, including various types of dipole unit cells, renormalized polarizability models, the fast Fourier transform (FFT) technique, and the fast multipole method (FMM).

7.4 Multiple Scattering Theory of Electromagnetic Waves

After introducing the analytical and numerical methods for handling the EM scattering of a single particle, in what follows, we will quickly review the rigorous theories for the treatment of multiple scattering of EM waves. These theories, such as the analytical wave theory and the Foldy–Lax equations (FLEs), were initially created for both classical (such as electrons) and quantum (such as EM and acoustic) waves [323, 456]. As a result, they approach problems involving multiple scattering in a highly generic way. We will also talk about how the FLEs and analytical wave theory are equivalent.

When thermal radiation propagates in DDM, it undergoes scattering in a very complicated way. To better understand microscopic effects originating from the EM interferences in the radiative transfer process, one should resort to the fundamental theories and methods for the treatment of EM scattering by a single particle as well as particle groups. In this section, we first summarize the single and multiple scattering theories of EM waves in DDM, which are directly based on the first principle of Maxwell's equations. On this basis, we proceed to an introduction on the relationship between Maxwell's equations and the RTE, and then discuss briefly the applicability of the RTE. It is worth noting that recently Doicu and Mishchenko [457–462] published a series of reviews summarizing the multiple scattering theory of EM waves in random media as well as the connection between this theory and the RTE, which contain many technical details that can be referred to.

Before we dive into the remainder of this chapter, several crucial postulations should be made. (1) We assume that the radiation must have sufficient spatial and temporal coherence to permit coherent effects [463, 464]. (2) We only consider the static multiple scattering problem in this chapter, by assuming that the scattering processes are much faster than the random movement of particles in the disordered media. To be more precise, the dynamic positional fluctuations of the random media during the photon scattering process should be very small compared to the wavelength, that is, $kv_p \ll 1/\tau$, where v_p is the velocity of the particle and τ is the time scale of an individual photon scattering event; otherwise, considerable inelastic effects like Doppler frequency shifts and decoherence will occur [465]. For nonresonant scattering, we have $\tau \sim l_s/c_0$, with c_0 denoting the speed of light, leading to a condition as $v_p/c_0 \ll 1/(kl_s)$. Since for most disordered media kl_s is substantially larger than 1, this condition gives a more stringent criterion than the conventional one, $v_p \ll c_0$ [466, 467]. Moreover, for resonant multiple scattering, owing to the time delay brought by resonances, τ becomes much larger [296], resulting in a much smaller upper limit for v_p. Therefore, under this assumption, at a given moment, the multiple wave scattering can be described by assuming that the scatterers are all fixed and solving the corresponding (quasi)instantaneous problem in the frequency domain without any considerations of the inelastic effects [468]. Apparently, for certain densely packed rigid DDM where all scatterers are stationary, there is no need to use this assumption. (3) On the basis of the second assumption, we further apply the ergodicity hypothesis for the random motions of scatterers, and therefore the time-averaged signals over a sufficiently long period of time can be replaced by the ensemble-averaged ones over all system states, such as positions, sizes, and orientations of the scatterers, with appropriate probability functions characterizing all the system states. This assumption is important for conventional (i.e., not ultrafast) detection techniques that usually take a long period of time, thus permitting us to study the dynamic problem statically [301]. Similarly, for those fixed DDM, there is no need to use this assumption. But it is always postulated in such stationary media that the ensemble average over all system states can be achieved by characterizing a large number of different samples from the same fabrication process or different zones in a single sample [469, 470]. This is important to eliminate strong statistical fluctuations. (4) We assume that there are not any quantum [471–473] or nonlinear effects [474–476], for both the radiation sources and the DDM. (5) Here we only work in three dimensions (3D) for generality and brevity, and the same problem in one dimensions (1D) and two dimensions (2D) can be largely simplified after symmetry considerations. See [299, 436, 477–481] for more details.

Analytic wave theory stemmed from the work by Frisch [482], which presented a Feynman diagrammatic representation and the Bethe–Salpeter equation technique, borrowed from quantum field theory (QFT), for the treatment of multiple scattering of waves. This method was then used and further developed by Ishimaru [435, 483], Barabanenkov [484, 485], Tsang and Kong [298, 486, 487], Lagendijk [296], Nieuwenhuizen [297], Sheng [299], Mishchenko et al. [301], and Wang and Zhao [313, 488], to name a few. In this section, we attempt to give a brief introduction to this theory and apply it by deriving several analytical models of the DSE in Section 7.4.6.

7.4.1 Dyson Equation

Let us start from the general case of an infinite nonmagnetic 3D medium, where the spatial distribution of permittivity $\varepsilon(\mathbf{r})$ is inhomogeneous and can be generally described as $\varepsilon(\mathbf{r})=1+\delta\varepsilon(\mathbf{r})$, where $\delta\varepsilon(\mathbf{r})$ is the fluctuational part of the permittivity due to the random morphology of the inhomogeneous medium. EM wave propagation in such media is described by the vectorial Helmholtz equation [296, 298, 305]:

$$\nabla\times\nabla\times\mathbf{E}(\mathbf{r})-k^2\varepsilon(\mathbf{r})\mathbf{E}(\mathbf{r})=0. \tag{7.4.1}$$

Let $k^2=\omega^2/c_0^2$, in which k is the wavenumber in the background medium, and $V(\mathbf{r})=k^2\delta\varepsilon(\mathbf{r})=\omega^2\delta\varepsilon(\mathbf{r})/c_0^2$ be the disordered "potential" inducing EM scattering, where c_0 is the speed of light in the background medium. Then we have an alternative form of the vectorial Helmholtz equation convenient for EM scattering problems in random media:

$$\nabla\times\nabla\times\mathbf{E}(\mathbf{r})-k^2\mathbf{E}(\mathbf{r})=V(\mathbf{r})\mathbf{E}(\mathbf{r}). \tag{7.4.2}$$

To solve the equation, we can introduce the dyadic Green's function (DGF) for this random medium that satisfies

$$\nabla\times\nabla\times\mathbf{G}(\mathbf{r},\mathbf{r}')-k^2\mathbf{G}(\mathbf{r},\mathbf{r}')=V(\mathbf{r})\mathbf{G}(\mathbf{r},\mathbf{r}')+\mathbf{I}\delta(\mathbf{r},\mathbf{r}'). \tag{7.4.3}$$

In the meanwhile, Green's function in the homogeneous background medium is[4]

$$\nabla\times\nabla\times\mathbf{G}_0(\mathbf{r},\mathbf{r}')-k^2\mathbf{G}_0(\mathbf{r},\mathbf{r}')=\mathbf{I}\delta(\mathbf{r},\mathbf{r}'), \tag{7.4.4}$$

where $\mathbf{I}$ is the identity matrix. Taking the Fourier transform with respect to $\mathbf{r}$ and $\mathbf{r}'$ to the reciprocal space in terms of the momentum vectors $\mathbf{p}$ and $\mathbf{p}'$ and letting $V(\mathbf{r},\mathbf{r}')=k^2\delta\epsilon(\mathbf{r})\delta(\mathbf{r}-\mathbf{r}')$ using the Dirac delta function, we can write down the solution for DGF in the disordered media as

$$\mathbf{G}(\mathbf{p},\mathbf{p}')=\mathbf{G}_0(\mathbf{p},\mathbf{p}')+\mathbf{G}_0(\mathbf{p},\mathbf{p}_2)\mathbf{V}(\mathbf{p}_2,\mathbf{p}_1)\mathbf{G}(\mathbf{p}_1,\mathbf{p}'), \tag{7.4.5}$$

where the dummy variables $\mathbf{p}_1$ and $\mathbf{p}_2$ will be integrated out and we do not write this integral explicitly here as well as below. This equation is known as the Lippmann–Schwinger equation [297, 299, 301]. By introducing the T-operator $\mathbf{T}$, Eq. (7.4.5) is transformed into the form

$$\mathbf{G}(\mathbf{p},\mathbf{p}')=\mathbf{G}_0(\mathbf{p},\mathbf{p}')+\mathbf{G}_0(\mathbf{p},\mathbf{p}_2)\mathbf{T}(\mathbf{p}_2,\mathbf{p}_1)\mathbf{G}_0(\mathbf{p}_1,\mathbf{p}'). \tag{7.4.6}$$

Based on Eqs. (7.4.5) and (7.4.6), it can be easily shown that the T-operator is given by

$$\mathbf{T}(\mathbf{p},\mathbf{p}')=\mathbf{V}(\mathbf{p},\mathbf{p}')+\mathbf{V}(\mathbf{p},\mathbf{p}_2)\mathbf{G}_0(\mathbf{p}_2,\mathbf{p}_1)\mathbf{T}(\mathbf{p}_1,\mathbf{p}'). \tag{7.4.7}$$

[4] In vacuum, it is the free-space Green's function, whose expression in the real domain is given in Eq. (7.4.34).

If the medium only contains one discrete scatterer, $\mathbf{T}(\mathbf{p},\mathbf{p}')$ is then known as the T-operator for the single scatterer. Obviously, for a random medium composed of many scatterers, Eq. (7.4.5) still applies. However, if each scatterer can be described by its own T-operator, it is more convenient to transform Eq. (7.4.5) into the form only involving the T-operators of the individual particles, rather than the "scattering potential" $\mathbf{V}$. This is most suitable for a random medium consisting of well-defined, discrete scatterers. Since the T-operator of the jth scatterer is analogously given by [298, 299]

$$\mathbf{T}_j(\mathbf{p},\mathbf{p}') = \mathbf{V}_j(\mathbf{p},\mathbf{p}') + \mathbf{V}_j(\mathbf{p},\mathbf{p}_2)\mathbf{G}_0(\mathbf{p}_2,\mathbf{p}_1)\mathbf{T}(\mathbf{p}_1,\mathbf{p}'), \tag{7.4.8}$$

where $\mathbf{V}_j(\mathbf{p},\mathbf{p}')$ is the scattering potential of the jth scatterer, which constitutes the scattering potential of the system simply as $\mathbf{V}(\mathbf{p},\mathbf{p}') = \sum_{j=1}^{N} \mathbf{V}_j(\mathbf{p},\mathbf{p}')$ [298]. After some manipulations, the T-operator of the full system is then given by

$$\begin{aligned}\mathbf{T}(\mathbf{p},\mathbf{p}') = &\sum_{i=1}^{N} \mathbf{T}_j(\mathbf{p},\mathbf{p}') + \sum_{i=1}^{N}\sum_{j\neq i}^{N} \mathbf{T}_i(\mathbf{p},\mathbf{p}_1)\mathbf{G}_0(\mathbf{p}_1,\mathbf{p}_2)\mathbf{T}_j(\mathbf{p}_2,\mathbf{p}') \\ &+ \sum_{i=1}^{N}\sum_{j\neq i}^{N}\sum_{l\neq j}^{N} \mathbf{T}_i(\mathbf{p},\mathbf{p}_1)\mathbf{G}_0(\mathbf{p}_1,\mathbf{p}_2)\mathbf{T}_j(\mathbf{p}_2,\mathbf{p}_3)\mathbf{G}_0(\mathbf{p}_3,\mathbf{p}_4)\mathbf{T}_j(\mathbf{p}_4,\mathbf{p}')\cdots,\end{aligned} \tag{7.4.9}$$

where $\mathbf{p}_3$ and $\mathbf{p}_4$ are also dummy variables to be integrated out. This formula can be understood as the sum of all multiple wave scattering paths at different orders, in which the T-operators of scatterers visited by these paths are connected by Green's functions or the propagators [298].

In this fashion, the Lippmann–Schwinger equation for a medium consisting of N discrete scatterers is rewritten as

$$\mathbf{G}(\mathbf{p},\mathbf{p}') = \mathbf{G}_0(\mathbf{p},\mathbf{p}') + \mathbf{G}_0(\mathbf{p},\mathbf{p}_2)\sum_{j=1}^{N} \mathbf{T}_j(\mathbf{p}_2,\mathbf{p}_1)\mathbf{G}_j(\mathbf{p}_1,\mathbf{p}'), \tag{7.4.10}$$

where Green's function with respect to each scatterer $\mathbf{G}_j(\mathbf{p},\mathbf{p}')$ is given by

$$\mathbf{G}_j(\mathbf{p},\mathbf{p}') = \mathbf{G}_0(\mathbf{p},\mathbf{p}') + \mathbf{G}_0(\mathbf{p},\mathbf{p}_2)\sum_{i=1,i\neq j}^{N} \mathbf{T}_i(\mathbf{p}_2,\mathbf{p}_1)\mathbf{G}_i(\mathbf{p}_1,\mathbf{p}'). \tag{7.4.11}$$

This equation is also known as FLEs for multiple scattering of classical waves [301, 322, 323], which will be discussed in Section 7.4.3.

Then, to obtain a statistically meaningful description of the random medium, it is necessary to take the ensemble average of the full system to eliminate the impact of a specific configuration. Taking ensemble average of Eq. (7.4.10), we obtain

$$\langle \mathbf{G}(\mathbf{p},\mathbf{p}')\rangle = \mathbf{G}_0(\mathbf{p},\mathbf{p}') + \mathbf{G}_0(\mathbf{p},\mathbf{p}')\langle \mathbf{T}(\mathbf{p},\mathbf{p}')\rangle \mathbf{G}_0(\mathbf{p},\mathbf{p}'), \tag{7.4.12}$$

where $\langle \mathbf{G}(\mathbf{p},\mathbf{p}')\rangle$ denotes the ensemble averaged amplitude of Green's function, and $\langle \mathbf{T}(\mathbf{p},\mathbf{p}')\rangle$ is the ensemble-averaged T-operator of the full system by invoking Eq. (7.4.9),

$$\langle \mathbf{T}(\mathbf{p},\mathbf{p}')\rangle = \left\langle \sum_{i=1}^{N} \mathbf{T}_j(\mathbf{p},\mathbf{p}') \right\rangle + \left\langle \sum_{i=1}^{N} \sum_{j\neq i}^{N} \mathbf{T}_j(\mathbf{p},\mathbf{p}_1)\mathbf{G}_0(\mathbf{p}_1,\mathbf{p}_2)\mathbf{T}_j(\mathbf{p}_2,\mathbf{p}') \right\rangle + \cdots. \tag{7.4.13}$$

After some manipulations of identifying and retaining only irreducible terms in the ensemble-averaged T-operator, we obtain the well-known Dyson equation for the coherent or mean component of the (electric) field as [296–298]

$$\langle \mathbf{G}(\mathbf{p},\mathbf{p}')\rangle = \mathbf{G}_0(\mathbf{p},\mathbf{p}') + \mathbf{G}_0(\mathbf{p},\mathbf{p}')\mathbf{\Sigma}(\mathbf{p},\mathbf{p}')\langle \mathbf{G}(\mathbf{p},\mathbf{p}')\rangle, \tag{7.4.14}$$

where $\mathbf{\Sigma}(\mathbf{p},\mathbf{p}')$ is the so-called self-energy (or mass operator) containing all irreducible multiple scattering expansion terms in the T-operator $\langle \mathbf{T}(\mathbf{p},\mathbf{p}')\rangle$. If we express the multiple wave scattering processes involving many particles into Feynman diagrams according to Eq. (7.4.13), where particles (represented by $\mathbf{T}_j$) are connected by the propagator ($\mathbf{G}_0$) and their relationships (including positional correlations between different particles and turning back to the same particle), then the irreducible terms stand for those multiple wave scattering diagrams that cannot be divided without breaking the innate particle connections, including the same particle or particle correlations. For more details on irreducible and reducible diagrams, see [297, 298, 489].

For a statistically homogeneous medium having translational symmetry after ensemble average, $\mathbf{\Sigma}(\mathbf{p},\mathbf{p}') = \mathbf{\Sigma}(\mathbf{p})\delta(\mathbf{p}-\mathbf{p}')$ and $\langle \mathbf{G}(\mathbf{p},\mathbf{p}')\rangle = \langle \mathbf{G}(\mathbf{p})\rangle \delta(\mathbf{p}-\mathbf{p}')$ [299]. In the momentum representation, the free-space DGF is $\mathbf{G}_0(\mathbf{p}) = -1/(k^2\mathbf{I} - p^2(\mathbf{I}-\hat{\mathbf{p}}\hat{\mathbf{p}}))$, and thus the averaged amplitude Green's function is

$$\langle \mathbf{G}(\mathbf{p})\rangle = -\frac{1}{k^2\mathbf{I} - p^2(\mathbf{I}-\hat{\mathbf{p}}\hat{\mathbf{p}}) - \mathbf{\Sigma}(\mathbf{p})}, \tag{7.4.15}$$

where $\hat{\mathbf{p}} = \mathbf{p}/p$ is the unit vector in the momentum space. Through this equation, self-energy $\mathbf{\Sigma}(\mathbf{p})$ provides a renormalization for the EM wave propagation in random media and determines the effective (renormalized) permittivity as [296]

$$\varepsilon_{\text{eff}}(\mathbf{p}) = \mathbf{I} - \frac{\mathbf{\Sigma}(\mathbf{p})}{k^2}. \tag{7.4.16}$$

For a statistically isotropic random medium, the obtained momentum-dependent effective permittivity tensor is decomposed into a transverse part and a longitudinal part as $\boldsymbol{\varepsilon}(\mathbf{p}) = \varepsilon^{\perp}(\mathbf{p})(\mathbf{I}-\hat{\mathbf{p}}\hat{\mathbf{p}}) + \varepsilon^{\parallel}(\mathbf{p})\hat{\mathbf{p}}\hat{\mathbf{p}}$, where $\varepsilon^{\perp}(\mathbf{p}) = 1 - \Sigma^{\perp}(\mathbf{p})/k^2$ and $\varepsilon^{\parallel}(\mathbf{p}) = 1 - \Sigma^{\parallel}(\mathbf{p})/k^2$ determine the effective permittivities of transverse and longitudinal modes in momentum space [296]. Therefore, by determining the poles of the amplitude Green's function, we can obtain the dispersion relation that corresponds to collective excitations of the disordered medium (i.e., like "the band structure" in periodic systems). An equivalent and frequently used method to find the collective excitations is to resort to the spectral function [296, 299, 490]:

$$\begin{aligned}\mathbf{S}(\omega,\mathbf{p}) = -\operatorname{Im}\langle\mathbf{G}(\omega,\mathbf{p})\rangle &= \frac{\operatorname{Im}\varepsilon^{\parallel}(\omega,\mathbf{p})/k^2}{[\operatorname{Re}\varepsilon^{\parallel}(\omega,\mathbf{p})]^2+[\operatorname{Im}\varepsilon^{\parallel}(\omega,\mathbf{p})]^2}\hat{\mathbf{p}}\hat{\mathbf{p}} \\ &+ \frac{\operatorname{Im}\varepsilon^{\perp}(\omega,\mathbf{p})/k^2}{[\operatorname{Re}\varepsilon^{\perp}(\omega,\mathbf{p})-p^2/k^2]^2+[\operatorname{Im}\varepsilon^{\perp}(\omega,\mathbf{p})]^2}(\mathbf{I}-\hat{\mathbf{p}}\hat{\mathbf{p}}).\end{aligned} \tag{7.4.17}$$

Since longitudinal modes are usually not propagating, here we mainly consider the transverse modes. By finding the (local) maxima of the transverse components of the spectral function in real momentum space, the transverse (propagating) mode's wavenumber can be calculated as $K^{\perp}=\sqrt{\varepsilon^{\perp}(\omega,p_{\max})}k$, where $p_{\max}$ is the momentum value that makes the spectral function maximal. Then the effective refractive index of the disordered medium is directly obtained from the real part of the effective wavenumber as

$$n_{\text{eff}} = \frac{\operatorname{Re}K^{\perp}}{k}, \tag{7.4.18}$$

and the extinction coefficient of this disordered medium is given by the imaginary part of the effective wavenumber:

$$\kappa_{\text{e}} = 2\operatorname{Im}K^{\perp}. \tag{7.4.19}$$

A scrutiny of the transverse component of the spectral function tells us that there is a difference between the real part of $K^{\perp}$ and $p_{\max}$. More precisely, it reaches maximum when $\operatorname{Re}\varepsilon^{\perp}(\omega,\mathbf{p})-p^2/k^2$ approaches zero, that is $(\operatorname{Re}K^{\perp})^2-(\operatorname{Im}K^{\perp})^2=p_{\max}^2$. For nonabsorbing, purely scattering media, the scattering coefficient κ_s is equal to the extinction coefficient κ_e, and this momentum mismatch profoundly stands for a propagating momentum shell broadening due to scattering scaling as κ_s. This broadening can be rather large for strongly scattering media [296], while for weakly scattering media, we can apply the on-shell approximation, namely, $\operatorname{Re}K^{\perp}\approx p_{\max}$, which is employed in a recent work by the present author [313].

Note that after the above treatments, we have implicitly assumed that the propagation of waves in disordered media is dominated by only one (transverse) mode, and thus a distinct wavenumber (propagation constant) can be well defined, so do the related radiative properties. However, it is not proved rigorously. Generally, the wave propagation behavior in disordered media is determined by the contributions of several modes with different wavenumbers and the spatial dispersion (nonlocality) is sometimes important, especially when it comes to the reflection and transmission at the boundary of a disordered medium with strong fluctuations at the scale of wavelength [491, 492]. Nevertheless, for conventional disordered media, we can safely adopt this single-mode treatment.

According to the above analysis of the Dyson equation, the first task to determine the radiative properties of DDM is to derive the self-energy. As a first-order perturbative approximation, ISA gives a self-energy that is simply the ensemble average of the sum of the T-operators of all scatterers per unit volume. If we only consider an ensemble of point dipole scatterers as the ideal case, the self-energy is given by [297, 493]

$$\Sigma_{\mathrm{ISA}} = n_0 t_0, \tag{7.4.20}$$

where $t_0 = -k^2\alpha$ is the T-operator of a single ED scatterer in the momentum representation, α is the dipole polarizability, and n_0 is the number density of particles. Note for the infinitely small scatterers, the self-energy is momentum independent.

7.4.2 The Bethe–Salpeter Equation

It is noted that the Dyson equation and the self-energy only provide a characterization for coherent EM field propagation in random media, that is, the first moment of the EM field $\langle \mathbf{G}(\mathbf{p}) \rangle$, while a more relevant quantity to our concern is the radiation intensity in disordered media that directly determines the phase function of each scattering process in terms of energy transport. This is exactly governed by the Bethe–Salpeter equation [296–298], which describes the second moment of the EM field $\langle \mathbf{G}\mathbf{G}^* \rangle$ in random media. In the operator notation, the Bethe–Salpeter equation is written as

$$\langle \mathbf{G}\mathbf{G}^* \rangle = \langle \mathbf{G} \rangle \langle \mathbf{G}^* \rangle + \langle \mathbf{G} \rangle \langle \mathbf{G}^* \rangle \mathbf{\Gamma} \langle \mathbf{G}\mathbf{G}^* \rangle, \tag{7.4.21}$$

where $\mathbf{\Gamma}$ is the irreducible vertex representing the renormalized scattering center for the incoherent part of radiation intensity due to random fluctuations of the disordered media. It can be understood as the differential scattering coefficient as well as (nonnormalized) the scattering phase function relevant in the RTE if the momentum shell broadening of the transport processes can be neglected, which means that the radiation intensity is concentrated on the momentum shell $p = K$. This is the case for random media containing dilute scatterers or weak scatterers. Moreover, since in most circumstances, one only needs to consider transverse EM waves, and the transverse component of the irreducible vertex can be written in the momentum representation as [296, 299, 494]

$$\Gamma^{\perp}(K\hat{\mathbf{p}}, K\hat{\mathbf{p}}') = (\mathbf{I} - \hat{\mathbf{p}}\hat{\mathbf{p}})\mathbf{\Gamma}(K\hat{\mathbf{p}}, K\hat{\mathbf{p}}')(\mathbf{I} - \hat{\mathbf{p}}'\hat{\mathbf{p}}') \tag{7.4.22}$$

under the *on-shell* approximation. For isotropic media, the scattering properties (the scattering coefficient and the phase function) do not rely on the incident direction but only on the solid angle Ω_s between the incident and scattering directions, which can be described by the polar scattering angle $\theta_s = \arccos(\hat{\mathbf{p}} \cdot \hat{\mathbf{p}}')$ and the azimuth angle φ_s. The choice of the latter depends on the definition of the local frame of spherical coordinates with respect to the incident direction $\hat{\mathbf{p}}'$. Therefore, the differential scattering coefficient with respect to the incident direction $\hat{\mathbf{p}}'$ can be obtained as [495]

$$\frac{\mathrm{d}\kappa_s}{\mathrm{d}\Omega_s} = \frac{1}{(2\pi)^2}\Gamma^{\perp}(K\hat{\mathbf{p}}, K\hat{\mathbf{p}}'). \tag{7.4.23}$$

From this equation, the scattering coefficient can be obtained by integrating over Ω_s (or equivalently, θ_s and ϕ_s).

The irreducible vertex and the self-energy are not independent of each other. For a nonabsorbing medium, the scattering coefficient should be equal to the extinction coefficient as required by the energy conservation, which leads to the following equation:

$$\frac{1}{(2\pi)^2}\int_0^{2\pi}\Gamma^{\perp}(K\hat{\mathbf{p}}, K\hat{\mathbf{p}}')\mathrm{d}\hat{\mathbf{p}} = -\frac{\mathrm{Im}\Sigma}{n_{\mathrm{eff}}k}. \tag{7.4.24}$$

This equation is also known as the Ward–Takahashi identity [299, 494], originally established in QFT [496]. Note there is a general form for this identity that does not require the on-shell approximation, which will not be shown here for simplicity [299].

For an ensemble of point dipole scatterers, the irreducible vertex has a simple expression under the ISA:

$$\mathbf{\Gamma}(K\hat{\mathbf{p}}, K\hat{\mathbf{p}}') = n_0|t_0|^2\mathbf{I}\otimes\mathbf{I}, \tag{7.4.25}$$

where $\otimes$ means the tensor product. By taking the transverse component of the irreducible vertex and integrating over the azimuth angle φ_s with the incident direction $\hat{\mathbf{p}}'$ fixed, the differential scattering cross coefficient with respect to the polar angle is

$$\frac{\mathrm{d}\kappa_s}{\mathrm{d}\theta_s} = \frac{1+\cos^2\theta_s}{4\pi}n_0|t_0|^2. \tag{7.4.26}$$

By combining Eqs. (7.4.26) and (7.4.20), one can directly examine that for non-absorbing scatterers, the Ward–Takahashi identity Eq. (7.4.24) is apparently fulfilled, which, in this case, is equivalent to the optical theorem in the scattering theory [285, 296, 435, 493].

Last but not least, we note that the Dyson and Bethe–Salpeter equations, and the self-energy and irreducible intensity vertex, are directly derived from the vectorial Helmholtz equation. This means that it is a first-principle method for EM waves and can treat, generally, any disordered medium. Therefore, for continuous heterogeneous media where no discrete scatterers exist, the self-energy and irreducible intensity vertex can still be calculated, for instance, the well-known bilocal approximation for the self-energy of Gaussian random media [298]. By contrast, this kind of random media cannot be conveniently treated by FLEs. And it should be borne in mind that the irreducible vertex is a much more complex quantity than the differential scattering coefficient, because the former describes all interference phenomena in the multiple wave scattering process and cannot always be simply interpreted as the differential scattering coefficient especially when the on-shell approximation does not apply.

7.4.3 The Foldy–Lax Equations

As mentioned in Section 7.4.2, FLEs are also general equations describing multiple scattering of both classical and quantum waves in disordered media containing discrete scatterers. This set of equations obtained its name from the early workers Foldy [322] and Lax [323]. Generally speaking, FLEs are the derivation of the analytic wave theory [298]. However, due to the explicit formalism and easy numerical implementation, FLEs have become a more widely used method than the analytic wave theory, especially in engineering. In this section, we briefly introduce the basics of FLEs along with the multipole expansion method and CDM, which are important for practical use. The readers can refer to the extensive review regarding FLEs for the treatment of DDM with formal definitions and general theoretical derivations by Mishchenko et al. [468].

The well-known FLEs depicting the multiple scattering processes of EM waves among N discrete scatterers read [298, 323, 447, 497]

$$\mathbf{E}_{\mathrm{exc}}^{(j)}(\mathbf{r}) = \mathbf{E}_{\mathrm{inc}}(\mathbf{r}) + \sum_{\substack{i=1 \\ i\neq j}}^{N} \mathbf{E}_{\mathrm{s}}^{(i)}(\mathbf{r}), \tag{7.4.27}$$

where $\mathbf{E}_{\mathrm{inc}}(\mathbf{r})$ is the incident electric field, $\mathbf{E}_{\mathrm{exc}}^{(j)}(\mathbf{r})$ is the electric component of the so-called exciting field impinging on the vicinity of particle j, and $\mathbf{E}_{\mathrm{s}}^{(i)}(\mathbf{r})$ is the electric component of partial scattered waves from particle i. We also show FLEs schematically in Fig. 7.9, where the scatterers are assumed to be spheres. Obviously, the scatterers can have arbitrary geometries. The physical significance of FLEs is straightforward. This series of equations describe that the exciting field impinging on the vicinity of particle j is the sum of the incident field $\mathbf{E}_{\mathrm{inc}}(\mathbf{r})$ and all partial scattered field from all other particles.

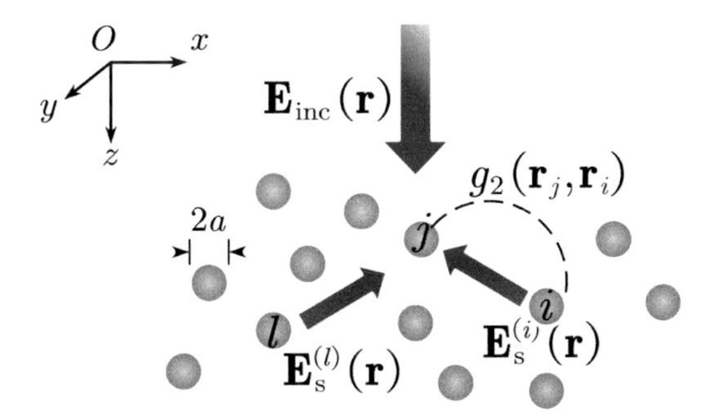

Figure 7.9 A schematic of FLEs for multiple scattering of EM waves in randomly distributed spherical particles in three dimensions. The particles are numbered as i, j, l, etc. The dashed line denotes $g_2(\mathbf{r}_j, \mathbf{r}_i)$, the pair distribution function between the two particles. The blue thick arrow indicates the propagation direction of the incident wave, while the red thin arrows stand for the propagation directions of the partial scattered waves from particles i to j and from l to j.

The Multipole (VSWF) Expansion of FLEs

To solve FLEs for a disordered medium consisting of spheres as shown in Fig. 7.9, it is convenient to expand the electric fields in VSWFs to utilize the spherical boundary condition of individual particles, following the way usually done for a single spherical particle in Mie theory. The expansion coefficients then naturally correspond to multipoles supported by the particles [298, 447, 498–500]. As a matter of fact, in the spirit of the EBCM, this expansion is still applicable for nonspherical particles if their T-matrix elements are given. Here we only consider spherical particles as a simple introduction. Using this technique, the exciting field $\mathbf{E}_{\rm exc}^{(j)}(\mathbf{r})$ is expressed as

$$\mathbf{E}_{\rm exc}^{(j)}(\mathbf{r})=\sum_{mnp} c_{mnp}^{(j)}\mathbf{N}_{mnp}^{(1)}(\mathbf{r}-\mathbf{r}_j). \tag{7.4.28}$$

Based on the expansion coefficients of the exciting field, the scattered field from particle i propagating to the arbitrary position $\mathbf{r}$ can be obtained through its T-matrix elements T_{mnp} as [298, 447]

$$\mathbf{E}_{\rm s}^{(i)}(\mathbf{r})=\sum_{mnp} c_{mnp}^{(i)}T_{mnp}\mathbf{N}_{mnp}^{(3)}(\mathbf{r}-\mathbf{r}_i). \tag{7.4.29}$$

Inserting Eqs. (7.4.28) and (7.4.29) into Eq. (7.4.27), we obtain

$$\sum_{mnp} c_{mnp}^{(j)}\mathbf{N}_{mnp}^{(1)}(\mathbf{r}-\mathbf{r}_j)=\mathbf{E}_{\rm inc}(\mathbf{r})+\sum_{\substack{i=1\\ i\neq j}}^{N}\sum_{mnp} c_{mnp}^{(i)}T_{mnp}\mathbf{N}_{mnp}^{(3)}(\mathbf{r}-\mathbf{r}_i). \tag{7.4.30}$$

To solve this equation, we need to translate the VSWFs centered at $\mathbf{r}_i$ to their counterparts centered at $\mathbf{r}_j$. Using the translation addition theorem for VSWFs (see [313]), Eq. (7.4.30) becomes

$$\begin{aligned}\sum_{mnp} c_{mnp}^{(j)}\mathbf{N}_{mnp}^{(1)}(\mathbf{r}-\mathbf{r}_j)=\mathbf{E}_{\rm inc}(\mathbf{r})+\sum_{\substack{i=1\\ i\neq j}}^{N}\sum_{mnp\mu\nu q}\\ c_{\mu\nu q}^{(i)}T_{\mu\nu q}A_{mnp\mu\nu q}^{(3)}(\mathbf{r}_j-\mathbf{r}_i)\mathbf{N}_{mnp}^{(1)}(\mathbf{r}-\mathbf{r}_j),\end{aligned} \tag{7.4.31}$$

where $A_{mnp\mu\nu q}^{(3)}(\mathbf{r}_j-\mathbf{r}_i)$ can translate the outgoing VSWFs centered at $\mathbf{r}_i$ to regular VSWFs centered at $\mathbf{r}_j$. We further expand the incident waves into regular VSWFs centered at $\mathbf{r}_j$ with the expansion coefficients $a_{mnp}^{(j)}$, use the orthogonal relation of VSWFs with different orders and degrees and obtain the following equation:

$$c_{mnp}^{(j)} = a_{mnp}^{(j)} + \sum_{\substack{i=1 \\ i \neq j}}^{N} \sum_{\mu\nu q} c_{\mu\nu q}^{(i)} T_{\mu\nu q} A_{mnp\mu\nu q}^{(3)}(\mathbf{r}_j - \mathbf{r}_i). \tag{7.4.32}$$

The matrix form of this equation can be directly exploited in the numerical calculation, provided the particles' positions are known.

Actually, Eq. (7.4.32) is the governing equation of the well-established Fortran code multiple-sphere T-matrix method (MSTM) developed by Mackowski and Mishchenko [447, 448] to exactly solve the multiple scattering of EM waves for a group of spherical particles. This code is further extended to optically active media [448, 501]. MSTM is widely used by many authors in the fields of thermal radiation transfer [312], astrophysics [502], and nanophotonics [304], together with a similar Fortran code developed by Xu, the generalized multiparticle Mie-solution (GMM) [500]. Other similar algorithms using the multipole expansion include Stout et al.'s recursive transfer matrix method [503] and Chew et al.'s recursive T-matrix method [504]. Notably, exploiting the Compute Unified Device Architecture (NVIDIA Corporation, Santa Clara, CA) acceleration feature of GPUs for parallel computing, Egel et al. [505] recently built a Matlab toolbox using the same governing equation, which can be at least twice as fast as the MSTM code, according to their testing cases.

For nonspherical particles, the generalization of this multipole expansion method of FLEs is then the multiple particles T-matrix method [506], which uses the full-form T-matrix of nonspherical particles. However, it should be noted that a severe limitation of the multiple-particle T-matrix method is its incapability to deal with elongated particles placed in the near field of each other or of an interface [507]. This is because the VSWF decomposition of the EM field is formally valid only in a uniform background beyond the smallest sphere that circumscribes the entire particle. Recent efforts to solve this limitation include [507–512], to name a few. As a variant, the fast multi-particle scattering algorithm [513–515] has been developed in recent years to accelerate the computation speed of the multiparticle scattering problem by using the combination of the integral equation technique to discretize each well-separated nonspherical particle [516] (or closely placed particle clusters in order to avoid the overlapping of the enclosing sphere due to the same reason above), the Debye scalar potential representation,[5] and the FMM [518–521]. This algorithm permits us to compute the EM field for ensembles containing several thousands of particles on a single CPU [513].

[5] Instead of using the VSWF representation in Eqs. (7.4.28)–(7.4.32), this scalar potential representation can reduce the complexity by avoiding the heavy use of vector algebra and vector translation functions [517].

The Coupled-Dipole Model

Since Eq. (7.4.32) is a general equation that includes all multipolar excitations in the spheres, a much simpler method that only considers the electric dipolar excitations is also frequently used. This is the well-known CDM. The governing equation of this method and the DDA method is the same, despite the fact that they are used for different purposes. The CDM, although very simplified, still preserves the essence of multiple scattering physics and, therefore, usually acts as a prototype model for studying many complicated mechanisms, for instance, the recurrent scattering mechanism [494]. Moreover, it is also very suitable for a group of small particles in which only dipolar modes are excited, not necessarily spheres. Along with the rapid development of nanofabrication, plasmonics, and nanophotonics, the CDM is now already verified experimentally for a variety of nanostructures, including 1D chains of metallic (typically gold and silver that SPPs) nanoparticles [522], ordered [523, 524] and disordered [525] 2D arrays of metallic nanoparticles, and 3D nanoparticle aggregates [526]. The nanoparticles can be nanorods [525], nanospheres [522], and many other geometries [527], only if they show an ED EM response. Moreover, for ultracold two-level atoms, in the limit of low excitation intensity of light, where atoms are not saturated, and assuming no Zeeman degeneracy of hyperfine sublevels or other internal quantum effects, for example, Sr atoms with zero electronic angular momentum [528], the response of a single two-level atom over light can also be treated as a linear point dipole. In this circumstance, the CDM is also suitable to describe the light–matter interaction in ultracold atomic clouds [529].

Since we are working in a single frequency/wavelength (the frequency domain), we will abbreviate the frequency dependence of all quantities appearing below. In vacuum or any homogeneous, isotropic host medium, the CDM has the following form [305]:

$$\mathbf{d}_j = \alpha \left[\mathbf{E}_{\text{inc}}(\mathbf{r}_j) + k^2 \sum_{i=1, i\neq j}^{N} \mathbf{G}_0(\mathbf{r}_j, \mathbf{r}_i)\mathbf{d}_i \right], \tag{7.4.33}$$

where α is the polarizability of the particle. $\mathbf{E}_{\text{inc}}(\mathbf{r}_j)$ is the incident field impinging on the jth particle. For instance, for a plane-wave illumination along the z axis, we have $\mathbf{E}_{\text{inc}}(\mathbf{r}_j) = \mathbf{E}_0 \exp(\mathrm{i}\mathbf{k}\cdot\mathbf{r}_j)$, with $\mathbf{k} = k\hat{\mathbf{z}}$. $\mathbf{d}_i$ being the excited dipole moment of the ith particle. $\mathbf{G}_0(\omega, \mathbf{r}_j, \mathbf{r}_i)$ is the free-space DGF and describes the propagation of the scattered field of the jth dipole to the ith dipole as [494]

$$\mathbf{G}_0(\mathbf{r}_j, \mathbf{r}_i) = \frac{\exp(\mathrm{i}kr)}{4\pi r}\left(\frac{\mathrm{i}}{kr} - \frac{1}{k^2r^2} + 1\right)\mathbf{I} + \frac{\exp(\mathrm{i}kr)}{4\pi r}\left(-\frac{3\mathrm{i}}{kr} + \frac{3}{k^2r^2} - 1\right)\hat{\mathbf{r}}\hat{\mathbf{r}} - \frac{\delta(\mathbf{r})}{3k^2}, \tag{7.4.34}$$

where the Dirac delta function $\delta(\mathbf{r})$ is responsible for the so-called local field in the scatterers [494], $\mathbf{I}$ is the identity matrix, and $\hat{\mathbf{r}}$ is the unit vector of $\mathbf{r} = \mathbf{r}_j - \mathbf{r}_i$. After the EM responses of all scatterers (namely, all dipole moments $\mathbf{d}_i$ with respect to a specific incident field) based on the above multiple wave scattering equations are calculated, the total scattered field of the random cluster of particles at an arbitrary position $\mathbf{r} \neq \mathbf{r}_j$, where $\mathbf{r}_j$ denotes the position of scatterers, is computed as

$$\mathbf{E}_s(\mathbf{r}) = k^2 \sum_{i=1}^{N} \mathbf{G}_0(\mathbf{r}, \mathbf{r}_j)\mathbf{d}_j, \tag{7.4.35}$$

where Green's function $\mathbf{G}_0(\mathbf{r}, \mathbf{r}_j)$ then describes the propagation of the scattered field of the jth dipole to a given position $\mathbf{r}$, similarly. And the total extinction cross section of the group of scatterers can also be calculated:

$$C_\mathrm{e} = k \sum_{j=1}^{N} \mathrm{Im}(\mathbf{d}_j \cdot \mathbf{E}^*_{\mathrm{exc},j}), \tag{7.4.36}$$

where $\mathbf{E}_{\mathrm{exc},j}$ is the exciting field imping on the jth particle and given by $\mathbf{E}_{\mathrm{exc},j} = \mathbf{d}_j/\alpha$. The total absorption cross section is calculated as

$$C_\mathrm{a} = k \sum_{j=1}^{N} \mathrm{Im}(\mathbf{d}_j \cdot \mathbf{E}^*_{\mathrm{exc},j} - \frac{k^3}{6\pi}|\mathbf{d}_j|^2). \tag{7.4.37}$$

Therefore, the total scattering cross section of the system of dipoles is directly given by $C_\mathrm{s} = C_\mathrm{e} - C_\mathrm{a}$.

7.4.4 The Connection between RTE and Electromagnetic Theory

The above equations are all related to the propagation of EM waves in the framework of Maxwell's equations, while the radiative properties are defined in the framework of the RTE that treats radiation as energy bundles (like classical particles). In this section, we attempt to briefly discuss how to establish the relationship between Maxwell's equations and the equation of radiative transfer, as well as the hydrodynamic limit and the diffusion equation.

As mentioned in Section 2.2, the RTE is derived phenomenologically in its initial stage from energy conservation considerations [530]. In the 1970s–1980s, there were continuous efforts to establish the relationship between the RTE and Maxwell's equations, and it was finally shown that this connection can be made by means of the analytical wave theory [435, 487, 531, 532]. To put it simply, the RTE can be derived by retaining the so-called ladder diagrams in the Bethe–Salpeter equation in which the field correlation function $\langle \mathbf{E}(\mathbf{r})\mathbf{E}^*(\mathbf{r}')\rangle$ is expressed into the specific intensity. Specifically, the ladder diagrams are constructed by connecting the multiple wave scattering trajectories (according to Eq. (7.4.13)) to their complex-conjugated counterparts with exactly the same ordering of scatterers, and in each of these trajectories, all scatterers are assumed to be visited only once. This procedure results in diagrams looking like a series of "ladder," which then give them the name. As a result, this ladder approximation only describes the transport of radiation intensity, and at the intensity (or more formally, the irreducible vertex) level, it is actually equivalent to the ISA [297]. All kinds of *microscopic and mesoscopic* wave interference effects occurring outside the scatterers are neglected in the RTE [296–299, 301, 468, 533]. It is also noted that according to the derivation procedure, the RTE describes the transport at length and time scales much larger than the wavelength and the period of light, and assumes weak scattering, that is, scattering/transport mean free path is much larger than the wavelength [296–299, 435]. Details of the derivation of the RTE using the diagrammatic technique in the analytic wave theory can be found in many monographs and papers, for example [297, 298, 298, 534, 535] and thus are not shown here. For a historical review about the relationship between the analytic wave theory and the RTE, see the Van de Hulst essay paper by Tsang [489]. Remarkably, Doicu and Mishchenko presented a series of reviews to describe how to derive the RTE from Maxwell's equations using different methods and discuss relevant interference effects [457–462], including the far-field FLEs [457] and the analytic wave theory (or directly dubbed "Dyson and Bethe–Salpeter equations") [458].

To show the connection between the RTE and Maxwell's equations more intuitively, in Fig. 7.10, a schematic diagram of different regimes in the theoretical treatment of radiative transfer, proposed by van Tiggelen et al. [533], is presented. In this figure, different regimes of radiative transfer are distinguished by three parameters, that is, the wavelength of radiation λ, the degree of disorder characterized by the mean free path l divided by the wavelength, and the length scale of radiative transfer quantified by the sample size L divided by the wavelength. It is clearly indicated by this figure that the RTE is most suitable for the situations in which $l \gg \lambda$ (weak scattering or, equivalently, weak disorder) and $L \gg \lambda$. When the disorder (or equivalently, scattering strength) is increased

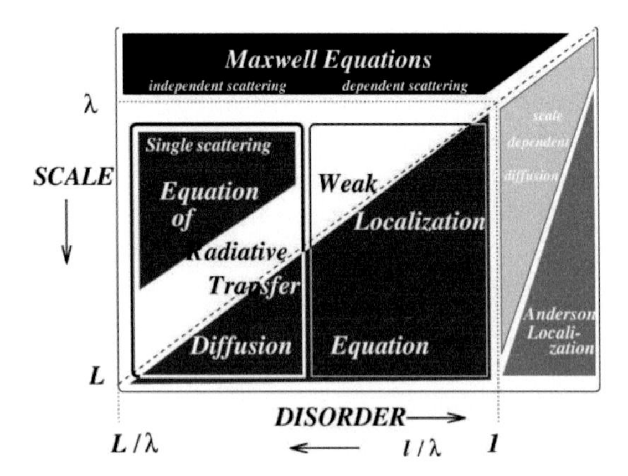

Figure 7.10 Schematic diagram of different regimes in the theoretical treatment of radiative transport. Reprinted figure with permission from [533]. Copyright 2000 by the American Physical Society.

with intermediate scattering strength (i.e., l/λ is sufficiently larger than unity but away from the weak scattering limit, i.e., l is at the scale of several λ's), mesoscopic interferences like the weak localization occur at the length scale of $L \gg \lambda$. Weak localization, also known as the coherent backscattering (CB), can be expressed as a series of most crossed diagrams [536–538] (or the so-called cyclical terms [301, 539]) in the diagrammatic representation of the irreducible intensity vertex in the Bethe–Salpeter equation, which cannot be considered in the RTE (see Section 7.4.7). Moreover, when the disorder continues to increase to result in $l \lesssim \lambda$, strong localization (or known as Anderson localization) can occur. And at small length scales and intermediate disorder (or scattering strength), the DSE is important, manifested as the wave interferences that take place at the length scale $L \sim \lambda$,

In a rigorous sense, the RTE cannot be applied in the dependent scattering regime since it is derived under the ISA, which is also implied by Fig. 7.10. However, it is quite useful in practice to retain the form of RTE by correcting the mesoscopic radiative properties by considering DSE. This is the key assumption of the present chapter and most reviewed references herein [298]. It is thus postulated that the DSE introduces some additional perturbative terms to the diagrammatic expansion of the RTE and results in a modification to the mesoscopic radiative properties. In this manner, the numerical techniques to solve the RTE in large-scale disordered media can be exploited, which is very valuable in practice. For instance, the dense media radiative transfer theory (DMRT), developed by Tsang et al. [298, 487, 489] in the spirit of this postulation, has been widely applied in the field of microwave remote sensing for the prediction of radiative transfer in dense media like terrestrial snow with densely packed ice grains, whose validity has been confirmed by extensive experimental measurements [540–543]. As a consequence, this postulation is also viable in many other practical applications like thermal engineering where moderate refractive index materials are used, usually away from EM resonances.[6] To the best of our knowledge and experiences, this is the only feasible approach to treat large-scale DDM when the mesoscopic radiative properties are significantly affected by the DSE.

[6] When there are EM resonances like Mie or some internal resonances in the densely packed scatterers, it is not quite clear whether this postulation can effectively work. By now there are very few works on resonant multiple scattering in dense DDM in the regime described by the RTE ($L \sim l$). Most works about resonant multiple wave scattering are either on dilute DDM, for example [296, 494, 544], or on the diffusive regime, for example [545, 546], to name a few. In particular, in highly scattering DDM in the diffusive transport regime, it is instructive to note that the diffusion coefficient should be significantly modified by introducing a position dependence to account for *mesoscopic* wave interferences like weak and strong localization mechanisms, and the theory agrees well with experiments and numerical simulations [547–552]. This may provide some implications on developing and justifying similar schemes in the RTE regime for microscopic interferences under resonant multiple wave scattering.

A well-known limit of the RTE is the celebrated diffusion equation of photons, which follows from the diffusive transport behavior of other classical particles in Brownian motion, like macroscopic heat transport by phonons [297, 299, 435]. This equation applies for highly scattering media with sufficiently low absorption, and the thickness of the sample should be much larger than the scattering mean free path, namely, $L \gg l_s$. As a result, the photons are scattered so many times before exiting the sample that they "forget" their initial transport direction, and therefore the long-time (long path-length) transport behavior can be described as isotropic, even if the scattering phase function itself is also anisotropic [297]. The resulting diffusive scale of photon migration mean free path is called the transport mean free path $l_{\mathrm{tr}} = l_s/(1-g)$. In this circumstance, we say that the radiative transfer enters the "diffusive regime," which is important and exhibits distinctions from the regime of the RTE, as will be shown in Section 7.4.5.

7.4.5 The Dependent Scattering Effect

General Considerations on the DSE

It is necessary to explore the underlying mechanisms in the DSE and develop corresponding analytical and semianalytical (SA) methods to predict the effect of dependent scattering mechanism on the mesoscopic radiative properties. In this section, we first describe some basic mechanisms involved in the DSE, including the far-field DSE, near-field DSE, recurrent scattering, structural correlations, and the effect of absorbing host media, and summarize relevant theoretical models that deal with them, most of which have closed-form analytical formulas. Then numerical methods to model the DSE are summarized, including the supercell method, the representative volume element method and the direct numerical simulation method.

The DSE, which can occur in the near field as well as in the far field among adjacent scatterers [297, 299, 553], will significantly affect the radiative properties in different length scales. Accordingly, the DSE can be roughly classified into two categories, that is, the far-field and near-field DSEs. Figure 7.11 schematically shows their differences. In the following, we will briefly review theoretical considerations on these two categories of dependent scattering mechanisms. And in particular, we also introduce the recurrent scattering mechanism, the role of structural correlations, and the absorption of the background medium on the DSE.

The Far-Field DSE

The far-field DSE, depicted in Fig. 7.11a, mainly considers the interference of scattered EM waves from different scatterers in the far field, while different scatterers can also interact with each other through far-field scattered waves. The main contribution to the far-field DSE, in conventional cases, stems from the interparticle correlations, since these correlations lead to constructive or destructive interferences among the far-field scattered waves. Therefore, theoretical

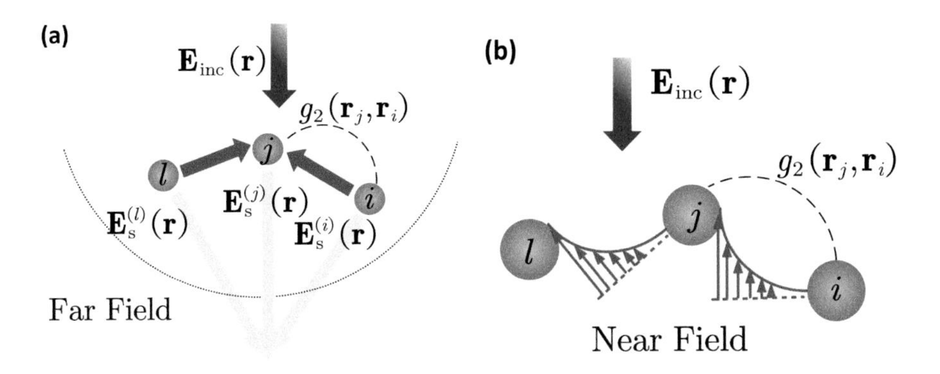

Figure 7.11 Schematic of the (a) far-field and (b) near-field DSEs in a discrete disordered medium. Here, for simplicity, only three scatterers are presented, which are numbered as i, j, and l. The dashed line denotes $g_2(\mathbf{r}_j, \mathbf{r}_i)$, the pair distribution function between the two scatterers. In (a), the dotted line indicates a schematic boundary of the far-field region outside of the entire medium, and the scatterers are also distributed in the far-field region of each other. The blue thick arrow indicates the incident wave $\mathbf{E}_{\text{inc}}$, while the red thick arrows stand for the partially scattered, propagating waves from particles i to j and from l to j. The yellow thick arrows represent scattered waves that propagate to the far field out of the medium $\mathbf{E}_{\text{s}}^{i}(\mathbf{r})$, $\mathbf{E}_{\text{s}}^{j}(\mathbf{r})$ and $\mathbf{E}_{\text{s}}^{l}(\mathbf{r})$. In (b), the scatterers are located in the near fields of each other, and the multiple wave scattering is dominated by the near-field tunneling of evanescent waves, as shown by the red thin arrows combined with a curve representing the exponential decay of wave amplitudes.

models on this mechanism usually relies on some description of the distribution of scatterers. More specifically, the structure factor, which is the Fourier transform of the two-particle correlation function, usually provides a first-order consideration of the far-field DSE, leading to the well-known interference approximation (ITA). In this method, the single scattering property of an individual scatterer is assumed to be not affected, namely, no interparticle interactions through scattered waves (denoted by the red thick arrows in Fig. 7.11a). As a result, the ITA is not able to fully capture the far-field DSE, especially when the packing density is large, structural correlations are strong, or the scatterers are highly scattering.

A conspicuous mechanism that is not accounted for in the first-order correction of the far-field DSE is the deformation of local EM field impinging on each scatterer, due to the scattered waves from adjacent scatterers. It means that the local incident field with respect to each scatterer is no longer the same as the externally incident field as the ITA and ISA assume. This mechanism becomes appreciable at moderate and high packing densities for strongly scattering particles. To tackle with this mechanism, many authors have introduced a homogenized environment with some effective refractive index surrounding each scatterer to modify the ITA model [288, 554, 555]. However, this type of method is not capable of explicitly demonstrating how the local incident field is altered by other scatterers. Moreover, they cannot consider some circumstances in which, for a plane-wave illumination, the local field can be deformed into neither plane-wave-like nor spherical-wave-like structure, resulting in the invalidity of the assumption of the existence of an effective refractive index for the surrounding background. The latter mechanism is seldom discussed, and will substantially affect the mesoscopic radiative properties, which was elucidated in [313] based on a dependent scattering model derived from the quasicrystalline approximation (QCA) by our group. But the QCA approach cannot tackle with the resonances well. The details of theoretical models mentioned here will be further discussed in Section 7.4.6.

The Near-Field DSE

As the concentration of scattering particles in disordered media rises, they are inclined to step into the near fields of each other [304, 556] (i.e., the clearance c between scatterers is comparable or even smaller than the wavelength, approximately, $kc \lesssim 1$). In this circumstance, near-field interaction (NFI) among scatterers, which is negligible when the scatterers are in the far fields of each other, can also contribute to radiation energy tunneling and thus transport properties. Specifically, the NFI indicates the EM coupling of different scatterers through their scattered near fields, which are mainly composed of EM wave components with large wavenumbers ($k > k_0$, where k_0 is the wavenumber of free-space radiation). Moreover, the NFI is purely vectorial, namely, containing both transverse and longitudinal components [296], while in the far-field DSE, the far-field scattered waves are always transverse spherical waves. This can be intuitively understood from the DGF in Eq. (7.4.34), in which the transverse component contains a term slowly decaying with the distance r as $1/r$ (i.e., a spherical wave) and the longitudinal component contains two terms decaying as $1/r^2$ and $1/r^3$.[7] As a consequence, the NFI leads to a more intricate picture of dependent scattering than the far-field interaction. Figure 7.11b schematically shows the possible tunneling of evanescent waves between nearby scatterers, where the red thin arrows combined with a curve representing the exponential decay of the amplitudes of evanescent waves.

The near-field DSE in densely packed DDM is still difficult to fully capture by now. Recently, there has been growing interest in addressing this mechanism thanks to the development of computational and experimental capabilities. It was numerically and experimentally shown by Naraghi et al. that the NFI can enhance the total transmission of disordered media by adding channels of transport [304]. It is also demonstrated that the longitudinal component of the NFI is a hindering factor for Anderson localization in three dimensions [557–559], while interestingly Silies et al. revealed that near-field coupling assists the formation of

[7] We have explicitly presented the expressions of transverse and longitudinal components in the DGF in Eqs. (7.4.51) and (7.4.52) for the convenience of calculation.

localized modes [560]. Pierrat et al. reported that NFI of a dipolar emitter with more than one particle creates optical modes confined to a small volume around it and gives rise to strong fluctuations in the local density of states (LDOS) [561]. Notably, Tishkovets et al. [562–566] carried out a series of theoretical analysis on the near-field DSE in clusters of densely packed scatterers with sizes comparable or smaller than the wavelength, where the near-field mutual shielding effect, the inhomogeneous field, and the negative values of the degree of linear polarization in the backscattering direction were discussed. Nevertheless, a clear elucidation of the near-field DSE and its influence on the mesoscopic radiative properties is still rare, although several phenomenological models are developed to qualitatively address the effect of NFI, which will be presented in Section 7.4.6. Note the methods based on the coherent-potential approximation (CPA) are believed to account for the NFIs to some degree in densely packed DDM, although in an implicit way through an effective refractive index.

In addition, since NFI is much stronger than far-field interaction, it is promising to utilize NFI to achieve extreme light–matter interaction. Therefore, the near-field DSE still needs to be systematically and quantitatively investigated. Recently, Shen and Dogariu [567] investigated the phase and effective interaction volume of a nanoparticle, which provided some important insights for the near-field DSE.

Recurrent Scattering Mechanism

Another mechanism in the DSE worth discussing is the recurrent scattering mechanism. When the scattering strength increases, the probability of a multiply scattered wave propagating back to a scatterer that it formerly visited also grows, leading to a closed-loop-like scattering trajectory. For very strongly scattering media, for example, cold atomic clouds near the atomic bare resonance and metallic nanoparticles near the localized surface plasmonic resonances, the influence of recurrent scattering is significant [305, 494]. However, the analytical calculation of the recurrent scattering mechanism is still very troublesome and can only be done for very simplified cases, for example, recurrent scattering between two point scatterers [494, 568, 569]. Figure 7.12 shows a typical recurrent scattering scheme between a pair of nearby scatterers [494], which can result in analytical expressions for the self-energy. In particular, there are two types of contributions to the self-energy, $\mathbf{\Sigma}^{(2,a)}$ and $\mathbf{\Sigma}^{(2,b)}$. The former describes two-scatterer processes in which the radiation incident on one scatterer eventually returns to the same one, and the latter describes all processes in which the radiation incident on one scatterer emerges from the second. In the diagrammatic representation of self-energy, this mechanism amounts to a group of self-connected diagrams. The calculation for this self-energy is presented in Section 7.4.6, which was conducted by Cherroret et al. [494]. It has been shown that for resonantly scattering particles, the impact of this two-scatterer recurrent scattering process becomes significant even in a dilute disordered medium ($4\pi n_0/k^3 \ll 1$) [494, 568, 569].

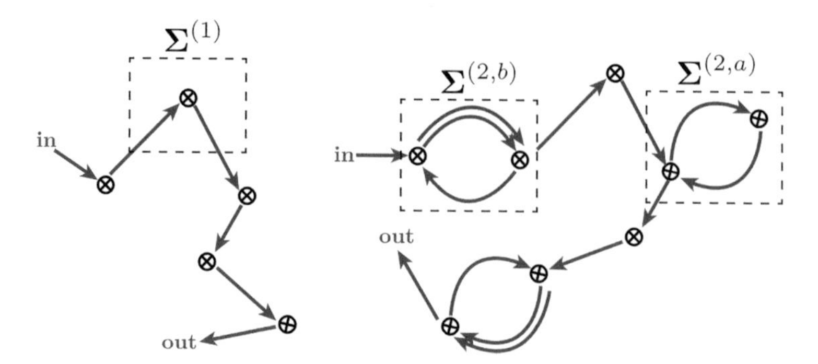

Figure 7.12 Schematic of the recurrent scattering mechanism in a random medium containing discrete scatterers. The ⊗ indicates a scatterer and the lines stand for the propagation of waves. Left: a multiple scattering path visiting independent scatterers, in which the propagating wave is never scattered more than once by the same scatterer. The contribution to the self-energy is $\mathbf{\Sigma}^{(1)}$, namely, the ISA. Right: a multiple scattering path involving the repeated scattering between pairs of scatterers. There are two types of contributions to the self-energy, $\mathbf{\Sigma}^{(2,a)}$ and $\mathbf{\Sigma}^{(2,b)}$. The former describes binary processes in which the radiation incident on one scatterer eventually returns to the same one, and the latter describes all processes in which the radiation incident on one scatterer emerges from the second. Reprinted figure with permission from [494]. Copyright 2016 by the American Physical Society.

The Role of Structural Correlations

As mentioned in Section 7.4.5, one important factor that leads to the dependent scattering mechanism is known as the structural correlations, which describe the possible reminiscence of order (usually short- or medium-ranged) existing in the spatial variation of the dielectric constant in disordered media [481]. They can give rise to definite phase differences among the scattered waves [288, 323, 324, 481, 570–572], which can well preserve over the ensemble average procedure. Hence, constructive or destructive interferences among the scattered waves occur and thus affect the transport properties of light remarkably. This is also called "partial coherence" by Lax [323, 324]. The most well-known type of structural correlation is the hard-sphere positional correlation in disordered media consisting of purely hard spheres without any additional interparticle interactions [570, 573]. This is due to the fact that hard spheres cannot deform or penetrate into each other, and the structural correlations emerge when the concentration of spheres is substantial (usually under a volume fraction of $f_v > 5\%$ [574]). Moreover, when the correlation length of particle positions is comparable with or smaller than the wavelength, the structural correlations play a very important role in determining the microscopic interferences and thus radiative properties [288, 570, 575].

Generally speaking, the structural correlations are not only affected by the packing density (or volume fraction) but also by the interaction potential between particles. In fact, for some types of specially designed structural correlations, even if the concentration of particles is not very high, the strong positional correlation will give rise to very significant interference phenomena, for instance, in the so-called short-range ordered hyperuniform media [571, 576, 577]. Several typical kinds of interaction potential among particles, for example, the surface adhesive potential [578, 579] and the interparticle Coulombic electrostatic potential [288, 580], can be realized experimentally. By controlling the interaction potential and thus the structural correlations, a flexible manipulation of the radiative properties of random media can be achieved [314].

The Effect of an Absorbing Host Medium

In most theoretical considerations of the multiple scattering of waves, the background medium is assumed to be nonabsorbing. This assumption holds for many applications of visible light in atmospheric sciences and optics. On the other hand, in many other applications, for example, infrared spectroscopy, the host medium may become absorptive. As a matter of fact, dependent scattering in an absorbing host medium is still an unsolved problem both theoretically and experimentally. As light scattering by a single spherical particle embedded in an absorbing background medium was not formally solved theoretically until the 2000s [581–585] and is still under intensive theoretical and experimental investigation [586–589], a full theory for the multiple scattering of EM waves in an absorbing host medium was still not well established. Such a theory is nowadays in need for many applications, for example, the use of highly scattering nanoparticles to enhance the light absorption of thin-film silicon solar cells [590, 591], as well as microsphere-enhanced subdiffraction optical imaging [592]. Recently, progresses on this problem were made by Mishchenko et al. [593, 594], who established the FLEs of multiple scatterers, solved the coherent EM field and derived the RTE in a weakly absorbing host media. Their derivation was based on the far-field approximation, namely, with no considerations of the DSE. Durant et al. [310, 595] were the first to investigate the DSE in an absorbing matrix. They developed an analytical formula using the diagrammatic expansion method (as will be shown to be an extension of Keller's approach) to predict the extinction coefficient, which considered the DSE and was verified by full-wave numerical simulations.

7.4.6 Theoretical Models of Radiative Properties Considering the DSE

In this section, we present some notable theoretical models that are used to predict the radiative properties of DDM with considerations of the DSE, some of which are already mentioned in Section 7.4.5. We will give a brief discussion on their physical significance, derivation procedure, advantages, and limitations. Most of the models can result in closed-form formulas for the extinction coefficient κ_e, while for scattering and absorption coefficients as well as the scattering phase function, many models did not have analytical expressions. This is because the latter quantities require additional derivations on the incoherent intensity (e.g., the irreducible vertex in the analytic wave theory), while deriving the extinction coefficient only needs the knowledge of the coherent field (e.g., the self-energy in the analytic wave theory), which is much easier to deal with.

Here we focus on the theoretical models developed for homogeneous spherical scatterers due to their high availability, which can be easily extended to the cases of other shapes, like multilayered spheres and cylinders in the 2D case. However, for scatterers of more complicated shapes like spheroids, cubes, and pyramids, no general closed-form theoretical models are available due to many difficulties, one of which is to obtain the solution of the pair distribution function (PDF) describing structural correlations. Moreover, we only consider a single species of particles, and the particles are disorderedly distributed in vacuum without loss of generality. In addition, for most models, we mainly give analytical expressions for ED with a concise derivation process to present the underlying physics, while for high-order multipolar excitations, we basically provide the formulas derived in the literature.

Interference Approximation

As discussed in Section 7.4.5, since the scatterers are randomly distributed in the medium and have finite sizes, it is crucial to take the interparticle correlations into account in the analysis of the dependent scattering mechanism [288, 570], especially when the volume concentration is substantial. The ITA [596], also known as the collective scattering (CS) approximation that takes the "collective scattering" due to structural correlations into account [304], is regarded as the first-order correction to the DSE. This method only considers the correlation between a pair of particles, which is described by the PDF, $g_2(\mathbf{r}_1,\mathbf{r}_2)$ (already schematically shown in Figs. 7.9 and 7.11a). Specifically, it is the conditional probability density function of finding a particle centered at the position $\mathbf{r}_1$ when a fixed particle is seated at $\mathbf{r}_2$. When assuming the random medium is statistically homogeneous and isotropic, the PDF only depends on the distance between the pair of particles, that is, $g_2(\mathbf{r}_1,\mathbf{r}_2)=g_2(|\mathbf{r}_1-\mathbf{r}_2|)$ [481, 573]. There are already several approximate analytical solutions of the PDF for some specific random systems, for example, [573, 578]. A well-known approximation for hard spheres is the Percus–Yevick (P–Y) approximation [573]. On the other hand, the PDF $g_2(r)$ between pairs of particles can also be obtained experimentally by analyzing the statistical correlations in the microscopic structures [597]. On this basis, the structure factor can be calculated through a Fourier transform process for the pair correlation function $h_2(r)=g_2(r)-1$ as

$$S(\mathbf{q})=1+n_0\int d\mathbf{r}h_2(r)\exp(-i\mathbf{q}\cdot\mathbf{r}). \tag{7.4.38}$$

In the ITA, this structure factor is then used as a correction to the ISA to derive the mesoscopic radiative properties. In this method, the single scattering property of an individual scatterer is assumed to be not affected. Then in a statistically homogeneous and isotropic disordered medium, the differential scattering coefficient in this method can be given by

$$\frac{d\kappa_{s,ITA}}{d\theta_s} = n_0 S(\mathbf{q}) \frac{dC_s}{d\theta_s}, \tag{7.4.39}$$

where $\frac{dC_s}{d\theta_s}$ is the single-particle differential scattering cross section. Here θ_s is the polar scattering angle, and the dependence on the azimuth angle is integrated out (namely, an azimuthal symmetry is assumed here), and $\mathbf{q}$ is chosen to be the difference between the scattered wavevector and the incident wavevector, that is, $q = 2k \sin\theta_s$. For spherical and homogeneous Mie scatterers (Eq. (7.3.55)), we have

$$\frac{d\kappa_{s,ITA}}{d\theta_s} = \frac{n_0 \pi}{k^2} S(2k \sin\theta_s)(|S_1(\theta_s)|^2 + |S_2(\theta_s)|^2), \tag{7.4.40}$$

where the scattering amplitudes $S_1(\theta_s)$ and $S_2(\theta_s)$ are given by the Mie theory. The total scattering coefficient is obtained by directly integrating the differential scattering coefficient over θ_s as $\kappa_{s,ITA} = \int_0^\pi \frac{d\kappa_{s,ITA}}{d\theta_s} \sin\theta_s d\theta_s$, and the scattering phase function and the asymmetry factor can also be accordingly computed. In the long wavelength limit $q \to 0$, the structure factor is given by [481, 573]

$$S(q=0) = \frac{(1-f_v)^4}{(1+2f_v)^2}, \tag{7.4.41}$$

which results in the well-known Twersky's formula for the scattering coefficient as [598]

$$\kappa_{s,Twersky} = n_0 C_s \frac{(1-f_v)^4}{(1+2f_v)^2}. \tag{7.4.42}$$

As discussed in Section 7.4.5, the ITA model accounts for the first-order far-field interference effect, which is the most widely used method [288, 329, 570, 572, 575, 599, 600]. For example, Tien et al. [307, 329, 330] implemented this approach to consider the DSE on radiative properties of packed beds containing spherical particles. Garcia et al. [302] also employed it and showed that the theoretical prediction for the total transmission of the photonic glass can achieve a much better agreement with the experimental result than ISA [302]. However, it was demonstrated in this experiment that a quantitative agreement is still not available, especially in the long wavelength range, where DSE is expected to be more prominent since the distance between particles becomes comparable or smaller than the wavelength, resulting in more particles involved in the multiple wave scattering process. Similarly, Conley et al. [600] calculated the decay rate of diffuse radiation transport in a 2D dielectric medium (refractive index is 3.5) containing randomly distributed circular air holes whose volume fraction is 20%, and they found the prediction of this model deviated from the numerically exact result significantly, where the discrepancy could reach an order of magnitude in some cases that exhibit strong positional correlations among the scatterers.

Local Field Correction (the Maxwell–Garnett Approximation)

The local field correction, also known as the Lorenz–Lorentz relation (LLR) takes the following form in 3D random media for the self-energy:

$$\Sigma_{\mathrm{LLR}} = \frac{n_0 t_0}{1 + n_0 t_0/(3k^2)} = -\frac{n_0 \alpha k^2}{1 - n_0 \alpha/3}. \tag{7.4.43}$$

This model is originally derived from the mean-field assumption in atomic and molecular optics (see [285, 317]) using a local field concept for an ideal cubic array of ED and therefore was usually regarded as not capable of considering the DSE in disordered media. On the contrary, it was shown by Lagendijk et al. using the diagrammatic expansion that this formula indeed takes positional correlations among randomly distributed point dipolar scatterers ($a \to 0$) into account (namely, nonoverlapping condition for different scatterers) into infinite scattering orders [601].[8] This can be understood by noting that Eq. (7.4.43) is alternatively rewritten as

$$\Sigma_{\mathrm{LLR}} = n_0 t_0 \left[1 + \left(-\frac{1}{3}\right) \frac{n_0 t_0}{k^2} + \left(-\frac{1}{3}\right)^2 \left(\frac{n_0 t_0}{k^2}\right)^2 + \left(-\frac{1}{3}\right)^3 \left(\frac{n_0 t_0}{k^2}\right)^3 + \cdots \right], \tag{7.4.44}$$

which is actually a summation of all scattering orders with the prefactors indicating two-particle, three-particle, and four-particle correlations, and so on.

According to Eq. (7.4.16), the effective permittivity is then given by $\varepsilon_{\mathrm{eff}} = 1 - \Sigma/k^2$, and equivalently, we have

$$\frac{\varepsilon_{\mathrm{eff}} - 1}{\varepsilon_{\mathrm{eff}} + 2} = \frac{1}{3} n_0 \alpha, \tag{7.4.45}$$

which is indeed the Clausius–Mossotti relation. If the polarizability α is calculated from the electrostatic approximation for a particle much smaller than the wavelength as $\alpha_{\mathrm{ES}} = 4\pi a^3 (\varepsilon_p - 1)/(\varepsilon_p + 2)$,[9] the well-known Maxwell–Garnett approximation is obtained:

$$\frac{\varepsilon_{\mathrm{eff}} - 1}{\varepsilon_{\mathrm{eff}} + 2} = f_v \frac{\varepsilon_{\mathrm{p}} - 1}{\varepsilon_{\mathrm{p}} + 2}. \tag{7.4.46}$$

Equation (7.4.46) is only valid for very small scatterers with only ED excitation with negligible scattering cross sections. When the size of the particle increases, the scattering becomes significant that can introduce additional imaginary parts into the effective permittivity. In this situation, the polarizability is expressed in the first-order electric Mie coefficient a_1 as [38]

$$\alpha_{\mathrm{ED}} = \frac{6\pi \mathrm{i}}{k^3} a_1 = \frac{6\pi \mathrm{i}}{k^3} \frac{m^2 j_1(mx)[x j_1(x)]' - j_1(x)[mx j_1(mx)]'}{m^2 j_1(mx)[x h_1(x)]' - h_1(x)[mx j_1(mx)]'}, \tag{7.4.47}$$

[8] Note in [601], the definition of "dependent scattering" is equivalent to recurrent scattering, in which the same scatterer is visited more than once, different from ours. Our definition of dependent scattering is a more general one.

[9] Note for nonabsorbing particles, this formula violates the optical theorem and thus cannot be used when the scattering is significant [493, 602].

where $m = \sqrt{\varepsilon_p}$ is the complex refractive index of the particle. In combination with Eq. (7.4.45), the model is called the extended Maxwell–Garnett theory (EMGT). Analogously, by considering the magnetic-dipole excitation described by the Mie coefficient b_1, this model can further be extended to result in an effective permeability [603]:

$$\frac{\mu_{\text{eff}} - 1}{\mu_{\text{eff}} + 2} = \frac{2\pi n_0 \mathrm{i}}{k^3} b_1. \tag{7.4.48}$$

Although this model was originally derived for the cubic lattice [603] and not formally derived for disordered media, it can be deduced from the full-wave equations describing an ensemble of coupled magnetic dipoles [604] in analogy of the procedure of [601] for magnetic field instead. Ruppin [605] presented an extensive summary and evaluation for these EMGTs. These formulas have recently received a lot of attention due to the invention of metamaterials and metasurfaces, for example, negative–refractive–index metamaterials from dielectric particles supporting both electric and magnetic dipoles [606].

Therefore, the effective refractive index is calculated as $n_{\text{eff}} = \text{Re}\sqrt{\varepsilon_{\text{eff}}\mu_{\text{eff}}}$, and the extinction coefficient is given by $\kappa_e = 2k\text{Im}\sqrt{\varepsilon_{\text{eff}}\mu_{\text{eff}}}$. Since this model is derived for an ensemble of point scatterers with infinitesimal exclusion volumes, it is not capable of considering the structural correlations for finite-size particles, which may lead to a broadening and shift for the resonances in the spectra [607]. This effect is taken into account by the QCA (see Section 7.4.6), which is actually equivalent to the local field correction in the point scatterer limit ($a \to 0$).

Keller's Approach

Developed by Keller et al. [608–610], this approach is a perturbative formula for the effective propagation constant up to the second order of number density n_0. In this approach, the self-energy for an ensemble of ED in the reciprocal space is given by

$$\mathbf{\Sigma}_{\text{Keller}}(\omega, \mathbf{p}) = n_0 t_0 \mathbf{I} + n_0^2 t^2 \int_{-\infty}^{\infty} \mathrm{d}^3\mathbf{r} \mathbf{G}_0(\mathbf{r}) h_2(\mathbf{r}) \exp(\mathrm{i}\mathbf{p} \cdot \mathbf{r}). \tag{7.4.49}$$

If we only consider the transverse waves and set $p = K$, namely, the effective propagation constant, we can obtain the self-energy under the on-shell approximation as

$$\begin{aligned}\Sigma_{\text{Keller}}^{\perp}(K) = n_0 t - \frac{n_0^2 t^2}{3k^2} + 4\pi n_0^2 t^2 \int r^2 \mathrm{d}r h_2(r) \\ \times \left[G_0^{\perp}(r) j_0(Kr) + \left(G_0^{\parallel}(r) - G_0^{\perp}(r) \right) \frac{j_1(Kr)}{Kr} \right],\end{aligned} \tag{7.4.50}$$

where the second term arises from the singular part of Green's function, indicating the local contact between particles, and the third term is the leading-order contribution of the two-particle correlations. $G_0^{\perp}(r)$ and $G_0^{\parallel}(r)$ are the transverse and longitudinal components of Green's function, which are given by

$$G_0^{\perp}(r) = -\left(\frac{\mathrm{i}}{kr} - \frac{1}{(kr)^2} + 1\right)\frac{\exp(\mathrm{i}kr)}{4\pi r} \tag{7.4.51}$$

and

$$G_0^{\parallel}(r) = 2\left(\frac{\mathrm{i}}{kr} - \frac{1}{(kr)^2}\right)\frac{\exp(\mathrm{i}kr)}{4\pi r}. \tag{7.4.52}$$

In this situation, it is noted that the self-energy and thus effective permittivity depend on the momentum (wavevector), which is known as the spatial dispersion or nonlocality [491, 595, 611].[10] The effective propagation constant (wavenumber) K can therefore be solved self-consistently by using the relation

$$K^2 = k^2 - \Sigma_{\mathrm{Keller}}^{\perp}(K). \tag{7.4.53}$$

For scalar waves, the formula becomes simpler without the singular part of Green's function [595]. More generally, by taking multipoles into account, the propagation constant is given by [595, 611, 612]

$$K^2 = k^2 + \mathrm{i}\frac{4\pi n_0 \mathrm{S}_k(0)}{k} + \left(\mathrm{i}\frac{4\pi n_0 \mathrm{S}_k(0)}{k}\right)^2 \frac{1}{K}\int_0^{\infty} \mathrm{e}^{\mathrm{i}kr}\sin(Kr) g_2(r)\mathrm{d}r, \tag{7.4.54}$$

where $S_k(0)$ is the forward-scattering amplitude with respect to the background medium with a wavenumber k. Hespel et al. [611] reported experimental measurements of the extinction coefficient in a suspension of PS spheres. It was found that the Keller model is in good agreement with the data provided that nonlocal effects are properly taken into account. Moreover, the local version using $p = k$ of this model can lead to substantial deviations from experimental results especially for large particles (size parameter ~ 1) and for high volume densities. They also examined the simple criterion establishing the regime of independent scattering previously introduced by Hottel et al. [327], which was shown to be not consistent with their experimental data. Notably, Derode et al. [613] tested this formula using acoustic waves in random media composed of metallic rods in water, where the scatterer densities were (6% and 14%), and the agreement was quite good. Chanal et al. [614] numerically examined Keller's formula for 2D random media with a wide range of particle sizes ($a/\lambda = 1/40, 1/20, 1/10, 1/5$) and volume fractions (up to 50%). The refractive index of the particles is 2.25 with varying imaginary parts ranging from 0 to 0.2. They found that this formula can achieve a good agreement with numerical results up to a volume fraction of 30%. Durant et al. further extended this model for absorbing host media [595] and verified it numerically [310].

Quasicrystalline Approximation

The above Eqs. (7.4.49)–(7.4.54) only involve the treatment of two-particle correlations. In much denser random media, high-order position correlations involving three or more particles simultaneously become important. However,

[10] Although the nonlocality is considered, this method still assumes that the wave propagation behavior is mainly determined by a single mode. See the discussion in Section 7.4.1.

there are no closed-form formulas for the correlation functions, and it is also difficult to analytically calculate third- and high-order diagrams in the analytic wave theory. Therefore, approximations on the correlations are necessary, among which the QCA is the mostly used [313, 324, 499]. In this method, three- and higher-order correlations are treated as a hierarchy of PDF, for example, $g_3(\mathbf{r}_1,\mathbf{r}_2,\mathbf{r}_3)=g_2(\mathbf{r}_1,\mathbf{r}_2)g_2(\mathbf{r}_2,\mathbf{r}_3)$, $g_4(\mathbf{r}_1,\mathbf{r}_2,\mathbf{r}_3,\mathbf{r}_4)=g_2(\mathbf{r}_1,\mathbf{r}_2)g_2(\mathbf{r}_2,\mathbf{r}_3)g_2(\mathbf{r}_3,\mathbf{r}_4)$, and so forth, where g_3 and g_4 indicate three-particle and four-particle distribution functions, respectively [324, 499, 565, 615]. This approximating method permits us to solve the propagation problem of coherent EM field (mean field) in random media in closed-form formulas. In this one respect, QCA is actually a perturbative approach only containing multiple wave scattering diagrams with cascading two-particle statistics, although it still takes infinite scattering orders into account [497, 616].

QCA was initially proposed by Lax [324] for both quantum and classical waves, and examined by exact numerical simulations as well as experiments to be satisfactorily accurate for the DSE in moderately dense random media [617, 618]. It is also widely used in the prediction of optical and radiative properties of disordered materials for applications in remote sensing [542], as well as thermal radiation transfer [303, 619]. More generally, its validity for ultrasonic wave propagation in acoustical random media is also frequently verified numerically and experimentally [620].

In the low-frequency limit for ED particles, the self-energy under QCA is expressed in a self-contained way in the reciprocal space as

$$\mathbf{\Sigma}(\mathbf{p})=n_0t_0\mathbf{I}+n_0t_0\int_{-\infty}^{\infty}\mathbf{G}_0(\mathbf{p})\,H_2(\mathbf{p})\,\mathbf{\Sigma}(\mathbf{p}), \tag{7.4.55}$$

where $H_2(\mathbf{q})$ is defined as the Fourier transform of the pair correlation function $h_2(r)$ as

$$H_2(\mathbf{q})=\int_{-\infty}^{\infty}\mathrm{d}^3\mathbf{r}h_2(r)\exp(-\mathrm{i}\mathbf{q}\cdot\mathbf{r}). \tag{7.4.56}$$

Note it differs from the structure factor in Eq. (7.4.38) by unity. Letting $\mathbf{\Sigma}(\mathbf{p})=\Sigma\mathbf{I}$ be $\mathbf{p}$ independent, which is applicable for small particles meaning the scattering properties are not spatially dispersive (i.e., the locality is assumed), we have

$$\Sigma\mathbf{I}=n_0t_0\mathbf{I}-\frac{n_0t_0\Sigma}{3k^2}\mathbf{I}+n_0t_0\Sigma\int_{-\infty}^{\infty}\mathrm{d}\mathbf{r}\mathrm{PV}\mathbf{G}_0(\mathbf{r})\,h_2(r), \tag{7.4.57}$$

where we have separated the singular part of Green's function and defined the PV of Green's function as $\mathrm{PV}\mathbf{G}_0(\mathbf{r})$. The integral is also transformed from the reciprocal domain to the space domain. Then the self-energy is solved as

$$\Sigma=\frac{n_0t_0}{1+n_0t_0/(3k^2)+2n_0t_0\int_0^{\infty}\mathrm{d}rr\exp(ikr)\,[g_2(r)-1]/3}. \tag{7.4.58}$$

This equation in the point scatterer limit ($a\to0$) is equivalent to the LLR formula in Section 7.4.6. Therefore, according to Eq. (7.4.16), the effective propagation constant K is given by

$$K^2 = k^2 - \frac{1}{1/(n_0 t_0) + 1/(3k^2) + 2\int_0^\infty \mathrm{d}r r \exp(ikr)\left[g_2(r) - 1\right]/3}. \tag{7.4.59}$$

After the effective propagation constant for the coherent wave is calculated, the differential scattering coefficient for the incoherent wave can be derived. It is determined by the irreducible intensity vertex $\mathbf{\Gamma}$. Again here we only consider two-particle statistics. The irreducible intensity vertex is solved as

$$\Gamma(\mathbf{p}, \mathbf{p}') = \left[n_0 |C|^2 + n_0^2 |C|^2 H_2(\mathbf{p} - \mathbf{p}')\right] \mathbf{I} \otimes \mathbf{I}, \tag{7.4.60}$$

where $C = \Sigma/n_0$. Afterward, we take the on-shell approximation, which implies the photons transport with a fixed momentum value $p = K$ and those excitations with other momentum values are negligible. It gives

$$\Gamma(K\hat{\mathbf{p}}, K\hat{\mathbf{p}}') = \left[n_0 |C|^2 + n_0^2 |C|^2 H_2(K\hat{\mathbf{p}} - K\hat{\mathbf{p}}')\right] \mathbf{I} \otimes \mathbf{I}. \tag{7.4.61}$$

Since the pair correlation function $H_2(K\hat{\mathbf{p}} - K\hat{\mathbf{p}}')$ only depends on the difference between $\hat{\mathbf{p}}$ and $\hat{\mathbf{p}}'$, the present medium is isotropic. And for unpolarized radiation transport, the azimuth symmetry is preserved. By taking the transverse component of the irreducible intensity vertex, the differential scattering coefficient can be obtained by integrating over the azimuth angle φ_s with the incident direction $\hat{\mathbf{p}}'$ fixed as [495]

$$\frac{\mathrm{d}\kappa_s}{\mathrm{d}\theta_s} = \frac{n_0 |C|^2 (1 + \cos^2\theta_s)}{4\pi} \left\{1 + n_0 H_2[2K \sin(\theta_s/2)]\right\}. \tag{7.4.62}$$

Therefore, the scattering coefficient can be calculated accordingly.

For spherical scatterers supporting high-order multipoles, the equations to calculate the effective propagation constant under QCA can also be developed [298]. Here we only present the main formulas as follows:

$$K - k = -\frac{i\pi n_0}{k^2} \sum_{n=1}^{N_{\max}} (2n+1)\left(T_n^{(M)} X_n^{(M)} + T_n^{(N)} X_n^{(N)}\right), \tag{7.4.63}$$

where $T_n^{(M)}$ and $T_n^{(N)}$ are the T-matrix elements and for spheres, $T_n^{(M)} = -b_n$ and $T_n^{(N)} = -a_n$. $X_n^{(M)}$, and $N_{\max}$ is the maximum expansion order for the multipolar modes. Here $X_n^{(N)}$ can be understood as the ensemble-averaged excitation amplitudes for the multipolar modes, which can be calculated as

$$\begin{aligned} X_v^{(M)} = -2\pi n_0 \sum_{n=1}^{N_{\max}} \sum_{p=|n-v|}^{|n+v|} & (2n+1)[L_p(k, K|d) + M_p(k, K|d)] \\ & \times [T_n^{(M)} X_n^{(M)} a(1, n| - 1, v|p) A(n, v, p) \\ & + T_n^{(N)} X_n^{(N)} a(1, n| - 1, v|p, p-1) B(n, v, p)], \end{aligned} \tag{7.4.64}$$

$$
\begin{aligned}
X_{\upsilon}^{(N)} = & -2\pi n_0 \sum_{n=1}^{N_{\max}} \sum_{p=|n-\upsilon|}^{|n+\upsilon|} (2n+1)[L_p(k,K_{\text{eff}}|d) + M_p(k,K|d)] \\
& \times [T_n^{(M)} X_n^{(M)} a(1,n|-1,\upsilon|p,p-1) B(n,\upsilon,p) \\
& + T_n^{(N)} X_n^{(N)} a(1,n|-1,\upsilon|p) A(n,\upsilon,p)],
\end{aligned}
\tag{7.4.65}
$$

where $L_p(k,K|D)$ and $M_p(k,K|d)$ are given as

$$
M_p(k,K_{\text{eff}}|d) = \int_d^{\infty} r^2 [g_2(r) - 1] h_p(kr) j_p(K_{\text{eff}} r)\, \mathrm{d}r \tag{7.4.66}
$$

and

$$
L_p(k,K|d) = -\frac{d^2}{K^2 - k^2} \times [k h_p{}'(kd) j_p(Kd) - K h_p(kd) j_p{}'(Kd)]. \tag{7.4.67}
$$

These formulas are derived by Tsang and Kong [298, 486, 541], and by applying the distorted Born approximation (DBA) [616], namely, considering the first-order scattering for a thin layer, the scattering phase function and scattering coefficient can be obtained [298]. This is called DMRT [298, 489], which indeed follows from the original theory for electron transport in disordered materials (more specifically, liquid metals) [621]. Recently, we have re-derived the QCA formulas for a random system containing dual-dipolar particles in which only electric and magnetic dipoles are excited [313]. Specifically, in terms of the intensity transport, we have obtained a Bethe–Salpeter-type equation for this system. By applying the far-field and on-shell approximations as well as Fourier transform techniques, we have finally obtained the scattering phase function and scattering coefficient, without resorting to the DBA because full multiple scattering series of radiation intensity is accounted for. Our treatment is based on more explicit arguments and can be easily extended to multipolar excitations.

Recurrent Scattering Models

The only recurrent scattering formula is derived by van Tiggelen et al. [494, 568, 569] for each pair of (uncorrelated) scatterers, as schematically depicted in Fig. 7.12. Recently, this formula was extended to consider the correlations between particles [622, 623]. For very small dipole scatterers, by assuming that $\boldsymbol{\Sigma}$ is $\mathbf{p}$-independent, we have the formula in the following form:

$$
\begin{aligned}
\boldsymbol{\Sigma}_{\text{rec}} = & n_0 t \mathbf{I} + n_0^2 t_0^2 \int \mathrm{d}^3\mathbf{r} \mathbf{G}_0(\mathbf{r}) h_2(\mathbf{r}) + n_0^2 t_0^3 \int \mathrm{d}^3\mathbf{r} \frac{\mathbf{G}_0^2(\mathbf{r})[1 + h_2(\mathbf{r})]}{\mathbf{I} - t^2 \mathbf{G}_0^2(\mathbf{r})} \\
& + n_0^2 t_0^4 \int \mathrm{d}^3\mathbf{r} \frac{\mathbf{G}_0^3(\mathbf{r})[1 + h_2(\mathbf{r})]}{\mathbf{I} - t^2 \mathbf{G}_0^2(\mathbf{r})}.
\end{aligned}
\tag{7.4.68}
$$

The transverse component of the self-energy is then given as

$$\begin{aligned}\Sigma_{\text{rec}}^{\perp} = & n_0 t_0 - \frac{n_0^2 t_0^2}{3k^2} + \frac{2n_0^2 t_0^2}{3} \int r\,\mathrm{d}r h_2(r) \exp(ikr) \\ & + 4\pi n_0^2 t_0^3 \int r^2 \mathrm{d}r \left[\frac{2}{3} \frac{G_0^{\perp 2}(r)}{1 - t_0^2 G_0^{\perp 2}(r)} + \frac{1}{3} \frac{G_0^{\parallel 2}(r)}{1 - t_0^2 G_0^{\parallel 2}(r)} \right] [1 + h_2(r)] \\ & + 4\pi n_0^2 t_0^4 \int r^2 \mathrm{d}r \left[\frac{2}{3} \frac{G_0^{\perp 3}(r)}{1 - t_0^2 G_0^{\perp 2}(r)} + \frac{1}{3} \frac{G_0^{\parallel 3}(r)}{1 - t_0^2 G_0^{\parallel 2}(r)} \right] [1 + h_2(r)]. \end{aligned} \tag{7.4.69}$$

Therefore, the effective propagation constant and the effective permittivity can be obtained by using the Dyson equation. Moreover, van Tiggelen et al. [494, 568, 569] also derived the corresponding irreducible intensity vertex for uncorrelated particles, while for correlated particles, by now no similar formulas are obtained.

This recurrent scattering model also belongs to the perturbative approach in the second order of the particle number density n_0 under the framework of the analytical wave theory, which is valid in very dilute random media, because in denser media, recurrent scattering between three or more particles might become prominent. More precisely, it was shown in [494, 568, 569] the criterion for the diluteness is $4\pi n_0/k^3 \ll 1$. Recently, Kwong et al. [622] used numerical calculations to examine the validity range of this model.

Coherent Potential Approximation and Its Modifications

The concept of coherent potential starts from very simple assumptions, which was first developed for disordered electronic systems [323, 624, 625]. Consider that in a renormalized (effective) disordered medium with an effective Green's function G_e (also called the "modified propagator" [323]), the Lippman–Schwinger equation (Eq. (7.4.6)) is rewritten in the operator form as

$$\mathbf{G} = \mathbf{G}_e + \mathbf{G}_e \overline{\mathbf{T}} \mathbf{G}_e, \tag{7.4.70}$$

where $\langle \overline{\mathbf{T}} \rangle$ is the T-operator of the full system in the effective medium. Taking the ensemble average of Eq. (7.4.70), we obtain

$$\langle \mathbf{G} \rangle = \mathbf{G}_e + \mathbf{G}_e \langle \overline{\mathbf{T}} \rangle \mathbf{G}_e. \tag{7.4.71}$$

According to the definition of the effective Green's function, we have $\langle \mathbf{G} \rangle = \mathbf{G}_e$, and thus the ensemble-averaged T-operator in the effective medium should be zero, that is,

$$\langle \overline{\mathbf{T}} \rangle = 0. \tag{7.4.72}$$

In this sense, in the effective medium, T-operator vanishes, leading to a zero scattering condition. By expanding the many-particle T-operator into the multiple wave scattering series of individual particles' T-operators, we have

$$\langle\overline{\mathbf{T}}\rangle=\left\langle\sum_{i=1}^{N}\overline{\mathbf{T}}_i\right\rangle+\left\langle\sum_{i=1}^{N}\sum_{j\neq i}^{N}\overline{\mathbf{T}}_i\mathbf{G}_e\overline{\mathbf{T}}_j\right\rangle+\left\langle\sum_{i=1}^{N}\sum_{j\neq i}^{N}\sum_{k\neq j}^{N}\overline{\mathbf{T}}_i\mathbf{G}_e\overline{\mathbf{T}}_j\mathbf{G}_e\overline{\mathbf{T}}_k\right\rangle\cdots=0, \tag{7.4.73}$$

where $\overline{\mathbf{T}}_j$ is the T-operator of the jth scatterer in the effective medium. And in this circumstance, the self-energy in the effective medium $\overline{\boldsymbol{\Sigma}}$ is also zero according to the Dyson equation.

At first sight, it seems that Eq. (7.4.73) involving the ensemble-averaged T-operators is still difficult to solve, and no difference is found compared with the original multiple wave scattering series for a disordered medium except for a modified Green's function. Indeed, the attractive point of the CPA lies in the zero scattering condition, because of which it is reasonable to assume that in the effective medium, all particles are weakly scattering and can be regarded as independent scatterers. As a consequence, it would be a good approximation by letting

$$\langle\overline{\mathbf{T}}\rangle\approx\left\langle\sum_{i=1}^{N}\overline{\mathbf{T}}_i\right\rangle=0, \tag{7.4.74}$$

and then the effective propagation constant can be solved self-consistently. This is the basic formula of the CPA for the calculation of radiative properties [299]. For uncorrelated disordered media, this formula is rather accurate up to the third order of the T-operator expansion [299]. The inaccuracies arise from its inability to capture microstructural correlations and recurrent scattering effects. In this circumstance, high-order techniques used in the previous models like QCA and recurrent scattering expansions can also be employed to further improve the accuracy. For example, Tsang and Kong [486] extended the QCA model by using the CPA approach and derived the QCA-CP model for an ensemble of spherical particles.

Recently, this approach has received substantial attention in the design and modeling of dielectric ordered and disordered metamaterials. For instance, Slovick et al. developed a generalized effective medium formula [626] based on the CPA (which they called the "zero-scattering condition") and proposed a design for a negative-index metamaterial using electric and magnetic dipolar excitations [627]. This model was further employed in the design of negative-index metamaterials with high-order multipoles (electric quadrupole) [628], which was numerically validated by the plane-wave expansion (PWE) method as well as finite-difference time-domain (FDTD) simulations.

Another important improvement for the basic CPA formula is the energy-based CPA (ECPA) proposed by Soukoulis et al. [629–631]. In order to account for the short-range correlations in disordered media consisting of densely packed spheres (volume fraction up to 0.6), Soukoulis et al. [629] first modified the CPA approach by considering a coated sphere as the basic scattering unit. However, for low volume fraction, this approach undesirably gives a phase velocity higher than the velocity of light near Mie resonances. Furthermore, Busch and Soukoulis

[630, 631] improved this coated CPA model by using the heuristic idea that in a random medium, the energy density should be uniform when averaged over the correlation length of the microstructure, as schematically shown in Fig. 7.13 with the coated sphere represented by the dashed lines. To calculate the effective dielectric constant $\bar{\varepsilon}$, the coated sphere of radius $R_c = a/f^{1/3}$ is embedded in a uniform medium. The self-consistent condition for the determination of $\bar{\varepsilon}$ is that the energy of a coated sphere is equal to the energy of a sphere with the radius R_c and the dielectric constant $\bar{\varepsilon}$, schematically illustrated in Fig. 7.13(b–c), that is,

$$\int_0^{R_c} \mathrm{d}^3\mathbf{r}\rho_E^{(1)}(\mathbf{r}) = \int_0^{R_c} \mathrm{d}^3\mathbf{r}\rho_E^{(2)}(\mathbf{r}), \tag{7.4.75}$$

whereas the energy density $\rho_E(\mathbf{r})$ is given by

$$\rho_E(\mathbf{r}) = \frac{1}{2}[\varepsilon(\mathbf{r})|\mathbf{E}(\mathbf{r})|^2 + \mu|\mathbf{H}(\mathbf{r})|^2]. \tag{7.4.76}$$

Since the energy densities (and surely the EM fields) implicitly depend on the effective permittivity, from Eqs. (7.4.75) to (7.4.76), the effective permittivity can be determined self-consistently. After this procedure, the self-energy Σ_{ECPA} of the random medium can be calculated with respect to the effective permittivity under the ISA. Thus, the scattering coefficient is given by [630]

$$\kappa_{\mathrm{s,ECPA}} = \frac{\sqrt{2}\mathrm{Im}\Sigma_{\mathrm{ECPA}}}{\left[(k_e^2 - \mathrm{Re}\Sigma_{\mathrm{ECPA}})^2 + \sqrt{(k_e^2 - \mathrm{Re}\Sigma_{\mathrm{ECPA}})^2 + (\mathrm{Im}\Sigma_{\mathrm{ECPA}})^2}\right]^{1/2}}, \tag{7.4.77}$$

where $k_e = k_0 n_{\mathrm{ECPA}}$ is the wavenumber in the effective medium and $n_{\mathrm{ECPA}} = \sqrt{\bar{\varepsilon}}$ is the effective refractive index.

The ECPA approach is most suitable for disordered media composed of extremely dense-packed monodisperse particles exhibiting Mie resonances, in which other dependent scattering models like QCA and recurrent scattering models do not apply or exhibit divergences in the calculation, and the solution of other CPA methods can disappear or jump abruptly or have multiple solutions [630]. As a notable example for the use of this approach, Maret et al. [546, 632] recently carried out a series of numerical and experimental works on photonic glasses containing densely assembled ($f_v \sim 0.5$) monodisperse spherical TiO_2 or PS nanoparticles near strong Mie resonances and showed the good predication capability of the ECPA approach for the light transport properties without any fitting parameters.[11] It was also demonstrated that this approach can quantitatively predict the reflectance spectra of these photonic glass samples with different thicknesses as structural color materials [633].

[11] Note in these works, different from the original ECPA paper [630], the ITA is used to calculate the transport mean free path with respect to the effective medium with n_{ECPA}. Maret et al. claimed that this model (ECPA combined with ITA) "takes into account resonant Mie scattering, short-range positional correlations, optical near-field coupling of randomly packed, spherical scatterers."

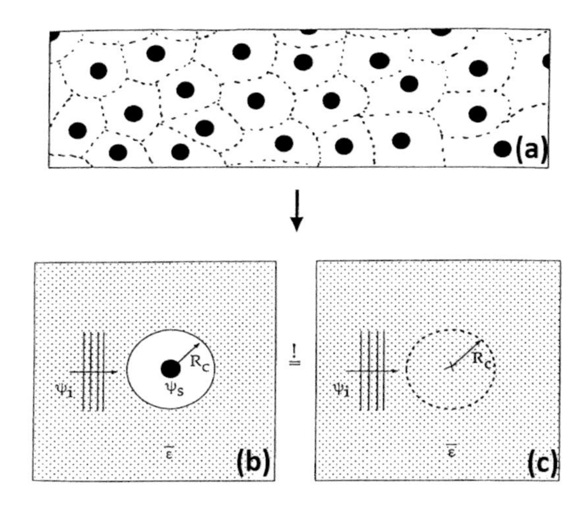

Figure 7.13 The energy-density CPA approach. (a) In a random medium composed of dielectric spheres, the basic scattering unit may be regarded as a coated sphere, as represented by the dashed lines. To calculate the effective dielectric constant $\bar{\varepsilon}$, a coated sphere of radius $R_c = R/f^{1/3}$ is embedded in a uniform medium. The self-consistent condition for the determination of $\bar{\varepsilon}$ is that the energy of (b) a coated sphere is equal to the energy of (c) a sphere with radius R_c and dielectric constant $\bar{\varepsilon}$. Reprinted figure with permission from [630]. Copyright 1995 by the American Physical Society.

Phenomenological Models for Near-Field DSE

Previous models are mostly based on the far-field approximations of EM scattering or include the NFI implicitly, for example, the CPA-based methods indeed account for the NFIs to some extent by using an effective refractive index. It is not easy to explicitly determine the role played by the NFI among scatterers. To do this, several authors developed phenomenological models that explicitly take the NFIs into consideration, for instance, Liew et al. [556] also proposed a similar model to predict the near-field DSE in densely packed PS spheres (volume fraction 64%) by using a near-field-dependent effective refractive index of the background. The value of the background refractive index n_b for a particle is obtained by averaging the actual refractive index surrounding the particle with a weighting factor from an exponentially decaying evanescent field as

$$n_b = \frac{\int_0^\infty n(r)\exp(-\beta r/\lambda) r^2 \mathrm{d}r}{\int_0^\infty \exp(-\beta r/\lambda) r^2 \mathrm{d}r}, \tag{7.4.78}$$

where r is the distance from the particle's surface, λ is the wavelength of light in vacuum, $n(r)$ is the ensemble-averaged refractive index distribution based on the packing geometry, and β is a fitting parameter that is expected to depend on the refractive indices of the particles and the local packing geometry. Their experimental results lead to a fitting parameter $\beta = 14.1 \pm 3.2$ with a relative standard error around 9%, indicating the applicability of this phenomenological model, as shown in Fig. 7.14b with the corresponding background refractive index presented in Fig. 7.14a. It was demonstrated that near-field effects together with the short-range order reduce the scattering strength by one order of magnitude in random close-packed structures. Peng and Dinsmore [634] proposed a similar model by modifying the ITA, which used an effective background medium with a refractive index n_{eff}, to consider the effect of neighboring particles. To obtain n_{eff}, they proposed a near-field coupling distance r_c of neighboring particles, which was estimated as $r_c = (n_s/n_b)\lambda/2$, where n_s and n_b are the refractive indices of particle and matrix materials. Then the effective index can be calculated from the volume-averaged refractive index within a spherical region of radius r_c surrounding a typical scatterer. This model agreed well with their experimental data of high-concentration films of randomly packed ZnS-PS core-shell microspheres. The calculated effective index by this model for closely packed PS spheres is also presented in Fig. 7.14a; see [556] for comparison.

Recently, Naraghi and Dogariu [304] proposed a phenomenological model that added an evanescent-wave-scattering correction to predict the near-field DSE. Their model of transport mean free path reads

$$l^*_{\text{CS+NF}} = \frac{1}{n_0 C_s (1-g)} + \overline{\left(\frac{P_{\text{NF}}}{n_0 C_{\text{NF}} (1-g_{\text{NF}})}\right)}, \tag{7.4.79}$$

where the first term on the RHS is exactly the transport mean free path using the ITA (which was called CS by the authors). In the second term, C_{NF} and g_{NF} are the scattering cross section and asymmetry factor of evanescent waves impinging on a spherical particle [635], and $P_{\text{NF}} = n_0 \lambda^3 \exp(-\kappa_{\text{NF}} d)$ is the probability function for evanescent wave transfer, where κ_{NF} is the characteristic attenuation coefficient of the evanescent waves. Because the decay rate of the evanescent waves depends on the incident angle, an average process $\overline{(...)}$ is taken over the angular domain defined by the refractive indices of the particle and its surrounding medium. Although this model qualitatively captures the physical significance of near-field DSE, especially in the high-concentration range, its prediction on the transport mean free path showed a substantial deviation from experimental data in the intermediate and large volume fraction range [304].

7.4.7 Other Related Interference Phenomena in Mesoscopic Physics

The study of radiative properties of disordered media is closely related to the field of mesoscopic physics, which also investigates the physics of propagation, scattering, and interference of quantum and classical waves in disordered materials. In this section, we will introduce some important phenomena that all arise from wave interferences in disordered materials, including the CB cone, the Anderson localization of light and the statistics as well as correlations induced by disorder. In fact, due to their close relation to the DSE, we have unavoidably mentioned these phenomena in the Far-Field DSE and the Near-Field DSE sections. The study of these phenomena can provide a different viewpoint that

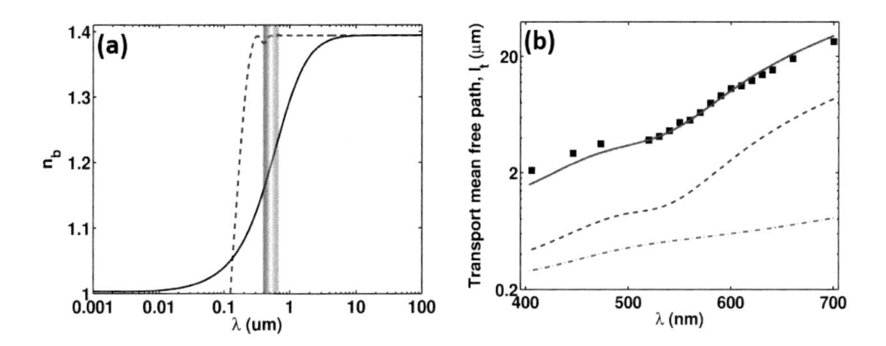

Figure 7.14 A phenomenological near-field DSE model. (a) Near-field effects on form factors can be included in an effective background refractive index n_b, whose value is calculated from Eq. (7.4.78). It approaches the refractive index of air at a short wavelength, and that of a homogenized medium at a long-wavelength. The wavelength range of the experimental measurement is highlighted with color. For comparison, the value of n_b obtained from [634] is plotted with the blue dashed line. (b) Measured (black square) and estimated (lines) transport mean free path l_{tr} vs. wavelength λ. The green dash-dots curve represents l_{tr} estimated without short-range order and near-field effects, the blue dashed line is with short-range order but no near-field effects and the red solid curve is with both. The experiment was done by measuring the CB cone. Reprinted from [556]. © 2011 Optical Society of America

focuses on the universal behavior of disordered media instead of the microscopic details of interparticle EM interactions and offer new methods to obtain more information about the radiative properties. We expect the brief discussion on this discipline can be helpful for the study of the DSE in DDM.

On the other hand, in atomic physics, understanding light propagation and scattering in disordered cold atomic clouds is crucial with important applications in quantum information science. Since cold atoms are extremely scattering near resonance, researchers tried to study the resonant multiple and dependent scattering phenomena of radiation in them, due to the advantages of cold atomic systems over conventional micro/nanoscale scattering media, including well-controlled systems and widely tunable parameters [528, 636]. Remarkably, for dilute cold atomic clouds, the RTE is also widely applied to the description of multiple scattering of photons [636, 637]. Moreover, the prominent collective effects like subradiance and superradiance due to the multiple wave scattering and interferences are also of fundamental importance [638]. In this section, we attempt to give a basic description of light–atom interactions and introduce two remarkable interference phenomena, including the breakdown of the mean-field optics and the significant role of structural correlation in atomic gases, in which multiple wave scattering plays a crucial role. Due to the strong light–matter interaction and high-accuracy experiments, these phenomena are much easier to observe and control in cold atomic gases than in conventional DDM. We expect

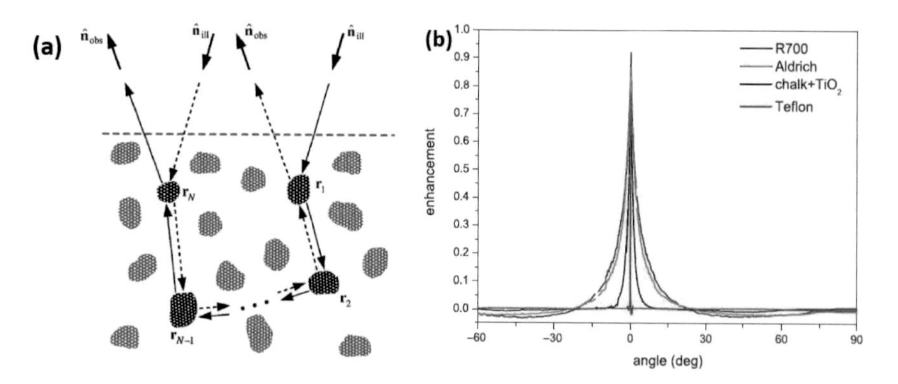

Figure 7.15 (a) The schematic of the mechanism of the coherent backscattering cone in a random medium containing discrete scatterers. The solid lines denote a random multiple scattering trajectory while dashed lines stand for its time-reversed counterparts. Reprinted from [642], Copyright 2016, with permission from Elsevier. (b) Experimental data of CB enhancement for different disordered media. Reprinted with permission from [643]. Copyright © EPLA, 2008.

that this section can establish a bridge among different communities and inspire further studies on dependent scattering.

Initially, mesoscopic physics mainly investigated the quantum transport phenomena involving electrons in disordered electronic materials with the consideration of the wave nature of electrons in mesoscopic-scale samples, which are smaller than the coherence length of electrons, leading to the emergence of quantum interference effects [553]. Later on, due to the easiness of experimental implementation and observation in optics, light transport in disordered dielectric materials offers a good platform for mimicking quantum transport behaviors of electrons, for instance, the direct observation of Anderson localization [639] and the straightforward measurement of intensity correlations and statistics [640, 641], etc.

Coherent Backscattering Cone

Even in the limit of extremely dilute media, the ladder approximation that leads to the RTE is not rigorously exact, because the well-known CB cone, called the weak localization phenomenon [301, 304, 644], can emerge. This is due to the constructive interference between each multiple scattering trajectory and its time-reversed counterpart in the exact backscattering direction, as shown in Fig. 7.15a. Note that the interference between each multiple scattering trajectory and its time-reversal counterpart is always constructive when time-reversal symmetry is conserved [297, 299, 301].

When the scattering strength becomes stronger (i.e., mean free path l becomes smaller), the angular width of the CB cone broadens as $(kl)^{-1}$, as shown in Fig. 7.15b, and thus affects the overall reflectance and transmittance (due to energy conservation) more significantly. As a consequence, for these media, the interference between each multiple scattering trajectory and its time-reversed counterpart should be taken into account, resulting in a series of most crossed diagrams [536–538] (or the so-called cyclical terms [301, 539]) in the diagrammatic representation of the irreducible intensity vertex in the Bethe–Salpeter equation. In practice, the conventional RTE is first used to solve the intensity distribution, and an extra calculation is made to obtain the contribution of the crossed diagrams based on the results of the first step (i.e., by using the property of time-reversal invariance) [301, 538, 644]. Moreover, besides using Ohm's law, another method to obtain the transport mean free path is to fit the analytical formula of the CB cone [304]. The peak enhancement factor, ideally, is 2. However, it strongly depends on the thickness and the absorption of the sample [299, 537, 645] and other factors that affect the coherence of multiple wave scattering.

CB is also frequently discussed in the field of astrophysics. The opposition effect is believed to be a manifestation of it [301, 468, 539], like the one exhibited by the rings of Saturn, which are composed of particles that are covered by small, submicron-sized H_2O grains.

Anderson Localization of Light

When the scattering strength continues to increase, making the Ioffe–Regel condition satisfied, that is, $kl \leq 1$ [296], Anderson localization can occur. In this regime, the wave packets are exponentially localized, and a halt of light diffusion is induced. Anderson localization can be formed by the constructive interference between a closed multiple scattering loops with its time-reversed counterpart given a strong scattering strength, as shown in Fig. 7.16a. By now, Anderson localization of light in one and two dimensions has been theoretically and experimentally confirmed [646–649]. For example, in Figs. 7.16(b–d), an experimental observation of the transition from ballistic transport to diffusion in two dimensions is presented, where the ensemble-averaged intensity follows a Gaussian distribution in space, and then Anderson localization, where the ensemble-averaged intensity is exponentially localized, with the increase of disorder in a 2D photonic lattice [650, 651]. Note the difference between the Anderson localization scheme and the recurrent scattering mechanism shown in Fig. 7.12, because the latter takes place in the microscopic scale while the former occurs in the mesoscopic scale.[12]

[12] However, in the Anderson localization regime, it is hard to tell whether the phenomenon is mesoscopic or microscopic because the wavelength and mean free path are comparable with each other. For convenience, we regard it as a mesoscopic phenomenon.

On the other hand, the Anderson localization of light in 3D is still under intensive theoretical and experimental pursuit but has not been unambiguously observed [545, 553, 652–656]. By measuring the thickness dependence of transmittance, it is possible to identify the onset of Anderson localization. During the transition from the diffusive transport regime to the Anderson localization regime, the total transmittance would follow $T \propto 1/L^2$ based on the prediction of the single-parameter scaling theory of localization [652, 657]. And in the Anderson localized regime, the relation becomes $T \propto \exp(-L/l_{\mathrm{loc}})$, where l_{loc} is the localization length that describes the spatial extent of localized modes. The first experimental claim on the 3D Anderson localization of light in GaAs powders [652] used this method, which, however, was refuted and shown to be the result of weak absorption [658, 659]. Later, to further search for 3D Anderson localization, time-resolved measurement was conducted in TiO_2 powders due to the advantage to separate scattering and absorption effects of this method [545, 660]. By quantifying the deviation of time-resolved transmission from the diffusion equation (more specifically, nonexponential decay at long times), these works also claimed the observation of 3D Anderson localization. However, this claim was later put in question [661] and finally refuted by the authors themselves [653], by demonstrating that the nondiffusive behavior was a consequence of fluorescent emission due to the impurities in the sample at high input laser powers. Therefore, by now there is no unambiguous evidence of the 3D Anderson localization of light.

Remarkably, recently it has been theoretically shown that 3D Anderson localization transition may be found in certain disordered structures with strong short-range correlations (i.e., hyperuniform networks) near the photonic band edges [552], following from the early theoretical proposal of John [662] but without any defect modes. Based on similar considerations, it has been numerically shown that 3D Anderson localization can occur in quasiperiodic structures (i.e., icosahedral quasicrystal) without any additional disorder [663], whereas they cannot be regarded as disordered materials due to the existence of long-range order. It would be interesting to examine these theoretical proposals experimentally, which can provide profound implications for wave physics.

Statistics and Correlations in Disordered Media

Due to the existence of disorder, the optical responses of disordered media vary from sample to sample, from position to position, and therefore it is more practical to investigate the statistics and correlations in their responses and obtain some general and global properties of them. These mesoscopic phenomena, including the statistical distribution of intensity speckles, spatial, and spectral correlations of intensity fluctuations, are comprehensively discussed in the mesoscopic physics community [299, 553]. Here we introduce several well-known statistic phenomena of light scattering in disordered media. For more details on this topic, see [640, 641].

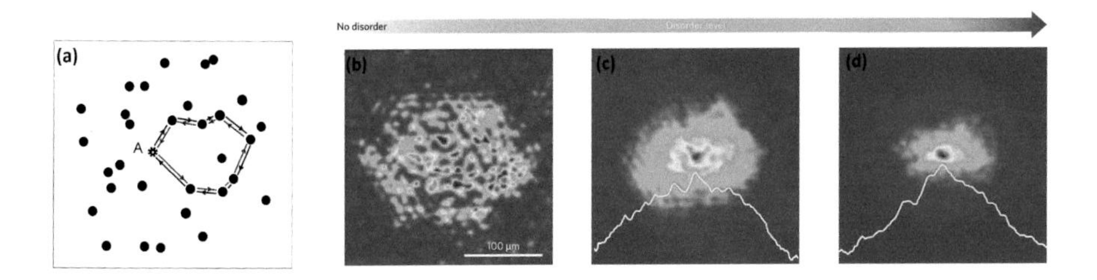

Figure 7.16 (a) The schematic diagram of the mechanism of Anderson localization in a disordered medium. Here the light source is denoted by a star symbol at position A and the spheres denote the scattering elements. A multiple scattering path that returns to the light source forms a loop. Its time-reversed counterpart propagates in exactly the opposite direction along this loop. The two paths thus will acquire exactly the same phase, and interfere constructively in A. Reprinted from [652], Copyright 1997, Springer Nature. (b–d) Experimental observation of the transition from (b) ballistic transport to (c) diffusion, and then to (d) Anderson localization with the increase of disorder in a 2D photonic lattice, in which the white thin lines denote the ensemble-averaged intensity profile in space. Reprinted with permission from [650, 651], Springer Nature.

The first is the amplitude and intensity distribution in the speckle pattern, which forms due to the complex interference of randomly scattered waves traveling different path lengths in disordered media. The amplitude follows a Rayleigh distribution, and the intensity is described by a negative exponential probability distribution as

$$P(I) = \frac{1}{\langle I \rangle_{\mathrm{c}}} \exp\left(-\frac{I}{\langle I \rangle_{\mathrm{c}}}\right). \tag{7.4.80}$$

This is generally known as Rayleigh statistics in random media, where $\langle I \rangle_{\mathrm{c}}$ is configuration averaged intensity and $P(I)$ is the probability distribution function of intensity. The condition for this statistics is that the speckle pattern should arise from the interferences of a large number of scattered waves with independently varying amplitudes and phases, and the phases should uniformly distributed in the range of 0–2π [553, 664]. Therefore, if these conditions are not fulfilled, for instance, when the scattering events are strongly correlated or the scattering strength is too weak to cover the phase range, or the scattering strength is too strong to exhibit Anderson localization, speckles can exhibit non-Rayleigh statistics. An experimental example of the breakdown of Rayleigh statistics and the emergence of super-Rayleigh statistics is given in Fig. 7.17. This is achieved by generating strongly correlated phases in the wavefront of incident light using a spatial light modulator (SLM), which can result in high-contrast speckle patterns that exhibit non-Rayleigh statistics.

Besides the statistic distribution of intensities, there are also universal correlations for intensities at different positions and frequencies. One frequently used intensity correlation function is defined for the intensity fluctuations in the speckle

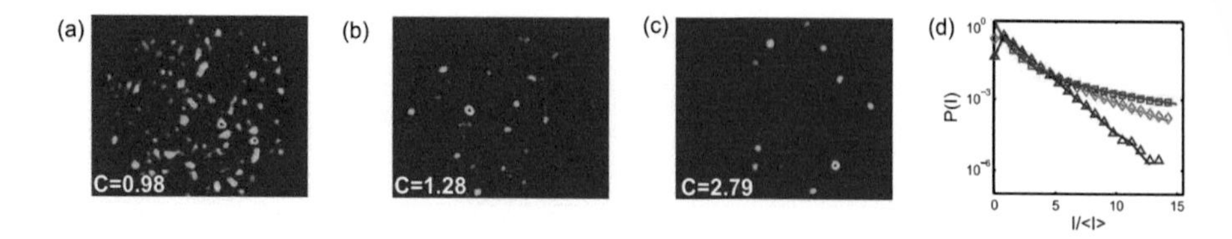

Figure 7.17 Speckle patterns and intensity distribution in disordered media. (a–c) Images of the speckle patterns with different statistics by controlling the contrast parameter $C=\sqrt{\langle I^2\rangle/\langle I\rangle^2-1}$ using a SLM to modulate the incident wavefront, and (d) the corresponding intensity distribution function. (a) A speckle pattern with standard Rayleigh statistics, which has a contrast of $C=0.98$ and a negative exponential intensity distribution [(d), blue triangles]. (b) A speckle pattern with super-Rayleigh statistics, which has a contrast of $C=1.28$, and an intensity distribution that decays slower than the negative exponential [(d), green diamonds]. (c) A speckle pattern with super-Rayleigh statistics with a higher contrast of $C=2.79$ [(d), red squares]. Reprinted figure with permission from [664]. Copyright 2014 by the American Physical Society.

patterns as $C=\langle\delta I\delta I'\rangle$, where $\delta I=I-\langle I\rangle$ is the fluctuation of short-range intensity and $\delta I'$ stands for the fluctuation of intensity at a different spatial or frequency position [466, 470, 665–667]. In this chapter, we mainly introduce spatial correlations. The correlation function C contains three components, which describe the short range (C_1), long range (C_2), and infinite range (C_3) correlations [667]. In particular, short-range spatial correlations describe the averaged size of a speckle spot, while multiple scattering can induce long-range spatial correlations. In the limit of weak scattering, the main contribution to the spatial correlation is the short-range one, which is given by the square of the field correlation function as $C_1(\mathbf{R})=|E(\mathbf{r})E^*(\mathbf{r}+\delta\mathbf{r})|^2$. Shapiro theoretically showed that [665]

$$C_1\propto\left(\frac{\sin(k\delta r)}{kr}\right)^2\exp\left(-\frac{\delta r}{l}\right). \tag{7.4.81}$$

Using a scanning near-field optical microscopy (SNOM), Emiliani et al. [667] directly measured the 2D short-range intensity correlation function of a disordered dielectric structure of microporous SiO_2 glass with randomly oriented and interconnected pores around 200 nm, which is shown in Fig. 7.18a. They found that Eq. (7.4.81) can indeed capture the short-range intensity correlation behavior. However, Carminati [668] theoretically demonstrated that in the deep near field, the spatial correlation length of the field correlation function heavily depends on the local microscopic environment and thus does not show any universal behavior as described by Eq. (7.4.81), shown in Fig. 7.18b. The dynamic fluctuations of speckle patterns stemming from the movement of scattering particles in random media enable a novel imaging method for soft materials, especially for biological materials called laser speckle contrast imaging [669], which is widely applied in the real-time imaging of blood flows in the retina, skin, brain, and many other

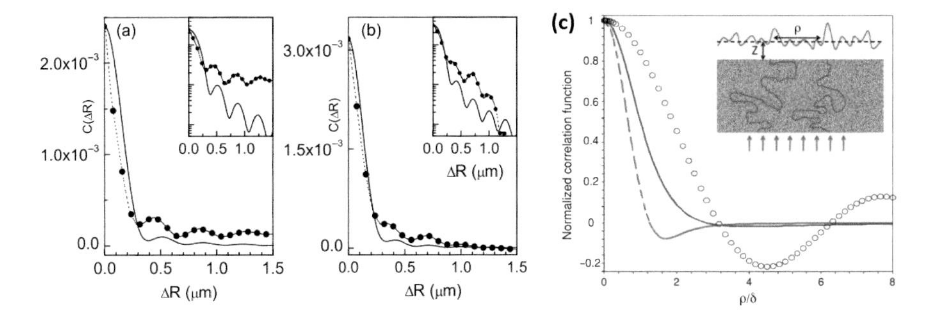

Figure 7.18 Spatial correlation function of intensity. (a–b) SNOM measurement of averaged radial profile of correlation function (dots), (a) probe wavelength of 780 nm, and (b) probe wavelength of 632 nm. The corresponding theoretical fits using Eq. (7.4.81) are shown in solid lines. Reprinted figure with permission from [667]. Copyright 2003 by the American Physical Society. (c) Normalized field spatial correlation function in a plane at a distance z vs ρ/δ. δ is a reference length scale. Black markers: Far-field regime. Blue solid line: near-field intermediate regime. Red dashed line: extreme near-field regime. Reprinted figure with permission from [668]. Copyright 2010 by the American Physical Society.

kinds of tissues and organs [670, 671]. Similar techniques also include the diffuse correlation spectroscopy [672, 673], which can also provide a route to measure radiative/optical properties.

In recent years, the rapid development of photonics further accelerates the advance of mesoscopic physics and gives rise to a brand new field called "disordered photonics" [674]. Fascinating achievements were made thanks to the availability of high-speed SLMs, including imaging, focusing, and multiplexing through disordered media. More specifically, it has received great attention recently that the statistics and correlations of the speckle patterns due to wave interference can be exploited to realize efficient spatial/temporal imaging and focusing through the scattering disordered media [675–679], show promising application to biological and medical imaging and laser therapy (like, deliver light to a specific position in a human body), as well as in-situ optical/infrared diagnosis of turbid or semitransparent coatings used in industry thermal barrier coatings. Moreover, engineering disorder itself to manipulate light transport is also a possible way, for instance, by controlling the fabrication process [680] or using spatially modulated pumping (all-optical SLM) [681]. It is of great interest to find its applications in thermal radiation control. A detailed introduction is, however, out of the scope of this article. For more details, see [674, 682].

As a short summary, in this section, we introduce and discuss the wave interference phenomena in mesoscopic physics, including the CB cone and Anderson localization, as well as the statistics and correlations of scattered waves. This field is closely related to the study of the DSE, and these phenomena are also frequently mentioned in the Far-Field DSE and the Near-Field DSE sections because the microscopic and mesoscopic interferences have influences on each other and cooperatively contribute to the radiative transfer processes. And the field of mesoscopic physics provides relevant theoretical and experimental tools for the study of radiative properties, like the measurement of the CB cone, the diagrammatic technique, and so on. We hope that this section can establish a bridge among different communities and inspire further studies on dependent scattering.

7.5 Theory of Near-Field Thermal Radiation

7.5.1 Introduction

This section seeks to present an overview of NFRHT that can occur between subwavelength objects or macroscopic objects separated by subwavelength gaps. Quantum mechanical analysis becomes more significant as emitting objects get smaller, albeit that is outside the scope of this book. Therefore, we do not talk about the implications of quantum mechanics and instead use EM wave analysis.

The concept of NFRHT is predicated on the idea that it is possible to take into account the observable effects that are caused by electron oscillations and/or lattice vibrations that occur between two objects. These oscillations are amplified in a medium either because of the energy of the medium, which may be determined by the temperature of the medium, or with the assistance of an external source (e.g., an incident laser beam). As a result, Maxwell's equations predict that fluctuations in an electron field (plasmons) or vibrating lattices (phonons) can tunnel through to the second medium. The traditional theory of thermal radiation did not take into account the role of the near field fluctuating until the subject of "radiant" energy transfer, which was raised in the 1960s, got the attention of scholars like Rytov [390]. Rytov brought to light the fact that the effects of a fluctuating field were neglected in the work that had been previously published about calculations of radiative energy transfer between bodies that were relatively close together. He formulated the relationship between the classical theory of thermal radiation and the fluctuating near fields, and then developed the FDT. The FE theory was derived from FDT. According to the FDT, the origin of thermal radiation is linked to the random motions of charges inside the medium at temperatures higher than 0 K. As a contemporary interpretation of Prevost's theory, any object that has a temperature that is finitely greater than absolute zero will emit a fluctuating EM field, which originates from an internal field caused by the oscillations of electrons, molecules, or lattices [7]. In this context, we are solely concerned with thermal emission and do not take into account the various other types of EM emissions.

Let us consider two nonmagnetic, isotropic, homogeneous plane plates that are parallel to one another and separated by a small gap that ranges from a few nanometers to a few tens of micrometers. Assume that the temperature of one of the plates is higher than that of the other. Although thermal radiation

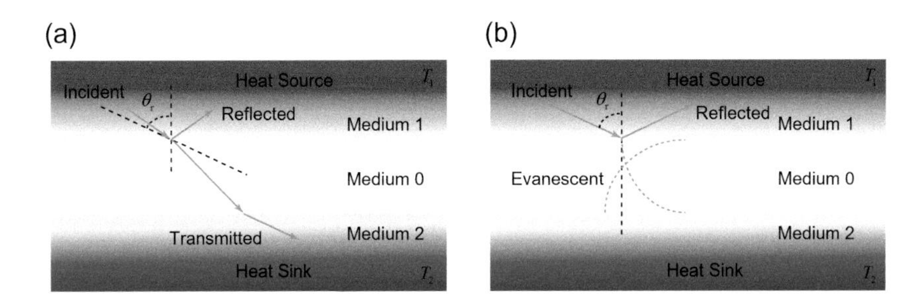

Figure 7.19 EM-wave propagation between two objects (a) for $\theta_i < \theta_c$; (b) $\theta_i > \theta_c$.

can transfer from one plane to the other in both direction, the direction of the net radiative transfer heat flux should be from the hot side to the cold side. The geometry is depicted in Fig. 7.19(a). If the angle at which the intensity beam of thermal radiation emitted from deep in the bulk of the hot medium strikes at its surface is less than the critical angle, the radiative energy will be transmitted out, as shown in the diagram. The critical angle in this context refers to the one that can be calculated using Snell's law when applied to interfaces as discussed in Section 7.1.5. Meanwhile, if the AOI of the radiative intensity beam is larger than the critical angle, there would be no transfer of energy, but TIR will cause evanescent waves to be present. As mentioned earlier, evanescent waves are naturally oscillating standing waves with no net energy carried. However, if the second medium is located within a near enough proximity to the first medium, the evanescent waves are able to tunnel the energy from the first medium to the second one. Since they experience an exponential decay with the distance from the surface (cf. Eq. (7.1.96)), the evanescent waves will be unable to transfer energy to the other medium located in a far field.

In general, the transmission of energy by EM waves can be broken down into two distinct categories: EM modes of propagation in the far field and the tunneling of near-field evanescent modes. These near-field modes are caused by evanescent waves that decay rapidly over length scales ranging from a few hundred nanometers to micrometers. Far-field modes are the principal carrier in energy transport in classic radiative heat transfer, and it is in this regime that Planck's blackbody law is stated. In other words, traditional radiative heat transfer is contributed by far-field modes. On the other hand, in situations where two bodies are separated by a small distance, it is feasible for evanescent modes to tunnel across the surface of the first medium and into the second medium. This phenomenon is illustrated in Fig. 7.19(b). As we will see in the following discussion, it is possible to demonstrate that the ensuing increase in radiative transfer can significantly exceed the limit imposed by Planck's blackbody law by orders of magnitude, especially when the existence of SPhPs or surface plasmons leads to strongly enhanced tunneling of surface waves.

When we talk about conventional radiative transfer, we are referring to the process of radiative exchange between things that are very far apart. It is necessary to quantify "closeness" or "proximity" between objects for the purpose of radiative transfer, and this should be done in respect to the dominant peak wavelength of thermal emission. The RTE and Planck's blackbody distribution can be used to describe the exchange of radiant energy between bodies that are physically separated by distances that are significantly greater than the dominant wavelength. In this scenario, it is presumed that transport is incoherent since thermal radiation is conceived of as a particle (or photon, while one must be careful about this description because it is distinct from the quantum of EM fields [301]), since the coherence length of a blackbody is of the same order of magnitude as the main wavelength of thermal emission [224], as in Chapter 6. This is the so-called far-field regime. It is easy to understand that the coherence can only last for a few nanoseconds or a few tens of nanometers at a time because random oscillations of charges in a particular thermal source produce the EM waves. In light of this, a blackbody source is considered to be incoherent when viewed from a far-field distance.

The traditional (or macroscopic) theory of radiative transfer is no longer applicable once the typical size between structures that are exchanging thermal radiation decreases to a size comparable to or below the dominant wavelength of thermal emission. This is because the wave nature must be taken into consideration in our calculations in order to account for the phenomenon. In Part I on macroscopic thermal radiation transfer, we did not take into account the phase and polarization associated with the EM waves. As a result of this simplification, we have been able to narrow the problem down to the movement of the so-called RTE photons, Monte Carlo, and ray-tracing methods can be applied for solving it. The spectral behavior of the emission (which is based on Planck's blackbody distribution) and the spectral radiative properties of the matter are taken into consideration when analyzing the wavelength effects in that scenario. When the wave aspect of the event is taken into consideration, however, the phase and polarization of the waves shed light on the complex physical mechanisms that underlie the radiative exchange phenomenon. When considering the situation, it is necessary to take into account not only the phase but also the polarization fluctuations of the signal. Therefore, in the near-field regime, radiative energy transfer can only be described by using Maxwell's equations, which describe more general behavior than the RTE formulation used in macroscopic radiative transfer theory. In the near-field domain, the solution of the EM wave equations takes into account all of the following wave phenomena: transmission, reflection, refraction, and diffraction. In addition, the RTE can also be derived from EM theory with appropriate approximations that neglect essential wave aspects, including microscopic and mesoscopic interference phenomena as introduced in Section 7.4.4 and in [298, 301, 459–462, 488].

Based on the EM description of thermal radiation, evanescent waves are produced by the random motions of charges and can be found on the surface of any substance that has a temperature greater than 0 K. Even if evanescent waves do not travel to the far field, it is still possible for energy transfer to occur through these modes if a second body is brought into the evanescent wave field of the substance that is emanating the waves. The mathematical analysis of the problem reveals that even though there is no normal component of the Poynting vector at the first interface, there is a nonzero component at the second interface, which indicates that there is the potential for a net energy exchange to take place. This mechanism is usually dubbed radiation tunneling, which is responsible for radiative heat transfer in the near field that is greater than the values anticipated by Planck's blackbody law.

As mentioned earlier, Rytov was a pioneer in the field of EM description of thermal radiation emission. He described emission as the field generated via chaotic motion of charges within a material, behaving like small radiant dipoles with random amplitudes [390]. Based on Rytov's theory, we are able to draw the conclusion that oscillating dipoles release waves that transport radiative energy away from the surface of a body that emits it (i.e., propagating waves). These evanescent waves are said to exist and fluctuate along the interface between two materials, which exponentially decay over about a wavelength that is normal to that interface [683].

As frequently mentioned, near-field effects become predominate when bodies are separated by less than a few hundreds of nanometers, while the typical wavelengths involved in heat radiation are of the order of a few micrometers. As a result, the term "nanoscale thermal radiation" is sometimes used to refer to near-field thermal radiation in the scientific literature [229, 230, 428]. Nanoscales are the only scales at which near-field effects may be observed for thermal radiation. We discover the near-field effects at a variety of length scales for the other EM wave spectra. For instance, the term "near field" refers to the EM field that exists within a distance range of 1 mm–1 cm when referring to microwaves [683], which is much easier to probe.

A source of thermal radiation is considered to be incoherent, as was mentioned earlier. In general, the temporal or spectral coherence of a radiative source is manifested through emission within a limited spectral band, while the spatial coherence of the source is manifested through emission within a narrow angular range [684–686]. In the far-field region, thermal radiation emission can be modeled as a broadband phenomenon with a quasi-isotropic angular distribution. The opposite of this example is a laser source that has a high degree of both spatial and temporal coherences. In other words, the radiation that is emitted from the laser source is centered on a single wavelength and has a restricted angular distribution. On the other hand, surface waves, also known as surface modes or surface polaritons, can cause thermal sources to exhibit significant spatial and temporal coherences in the near field [335, 684].

Surface waves are usually referred to as hybrid modes that result from the (strong) coupling of an EM field with the mechanical oscillations of energy carriers within the matter. In the context of thermal radiation heat transfer, the term "hybrid mode" refers to the interaction between EM waves and collective oscillations of electrons or lattice vibrations. In particular, a SPP indicates the hybrid mode of collective motion that results from the interaction of free electrons and EM radiation. This mode can be observed in metals and doped semiconductors. In a similar manner, the hybrid mode of lattice vibrations (transverse optical phonons) and an EM field is referred to as an SPhP that is commonly found in polar crystals [419]. These surface polaritons, as particular examples of evanescent waves, travel through an interface between two materials and decay exponentially in the direction of surface normal [156]. According to Henkel et al. [685], surface polaritons have a significant impact on the coherence properties that are present in the near field of a thermal radiation source. In point of fact, radiative heat transfer between closely spaced bodies that support surface waves not only exceeds the amount that was predicted by Planck's blackbody law, but it also has the potential to become quasi-monochromatic due to the high degree of spectral coherence exhibited by these surface waves, as discussed in [684]. Exciting surface waves with a high degree of spatial coherence by, for example, a grating is another method that can be used to achieve highly directional thermal sources in the far field, as demonstrated in [335, 687]. Due to this high degree of spatial coherence, one can see laser-like emission from a thermal source even when at a large distance from it.

Because of recent developments in nanotechnology and nanofabrication techniques, near-field radiation heat transfer is no longer merely a theoretical phenomenon. It is now a real-world occurrence. NFRHT is now promising in applications such as thermal management of MEMS and nanoelectromechanical systems devices, nanoscale-gap thermophotovoltaic (TPV) power generation [385, 688–690], tuning of far- and near-field thermal radiation emission [340, 691–695], thermal rectification [696], near-field thermal microscopy and spectroscopy [697–699], and advanced nanofabrication techniques [383, 387]. The rich physics behind NFRHT lies at the interfaces of quantum mechanics, electrodynamics, and statistical thermodynamics, and thus is quite complicated and fascinating. Our aim in this section is to summarize the theoretical foundations of NFRHT. For the purpose of engineering applications, this section is more apt to give an intuitive, rather than a mathematical, approach to describe the physical phenomena that lie at the heart of this burgeoning discipline. For readers who are interested in acquiring additional information, the appropriate sources are cited throughout the text.

In the following sections, we will provide an overview of the EM treatment of thermal radiation, as well as a concise introduction to the ideas of the density of EM states (DOS) and coherence, both of which are necessary to comprehend the physical principles underlying NFRHT. After that, a discussion of evanescent waves and surface polaritons will follow. Afterward, an explanation of the

NFRHT that occurs between things that are closely spaced is provided by flux calculations performed in a 1D layered architecture.

7.5.2 Electromagnetic Treatment of Thermal Radiation and the Fluctuational Electrodynamics

The presence of evanescent and surface waves, as well as the fact that incoherent transport can no longer be assumed, both contribute to the fact that near-field thermal radiation is distinct from its counterpart that occurs in the far field. In this section, we will present the theoretical framework for describing near-field thermal radiation, as well as explain some fundamental principles that will assist in better comprehending the underlying physics.

When performing calculations in the far field for a variety of technical problems, we make the assumption that the typical size of objects exchanging radiant energy is significantly greater than the dominant wavelength of thermal radiation. In these circumstances, one need not take into account the wave nature of the radiation. As a result, the effects of diffraction and interference are considered to be minimal, and the blackbody distribution proposed by Planck is used to describe the emission energy of thermal radiation. The use of an absorption coefficient, which is nothing more than a proportionality constant, is what is done next to calculate the amount of radiation that an object absorbs. In order to calculate the total amount of energy absorbed by an object, this factor must be multiplied by the intensity with which it was hit. In a similar manner, scattering is represented by employing an *ad hoc* proportionality constant known as the scattering coefficient. This constant is used to quantify the amount of energy that is lost when a beam strikes an object and then scatters off of it. The scattering phase function, on the other hand, is a probability distribution function that depicts how the scattered radiation field is distributed in all solid angles.

As noted earlier, it is possible to distinguish between the far-field and the near-field regimes by making use of a critical length scale. This length scale is traditionally determined by Wien's law and corresponds to the peak wavelength of thermal emission governed by Planck's blackbody distribution: $\lambda_{\max}T = 2897.8\ \mu\text{m}\cdot\text{K}$. This length scale is widely used; however, it is still only an approximation and more precise values have been given for specific instances.

In this sense, we need to employ the EM description of thermal radiation that is based on the macroscopic Maxwell equations because radiative transport is coherent in the near field. These equations provide a description of the connection that exists between the fields, the sources, and the properties of the material. This set of equations accounts for the phenomenon of radiation being absorbed by the medium through the use of the imaginary part of the dielectric function. It is also possible to calculate the scattering of EM waves directly using Maxwell's equations. One way to do this is to assume, for instance, that the total field is the superposition of the incident field and the scattered field, as discussed in

Section 7.4. On the other hand, Maxwell's equations do not take into account the emission of thermal radiation. This connection between Maxwell's equations and the emission of heat radiation is made possible by Rytov's field equation theory [390], which will be covered in the following section.

The addition of a term that accounts for the emission of thermal radiation into Maxwell's equations is not a simple process. Starting from a quantum mechanical point of view allows for the most comprehensive explanation of thermal radiation emission. According to this explanation, the emission of thermal photon results from the transition of an elementary energy carrier (electrons, molecules, phonons, etc.) from a higher energy level to a lower energy level due to thermal fluctuations [700]. In order to make the connection between this phenomenon and EM waves, we need to take into account the emission of thermal radiation from the perspective of electrodynamics. In order to properly conduct the analysis, it is necessary to take into consideration both the far-field and near-field components of radiation (namely, propagating and evanescent waves).

Thermal excitation generates a chaotic motion of charged particles within an object at any temperature ($T > 0$ K), which induces oscillating dipoles. Because it is caused by random thermal motion, the fluctuating EM field that is produced is referred to as the thermal radiation field. It should be noted that the variations in the field are caused by thermal fluctuations of the volume densities of charges and currents on a macroscopic level. To put it another way, the EM field thermally generated is not the addition of the fields of the individual charges in the matter; rather, it is a field produced by sources that are also macroscopic (volume densities of charge and current). The FE theory can be applied to any type of medium as long as the medium is in a state of local thermodynamic equilibrium. In this way, an equilibrium temperature can be specified at any specific position inside the body at any given time. It may also be applied to conditions that are not in equilibrium in certain circumstances, depending on the amount of energy that is emitted by the body. In FE, thermal fluctuations in an object around an equilibrium temperature T would lead to random fluctuations of current, which thus become the radiating source of EM waves. To obtain the EM fields radiated by thermal fluctuations of current in the object, a fluctuating current density distribution $\mathbf{J}^{\mathrm{r}}$ should be introduced in Maxwell's equations in a nonmagnetic, inhomogeneous medium [391]:

$$\nabla \times \mathbf{E}(\mathbf{r},\omega) = \mathrm{i}\omega\mu_0 \mathbf{H}(\mathbf{r},\omega), \tag{7.5.1}$$

$$\nabla \times \mathbf{H}(\mathbf{r},\omega) = -\mathrm{i}\omega\varepsilon_0\varepsilon_r \mathbf{E}(\mathbf{r},\omega) + \mathbf{J}^{\mathrm{r}}(\mathbf{r},\omega), \tag{7.5.2}$$

where we assume the time-harmonic fields with the convention of $\exp(-\mathrm{i}\omega t)$ without loss of generality. Here $\mathbf{J}^{\mathrm{r}}(\mathbf{r},\omega)$ can be regarded as the random source that radiates EM field originating from thermal fluctuations. The presence of the random source makes this equation have the character of a Langevin equation for the EM field. Note by introducing random currents, Maxwell's equations in this situation become stochastic in nature, and are thus called the "stochastic Maxwell equations."

In the stochastic Maxwell's equations, the mean value of this random current density distribution, $\langle \mathbf{J}(\mathbf{r},\omega)\rangle$, is zero, in which $\langle\cdot\rangle$ means taking the ensemble average, indicating that the mean radiated field also vanishes. Note here that the ergodic hypothesis is used. By saying ergodicity, we mean the averaging over time can be replaced by an ensemble average [701]. The underlying idea is that the time average of their properties is equal to the average over the entire space of all states for certain systems. Therefore, the statistical properties of the random current density distribution are described by its two-point correlation function, providing the (quantum and thermal) average of the product of components of $\mathbf{J}(\mathbf{r},\omega)$ at two different points $\mathbf{r}$ and $\mathbf{r}'$ inside the medium. More specifically, the FDT for ensemble-averaged random current density due to thermal fluctuations is given by

$$\left\langle J_\alpha^{\mathrm{r}}\left(\mathbf{r}',\omega\right) J_\beta^{\mathrm{r}^*}\left(\mathbf{r}'',\omega'\right)\right\rangle = \frac{\omega\varepsilon_0\varepsilon''(\mathbf{r},\omega)}{\pi}\Theta(\omega,T)\delta\left(\mathbf{r}'-\mathbf{r}''\right)\delta\left(\omega-\omega'\right)\delta_{\alpha\beta}, \quad (7.5.3)$$

where the subscripts α and β represent orthogonal components denoting the state of polarization of the source. And $\Theta(\omega,T)$ stands for the mean energy of a Planck oscillator at frequency ω and temperature T:

$$\Theta(\omega,T) = \frac{\hbar\omega}{\exp\left(\hbar\omega/k_{\mathrm{b}}T\right)-1}. \quad (7.5.4)$$

In Eq. (7.5.3), the Dirac function $\delta\left(\mathbf{r}'-\mathbf{r}''\right)$ implies that the dielectric constant is local (namely, the fluctuations of the imaginary part of the dielectric constant at two different spatial points are only correlated in the limit $\mathbf{r}'' \to \mathbf{r}'$). The other Dirac function regarding the angular frequency $\delta\left(\omega-\omega'\right)$ implies that the spectral components with different frequencies are totally uncorrelated (namely, white noise), and $\delta_{\alpha\beta}$ accounts for the assumption of isotropic media. From Eq. (7.5.3), the relationship between the statistics of fluctuating current density distribution and the temperature T is thus established [391].

Now let us proceed to solve the EM field from the stochastic Maxwell's equations, for which the method of DGF, as mentioned in Section 7.4, can be used. Using the method of potentials, the electric and magnetic fields can be obtained as

$$\mathbf{E}(\mathbf{r},\omega) = \mathrm{i}\omega\mu_0 \int_V \mathrm{d}V' \mathbf{G}^{\mathrm{E}}\left(\mathbf{r},\mathbf{r}',\omega\right)\cdot\mathbf{J}^{\mathrm{r}}\left(\mathbf{r}',\omega\right), \quad (7.5.5)$$

$$\mathbf{H}(\mathbf{r},\omega) = \int_V \mathrm{d}V' \mathbf{G}^{\mathrm{H}}\left(\mathbf{r},\mathbf{r}',\omega\right)\cdot\mathbf{J}^{\mathrm{r}}\left(\mathbf{r}',\omega\right), \quad (7.5.6)$$

where $\mathbf{G}^{\mathrm{E}}\left(\mathbf{r},\mathbf{r}',\omega\right)$ and $\mathbf{G}^{\mathrm{H}}\left(\mathbf{r},\mathbf{r}',\omega\right)$ are the electric and magnetic DGFs with $\mathbf{r}$ and $\mathbf{r}'$ representing field and source points, respectively. In free space, the electric DGF is given by [225]

$$\mathbf{G}^{\mathrm{E}}\left(\mathbf{r},\mathbf{r}',\omega\right) = \left[\mathbf{I}+\frac{1}{k^2}\nabla\nabla\right] G_0\left(\mathbf{r},\mathbf{r}',\omega\right), \quad (7.5.7)$$

where G_0 is the scalar Green's function as given in Eq. (7.3.18), and the dyadic $\mathbf{I}$ is a 3×3 identity matrix. The magnetic DGF can be derived from the electric DGF by $\mathbf{G}^{\mathrm{H}}(\mathbf{r},\mathbf{r}',\omega)=\nabla\times\mathbf{G}^{\mathrm{E}}(\mathbf{r},\mathbf{r}',\omega)$. From Eqs. (7.5.5) and (7.5.6), it can be straightforwardly understood that the fields at the location of $\mathbf{r}$ are proportional to the sum of randomly fluctuating currents $\mathbf{J}^{\mathrm{r}}$ distributed at different $\mathbf{r}'$ locations within the emitting object. As mentioned earlier, however, the averaged radiated fields, $\langle\mathbf{E}\rangle$ and $\langle\mathbf{H}\rangle$, vanish due to the fact that $\langle\mathbf{J}^{\mathrm{r}}\rangle=0$.

By Eq. (7.5.3), it is expected that the ensemble average of the field correlations must not be zero. The most relevant quantity involving field correlations is the Poynting vector describing the energy transport of radiation. Therefore, here let us calculate the time-averaged Poynting vector (cf. Eq. (7.1.60)):

$$\langle\mathbf{S}(\mathbf{r},\omega)\rangle=4\times\frac{1}{2}\mathrm{Re}\left\{\langle\mathbf{E}(\mathbf{r},\omega)\times\mathbf{H}^*(\mathbf{r},\omega)\rangle\right\}, \tag{7.5.8}$$

in which a prefactor of 4 appears compared to Eq. (7.1.60), since only positive frequencies are taken into account in the Fourier decomposition of the time-dependent fields into frequency-dependent quantities, and thus due to the presence of cross product, time-averaged Poynting vector will involve those terms like $\langle E_m(\mathbf{r},\omega)H_n^*(\mathbf{r},\omega)\rangle$ with $m\neq n$, where the subscripts m and n are orthogonal components representing the polarization of the fields with $(m\neq n)$. These terms can be expressed in the form of correlation functions of the current density distribution by using Eqs. (7.5.5) and (7.5.6) and then invoking the FDT Eq. (7.5.3):

$$\begin{aligned}\langle E_m(\mathbf{r},\omega)H_n^*(\mathbf{r},\omega)\rangle&=\mathrm{i}\omega\mu_0\int_V \mathrm{d}V'\int_V \mathrm{d}V''G^{\mathrm{E}}_{m\alpha}(\mathbf{r},\mathbf{r}',\omega)\,G^{\mathrm{H}^*}_{n\beta}(\mathbf{r},\mathbf{r}',\omega)\\&\quad\times\left\langle J^{\mathrm{r}}_{\alpha}(\mathbf{r}',\omega)\,J^{\mathrm{r}^*}_{\beta}(\mathbf{r}'',\omega)\right\rangle\\&=\frac{k_0^2\Theta(\omega,T)}{\pi}\mathrm{i}\varepsilon''(\omega)\int_V \mathrm{d}V'G^{\mathrm{E}}_{m\alpha}(\mathbf{r},\mathbf{r}',\omega)\,G^{\mathrm{H}^*}_{n\alpha}(\mathbf{r},\mathbf{r}',\omega),\end{aligned} \tag{7.5.9}$$

in which the subscripts m and n indicate the polarization states of the fields observed at $\mathbf{r}$, while α and β describe polarization states of the source at $\mathbf{r}'$. In the last line Eq. (7.5.9), we should note it is a tensor sum, which means, for instance, the set of indices $m\alpha$ represents a summation performed over all orthogonal components (e.g., $G^{\mathrm{E}}_{mx}G^{\mathrm{H}^*}_{nx}+G^{\mathrm{E}}_{my}G^{\mathrm{H}^*}_{ny}+G^{\mathrm{E}}_{mz}G^{\mathrm{H}^*}_{nz}$). By this stage, we have established the relationship between the radiative heat flux and the temperature of the object under investigation based on the framework of FE. The DGFs should then be determined by classical EM analysis for specific systems, which would be quite different from and usually more complex than the free-space expression (Eq. (7.5.7)) if, for example, the medium is inhomogeneous with micro- and nanostructures [692, 702, 703].

Now let us discuss the concept of DOS, which is vital to the interpretation of the underlying physics of NFRHT. The (optical) DOS represents the number of possible photonic modes, per unit frequency and per unit volume, whose expression in vacuum is given by

$$N(\omega) = \frac{\omega^2}{\pi^2 c_0^3}, \tag{7.5.10}$$

which can be simply obtained by quantizing the EM field in an imaginary cavity in free space [700]. Then the energy density of thermal radiation from a blackbody at a given frequency, $U(\omega)$, can be calculated as the product of the DOS given by Eq. (7.5.10) and the mean energy of a Planck oscillator [704]

$$U(\omega) = \frac{\hbar\omega^3}{\pi^2 c_0^3 \left[\exp(\hbar\omega/k_{\mathrm{B}}T) - 1\right]}. \tag{7.5.11}$$

From this equation, we can then derive Planck's law of blackbody radiation for emissive power. Details can be found following in [224].

We anticipate a rise in the DOS due to the presence of evanescent and surface waves in the near field of a thermal source at a temperature of T. We calculate the local DOS (LDOS) at a particular position $\mathbf{r}$ in space since these waves are very sensitive to the distance from the thermal source (c.f. Eq. (7.1.96)) [704]. The energy density in a vacuum can be determined by adding the electric and magnetic energies of the source as [704]

$$\langle U_\omega(\mathbf{r},\omega,T)\rangle = 4 \times \frac{1}{4}\left[\varepsilon_0 \left\langle |\mathbf{E}(\mathbf{r},\omega)|^2 \right\rangle + \mu_0 \left\langle |\mathbf{H}(\mathbf{r},\omega)|^2 \right\rangle\right], \tag{7.5.12}$$

where a factor of 4 appears again due to the same reason as for Eq. (7.5.8). It is possible to construct an explicit equation for the energy density, such as Eq. (7.5.9) for the Poynting vector, by inserting the electric and magnetic field formulas and applying the FDT, as shown in [704]. The energy density that is given by Eq. (7.5.12) is relative to the energy density of the vacuum. This is because fluctuations in the vacuum are ignored when calculating the mean energy of a Planck oscillator because they do not have an impact on the calculations for radiative heat flux. The LDOS $N(r,\omega)$ in the near field can be computed by dividing the result of Eq. (7.5.12) by $\Theta(\omega,T)$.

7.5.3 Role of Evanescent and Surface Waves in NFRHT

Now let us discuss the role of evanescent and surface waves in near-field thermal radiation. an evanescent wave can be created in the process of TIR at an AOI larger than the critical angle θ_c [6, 225, 285]. Consider Fig. 7.20 where medium 0 is vacuum, medium 1 is a heat source at temperature T_1 with a refractive index n_1, and medium 2 is a heat sink at temperature $T_2 < T_1$ with a refractive index n_2. By FDT, thermal radiation is emitted throughout the volume of medium 1 and medium 2. Without loss of generality, we can assume that the EM waves emitted by these objects are propagating in the xOz plane only. For the waves with a large AOI compared to θ_c, they become evanescent waves due to TIR from medium 1. When the distance d between them is comparable or smaller than the

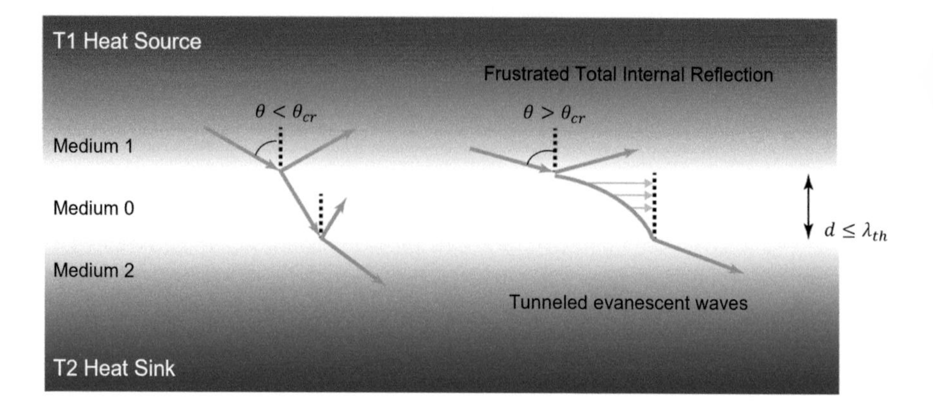

Figure 7.20 Schematic of near-field thermal radiation transfer through evanescent waves.

dominant wavelength of thermal radiation, these evanescent waves can tunnel into medium 2. In particular, the evanescent waves transmitted from medium 1 can excite the charges within medium 2 and dissipate the energy through Joule heating [705, 706]. This is the near-field radiative This mode of energy transfer is, therefore, NFRHT, or usually referred to as radiation tunneling, or frustrated TIR, since in this circumstance Poynting vector due to evanescent waves is no longer zero.

As a result of this additional transfer of energy in the nearby field, the radiative flux ends up being greater than the values indicated by Planck's distribution. The maximum radiative heat transfer takes place at the limit $d \to 0$ for the case of lossless dielectric materials with refractive indices n_1, and its achievable value is n_1^2 times the values predicted between blackbodies. This limit results from the fact that the blackbody intensity in a lossless material is proportional to n_1^2. However, when dealing with materials that support surface waves, such as metals, doped semiconductors, and polar crystals, above simple estimation for NFRHT is insufficient. As was covered earlier in this chapter, the process of thermal radiation emission should be understood from the perspective of electrodynamics, in which oscillating dipoles generate propagating and evanescent waves. In point of fact, in the case when materials are able to support surface waves or surface polaritons, the biggest contributing wavevector k_x that is parallel to the interface is able to significantly surpass the limit $n_1 k_0$ in the consideration of TIR.

As mentioned earlier, surface waves, also known as surface polaritons, are hybrid modes that are created when a mechanical oscillation couples with an EM field. The out-of-phase longitudinal oscillations of free electrons (also known as plasma oscillations) relative to the positive ion cores in a metal or a doped semiconductor cause dipoles, which in turn generate an EM field [156, 706]. An SPP is a component of the radiation spectrum in the near-field. In a manner analogous, the out-of-phase oscillations of transverse optical phonons in polar crystals, such as SiC, generate an EM field, and the near-field component of this field is referred to as an SPhP. Surface waves travel along the interface between two media, and while they do so, evanescent fields decay exponentially in the direction along the surface normal.

For the purpose of explaining the physics of surface waves, we will again use the plane interface shown in Fig. 7.20 as an example. In this example, both media 1 and 2 are infinite along the x- and y-directions. The frequency-dependent dielectric function of the media 1 is given by the expression $\varepsilon_1(\omega)$ when $z<0$, and the frequency-dependent dielectric function of the medium 2 is assumed to 1 (vacuum) when $z>0$. The presumptions that are laid out in Section 7.5.2 are applicable to medium 1. It is assumed that surface waves are only moving in the direction of x, and as a result, they do not transfer any energy outside of the medium if there is no other object in close vicinity to them. This assumption is made without sacrificing generality.

By analyzing the dispersion equations, which describe the relationship between the periodicity of the wave in time (represented by the angular frequency ω) and its periodicity in space (represented by the wavevector k_x in this case), it is possible to gain an understanding of the effect that surface waves have on the NFRHT. Solving Maxwell's equations at the interface 1–2 in a manner that is distinct for the TE- and TM-polarized waves enables one to ascertain the existence of such a dispersion relation. Another method of getting the dispersion relation is to find the poles of the Fresnel reflection coefficients at the interface 1–2. The conditions under which the Fresnel reflection coefficients can approach infinity correspond to their poles. This straightforward approach is one that we use in this context, with additional details available in [392].

The Fresnel reflection coefficients in TE and TM polarizations in terms of wavevectors are given by [707]

$$r_{12}^{\mathrm{TE}} = \frac{k_{z1} - k_{z2}}{k_{z1} + k_{z2}}, \tag{7.5.13a}$$

$$r_{12}^{\mathrm{TM}} = \frac{k_{z1} - \varepsilon_1(\omega) k_{z2}}{k_{z1} + \varepsilon_1(\omega) k_{z2}}. \tag{7.5.13b}$$

These formulas are the same as Eq. (7.1.92), in which the perpendicular and parallel polarizations correspond to TE and TM polarizations respectively. In TE polarization, the Fresnel reflection coefficient will diverge if the condition $k_{z1} + k_{z2} = 0$ is met. Since we are concerned with the dispersion relation of surface waves when there is an exponentially decaying field along the z direction in both medium 1 and medium 2, here we only deal with surface waves if both k_{z1} and k_{z2} are considered to be pure imaginary values. However, since the imaginary part of the z component of the wavevector is always positive, TE polarization does not support the existence of surface waves. If the materials are magnetic, however, it is possible to demonstrate that surface polaritons can occur in TE polarization [392] (cf. Eq. (7.1.92)).

On the other hand, for TM polarization, the Fresnel reflection coefficient approaches infinity if the condition $k_{z1}+\varepsilon_1(\omega)k_{z2}=0$ is satisfied. As a consequence of the fact that the real part of the dielectric function of medium 1 can be negative (for metals and polar crystals, cf. Section 7.2.1), surface waves can exist in this scenario. Following from $k_{zj}=\sqrt{\varepsilon_j k_0{}^2-k_x^2}$, the condition obtained in terms of k_x is

$$k_x = k_0\sqrt{\frac{\varepsilon_1(\omega)}{\varepsilon_1(\omega)+1}}. \tag{7.5.14}$$

Let us emphasize that this dispersion relation at the interface 1–2 poses two conditions for the presence of surface polaritons. Given that surface waves travel along the interface 1–2, the wavevector k_x needs to be a real number in order for it to be valid. In addition, because surface polaritons are evanescent waves moving in the direction of z, the wavevector k_x must be larger than the value of $k0$ that is found in a vacuum. Surface waves are created when the term contained within the square root in Eq. (7.5.14) has a value that is greater than unity. This can occur when $\varepsilon_1(\omega)<-1$ [392].

In order to demonstrate how the dispersion relation of surface waves is obtained, we will assume that medium 1 is a polar crystal SiC that is capable of supporting SPhPs in the mid-infrared region. It is possible to model the dielectric function of SiC as a damped harmonic oscillator using the Lorentz model (c.f. Eq. (7.2.7)):

$$\varepsilon_1(\omega)=\varepsilon_\infty\left[\frac{\omega^2-\omega_{\mathrm{LO}}^2+\mathrm{i}\Gamma\omega}{\omega^2-\omega_{\mathrm{TO}}^2+\mathrm{i}\Gamma\omega}\right], \tag{7.5.15}$$

where ε_∞ is the high-frequency dielectric constant, Γ is the damping factor, and ω_{LO} and ω_{TO} are the frequencies of longitudinal and transverse optical phonons, respectively. For SiC, these properties are taken to be $\varepsilon_\infty=6.7$, $\omega_{\mathrm{LO}}=966\ \mathrm{cm}^{-1}$, $\omega_{\mathrm{TO}}=790\ \mathrm{cm}^{-1}$, and $\Gamma=5\ \mathrm{cm}^{-1}$ [708]. For convenience for plotting the dispersion relation, the loss in the dielectric function of SiC is ignored (viz. $\Gamma=0$). The dispersion relation at the interface 1–2 is illustrated in Fig. 7.21, in which the light line in vacuum $k_x=k_0$, and the frequencies of transverse and longitudinal optical phonons are also shown.

It is vital to determine the zones in which the waves are either propagating or evanescent in vacuum in order to have a better understanding of the dispersion relation that is given in Fig. 7.21. In a vacuum, the equation that describes the z component of the wavevector is written as $k_{z2}=\sqrt{k_0^2-k_x^2}$. If $k_x\le k_0$, then the z component of the wavevector is a real number with no imaginary components; thus, the wave is propagating. As a result, these modes within the light line in vacuum correspond to propagating waves. On the other hand, when $k_x>k_0$, k_{z2} becomes a pure imaginary value, indicating that the wave is evanescent. As a result, the portion of the dispersion relation shown on the RHS of the light line in vacuum in Fig. 7.21 belongs to evanescent waves.

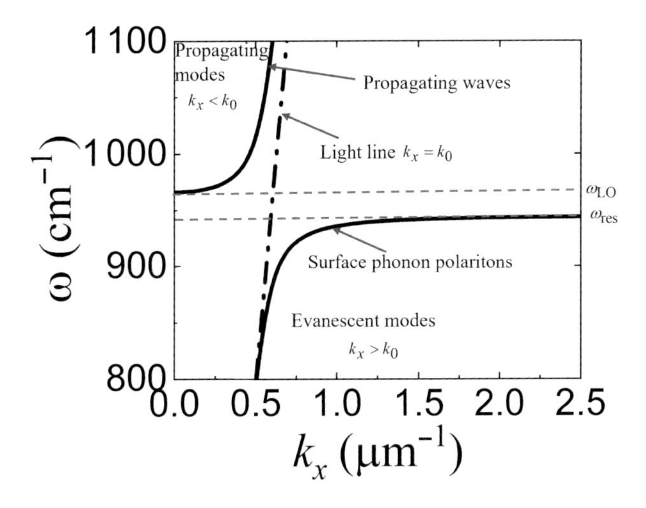

Figure 7.21 SPhP dispersion relation at SiC–vacuum interface via Eqs. (7.5.14) and (7.5.15).

Let us consider the dispersion relation within the light line that does not belong to the SPhP, which corresponds to the solution of Eq. (7.5.14) that does not meet the criterion that the term beneath the square root is greater than unity. This can be seen by looking at the real part of the dielectric function of SiC, given by Eq. (7.5.15). In fact, for frequencies greater than ω_{LO} or smaller than ω_{TO}, the real part of the dielectric function of SiC is larger than -1. On the other hand, we can note the dispersion relation outside of the light line (lower branch in Fig. 7.21) has a wavenumber $k_x > k_0$, which corresponds to an SPhP, and it is included in the frequency range between ω_{LO} and ω_{TO}. By evaluating Eq. (7.5.15), it is revealed that the real part of the dielectric function in this frequency range is less than -1. When the real part of the dielectric function is equal to -1, it approaches an asymptote of ω_{res}, which is defined as the resonant frequency of surface polariton at a single interface and is solved to be $\omega_{\mathrm{res}} = \sqrt{(\varepsilon_\infty \omega_{\mathrm{LO}}^2 + \omega_{\mathrm{TO}}^2)/(1+\varepsilon_\infty)}$ from Eq. (7.5.15) with $\Gamma = 0$. In this circumstance, the denominator of Eq. (7.5.14) becomes zero, leading to an infinite k_x, as can be observed from Fig. 7.21 near the frequency ω_{res}.

As mentioned earlier, LDOS represents the number of EM modes per unit frequency and per unit volume, and thus LDOS is directly proportional to $|\mathrm{d}k_x/\mathrm{d}\omega|$. At the resonance frequency, the value of $|\mathrm{d}k_x/\mathrm{d}\omega|$ can approach infinity, leading to a divergence of LDOS and energy density. As a result, we anticipate that the radiative heat exchange between the materials that support surface waves will be much higher than the values that are anticipated by Planck's distribution. In addition, the fact that a significant increase occurs close to a specific frequency ω_{res} hints at the presence of temporal or spectral coherence in the near field. It has been shown that the existence of surface waves has a significant impact on the spatial coherence of the near field of a thermal source [6, 684, 685]. Since the mechanical oscillations within the material (plasma oscillations or lattice vibrations) transmit their spatial coherency to the emitted EM field, a high degree of spatial coherence can be found very close to an emitting material that supports surface waves. Therefore, the excitation of surface waves in the far field causes thermal emission in a limited spectral band and a narrow angular range.

To directly observe surface waves or surface polaritons, techniques developed by near-field optics are usually employed [225]. A number of approaches have been developed to excite surface polaritons by means of TIR experienced by an external radiation beam. The two mostly used approaches are known as the Kretschmann and Otto configurations [156, 225]. The scenario is different when it comes to thermal radiation, as surface waves are excited by the random fluctuation of charges within the emitting material. SPhPs in SiC can be thermally excited as $\Theta(\omega, T)$ reaches its peak value around 10 μm at 300 K, as predicted by Wien's law. These temperatures are involved in thermal radiation applications at typical temperatures.

The SPP resonant frequency is around 9.69×10^{15} rad/s, which corresponds to a wavelength of approximately 0.194 μm. If medium 1 is a bulk region of gold, then the SPP resonant frequency is approximately 9.69×10^{15} rad/s.

As a remark, now let us mention the case of SPPs. If medium 1 in Fig. 7.20 is a bulk region of gold, then the SPP resonant frequency is around 9.69×10^{15} rad/s, equivalently to a wavelength around 0.194 μm. By using the Drude model Eq. (7.2.9) for gold, similar calculations can be done to obtain the dispersion relation. Because $\Theta(\omega, T)$ is relatively small at this SPP frequency for typical thermal radiation temperatures between 300 and 2000 K, the energy density at resonance is quite low. Therefore, SPhPs that have resonance in the infrared spectrum are typically more intriguing than SPPs from the point of view of thermal radiation [706]. However, substances such as doped silicon are able to support SPPs in the infrared, and as a result, they can behave similarly to polar crystals that support SPhPs [709].

7.5.4 Calculation of Near-Field Radiative Heat Transfer

Despite the fact that the study of near-field thermal radiation is still in its infancy, especially experimental studies, the challenge of calculating the near-field radiative heat flux was tackled in the late 1960s between objects with cryogenic temperatures, in which other modes of heat transfer are at a minimum [406, 710], and the dominant wavelength of thermal radiation is quite large. In these early works, thermal radiation near-field calculation results were presented between two bulk materials that were separated by a vacuum gap. In order to explain the results of thermal radiation heat transfer, they made use of the analogy with TIR, although this did not take into consideration all of the evanescent modes. In the 1970s, Polder and Van Hove presented the first correct radiative heat flux calculations between two bulk materials using FE [393].

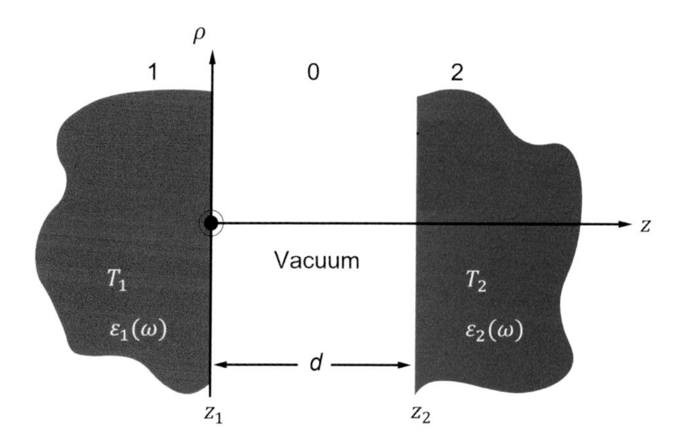

Figure 7.22 Schematic representation of two semi-infinite materials separated by a vacuum gap of thickness d.

Now let us discuss the NFRHT between two bulk materials separated by a vacuum gap of thickness d. According to Eq. (7.5.9), we need to first get the expressions of DGF for this case. For the two parallel semi-infinite media schematically shown in Fig. 7.22, the DGF can be expressed in terms of the 2D spatial Fourier transform in the cylindrical coordinate $\mathbf{r} = z\hat{\mathbf{z}} + \rho\hat{\boldsymbol{\rho}}$ as [436]

$$\mathbf{G}(\mathbf{r}, \mathbf{r}', \omega) = \frac{1}{4\pi^2}\int_{-\infty}^{\infty}\int_{-\infty}^{\infty} \mathbf{g}(k_\rho, z, z', \omega)\, \mathrm{e}^{\mathrm{i}k_\rho(\rho-\rho')}\mathrm{d}k_x\, \mathrm{d}k_y, \tag{7.5.16}$$

in which $k_\rho = \sqrt{k_x^2 + k_y^2}$ denotes the in-plane wavevector, and the full wavevector can be written as $\mathbf{k}_j = k_\rho\hat{\boldsymbol{\rho}} + k_{zj}\hat{\mathbf{z}}$, with $k_{zj} = \sqrt{\varepsilon_j\omega^2/c^2 - k_\rho^2}$. By inserting Eq. (7.5.16) into Eq. (7.5.9), the integration over $\mathrm{d}x\mathrm{d}y$ becomes an integration over $\mathrm{d}k_x\mathrm{d}k_y$. And for isotropic materials, this integration in the reciprocal domain can be converted into an integration in cylindrical coordinates (k_ρ, θ). For two parallel semi-infinite media, the DGF can be solved as [392, 706, 711]:

$$\mathbf{G}(\mathbf{r}, \mathbf{r}', \omega) = \frac{\mathrm{i}}{4\pi}\int_0^{\infty} \frac{k_\rho\, \mathrm{d}k_\rho}{k_{z1}}\left(\hat{\mathbf{s}}t_s\hat{\mathbf{s}} + \hat{\mathbf{p}}_2 t_p\hat{\mathbf{p}}_1\right)\mathrm{e}^{\mathrm{i}\left(k_{z2}z - k_{z1}z'\right)}\mathrm{e}^{\mathrm{i}k_\rho\left(r-r'\right)}, \tag{7.5.17}$$

where t_s and t_p are the transmission coefficients from medium 1 to medium 2 for s and p polarizations, and $\hat{\mathbf{s}}$ and $\hat{\mathbf{p}}$ are unit vectors denoting the polarization directions, given by $\hat{\mathbf{s}} = \hat{\boldsymbol{\rho}} \times \hat{\mathbf{z}}$, $\hat{\mathbf{p}}_1 = (k_\rho\hat{\mathbf{z}} - k_{z1}\hat{\boldsymbol{\rho}})/k_1$, and $\hat{\mathbf{p}}_2 = (k_\rho\hat{\mathbf{z}} - k_{z2}\hat{\boldsymbol{\rho}})/k_2$. If we are interested in computing the thermal radiation from a medium to vacuum, then the Fresnel transmission coefficients between the medium and vacuum can be substituted for t_s and t_p. After that, using this DGF and performing integration over the z direction in either region 1 or 2, one can get the Poynting vector and energy density at a specific position emitted from medium 1 or medium 2 according to Eq. (7.5.9). Note that the term $\mathrm{e}^{\mathrm{i}k_\rho\left(r-r'\right)}$ will be canceled as it is multiplied by its complex conjugate in the calculation of Poynting vector and energy density [712].

Since the z component of the time-averaged Poynting vector can be expressed as

$$\langle S_z(\mathbf{r},\omega)\rangle = \frac{1}{2}\operatorname{Re}\left[\langle E_x(\mathbf{r},\omega)H_y^*(\mathbf{r},\omega)\rangle - \langle E_y(\mathbf{r},\omega)H_x^*(\mathbf{r},\omega)\rangle\right], \tag{7.5.18}$$

by invoking the FDT with some manipulations, we can obtain the spectral energy flux from medium 1 to medium 2:

$$q''_{\omega,1\to 2} = \frac{\Theta(\omega,T_1)}{4\pi^2}\int_0^\infty \xi_{12}(\omega,k_\rho)k_\rho\,\mathrm{d}k_\rho, \tag{7.5.19}$$

in which

$$\xi_{12}(\omega,k_\rho) = \xi_{12}^p(\omega,k_\rho) + \xi_{12}^s(\omega,k_\rho) \tag{7.5.20}$$

is the total energy transmission coefficient or called photon tunneling probability. It is the sum of the energy transmission coefficients for p and s polarizations, which can be calculated as

$$\xi_{12}^p(\omega,k_\rho) = \frac{16\operatorname{Re}(\varepsilon_1 k_{z1}^*)\operatorname{Re}(\varepsilon_2 k_{z2}^*)\left|k_{z0}^2 \mathrm{e}^{2\mathrm{i}k_{z0}d}\right|}{\left|(\varepsilon_1 k_{z0}+k_{z1})(\varepsilon_2 k_{z0}+k_{z2})(1-r_{01}^p r_{02}^p \mathrm{e}^{2\mathrm{i}k_{z0}d})\right|^2}, \tag{7.5.21a}$$

$$\xi_{12}^s(\omega,k_\rho) = \frac{16\operatorname{Re}(k_{z1})\operatorname{Re}(k_{z2})\left|k_{z0}^2 \mathrm{e}^{2\mathrm{i}k_{z0}d}\right|}{\left|(k_{z0}+k_{z1})(k_{z0}+k_{z2})(1-r_{01}^s r_{02}^s \mathrm{e}^{2\mathrm{i}k_{z0}d})\right|^2}, \tag{7.5.21b}$$

both of which vary in the range of $[0,1]$. Here, r_{01} is the Fresnel reflection coefficient at the interface between vacuum and medium 1, with superscripts s and p denoting the polarization as before, $k_0 = \omega/c$ is the wavevector in vacuum. By Fresnel formulas (Eq. (7.1.92)), we have $r_{01}^s = (k_{z0}-k_{z1})/(k_{z0}+k_{z1})$ and $r_{01}^p = (\varepsilon_1 k_{z0}-k_{z1})/(\varepsilon_1 k_{z0}+k_{z1})$. Similar definitions and expressions also apply for r_{02}. The spectral energy flux from medium 2 to medium 1, $q''_{\omega,2\to 1}$, can be obtained by considering the reciprocity of the energy transmission coefficients, that is, $\xi_{12}^j(\omega,k_\rho) = \xi_{21}^j(\omega,k_\rho)$, and by $\Theta(\omega,T_1)$ with $\Theta(\omega,T_2)$. As a consequence, the net total energy (heat) flux can be readily obtained as

$$q''_{\text{net}} = \frac{1}{4\pi^2}\int_0^\infty\int_0^\infty [\Theta(\omega,T_1)-\Theta(\omega,T_1)]\,\xi_{12}(\omega,k_\rho)k_\rho\,\mathrm{d}k_\rho\,\mathrm{d}\omega. \tag{7.5.22}$$

From this equation, we can separate the energy flux into the contributions of propagating waves and evanescent waves. For propagating waves with $k_\rho < k_0$, let us consider the energy transmission coefficient for the j-polarization ($j = p, s$):

$$\xi^j_{\text{prop}}(\omega, k_\rho) = \frac{\left(1-\rho^j_{01}\right)\left(1-\rho^j_{02}\right)}{\left|1 - r^j_{01} r^j_{02} e^{2ik_{z0}d}\right|^2}, \quad k_\rho < k_0, \tag{7.5.23}$$

where r_{01} and r_{02} are the Fresnel coefficients from vacuum to medium 1 and medium 2, respectively, and $\rho_{01} = r_{01} r^*_{01}$ and $\rho_{02} = r_{02} r^*_{02}$ are the corresponding reflectivities. In the far field and incoherent limit $d \gg \lambda$, we can obtain

$$\left|1 - r^{p,s}_{01} r^{p,s}_{02} e^{2ik_{z0}d}\right|^2 \to \left(1 - \rho^{p,s}_{01} \rho^{p,s}_{02}\right). \tag{7.5.24}$$

From this equation, the inverse of Eq. (7.5.23) can be recast into the following form:

$$\frac{1 - \rho^{p,s}_{01} \rho^{p,s}_{02}}{\left(1 - \rho^{p,s}_{01}\right)\left(1 - \rho^{p,s}_{02}\right)} = \frac{1}{\varepsilon'^{p,s}_{\omega,1}} + \frac{1}{\varepsilon'^{p,s}_{\omega,2}} - 1, \tag{7.5.25}$$

and therefore the total energy flux in the far-field limit is given by

$$q''_{\text{net,FF}} = \frac{1}{4\pi^2 c^2} \int_0^\infty \int_0^{\pi/2} \left[\Theta(\omega, T_1) - \Theta(\omega, T_2)\right] \omega^2 \times \left(\frac{1}{1/\varepsilon'^{p}_{\omega,1} + 1/\varepsilon'^{p}_{\omega,2} - 1} + \frac{1}{1/\varepsilon'^{s}_{\omega,1} + 1/\varepsilon'^{s}_{\omega,2} - 1}\right) \cos\theta \sin\theta \, d\theta \, d\omega, \tag{7.5.26}$$

in which we have replaced the integration over k_ρ by using the definition of incident angle, $\sin\theta = k_\rho / k_0$. It is found this formula is similar to the equation of energy exchange between two large planes in macroscopic thermal radiation heat transfer [7]. Although the contributions of both polarizations are included in the expression of total energy flux, it is important to conduct the integration of the two polarizations independently, as shown in this equation. If the emissivities of the two surfaces have different dependencies on the polar angle and the polarization state, then averaging over the two polarizations to obtain the directional emissivity of each surface may result in substantial errors in the calculation.

For evanescent waves in vacuum with in-plane wavevectors $k_\rho > k_0$, the energy transmission coefficient ξ for the j-polarization ($j = p, s$) becomes

$$\xi^j_{\text{evan}}(\omega, k_\rho) = \frac{4\,\text{Im}\left(r^j_{01}\right)\text{Im}\left(r^j_{02}\right) e^{-2\gamma_0 d}}{\left|1 - r^j_{01} r^j_{02} e^{-2\gamma_0 d}\right|^2}, \quad k_\rho > k_0, \tag{7.5.27}$$

where $\gamma_0 = -ik_{z0} = \sqrt{k_\rho^2 - k_0^2}$. It is seen that as the distance d increases, the transmission coefficient decreases in an exponential fashion, and vanishes when the separation is substantially larger than the wavelength of radiation. On the other hand, when the separation d is substantially smaller than the dominant wavelength of thermal radiation, the contribution of near-field evanescent waves dominates, especially for metals, doped semiconductors, or polar crystals in which polaritonic modes can be excited in the near field. Since the in-plane wavevectors in these scenarios are quite large as was noted earlier, we can deduce approximate expressions for the transmission coefficient. In particular, for $k_\rho \gg k_0$, we can safely assume $k_{z1} \approx k_{z2} \approx k_{z0} \approx ik_\rho$ by definition. As mentioned earlier, in this case, the contribution of TE waves is negligible since r_{01}^s and r_{02}^s are vanishingly small. In addition to this, the Fresnel reflection coefficients for TM waves can be approximated as $r_{01}^p \approx (\varepsilon_1 - 1)/(\varepsilon_1 + 1)$ and $r_{02}^p \approx (\varepsilon_2 - 1)/(\varepsilon_2 + 1)$, which are thus independent of k_ρ. It follows that

$$\xi_{\text{evan}}(\omega, k_\rho) \approx \frac{4\,\text{Im}(r_{01}^p)\,\text{Im}(r_{02}^p)\,\text{e}^{-2k_\rho d}}{|1 - r_{01}^p r_{02}^p \text{e}^{-2k_\rho d}|^2}. \tag{7.5.28}$$

Inserting this formula into Eq. (7.5.19), the spectral heat flux from medium 1 to medium 2 in the limit of $d \to 0$ is given by

$$q''_{\omega,1\to2} \approx \frac{\Theta(\omega, T_1)}{\pi^2 d^2} \frac{\text{Im}(\varepsilon_1)\,\text{Im}(\varepsilon_2)}{|(\varepsilon_1 + 1)(\varepsilon_2 + 1)|^2} \int_{x_0}^{\infty} \left| 1 - \frac{(\varepsilon_1 - 1)(\varepsilon_2 - 1)}{(\varepsilon_1 + 1)(\varepsilon_2 + 1)} \text{e}^{-x} \right|^{-2} x\text{e}^{-x}\, \text{d}x \tag{7.5.29}$$

with $x_0 = 2k_0 d$. As a consequence, the spectral (as well as total) heat flux will be inversely proportional to d^2 in the extreme near-field limit. Moreover, when

$$\left| \frac{(\varepsilon_1 - 1)(\varepsilon_2 - 1)}{(\varepsilon_1 + 1)(\varepsilon_2 + 1)} \right| \ll 1, \tag{7.5.30}$$

the integral in Eq. (7.5.29) approaches 1, and in this scenario we can obtain the net spectral heat flux as

$$q''_{\omega,1\to2} - q''_{\omega,2\to1} \approx \frac{1}{\pi^2 d^2} \frac{\text{Im}(\varepsilon_1)\,\text{Im}(\varepsilon_2)}{|(\varepsilon_1 + 1)(\varepsilon_2 + 1)|^2} [\Theta(\omega, T_1) - \Theta(\omega, T_2)]. \tag{7.5.31}$$

Besides the NFRHT calculations presented here for the simplest scenarios of two bulk media, a large number of analytical and numerical computations for NFRHT between a wide range of materials and geometries have been conducted by researchers, for instance, between thin films [693, 713], between a dipole and a surface [711], between two dipoles [714, 715], between two large spheres [394], between a sphere and a surface [395], between a dipole and a structured surface [716], between 2D materials and their composites [717, 718], and among a group of objects (the so-called many-body radiative heat transfer) [719, 720]. Numerical approaches have been developed for the treatment of complex geometries, for instance, the FDTD method, BEM, and the T-DDA method [396, 397, 400, 721, 722].

Now let us remark the calculation of LDOS in the near field. It is possible to represent the LDOS in vacuum close to the surface of medium 1 into the sum of two terms, which are,

$$D(z, \omega) = D_{\text{prop}}(\omega) + D_{\text{evan}}(z, \omega), \tag{7.5.32}$$

where

$$D_{\text{prop}}(\omega) = \int_0^{k_0} \frac{\omega}{2\pi^2 c^2 k_{z0}} \left(2 - |r_{01}^s|^2 - |r_{01}^p|^2 \right) k_\rho \text{d}k_\rho \tag{7.5.33}$$

and

$$D_{\text{evan}}(z,\omega)=\int_{k_0}^{\infty}\frac{e^{-2z\gamma_0}}{2\pi^2\omega\gamma_0}\left[\text{Im}\left(r_{01}^s\right)+\text{Im}\left(r_{01}^p\right)\right]k_\rho^3\,\mathrm{d}k_\rho. \tag{7.5.34}$$

Note that for obtaining the above expressions, we should assume the effect of medium 2 is negligible, which means it is far apart from medium 1. See [723] for the calculation of LDOS in the vacuum gap with consideration of the effect of medium 2. From Eq. (7.5.33), the contribution of propagating waves is not a function of the distance from the medium surface z and therefore it will persist in both near and far fields. On the other hand, according to Eq. (7.5.34), the contribution of evanescent waves demonstrates an exponential decay with z, which will vanish in the far field limit. Therefore in the far field, thermal emission is only contributed by the propagating waves, as expected.

In point of fact, Eq. (7.5.33) incorporates terms that are associated with the directional-spectral emissivity. These terms are as follows: $\varepsilon_{\omega,1}^{\prime s}=1-\rho_{01}^s$ and $\varepsilon_{\omega,1}^{\prime p}=1-\rho_{01}^p$. As it moves closer and closer to the surface, the contribution of evanescent waves at the surface may begin to dominate when the value of $\text{Im}\left(r_{01}^p\right)$ is substantial. This is particularly true in the scenario in which SPhPs can be excited. There is the potential for extraordinarily high energy densities to exist in close proximity to the surface at a specific frequency [704, 706, 711].

In the extreme near field limit ($z\to 0$), we can also obtain approximate formulas for LDOS. By taking $k_\rho \gg k_0$, Eq. (7.5.34) can be recast into

$$D_{\text{evan}}\left(z,\omega\right)\approx\frac{1}{\pi^2\omega}\frac{\text{Im}\left(\varepsilon_1\right)}{\left|\varepsilon_1+1\right|^2}\int_{k_0}^{\infty}\mathrm{e}^{-2\beta z}k_\rho^2\,\mathrm{d}k_\rho. \tag{7.5.35}$$

It follows that

$$D_{\text{evan}}\left(z,\omega\right)\approx\frac{1}{4\pi^2\omega z^3}\frac{\text{Im}\left(\varepsilon_1\right)}{\left|\varepsilon_1+1\right|^2} \tag{7.5.36}$$

in which only the leading term in the integration is kept [704].

According to this equation, the near-field LDOS grows with z^{-3} as z gets smaller. Concerns have been raised about whether the FE theory would be valid in the extreme near field and how the near-field radiation would transition into thermal conduction. Generally speaking, the locality and homogeneity assumptions should be considered valid up until the separation spacing approaches around $1-2$ nm [384, 405]. When the separation continues to reduce, a nonlocal model for the permittivity should be used to modify the result of FE [401, 402]. Moreover, when the separation approaches the interatomic distance (smaller than 1 nm), the tunneling of acoustic phonons begins to play a substantial role, as demonstrated by recent numerical simulations [209, 403, 405]. This means there is a transition regime with contributions of both radiation and conduction. Recent tests carried out in a variety of laboratories at distances of around 1 nm or less indicate that the fundamental issues regarding the transition from radiation to conduction and the validity of FE theory at the extreme near field are still unresolved [421, 422, 724].

7.6 Summary

In this chapter, we have presented an overview of the theoretical foundations of MNTR. Firstly, we introduce the fundamentals of EM theory, including the macroscopic Maxwell's equations, constitutive relations, boundary conditions, Fresnel's formulas of reflection and refraction, etc. This is then followed by a general description of optical interactions in solids and related permittivity models, including the Drude–Lorentz model, Drude model, Clausius–Mossotti and Kramers–Kronig relations. Then, the theory of EM wave scattering is discussed, including the basic definitions and general properties of scattering cross section, integral representations of the scattering problem, Rayleigh scattering, Mie theory and the T-matrix method. We further introduce the theory of multiple wave scattering, which is actually very important for understanding the connection between the RTE and Maxwell's equations. We have also discussed the DSE comprehensively, with a summary of different theoretical models. Finally, theoretical foundations of NFRHT are given, including the FDT, EM DOS, surface polaritons, and so on. This is further followed by detailed calculations performed on two bulk media separated by a vacuum gap with relevant discussions.

8 Numerical Methods for Micro/Nanoscale Thermal Radiation

In this chapter, we aim to give a brief introduction to various numerical methods that are commonly used to tackle MNTR problems. First, a general overview of numerical methods for solving Maxwell's equations is given, which is actually from computational EMs and photonics. The two most widely used methods in MNTR studies are introduced in detail, including the FDTD method and the modal method (MM). The principles, boundary conditions, and implementation are described. Then comes a brief discussion on the simulation of EM wave propagation in disordered media, in order to delineate the connection of radiative transfer and EM wave theory. We further proceed to the specific algorithms dealing with near-field thermal radiation problems, in which fluctuating currents should be taken into account.

8.1 An Overview of Numerical Methods

Since the EM theory is generally valid at micro- and nanoscales, and the inclusion of fluctuational currents into the EM theory results in the FE, the numerical calculation of MNTR transport is, to a large extent, equivalent to solving Maxwell's equations. Therefore, a variety of numerical EM methods are viable for the prediction and modeling of thermal radiation properties of micro/nanostructures in both far- and near fields. Following the rapid development of photonics, more and more numerical algorithms have been developed, which can be directly used in MNTR. Now let us have a brief overview of several numerical EM methods frequently used in photonics and MNTR.

The FDTD method can be considered as a universal approach for directly solving Maxwell's equations as they look in the time domain, without any further approximation or specification of the processes, as the plane-wave solution and harmonic time dependence have done. As a finite-difference solution, this method needs to be applied to a grid of discretized space and time, in order to transform the differential equations into algebraic ones. The Yee grid, also known as the Yee mesh, is the most well known and widely used of these types of grids. This type of grid organizes the six different components of electric and magnetic fields in a specific order, with the goal of having only tangential field components on the unit cell interfaces. As a result, we can obtain an explicit algorithm that updates the EM fields when the time is increased by a small step. As a result, by utilizing this updating technique, we are able to track all of the peculiarities of the evolution of fields, which take place during the propagation, scattering, absorption, and any other events that can be observed with EM waves. In this manner, the FDTD approach can be regarded as a method to simulate real-world optical experiments, which is currently utilized rather frequently as a benchmark solution for the purpose of testing and verifying the results of real optical devices and other numerical methods. In addition, because it is implemented in the time domain, the FDTD method can be associated with physical phenomena that are not of the same kind as traditional electromagnetism, such as the kinetics of atoms or molecules, the dissipation of heat, and diffusion, among many other things. One of the limitations of the FDTD method is that it is necessary to update the field components at all points in the numerical space, even in areas of the space that are not being studied. This drawback, however, is compensated for by the linear scaling rule of the approach with the sizes, the relatively simple meshing of the entire domain, and the potential of effective parallelization of the computation.

When calculating the eigenmodes of straight waveguides or micro- and nanocavities in the frequency domain, the finite-difference frequency-domain (FDFD) approach can be employed effectively to provide accurate results. The simplicity of the procedure, in addition to its universal applicability, is the primary benefit of the approach, in which the application of periodic and absorbing boundary conditions (ABCs) is straightforward. The most significant shortcoming of the method is that its efficiency is inferior to other sophisticated approaches like the FEM.

The MM is also a frequency-domain approach with the harmonic wave ansatz. In this method, the calculation domain in geometry is sliced into layers that are uniform along a propagation axis (which is usually taken to be the z axis as a convention). The eigenmodes of each layer are first calculated, and then the field is expanded as a linear combination of these eigenmodes, while the z dependency of the eigenmodes can be characterized analytically through the use of propagation constants. In this manner, the thickness of each layer has no impact on the time consumption required for the computation. To address the scattering that occurs at the interfaces between the various layers, a mode-matching technique is used in the MM, in which reflection and transmission matrices are assigned to each interface. One of the benefits of the MM is that it provides direct access to the propagation constants of the eigenmodes as well as their coupling (or intermodal scattering). This gives the opportunity to investigate certain optical modes of interest and gets more physical insights. In addition, the MM naturally makes use of the periodicity by employing the Bloch mode formalism. This is a significant benefit. As a result, the MM is a sensible option for handling periodic geometries like gratings and PCs. In the plane that is lateral to the propagation axis, the MM naturally supports closed and periodic boundary conditions. In order to simulate open geometries, it is necessary to implement ABCs.

Another frequency-domain approach is the Green's function integral equation method. Different from the finite-difference methods, this approach utilizes equivalent integral equations. In these equations, the total field at any position is directly related to an overlap integral between a Green's function and the field inside or on the surface of a scattering object in a reference geometry. As mentioned earlier, Green's function represents the field that is produced by a point source in the reference geometry. By solving the self-consistent equations, one can determine the field that exists either inside or on the surface of the scattering object. As can be seen in the integral equation for the scattering problem in Section 7.3, the radiation that is emitted from a specific part in a scatterer is caused by the total field, which is the sum of the externally incident field and the contributions to the partial fields emitted by other components of the scatterer. This approach has a number of advantages, one of which is that the numerical problem can be simplified by focusing on either the surface or the inside of the scatterer. There is no need to calculate the field in an area that is outside the scatterer or artificially implement the perfectly matching or absorbing layers. In fact, boundary conditions outside the scatterer are automatically accounted for by the use of Green's function. In addition, if a scatterer that is placed on a structured material is considered rather than the same scatterer in free space, it is possible to replace Green's function in free space with the one for this structured material and repeat the same calculation in free space for the scatterer to obtain the final result. In this sense, this method provides a unified framework for dealing with numerical calculations in different EM environments, although the calculation of Green's function for media with complex structures is also a demanding task itself.

The FEM is also a popular frequency-domain method. There are two primary stages involved in the execution of the FEM. First, Maxwell's equations are transformed into a so-called variational form that incorporates integral expressions on the computational domain. This step is necessary before moving on to the next step. Second, the construction of the solution space, which should include a fair approximation of the exact solution, should be carefully done. This solution space is generated by first subdividing the computational domain into small geometric patches and then supplying many polynomials on each patch to approximate the solution. The patches and the local polynomials that are defined on them are referred to as finite elements. In two dimensions, the most typical examples of finite elements are triangles and rectangles. In three dimensions, the most typical examples of finite elements are tetrahedrons and cuboids, along with constant, linear, quadratic, and cubic polynomials. It is necessary to piece together these locally defined polynomial spaces in order to maintain the tangential continuity of the electric and magnetic fields across the boundaries of adjacent patches. In FEM, it is possible to treat complex geometrical shapes, like curved geometries, without using geometrical approximations. Moreover, the finite element mesh can be adjusted easily to fit the behavior of the solution, for instance, the treatment of singularities at corners. Furthermore, by using high-order approximations, the FEM can also ensure rapid convergence of the numerical solution to the exact solution.

A very large number of numerical methods and variants of those methods exist for solving Maxwell's equations, and it is not possible to consider all methods in one chapter. We have chosen the methods that we believe are predominantly used in MNTR. Methods not covered here include, among others, the multiple multipole method, the beam propagation method, the finite integral and finite volume methods, the method of lines, and the PWE method.

8.2 Finite-Difference Time-Domain Method

In this section, we will discuss the FDTD method. First introduced by Yee [725] in 1966 and later developed by Taflove and others [726–729], the FDTD method is considered to be one of the most general numerical methods in the fields of optics, electromagnetism, and MNTR. As mentioned earlier, it directly solves Maxwell's equations without resorting to any further approximations or derivations, it can therefore be considered a brute force approach. In this sense, the FDTD method can be regarded as a method to simulate real-world optical experiments, which is currently utilized rather frequently as a benchmark solution for the purpose of testing and verifying the results of real optical devices and other numerical methods. We begin by introducing finite-difference techniques and then proceed to explore particular features of optical waves as they propagate on the mesh. One of these features is numerical dispersion. We provide a concise overview of the practical considerations involved in the implementation of the FDTD numerical scheme, including the selection of grid sizes, the optimization of time steps to ensure the FDTD model maintains its stability, and the truncation of a numerical domain that has open boundary conditions. Both plasmonics and metamaterials are the fields that require a significant amount of involvement from frequency-dispersive materials like metals. For the sake of accuracy, we will propose several unique techniques for describing these types of materials. Since there is a solid collection of original works that are devoted to the FDTD method [730, 731], we made the conscious decision to restrict this chapter's coverage to the features of this domain that are the most important.

8.2.1 Finite-Difference Approximations of Derivatives

The finite-difference methods are widely used in the numerical simulation of optical and radiative properties of micro/nanostructures. There are no closed-form analytical solutions accessible for the problem of solving PDEs that represent boundary value problems, except for the scenarios of simple and well-defined geometrically. In this sense, numerical schemes provide the optimal and frequently most cost-effective solution for problems involving complicated geomet-

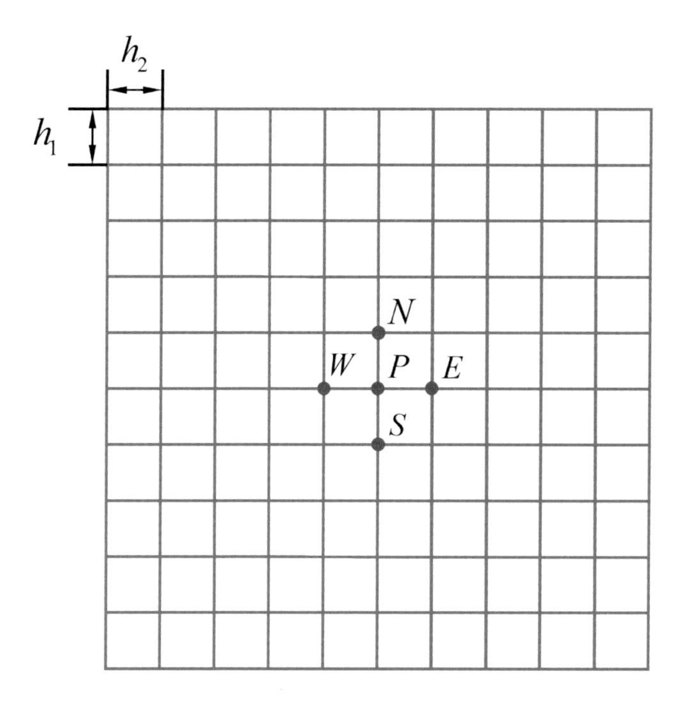

Figure 8.1 Rectangular mesh and five-point finite-difference operator.

rical shapes, with variable material qualities and frequently mixed boundary conditions.

The origin of the finite-difference method may be traced back to Gauss. This approach is considered to be one of the earliest numerical techniques. The finite-difference method is the simplest form of the point-value technique, namely, the method used for finding solutions at discrete points. This method is also known as the way of constructing a five-point star. The idea that underpins this method is to replace the PDEs of the field issue with many difference approximations. This ensures that the difference expressions get as close as possible to representing the derivatives. It is also known as the five-point star method. Let us consider the partial derivative of a function ϕ in the x direction, which can be expressed as

$$\frac{\partial \phi}{\partial x} = \lim_{\Delta x \to 0} \frac{\phi(x+\Delta x, y, z) - \phi(x, y, z)}{\Delta x}, \tag{8.2.1}$$

where Δx is the separation distance between adjacent points in the x direction. If Δx is sufficiently small, the first- and second-order partial derivatives of the function at point P shown in Fig. 8.1 can be given as

$$\frac{\partial \phi}{\partial x} \approx \frac{\phi(x+\Delta x, y, z) - \phi(x, y, z)}{\Delta x}, \tag{8.2.2}$$

$$\frac{\partial^2 \phi}{\partial x^2}|_p \approx \frac{\phi_w + \phi_e - 2\phi_p}{\Delta x}, \tag{8.2.3}$$

and

$$\frac{\partial^2 \phi}{\partial y^2}|_P \approx \frac{\phi_n + \phi_s - 2\phi_p}{\Delta y}, \tag{8.2.4}$$

where ϕ_j, $j = W, N, E, P, S$, indicates the function values at different points. Note it is more rigorous to consider Taylor's expansion series and investigate the order of truncation errors for finite-difference approximation, as discussed in [732]. A more general discussion on finite-difference methods from a mathematical point of view can be found in, for example, [733, 734].

8.2.2 The FDTD Scheme

Implementation of the FDTD method in three dimensions starts from the time-dependent Maxwell's equations for curls, Eqs. (7.1.2) and (7.1.4), and then the method discretizes them on the homogeneous Yee grid (or Yee mesh) composed of the unit Yee cell. The so-called Yee cell, as shown in Fig. 8.2, was first proposed by Kane Yee in 1966 [725]. The Yee grid is at the heart of the FDTD approach because it allows for a leapfrog time update of the fields to be performed on a staggered mesh. The components of the magnetic field are specified in the centers of the cube faces, while the components of the electric field are defined in the middle of the cube's edges. This setup is carried out with the intention of arranging the continuity of the field components (tangential components of **E** and **H**) at the interfaces.

Let us start from Maxwell's equations for isotropic and homogeneous media in the absence of charges and currents, and apply the material (Eqs. (7.1.6) and (7.1.7)):

$$\nabla \times \mathbf{E} = -\mu \frac{\partial \mathbf{H}}{\partial t} \tag{8.2.5}$$

and

$$\nabla \times \mathbf{H} = \varepsilon \frac{\partial \mathbf{E}}{\partial t}, \tag{8.2.6}$$

which lead to six scalar equations as follows:

$$\frac{\partial H_x}{\partial t} = \frac{1}{\mu}\left(\frac{\partial E_y}{\partial z} - \frac{\partial E_z}{\partial y}\right), \tag{8.2.7a}$$

$$\frac{\partial H_y}{\partial t} = \frac{1}{\mu}\left(\frac{\partial E_z}{\partial x} - \frac{\partial E_x}{\partial z}\right), \tag{8.2.7b}$$

$$\frac{\partial H_z}{\partial t} = \frac{1}{\mu}\left(\frac{\partial E_x}{\partial y} - \frac{\partial E_y}{\partial x}\right), \tag{8.2.7c}$$

$$\frac{\partial E_x}{\partial t} = \frac{1}{\varepsilon}\left(\frac{\partial H_z}{\partial y} - \frac{\partial H_y}{\partial z}\right), \tag{8.2.7d}$$

$$\frac{\partial E_y}{\partial t} = \frac{1}{\varepsilon}\left(\frac{\partial H_x}{\partial z} - \frac{\partial H_z}{\partial x}\right), \tag{8.2.7e}$$

$$\frac{\partial E_z}{\partial t} = \frac{1}{\varepsilon}\left(\frac{\partial H_y}{\partial x} - \frac{\partial H_x}{\partial y}\right). \tag{8.2.7f}$$

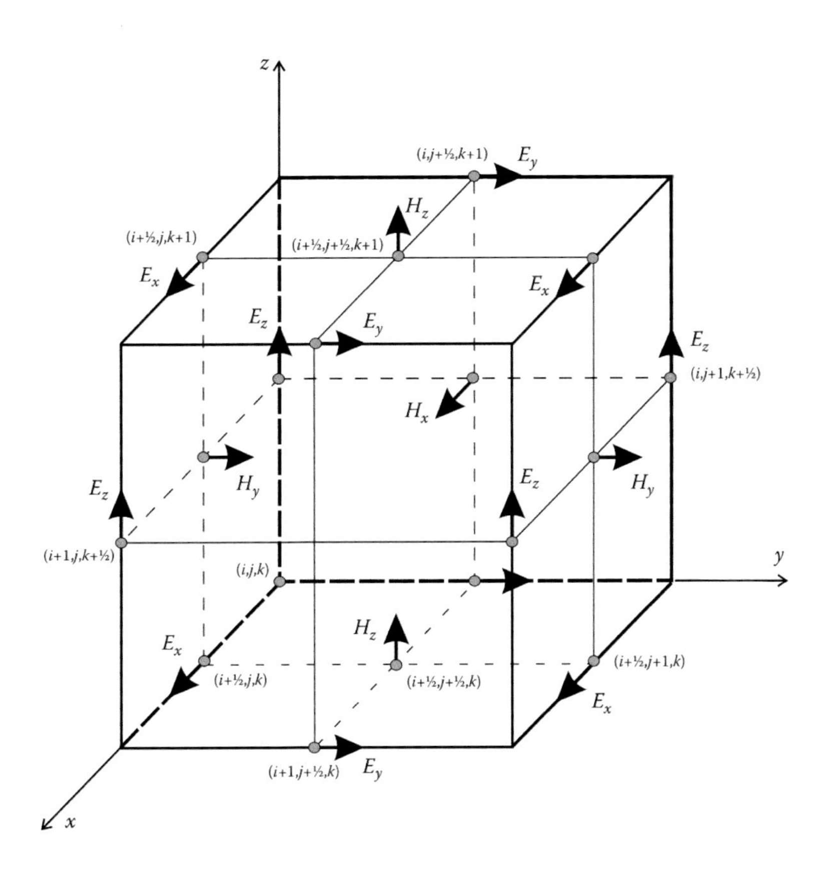

Figure 8.2 The Yee grid in three dimensions. Not all of the field components in the Yee cell are shown for the sake of clarity of presentation.

The finite-difference formalism is then obtained by utilizing the central finite difference approximation for space and temporal derivatives that are accurate to the second order. This is then applied to the Yee cell, and the following expressions are obtained:

$$E_x|_{i+1/2,j,k}^{n+1/2} = E_x|_{i-1/2,j,k}^{n-1/2} + \frac{\delta t}{\varepsilon_0 \varepsilon_{i+1/2,j,k}} \times \left(\frac{H_z|_{i+1/2,j+1/2,k}^{n} - H_z|_{i+1/2,j-1/2,k}^{n}}{\delta y} - \frac{H_y|_{i+1/2,j,k+1/2}^{n} - H_y|_{i+1/2,j,k-1/2}^{n}}{\delta z} \right), \tag{8.2.8}$$

$$E_y|_{i,j+1/2,k}^{n+1/2} = E_y|_{i,j+1/2,k}^{n-1/2} + \frac{\delta t}{\varepsilon_0 \varepsilon_{i,j+1/2,k}} \times \left(\frac{H_x|_{i,j+1/2,k+1/2}^{n} - H_x|_{i,j+1/2,k-1/2}^{n}}{\delta z} \frac{H_z|_{i+1/2,j+1/2,k}^{n} - H_z|_{i-1/2,j+1/2,k}^{n}}{\delta x} \right), \tag{8.2.9}$$

$$E_z|_{i,j,k+1/2}^{n+1/2} = E_z|_{i,j,k+1/2}^{n-1/2} + \frac{\delta t}{\varepsilon_0 \varepsilon_{i,j,k+1/2}} \times \left(\frac{H_y|_{i+1/2,j,k+1/2}^{n} - H_y|_{i-1/2,j,k+1/2}^{n}}{\delta x} - \frac{H_x|_{i,j+1/2,k+1/2}^{n} - H_x|_{i,j-1/2,k+1/2}^{n}}{\delta y} \right), \tag{8.2.10}$$

$$H_x|_{i,j+1/2,k+1/2}^{n+1} = H_x|_{i,j+1/2,k+1/2}^{n} + \frac{\delta t}{\mu_0 \mu_{i,j+1/2,k+1/2}} \times \left(\frac{E_y|_{i,j+1/2,k+1}^{n+1/2} - E_y|_{i,j+1/2,k}^{n+1/2}}{\delta z} - \frac{E_z|_{i,j+1,k+1/2}^{n+1/2} - E_z|_{i,j,k+1/2}^{n+1/2}}{\delta y} \right), \tag{8.2.11}$$

$$H_y|_{i+1/2,j,k+1/2}^{n+1} = H_y|_{i+1/2,j,k+1/2}^{n} + \frac{\delta t}{\mu_0 \mu_{i+1/2,j,k+1/2}} \times \left(\frac{E_z|_{i+1,j,k+1/2}^{n+1/2} - E_z|_{i,j,k+1/2}^{n+1/2}}{\delta x} - \frac{E_x|_{i+1/2,j,k+1}^{n+1/2} - E_x|_{i+1/2,j,k}^{n+1/2}}{\delta z} \right), \tag{8.2.12}$$

$$H_z|_{i+1/2,j+1/2,k}^{n+1} = H_z|_{i+1/2,j+1/2,k}^{n} + \frac{\delta t}{\mu_0 \mu_{i+1/2,j+1/2,k}} \times \left(\frac{E_x|_{i+1/2,j+1/2,k}^{n} - E_x|_{i+1/2,j,k}^{n+1/2}}{\delta y} - \frac{E_y|_{i+1,j+1/2,k}^{n+1/2} - E_y|_{i,j+1/2,k}^{n+1/2}}{\delta x} \right). \tag{8.2.13}$$

We can obtain a set of explicit formulas to update the electric (magnetic) fields at the next time moment by using the electric (magnetic) fields at the current moment and the magnetic (electric) fields at a shifted half-time step from the above equations.

8.2.3 Numerical Dispersion, Accuracy, and Stability

In finite-difference methods, all continuous functions and operations, such as derivatives, limits, and integrals, are approximated by discrete functions, which are then calculated on a grid. As a consequence, the accuracy of such approximations depends heavily on the grid. Even the propagation of a plane wave in vacuum, when described using a grid, will have dispersion and anisotropy as a direct consequence of this. These undesired characteristics, which are caused by the tabulation or discretization of functions, are usually given names such as numerical dispersion and numerical anisotropy, among many others. As a result, it is reasonable for us to assume that several of the well-known features of the trial solutions of Maxwell's equations will need to be carefully corrected. In this section, we will examine the consequences that discrete functions defined on a grid can have in the numerical FDTD method. We will focus on the fundamentals of stability analysis, and numerical dispersion, and examine the FDTD approach to ensure that the field-updating scheme exhibits divergence-free behavior.

As noted earlier, the solution of Maxwell's equations in a linear, isotropic, and homogeneous medium in the absence of sources is the well-known traveling harmonic plane-wave solution. If we express the field on discrete grids $(x,y,z,t)\rightarrow(a\delta x, b\delta y, c\delta z, n\delta t)$, with a, b, c, and n being the integers, we can obtain

$$\mathbf{E}(x,y,z,t)=\mathbf{E}(a,b,c,n)=\mathbf{E}_0\exp\left[\mathrm{i}(-\omega n\delta t+k_x a\delta x+k_y b\delta y+k_z c\delta z)\right] \quad (8.2.14)$$

and similar to the magnetic field. Note we use the $\mathrm{e}^{-\mathrm{i}\omega t}$ convention. A central difference time operator $\mathcal{D}_t^c$ applied on the field $\mathbf{E}$ in point (a,b,c,n) of such grid gives [735]

$$\begin{aligned}\mathcal{D}_t^c\mathbf{E}(x,y,z,t)\big|_{abc}^n &= \frac{\mathbf{E}_{abc}^{n+1/2}-\mathbf{E}_{abc}^{n-1/2}}{\delta t}=\mathbf{E}(a,b,c,n)\frac{\mathrm{e}^{-\mathrm{i}\omega\delta t/2}-\mathrm{e}^{\mathrm{i}\omega\delta t/2}}{\delta t}\\ &=\mathbf{E}(a,b,c,n)\frac{-2\mathrm{i}\sin(\omega\delta/2)}{\delta t}.\end{aligned} \quad (8.2.15)$$

In analogy, a spatial central difference gives

$$\mathcal{D}_x^c\mathbf{E}(x,y,z,t)\big|_{abc}^n=\mathbf{E}(a,b,c,n)\frac{2\mathrm{i}\sin(k_x\delta x/2)}{\delta x} \quad (8.2.16)$$

and corresponding expressions for other coordinates. Therefore, the six scalar equations constitute a homogeneous system of linear equations, expressed in the matrix form:

$$\begin{pmatrix} -T & 0 & 0 & 0 & Z/\varepsilon_0 & -Y/\varepsilon_0\\ 0 & -T & 0 & -Z/\varepsilon_0 & 0 & X/\varepsilon_0\\ 0 & 0 & -T & Y/\varepsilon_0 & -X/\varepsilon_0 & 0\\ 0 & -Z/\mu_0 & Y/\mu_0 & -T & 0 & 0\\ Z/\mu_0 & 0 & -X/\mu_0 & 0 & -T & 0\\ -Y/\mu_0 & X/\mu_0 & 0 & 0 & 0 & -T \end{pmatrix}\begin{pmatrix}E_x\\E_y\\E_z\\H_x\\H_y\\H_z\end{pmatrix}=0, \quad (8.2.17)$$

where $T=\sin(\omega\delta t/2)/\delta t$, $X=\sin(k_x\delta x/2)/\delta x$, $Y=\sin(k_y\delta y/2)/\delta y$, $Z=\sin(k_z\delta z/2)/\delta z$. A nontrivial solution can be found for this matrix equation if and only if the matrix's determinant is zero, from which the following condition is obtained:

$$\frac{1}{c^2\delta t^2}\sin^2\frac{\omega\delta t}{2}=\left(\frac{1}{\delta x}\sin\frac{k_x\delta x}{2}\right)^2+\left(\frac{1}{\delta y}\sin\frac{k_y\delta y}{2}\right)^2+\left(\frac{1}{\delta z}\sin\frac{k_z\delta z}{2}\right)^2. \quad (8.2.18)$$

This equation suggests the existence of a numerical artifact known as numerical dispersion, which is caused by the lattice discretization process. This particular type of dispersion can result in pulse distortion, fake anisotropy, and spurious refraction at the boundaries of different grids, even when it deals with such a simple problem of wave propagation in vacuum or a homogeneous dielectric medium. From this equation, it is evident that the values of temporal and spatial grid spacings have a significant impact on this numerical artifact. Because it determines the landscape and the dynamical resolution of fields' evolution, the constraints on the time step δt are of particular relevance. A too small δt will make the overall computation process more time-consuming, while a too large δt will probably breach the causal relationship between the fields' updates at various grid points. We need a numerical scheme stability study to evaluate the constraints that must be placed on the allowable time-step range.

Let us presume that the frequency can be extended to the complex plane with $\omega = \omega' + i\omega''$. In this manner, Eq. (8.2.18) can be solved formally. We can define a quantity ξ as

$$\xi = c\delta t \sqrt{\left(\frac{1}{\delta x}\sin\frac{k_x\delta x}{2}\right)^2 + \left(\frac{1}{\delta y}\sin\frac{k_y\delta y}{2}\right)^2 + \left(\frac{1}{\delta z}\sin\frac{k_z\delta z}{2}\right)^2}, \tag{8.2.19}$$

and therefore the numerical dispersion relation turns into

$$\sin^2\left(\frac{\omega\delta t}{2}\right) = \xi^2, \tag{8.2.20}$$

from which the frequency can be obtained as

$$\omega = \frac{2}{\delta t}\sin^{-1}\xi. \tag{8.2.21}$$

It is necessary to restrict the angular frequency ω to real values in order to achieve the goal of having a traveling wave solution that can propagate without experiencing losses eternally. Therefore, we should have $|\xi| \leq 1$ as required by condition $\mathrm{Im}(\omega) = 0$. As a result, we can deduce the following conclusion:

$$\delta t \leq \frac{1}{c}\left\{\left(\frac{1}{\delta x}\right)^2 + \left(\frac{1}{\delta y}\right)^2 + \left(\frac{1}{\delta z}\right)^2\right\}^{-1/2}. \tag{8.2.22}$$

This equation is called the Courant stability criterion, often known as the Courant–Friedrichs–Lewy condition. It does this by effectively restricting δt in the FDTD method by using the fixed grid. This is accomplished by setting restrictions for the sizes of grid intervals in space and time. We can thus obtain $\delta t < \delta x/(\sqrt{3}c)$ as the stability criterion for the cubic 3D mesh, and more broadly, for meshes with n dimensions, $\delta t < \delta x/(\sqrt{n}c)$ is the stability criterion. Otherwise, we have a $|\xi| > 1$, showing that electric and magnetic fields are able to increase without restriction.

Thus far we have been focusing on two different curls of Maxwell's equations. Because of the curl features of the EM field, continuous EM fields in isotropic homogeneous dielectrics naturally obey the other two divergence equations. However, discrete functions defined on grid points do not necessarily have this attribute automatically ensured for them. Therefore, a reliable numerical approach for solving Maxwell's equations needs to be able to fulfill discrete divergence equations as well. Because the divergence operator $\nabla \cdot \mathbf{E}$ cannot be canceled out in a vacuum in this case, this indicates that there are physically existing free electric charges. If we take into account the characteristics of EM fields, then the production of these spurious charges could result in instability in numerical algorithms and nonphysical outcomes. In point of fact, the FDTD method that employs the Yee cell with a leapfrog updating scheme is fortunately divergence-free. To demonstrate this, let us consider a unit Yee cell and prove the following equation is valid:

$$\frac{\partial}{\partial t}\oint_S \mathbf{E}\cdot\mathbf{n}\mathrm{d}S = 0. \tag{8.2.23}$$

For a cubic Yee cell centered at $i+1/2, j+1/2, k+1/2$, as shown in Fig. 8.2, the LHS of the above equation can be written as

$$\begin{aligned}\frac{\partial}{\partial t}\oint_S \mathbf{E}\cdot\mathbf{n}\mathrm{d}S = &\left(\left.\frac{\partial E_x}{\partial t}\right|_{i+1,j+1/2,k+1/2} - \left.\frac{\partial E_x}{\partial t}\right|_{i,j+1/2,k+1/2}\right)\delta y\delta z\\ &+\left(\left.\frac{\partial E_y}{\partial t}\right|_{i+1/2,j+1,k+1/2} - \left.\frac{\partial E_y}{\partial t}\right|_{i+1/2,j,k+1/2}\right)\delta x\delta z\\ &+\left(\left.\frac{\partial E_z}{\partial t}\right|_{i+1/2,j+1/2,k+1} - \left.\frac{\partial E_z}{\partial t}\right|_{i+1/2,j+1/2,k}\right)\delta x\delta y.\end{aligned} \tag{8.2.24}$$

According to the construction of the Yee cell, partial time derivatives of field $\mathbf{E}$ can be expressed through the central difference scheme by using the magnetic field components $\mathbf{H}$, yielding

$$\begin{aligned}\varepsilon_0\left.\frac{\partial E_x}{\partial t}\right|_{i+1,j+1/2,k+1/2} =& \frac{H_{z,i+1,j+1,k+1/2} - H_{z,i+1,j,k+1/2}}{\delta y}\\ &- \frac{H_{y,i+1,j+1/2,k+1} - H_{y,i+1,j+1/2,k}}{\delta z},\end{aligned} \tag{8.2.25}$$

$$\begin{aligned}\varepsilon_0\left.\frac{\partial E_x}{\partial t}\right|_{i,j+1/2,k+1/2} =& \frac{H_{z,i,j+1,k+1/2} - H_{z,i,j,k+1/2}}{\delta y}\\ &- \frac{H_{y,i,j+1/2,k+1} - H_{y,i,j+1/2,k}}{\delta z},\end{aligned} \tag{8.2.26}$$

$$\begin{aligned}\varepsilon_0\left.\frac{\partial E_y}{\partial t}\right|_{i+1/2,j+1,k+1/2} =& \frac{H_{x,i+1/2,j+1,k+1} - H_{x,i+1/2,j+1,k}}{\delta z}\\ &- \frac{H_{z,i+1,j+1,k+1/2} - H_{z,i,j+1,k+1/2}}{\delta x},\end{aligned} \tag{8.2.27}$$

$$\varepsilon_0 \frac{\partial E_y}{\partial t}\bigg|_{i+1/2,j,k+1/2} = \frac{H_{x,i+1/2,j,k+1} - H_{x,i+1/2,j,k}}{\delta z} - \frac{H_{z,i+1,j,k+1/2} - H_{zi,j,k+1/2}}{\delta x}, \tag{8.2.28}$$

$$\varepsilon_0 \frac{\partial E_z}{\partial t}\bigg|_{i+1/2,j+1/2,k+1} = \frac{H_{y,i+1,j+1/2,k+1} - H_{y,i,j+1/2,k+1}}{\delta x} - \frac{H_{z,i+1/2,j+1,k+1} - H_{z,i+1/2,j,k+1}}{\delta y}, \tag{8.2.29}$$

$$\varepsilon_0 \frac{\partial E_z}{\partial t}\bigg|_{i+1/2,j+1/2,k} = \frac{H_{y,i+1,j+1/2,k} - H_{y,i,j+1/2,k}}{\delta x} - \frac{H_{z,i+1/2,j+1,k} - H_{z,i+1/2,j,k}}{\delta y}. \tag{8.2.30}$$

Inserting all these expressions into Eq. (8.2.24), we can find the RHS indeed turns out to be zero. As a result, we have demonstrated that the overall flux of the electric field does not evolve over the course of time. This means the complete absence of electric charges in a vacuum at the time zero. As a result of this, we have come to the conclusion that the unit Yee cell does not produce any fictitious charges, and hence the divergence of the electric field on the grid is equal to zero.

8.2.4 Absorbing Boundary Conditions

Many problems in numerical EMs are open-boundary problems, which means that an unlimited amount of memory for the storage of data and an equally huge amount of processing time for computations are required if the computational domain is not artificially truncated. In order to reasonably truncate the computational domain, in FDTD, ABCs are frequently used. Due to the presence of these conditions, the signal will be able to propagate forward without causing any reflection at the boundary. When a wave that has been incident on a boundary continues to propagate in just one direction after it has left the boundary, we say that the boundary is reflectionless. It is common knowledge that the Dirichlet and Neumann boundary conditions, which are respectively characterized by the expressions $\phi=0$ and $\partial\phi/\partial n=0$, with ϕ being a field function, provide perfect reflections at the border. One might ponder whether or not the proper specification of ϕ at the boundary could lead to the production of zero reflection. Let us talk about how the proper boundary conditions can be defined to describe a boundary that does not reflect anything. The presented analysis is originated from Enquist and Majda [736].

Let us first consider the 1D case. The 1D wave equation propagating in the z direction and polarized over the x direction in vacuum is given by

$$\frac{\partial^2 E_x}{\partial z^2} - \frac{1}{c^2}\frac{\partial^2 E_x}{\partial t^2} = 0. \tag{8.2.31}$$

We can write this equation in the operator form by defining the operator $\hat{G}_{zt}$:

$$\hat{G}_{zt}E_x \equiv \left(\partial_{zz}^2 - \frac{1}{c_0^2}\partial_{tt}^2\right)E_x = 0. \tag{8.2.32}$$

The operator $\hat{G}_{zt}$ can be further factorized as

$$\hat{G}_{zt} = \hat{G}_{zt}^{+}\hat{G}_{zt}^{-} = \left(\partial_z + \frac{1}{c_0}\partial_t\right)\left(\partial_z - \frac{1}{c_0}\partial_t\right). \tag{8.2.33}$$

It can be seen that each of these two operators describes one-way the z propagation, in which $\hat{G}_{zt}^{+}$ governs the propagation in the positive direction and $\hat{G}_{zt}^{-}$ represents the propagation in the negative direction. Hence, the reflectionless ABCs can be straightforwardly written as

$$\hat{G}_{zt}^{-}H_y = 0, \quad \text{At } z = z_1, \tag{8.2.34a}$$

$$\hat{G}_{zt}^{+}E_x = 0, \quad \text{At } z = z_N, \tag{8.2.34b}$$

where z_1 and z_N are the left and right boundary nodes, respectively. By transforming the operator equations into finite-difference equations, we can obtain practically useful boundary conditions. For instance, Eq. (8.2.34b) can be written as

$$E_{x,N+1}^{n+1} = E_{x,N}^{n} + \left(\frac{c\delta t - \delta x}{c\delta t + \delta x}\right)\left(E_{x,N}^{n+1} - E_{x,N+1}^{n}\right). \tag{8.2.35}$$

The condition is more complicated in two dimensions, as it is more difficult to factorize the 2D wave equation that is given by

$$\hat{G}_{xyt}\mathbf{E} \equiv \left(\partial_{xx}^2 + \partial_{yy}^2 - \frac{1}{c_0^2}\partial_{tt}^2\right)\mathbf{E} = 0. \tag{8.2.36}$$

In this equation, the factorization results in the partial derivatives becoming intertwined with one another. In addition, different field components are involved, each of which presents a different continuity condition at the interfaces. For attacking this problem, Mur proposed a more efficient and optimal implementation for the ABC, which is now called Mur's ABC. For more details about analytical ABC, see [731].

Berenger offered a different version of the ABC strategy in 1994 [737]. He came up with the term perfectly matched layer (PML) for this ABC. The PML ABCs have quickly become the most effective and widely used technique for organizing nonreflecting (open) boundaries in both the time and frequency-domain methods. The fundamental concept of PML can be demonstrated with a straightforward illustration of a plane interface between two media, each of which is defined by a set of material parameters denoted by the symbols $\varepsilon_1, \varepsilon_2, \mu_1, \mu_2$, with a plane wave normally incident on the interface. The Fresnel formula for calculating the amplitude of the reflected wave can be expressed into the impedance as

$$r = \frac{Z_1 - Z_2}{Z_1 + Z_2}, \tag{8.2.37}$$

where $Z_1 = \sqrt{\mu_1/\varepsilon_1}$ and $Z_2 = \sqrt{\mu_2/\varepsilon_2}$. To make the interface reflectionless, which means $r = 0$, we obtain the impedance matching condition:

$$Z_1 = Z_2. \tag{8.2.38}$$

The absence of reflection indicates that the impinging energy is totally carried into medium 2. In medium 2, the energy can be absorbed or transported as in an infinitely transparent medium. The latter possibility cannot be implemented since the computational domain should be truncated. Hence, the former option is practically chosen. This is the principle of the PML, which truncates the computational domain in a specific direction and allows EM waves to enter it without any reflection. The electromagnet energy is then fully dissipated in the PML due to its absorption.

Let us recall the definition of the complex permittivity using the $\mathrm{e}^{-\mathrm{i}\omega t}$ convention:

$$\varepsilon(\omega) = \varepsilon\left(1 + \frac{\mathrm{i}\sigma_e}{\omega\varepsilon}\right) \tag{8.2.39}$$

and

$$\mu(\omega) = \mu\left(1 + \frac{\mathrm{i}\sigma_m}{\omega\mu}\right), \tag{8.2.40}$$

where ε and μ are the real constants and σ_e and σ_m are the conventional electric conductivity and magnetic conductivity, respectively, also assumed to be real constants. If medium 1 is nonabsorbing, from the impedance matching condition we can obtain

$$\sqrt{\frac{\mu_1}{\varepsilon_1}} = \sqrt{\frac{\mu_2\left(1 + \frac{\mathrm{i}\sigma_m}{\omega\mu_2}\right)}{\varepsilon_2\left(1 + \frac{\mathrm{i}\sigma_e}{\omega\varepsilon_2}\right)}}. \tag{8.2.41}$$

Moreover, we can assume that the real parts of complex permittivity and permeability are equal for the two media, namely

$$\varepsilon_1 = \varepsilon_2 = \varepsilon \tag{8.2.42}$$

and

$$\mu_1 = \mu_2 = \mu. \tag{8.2.43}$$

Combining the above equations, we can obtain

$$\frac{\sigma_m}{\mu} = \frac{\sigma_e}{\varepsilon}. \tag{8.2.44}$$

The above conditions can result in a plane-wave transport into the PML without any reflection. Now let us consider how this wave will propagate inside the PML. In medium 1, the wavenumber is given by $k_1 = \omega\sqrt{\varepsilon_1\mu_1}/c$, which is a real number. In medium 2, the wavenumber is

$$k_2 = \frac{\omega}{c}\sqrt{\varepsilon_2\mu_2\left(1 + \frac{\mathrm{i}\sigma_m}{\omega\mu_2}\right)\left(1 + \frac{\mathrm{i}\sigma_e}{\omega\varepsilon_2}\right)} = k_1\left(1 + \frac{\mathrm{i}\sigma_e}{\omega\varepsilon_1}\right). \tag{8.2.45}$$

It can be seen that in this situation, the real parts of the wavenumbers in the first and the second media are equal. This implies that the EM waves at the interface are temporally and spatially matched. The nonzero imaginary part of k_2 indicates that the wave amplitude decays exponentially in the PML and thus is absorbed:

$$\exp\left(\mathrm{i}k_2''z\right) = \exp\left(\mathrm{i}\frac{\mathrm{i}\sigma_e k_1}{\omega\varepsilon_1}z\right) = \exp\left(-\frac{\sigma_e k_1}{\omega\varepsilon_1}z\right). \tag{8.2.46}$$

The FDTD method is a time-domain method, and the field-updating schemes comprise time-domain permittivities and permeabilities, while the above Eqs. (8.2.43) and (8.2.43) for the permittivity and permeability are given in the frequency domain. In order to reconcile these, it is essential to convert the material functions of the PMLs into the time-domain representation. Let us first consider the 1D case. Given the existence of frequency-independent sources σ_e in the PML, using the central difference scheme, we can obtain

$$\frac{E_i^{n+1/2} - E_i^{n-1/2}}{\delta t} = -\frac{1}{\varepsilon_i}\left(\frac{H_{i+1/2}^n - H_{i-1/2}^n}{\varepsilon_0 h} + \sigma_{e,i}E_i^n\right), \tag{8.2.47}$$

with h being the spatial step, in which the E_i^n in the last term in brackets is evaluated exactly at the coordinate (i,n) since it is not involved with any finite-difference operation. As in the Yee cell, the electric field is defined only on half-integer time points; therefore, we should express E_i^n into the field values on these fields, which thus can be approximated as

$$E_i^n = \frac{E_i^{n+1/2} + E_i^{n-1/2}}{2}. \tag{8.2.48}$$

Combining the above two equations leads to a formal updating scheme for $E_i^{n+1/2}$ as

$$E_i^{n+1/2} = \frac{2\varepsilon_i - \sigma_i\delta t}{2\varepsilon_i + \sigma_i\delta t}E_i^{n-1/2} - \frac{2\delta t}{h\varepsilon_0\left(2\varepsilon_i + \sigma_{e,i}\delta t\right)}\left(H_{i+1/2}^n - H_{i-1/2}^n\right). \tag{8.2.49}$$

A similar formula can be obtained for $H_{i+1/2}^{n+1}$ by using the curl of the magnetic field:

$$\begin{aligned} H_{i+1/2}^{n+1} = {} & \frac{2\mu_{i+1/2} - \sigma_{m,i+1/2}\delta t}{2\mu_{i+1/2} + \sigma_{m,i+1/2}\delta t}H_{i+1/2}^n \\ & - \frac{2\delta t}{h\mu_0\left(2\mu_{i+1/2} + \sigma_{m,i+1/2}\delta t\right)}\left(E_{i+1}^{n+1/2} - E_i^{n+1/2}\right). \end{aligned} \tag{8.2.50}$$

In the 2D and 3D scenarios, the notion of perfect matching is not as straightforward as one might think, particularly for oblique incidence. According to what was demonstrated by Gedney [738], it is necessary to satisfy particular tensorial criteria, the nature of which is determined by the orientation of the interfaces. The most difficult parts to work with are the edges and corners of the computational domain since this is where PMLs that were formulated in separate directions interweave. Details of the explicit schemes for the time-domain implementation of 2D and 3D PMLs can be found in [739–741].

8.2.5 Considerations on Material Dispersion

As mentioned in Chapter 7, when there is a frequency dispersion of material properties, the constitutive equations will be given by the convolution formulas in the time domain. In this scenario, the displacement vector, for instance, is determined by electric fields in all previous time moments, indicating in some way that the medium traces the history of electric-field values. In certain specific instances, it is possible to considerably simplify the convolution expression while avoiding the necessity of doing backward temporal integration. Instead, an additional differential equation is generated, which is also numerically integrated through an explicit finite-difference scheme coupled with the main FDTD scheme. This additional differential equation is then solved using the FDTD. These cases are usually described by well-known models for the consideration of the frequency-dependent material properties. In this section, we take a look at how the Debey, Drude, and Lorentz models are often handled in FDTD.

The Debye relaxation model is used to describe the EM response of a material with polar molecules, which can be considered as a group of dipoles undergoing perturbation by an external electrical field. In this model, the susceptibility of the material in the frequency domain is given by [742]

$$\chi_D(\omega) = \frac{\chi_0}{1 - i\omega\tau}, \tag{8.2.51}$$

where τ is the relaxation time, and χ_0 is the constant related to the static orientational polarizability of molecules. By using this expression, the polarization current is obtained as

$$\mathbf{J}_p(\omega) = -i\omega\varepsilon_0 \frac{\chi_0}{1 - i\omega\tau} \mathbf{E}(\omega), \tag{8.2.52}$$

or equivalently,

$$\mathbf{J}_p(\omega) - i\omega\tau \mathbf{J}_p(\omega) = -i\omega\varepsilon_0\chi_0 \mathbf{E}(\omega). \tag{8.2.53}$$

This equation can be transformed into a time domain as (by replacing $-i\omega$ by $\partial/\partial t$)

$$\mathbf{J}_p(t) + \tau \frac{\partial \mathbf{J}_p(t)}{\partial t} = \varepsilon_0\chi_0 \frac{\partial \mathbf{E}(t)}{\partial t}. \tag{8.2.54}$$

If the derivative of the electric field is known, we can solve the time-dependent polarization current from Eq. (8.2.52). This kind of equation is usually called the auxiliary differential equation (ADE) in dispersive media for the FDTD method [743, 744]. For seeking a numerical solution, the derivatives in this equation can be replaced by finite differences and combined with the FDTD field-updating scheme. By using the second-order central difference scheme to the time derivatives centered at the time moment $n+1/2$, we obtain

$$\frac{\mathbf{J}_p^{n+1}+\mathbf{J}_p^n}{2}+\tau\frac{\mathbf{J}_p^{n+1}-\mathbf{J}_p^n}{\delta t}=\varepsilon_0\chi_0\frac{\mathbf{E}^{n+1}-\mathbf{E}^n}{\delta t}. \tag{8.2.55}$$

Thus, the updating scheme of the polarization current is obtained as

$$\mathbf{J}_p^{n+1}=\frac{2\tau-\delta t}{2\tau+\delta t}\mathbf{J}_p^n+\frac{2\varepsilon_0\varepsilon_d}{2\tau+\delta t}\left(\mathbf{E}^{n+1}-\mathbf{E}^n\right). \tag{8.2.56}$$

By invoking Ampere's law and implementing the finite-difference scheme in the time moment $n+1/2$, we have

$$\nabla\times\mathbf{H}^{n+1/2}=\varepsilon_0\frac{\left(\mathbf{E}^{n+1}-\mathbf{E}^n\right)}{\delta t}+\mathbf{J}_p^{n+1/2}+\sigma\frac{\mathbf{E}^{n+1}+\mathbf{E}^n}{2}, \tag{8.2.57}$$

in which $\mathbf{J}_p^{n+1/2}$ is not explicitly known and thus can be further approximated as

$$\mathbf{J}_p^{n+1/2}=\frac{\mathbf{J}_p^{n+1}+\mathbf{J}_p^n}{2}. \tag{8.2.58}$$

An explicit expression for it can be obtained by inserting the updating expression (8.2.56). Equations (8.2.55)–(8.2.58) can be used to explicitly update $\mathbf{E}^{n+1}$. Therefore, we have obtained the updating schemes for the fields and polarization current for a dispersive material described by the Debye model.

The Drude model for the permittivity of metals is already introduced in Section 7.2.1. The susceptibility in the Drude model (taking $\varepsilon_\infty=1$) is expressed as

$$\chi_{\text{Drude}}(\omega)=-\frac{\omega_p^2}{\omega^2+\mathrm{i}\gamma\omega}=-\frac{\omega_p^2}{\mathrm{i}\omega(-\mathrm{i}\omega+\gamma)}. \tag{8.2.59}$$

The polarization current is thus given by

$$\mathbf{J}_p(\omega)=\frac{-\mathrm{i}\omega\varepsilon_0\omega_p^2}{-\mathrm{i}\omega(-\mathrm{i}\omega+\gamma)}\mathbf{E}(\omega), \tag{8.2.60}$$

which results in an equation in the frequency domain as

$$-\mathrm{i}\omega\mathbf{J}_p(\omega)+\gamma\mathbf{J}_p(\omega)=\varepsilon_0\omega_p^2\mathbf{E}(\omega). \tag{8.2.61}$$

This equation can be transformed into the time domain, leading to an ADE as

$$\frac{\partial\mathbf{J}_p(t)}{\partial t}+\gamma\mathbf{J}_p(t)=\varepsilon_0\omega_p^2\mathbf{E}(t). \tag{8.2.62}$$

We can discretize this equation on a grid as

$$\frac{\mathbf{J}_p^{n+1}-\mathbf{J}_p^n}{\delta t}+\frac{\gamma}{2}\left(\mathbf{J}_p^{n+1}+\mathbf{J}_p^n\right)=\varepsilon_0\omega_p^2\frac{\mathbf{E}^{n+1}+\mathbf{E}^n}{2}. \tag{8.2.63}$$

Therefore, we can obtain the updated scheme for the polarization current as

$$\mathbf{J}_p^{n+1}=\frac{2-\gamma\delta t}{2+\gamma\delta t}\mathbf{J}_p^n+\frac{\varepsilon_0\omega_p^2\delta t}{2+\gamma\delta t}\left(\mathbf{E}^{n+1}+\mathbf{E}^n\right). \tag{8.2.64}$$

The same approximation as Eq. (8.2.58) can be taken to express $\mathbf{J}_p^{n+1/2}$ in the above formula (Eq. (8.2.64)). Therefore, following a similar updating procedure for the case of the Debye model, we can obtain the complete updating scheme of the Drude model.

The scenario for the Lorentz model is a little bit more complicated. Let us consider the polarization current in the frequency domain

$$\mathbf{J}_p(\omega) = -\mathrm{i}\omega\varepsilon_0 \frac{\omega_l^2}{\omega_0^2 - \mathrm{i}\gamma_l\omega - \omega^2}\mathbf{E}(\omega). \tag{8.2.65}$$

Similarly, the frequency-domain equation becomes

$$\left(\omega_0^2 - \mathrm{i}\omega\gamma_l - \omega^2\right)\mathbf{J}_p(\omega) = \mathrm{i}\omega\varepsilon_0\omega_l^2\mathbf{E}(\omega), \tag{8.2.66}$$

which is then transformed into the time domain, which gives the ADE as

$$\omega_0^2\mathbf{J}_p(t) + \gamma_l\frac{\partial \mathbf{J}_p(t)}{\partial t} + \frac{\partial^2\mathbf{J}_p(t)}{\partial t^2} = \varepsilon_0\omega_l^2\frac{\partial \mathbf{E}(t)}{\partial t}. \tag{8.2.67}$$

It is noted that this equation contains the second-order derivative. Therefore, the centering point of the finite-difference approximation should be n, which leads to

$$\omega_0^2\mathbf{J}_p^n + \gamma_l\frac{\mathbf{J}_p^{n+1} - \mathbf{J}_p^{n-1}}{2\delta t} + \frac{\mathbf{J}_p^{n+1} - 2\mathbf{J}_p^n + \mathbf{J}_p^{n-1}}{\delta t^2} = \varepsilon_0\omega_l^2\frac{\mathbf{E}^{n+1} - \mathbf{E}^{n-1}}{2\delta t}. \tag{8.2.68}$$

Hence, we get the updating scheme for the polarization current:

$$\mathbf{J}_p^{n+1} = 2\frac{2 - \omega_0^2\delta t^2}{2 + \gamma_l\delta t}\mathbf{J}_p^n + \frac{\gamma_l\delta t - 2}{2 + \gamma_l\delta t}\mathbf{J}_p^{n-1} + \frac{\varepsilon_0\omega_l^2\delta t}{2 + \gamma_l\delta t}\left(\mathbf{E}^{n+1} - \mathbf{E}^{n-1}\right). \tag{8.2.69}$$

To obtain $\mathbf{J}_p^{n+1/2}$, we also need to use the approximation (8.2.58). Then we can use similar updating procedures like those in the media described by the Debye and Drude models. Due to the presence of the second-order derivatives, the updating scheme needs to store four additional arrays of quantities, namely $\mathbf{E}_n$, $\mathbf{E}_{n-1}$, $\mathbf{J}_p^n$, and $\mathbf{J}_p^{n-1}$.

We note that instead of deriving the updating scheme of the polarization current $\mathbf{J}_p = \partial\mathbf{P}/\partial t$, it is feasible to consider the material dispersion directly using the polarization vector $\mathbf{P}$ [735].

To summarize, we have briefly summarized the FDTD method, which is by now the most universal method that can directly solve Maxwell's equations in the time domain. The discussion presented in this section, is only an preliminary introduction of this numerical method, with a large amount of important themes uncovered, including the total field/scattered field numerical scheme, subgridding and nonuniform meshing, and near-field–far-field transformation. It is vital to note that the FDTD method, by virtue of its algorithm, is well suited for parallel computing, which results in a significant acceleration of the calculations. For more information, the readers can refer to relevant monographs [730, 731, 745].

8.3 The Modal Method

8.3.1 General Introduction

The term "modal method" (MM) refers to a series of mode expansion approaches in the frequency domain that have been developed over the course of the 1980s to the 2000s. These methods were at first created as a grating theory and given the titles rigorously coupled waveguide analysis (RCWA) [746] and Fourier modal method (FMM) [747] in order to examine the diffraction of periodic structures. After that, a mode expansion technique known as the eigenmode expansion technique was used for the analysis of rotationally symmetric structures such as VCSELs and micropillar cavities [748, 749]. In recent years, the method has been employed to model open geometries without rotational or translational symmetries, such as the PC slab, thus called the aperiodic Fourier modal method (a-FMM) [750, 751]. There are just a few key conceptual differences between the various methodologies, and these differences are related to the symmetries and boundary conditions of the geometry that is being investigated, in addition to the technique that is being used to calculate the modes.

When using the mode expansion technique, the geometry needs to be segmented into layers having a relative permittivity profile that is uniform along a propagation axis, often the z axis. Then we determine the transverse eigenmodes of these layers using the wave equation, which are commonly referred to as eigenmodes. After this, the optical field in each layer is expanded on the corresponding eigenmodes, and that mode matching at the interface is used to connect the fields on each side of the interface between neighboring layers. We should also note the spatial discretization that is utilized in approaches based on finite differences and that is used in techniques based on mode expansion are not the same. The majority of numerical methods for EMs involve the incorporation of a spatial discretization of the computational domain that is subwavelength in nature. On the other hand, the MM only divides the geometry into layers that have the same relative permittivity profile throughout. The use of such a subdivision offers computing benefits in the context of structures composed mostly of sections that are uniform or periodic. The MM provides direct access to various fascinating physical quantities, such as mode profiles, effective indices, and scattering coefficients, as we will see in the following discussion. In spite of several decades' worth of intensive development, commercial software packages that implement the method are still difficult to come by [752], and there are only a small number of open-source software packages available [753, 754] with relatively limited functionality. Thankfully, the approach is not overly complicated, and a student with fundamental programming skills should be able to construct the method without any trouble. This section discusses the basics and also briefly introduces some advanced extensions of the MM, such as ABCs and the Bloch mode formalism.

The idea of expanding on eigenmodes is the first thing we cover here. After that, we will examine the simplest possible scenario, which is the 1D geometry.

During this step, we will become familiar with the idea of subdividing the geometry into uniform layers along a propagation axis. After that, we will broaden the scope of the formalism to include 2D structures and expand on multiple lateral eigenmodes. Formalisms for managing the specific example of a periodic structure are the subject of our final discussion. In this section, we will assume that the materials are nonmagnetic, and we will use the $\mathrm{e}^{-\mathrm{i}\omega t}$ convention for the time dependency.

The determination of eigenmodes, which acts as a basis set for the expansion of the optical field, is the first step in the MM. In order to characterize them, the relative permittivity profile is first split along the z axis into the layers that have translational symmetry. The permittivity profile of a random glass sample is depicted in Fig. 8.3. Along the z axis, the sample itself can be divided into three layers that are uniformly numbered from 2 to 4. The entire structure is made up of a total of five layers, including the first and fifth layers of the air that surrounds the sample.

We will now consider the geometry with a permittivity profile in the (x, y) plane that corresponds to the initial geometry but with uniformity along the z axis. This will be done for each layer. As an illustration, the (x, z) profile that corresponds to layer 2 in Fig. 8.3(a) is depicted in Fig. 8.3(b). When we talk about the eigenmodes of a particular layer in the text that follows, it is important to understand that we are referring to the solutions to the associated geometry that have uniformity along the z axis (namely, infinitely long). This can be seen in Fig. 8.3(b). The layers in the entire geometry do have finite thickness; however, the nonuniformity that results from this will be dealt with in a separate step using a recursive matrix formalism.

Let us repeat Eq. (7.1.35) here for a nonmagnetic medium as

$$\nabla^2\mathbf{E} - \varepsilon\mu\frac{\partial^2\mathbf{E}}{\partial t^2} + \nabla\left[\mathbf{E}\cdot\nabla(\ln\varepsilon)\right] = 0. \tag{8.3.1}$$

Based on the plane harmonic wave assumption, we can write the electric field in geometry with z invariance as (neglecting the $\mathrm{e}^{-\mathrm{i}\omega t}$ term)

$$\mathbf{E} = \mathbf{e}(\mathbf{r}_\perp)\mathrm{e}^{\mathrm{i}\beta z}, \tag{8.3.2}$$

where β is the so-called propagation constant that is the only parameter controlling the z dependence of the electric field. Inserting this ansatz into Eq. (8.3.1), we obtain an eigenvalue problem equation:

$$\nabla_\perp^2\mathbf{e} + (\nabla_\perp + \mathrm{i}\beta\hat{\mathbf{z}})\left(\mathbf{e}\cdot\nabla\ln\varepsilon\left(\mathbf{r}_\perp\right)\right) + \varepsilon\left(\mathbf{r}_\perp\right)k_0^2\mathbf{e} = \beta^2\mathbf{e}, \tag{8.3.3}$$

where $\nabla_\perp$ represents the transverse part of the nabla operator. This is a not conventional eigenvalue problem; it is a second-order eigenvalue problem that is more difficult to solve numerically than the conventional one. Because the z derivative of ε is vanishing, the scalar equations for the in-plane components $\mathbf{e}_\perp$

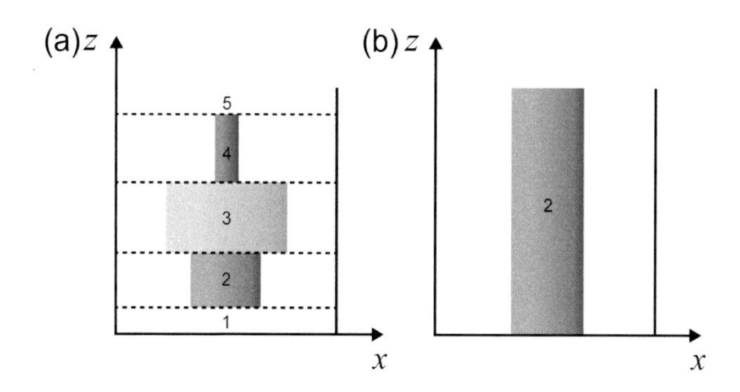

Figure 8.3 (a) The division of an example geometry into five layers. (b) The profile for layer 2 with uniformity along the z axis.

are not coupled to e_z. As a result, it suffices to solve the equation for the in-plane components:

$$\nabla_\perp^2 \mathbf{e}_\perp + \nabla_\perp \left[\mathbf{e}_\perp \cdot \nabla \ln \varepsilon\left(\mathbf{r}_\perp\right)\right] + \varepsilon\left(\mathbf{r}_\perp\right) k_0^2 \mathbf{e}_\perp = \beta^2 \mathbf{e}_\perp, \tag{8.3.4}$$

whose solution is, therefore, the eigenmode. This equation is the master equation for the MMs. By solving it, we are able to derive eigenmodes denoted by $\mathbf{e}_{\perp m}(\mathbf{r}_\perp)$ and eigenvalues that are the squares of the propagation constants denoted by β_m, where m is the mode number. We can also define an effective index written as $n_{\mathrm{eff,m}} = \beta_m / k_0$ to describe the propagation of the eigenmode. These eigenmodes serve as a basis set upon which the optical field in the layer is expanded. Note it has been demonstrated [755] that the eigenmodes obtained by solving Eq. (8.3.4) constitute a complete set for geometries that have a real-valued relative permittivity profile and uniformity along the y axis. However, it is not so clear whether or not the completeness extends to general geometries with complex permittivity profiles. Up to this point, there has been no evidence to suggest that this is not the case. As a result, the completeness of the eigenmodes is typically assumed, despite the absence of the formal proof.

Therefore, the eigenmodes allow us to expand an arbitrary optical field as

$$\mathbf{E}_\perp(\mathbf{r}) = \sum_{m=1}^{\infty} a_m \mathbf{e}_{\perp m}\left(\mathbf{r}_\perp\right) \mathrm{e}^{\mathrm{i}\beta_m z} + \sum_{m=1}^{\infty} b_m \mathbf{e}_{\perp m}\left(\mathbf{r}_\perp\right) \mathrm{e}^{-\mathrm{i}\beta_m z}, \tag{8.3.5}$$

where the expansion coefficients a_m and b_m correspond to the eigenmodes traveling in the forward and backward directions, respectively. Theoretically speaking, the summations would be performed over an infinite set of eigenmodes. However, in order to do calculations in a realistic setting, a truncation would be required that would result in a summation being performed over a finite number N of modes. Note that the $\mathbf{e}_\perp$ are the only components of the electric field involved in the Eqs. (8.3.2)–(8.3.5). The remaining components of the six-component EM field can be derived from the $\mathbf{e}_\perp$. Another choice is to consider the eigenvalue problem for the magnetic field, which computes the lateral magnetic field $\mathbf{h}_\perp$, and then deduces the other field components from the $\mathbf{h}_\perp$ components. Each of these methods is equivalent to each other.

8.3.2 One-Dimensional Geometry

Before moving on to investigate structures in 2D and 3D dimensions, we will first acquaint ourselves with the recursive matrix formalisms that describe the propagation of the field along the z axis in multilayer structures in simple 1D geometry. In the geometry of one dimension, the refractive index $n(z)$ changes as it moves down the z axis of propagation, but it is unaffected by the lateral x and y coordinates. In the effective index approximation, the 1D theory is not only straightforward but also very effective at describing how light travels through fiber Bragg gratings and VCSELs. Figure 8.4 illustrates one example of this kind of refractive-index profile. Following the MM, we will proceed by subdividing the refractive-index profile $n(z)$ into layers q of a constant index n_q $(q=1,2,3,\ldots)$, as shown in Fig. 8.4(a), and we will use a recursive matrix formalism to relate the fields in the various layers, as will be described in more detail in this section later. Considered at this time is a uniform layer q, as depicted in Fig. 8.4(b). In this geometry, the refractive index does not rely on position; hence, it is the simplest conceivable geometry. The eigenmode problem for the layer q reduces to

$$\nabla_\perp^2 \mathbf{e}_q + n_q^2 k_0^2 \mathbf{e}_q = \beta^2 \mathbf{e}_q. \tag{8.3.6}$$

We are going to work out the solution to this equation for a $\mathbf{r}_\perp$-dependent field that is propagating along the z axis. In this particular scenario, there is only one lateral eigenmode $\mathbf{e}_q$ in the layer q, and our expansion coefficient sets in Eq. (8.3.5) for the forward and backward traveling eigenmodes therefore simply reduce to the scalar ones a_q and b_q, respectively. The lateral eigenmode profile can be written simply as $\mathbf{e}_q = C\hat{\mathbf{u}}$, where C is a constant and $\hat{\mathbf{u}}$ is a unit vector in the (x,y) plane. This is because the lateral field components in the 1D case do not couple.

From Eq. (8.3.6), a simple dispersion relation can be given

$$n_q^2 k_0^2 = \beta_q^2, \tag{8.3.7}$$

and inside the layer q, the scalar electric field along the propagation axis z, $E_q(z)$, is expressed as

$$E_q(z) = a_q e^{i\beta_q z} + b_q e^{-i\beta_q z}, \tag{8.3.8}$$

where $\beta_q = n_q k_0$ and the full vectorial field is thus $\mathbf{E}_q(z) = E_q(z)\hat{\mathbf{u}}$.

This equation is a sum of a traveling field in both the forward and backward directions, where the variables a_q and b_q are constants that describe the phases and amplitudes of the forward and backward contributions, respectively. It constitutes a complete analytical solution, which indicates that once the coefficients a_q and b_q have been determined, the field in all of the spatial coordinates z can

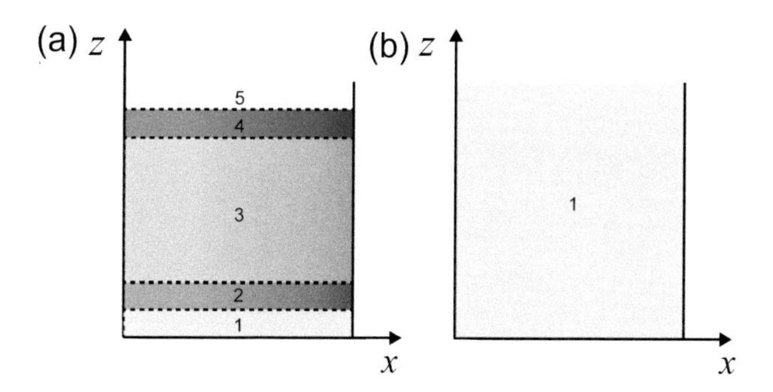

Figure 8.4 (a) A geometry consisting of layers with refractive-index profile $n(z)$, represented by different colors, depending only on the z coordinate. (b) The single-layer geometry.

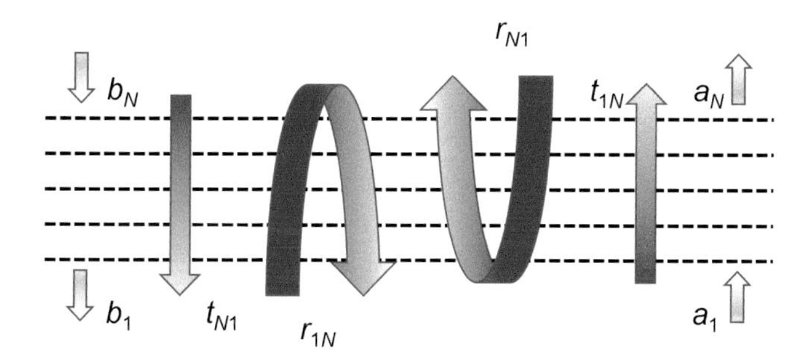

Figure 8.5 Forward and backward traveling light in layers 1 and N in an N-layer geometry. The fields are related by the reflection and transmission coefficients r_{1N}, t_{1N}, r_{N1}, and t_{N1}.

be immediately derived. We therefore just need to find the coefficients a_q and b_q, and after that, we can utilize the analytical dependence on z to compute the EM field. Moreover, we can also determine the intensity I, defined as the power flow per unit area for a 1D wave $a_q e^{in_q k_0 z}$, which can be calculated as follows: $I = nq|a_q|^2/(2\mu_0 c)$.

When dealing with geometries that are composed of numerous layers, we use a recursive matrix formalism to determine how the fields in each of the distinct layers are related to one another. Following this, we shall discuss two such formalisms, namely the transfer or T-matrix formalism [285] and the S-matrix formalism [756]. Consider the 1D N-layer geometry shown in Fig. 8.5, which is illuminated by two incoming fields: a forward traveling field a_1 in layer 1 and a backward traveling field b_N in layer N. These fields are moving in opposite directions. The scattering of the incoming fields results in the production of the outgoing fields b_1 and a_N, and this scattering can be characterized by the reflection and transmission coefficients defined by

$$r_{1N} = \frac{b_1}{a_1} \quad (b_N = 0), \tag{8.3.9}$$

$$t_{1N} = \frac{a_N}{a_1} \quad (b_N = 0), \tag{8.3.10}$$

$$r_{N1} = \frac{a_N}{b_N} \quad (a_1 = 0), \tag{8.3.11}$$

$$t_{N1} = \frac{b_1}{b_N} \quad (a_1 = 0), \tag{8.3.12}$$

and therefore the outgoing fields can be expressed as

$$a_N = t_{1N} a_1 + r_{N1} b_N \tag{8.3.13}$$

and

$$b_1 = r_{1N} a_1 + \mathbf{t}_{N1} b_N. \tag{8.3.14}$$

From Eqs. (8.3.13) and (8.3.14), we can obtain the S-matrix formalism that establishes the relationship between the outgoing scattered fields and the incoming fields. By using a matrix notation, we have

$$\begin{bmatrix} a_N \\ b_1 \end{bmatrix} = \mathbf{S} \begin{bmatrix} a_1 \\ b_N \end{bmatrix}, \tag{8.3.15}$$

where the matrix elements of $\mathbf{S}$ are obtained directly from Eqs. (8.3.13) and (8.3.14) as

$$\mathbf{S} = \begin{bmatrix} S_{11} & S_{12} \\ S_{21} & S_{22} \end{bmatrix} = \begin{bmatrix} t_{1N} & r_{N1} \\ r_{1N} & t_{N1} \end{bmatrix}. \tag{8.3.16}$$

Having the ability to relate the forward and backward traveling fields at one place in the geometry with those at another location in the geometry is, on the other hand, quite convenient. This is accomplished by the use of the T-matrix formalism, in which the fields on either side of the structure are connected as follows:

$$\begin{bmatrix} a_N \\ b_N \end{bmatrix} = \mathbf{T} \begin{bmatrix} a_1 \\ b_1 \end{bmatrix}. \tag{8.3.17}$$

Similarly, the T-matrix elements can also be derived from Eqs. (8.3.13) and (8.3.14) by rearranging terms that are

$$T = \begin{bmatrix} T_{11} & T_{12} \\ T_{21} & T_{22} \end{bmatrix} = \frac{1}{t_{N1}} \begin{bmatrix} t_{1N} t_{N1} - r_{1N} r_{N1} & r_{N1} \\ -r_{1N} & 1 \end{bmatrix}. \tag{8.3.18}$$

The S and T matrices are defined in Eqs. (8.3.16) and (8.3.18) based on the knowledge of the reflection and transmission coefficients, respectively. On the other hand, once we have computed the S or the T-matrix for an arbitrary geometry, we are able to infer the properties of reflection and transmission for

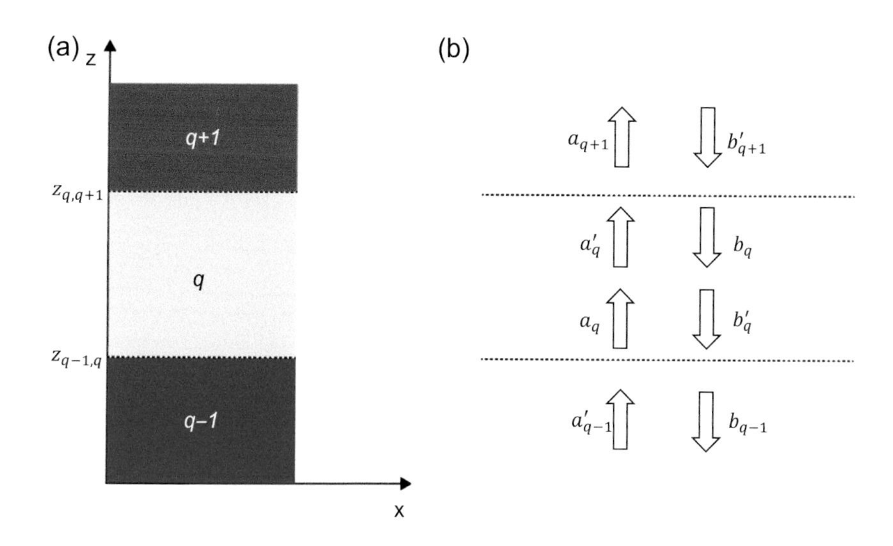

Figure 8.6 (a) A multilayer geometry. (b) The corresponding reference coordinate positions of the field expansion coefficients.

that geometry from the S and T matrices. In what follows, we will demonstrate that the T-matrix formalism is ideally suited for dealing with 1D geometries, whereas the S-matrix formalism is obligatory for dealing with 2D and 3D geometries.

By this stage, the expansion coefficients a_q and b_q for the layer q have been determined in relation to the $z=0$ plane, so that $E_q(0)$ equals a_q+b_q. However, when we take into consideration geometries with multiple layers, it will be beneficial to construct expansion coefficients relative to various points z contained within the layer q. These coefficients will be denoted by the notations a_q^z and b_q^z, respectively. For expansion coefficients that are determined in relation to the positions of the interfaces, in particular, we will implement the notation that is outlined in Fig. 8.6. From this point forward, we will define the coefficients a_q and b'_q relative to the z coordinate $z_{q-1,q}$ of the interface between the layers $q-1$ and q, and in the same way, we will define the coefficients a'_q and b_q relative to the z coordinate $z_{q,q+1}$ of the interface between the layers q and $q+1$.

We are able to describe the optical field in a 1D geometry using either the S- or the T-matrix formalism, and with any number of layers that we choose. First, using the T-matrix formalism, we will create the propagation matrix $\mathbf{P}_q$ for a uniform layer q with the length $L_q=z_{q,q+1}-z_{q-1,q}$ and the refractive index n_q that is depicted in Fig. 8.7. The decision that was made regarding the direction of the z axis will be justified later. The fields on either side of a uniform layer are related to one another by the propagation matrix, $\mathbf{P}_q$, as follows:

$$\begin{bmatrix} a'_q \\ b_q \end{bmatrix} = \mathbf{P}_q \begin{bmatrix} a_q \\ b'_q \end{bmatrix}, \tag{8.3.19}$$

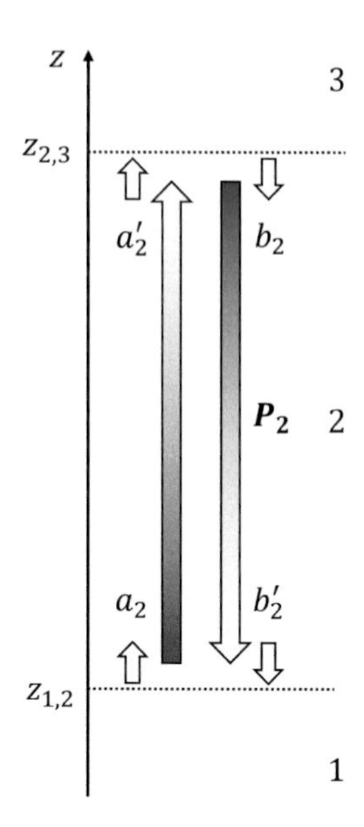

Figure 8.7 Propagation of light inside a 1D uniform layer.

and the matrix elements of the propagation matrix are given by

$$\mathbf{P}_q = \begin{bmatrix} e^{in_q k_0 L_q} & 0 \\ 0 & e^{-in_q k_0 L_q} \end{bmatrix}. \tag{8.3.20}$$

In a broader sense, the propagation matrix $\mathbf{P}_q(\Delta z)$ can be utilized in the interior of the layer q to associate the field coefficients a_q^z and b_q^z defined at z with the coefficients $a_q^{z'}$ and $b_q^{z'}$ defined at z' by substituting the distance $\Delta z = z' - z$ for the layer length L_q in Eq. (8.3.20). However, in order to keep the notation as short as possible, we will omit the argument (L_q) when referring to the propagation matrix $\mathbf{P}_q(L_q)$ over the full layer length L_q.

After determining the interface matrix $\mathbf{T}_{q,q+1}$, which relates the fields on each side of an interface between the layers q and $q+1$, as well as the propagation matrix $\mathbf{P}_q$, which relates the fields at different positions within the uniform layer q, we are now prepared to treat a structure with an arbitrary number of layers. The relation between the expansion coefficients can subsequently be derived for a three-layer geometry like the one shown in Fig. 8.8 by simply cascading the propagation and interface matrices, so that

$$\begin{bmatrix} a_3^z \\ b_3^z \end{bmatrix} = \mathbf{T}_\mathrm{t} \begin{bmatrix} a_1^0 \\ b_1^0 \end{bmatrix} = \mathbf{P}_3(z - z_{23})\,\mathbf{T}_{23}\mathbf{P}_2\mathbf{T}_{12}\mathbf{P}_1(z_{12}) \begin{bmatrix} a_1^0 \\ b_1^0 \end{bmatrix}. \tag{8.3.21}$$

And as we can see, the decision that was made regarding the orientation of the z axis in Figs. 8.7 and 8.8 make it possible for there to be a correspondence between the order of the geometry segments and the order of the T-matrix multiplication. The field at any given location in the geometry can be calculated

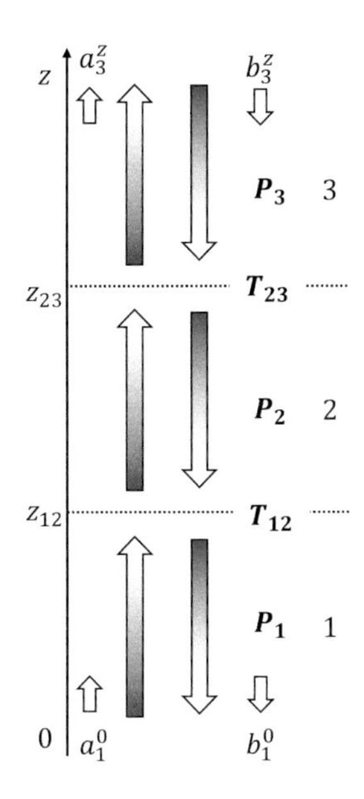

Figure 8.8 Propagation of light described by $\overline{\overline{T}}$ and $\overline{\overline{P}}$ matrices for a three-layer geometry.

by following this approach, which involves the cascading of the T and P matrices. In addition, the reflection and transmission characteristics of the whole structure can be computed by looking at the total T-matrix $\mathbf{T}_t$.

8.3.3 The Plane-Wave Expansion Method

We are now going to look at geometries in which the permittivity profile is dependent not only on the z coordinate but also on the lateral x coordinate. Similar to the 1D case, the first thing to do is to partition the geometry into the q layers that are uniform along the z axis, as shown in Fig. 8.3(a). After that, we compute the eigenmodes for each layer while assuming that the permittivity profile is independent of z. The EM fields will be expanded on these eigenmodes, and a recursive matrix formalism is used to relate the fields in distinct layers and to deal with the z dependence. Due to the fact that the inclusion of evanescent field components in the expansion equation (Eq. (8.3.5)) causes numerical instability in the T-matrix formalism when applied to the 2D problem, we will only be using the S-matrix formalism from this point forward.

There are a few different approaches that can be taken to compute the eigenmodes, and each one has both the pros and cons. The finite-difference method discussed in Section 8.2 can achieve this goal, and in this section, we will elaborate on the PWE method, which is commonly utilized in the FMM. When the lateral profile has translational symmetry in the 2D geometry, exact analytical descriptions of the eigenmodes are accessible. These eigenmodes can be derived using an SA approach. In the following, we will compute the eigenmodes in two dimensions by finding the solutions to the eigenvalue problems, which can be derived from the eigenvalue problem for in-plane EM fields, Eq. (8.3.4), for a special class of solutions. This class of solutions $(\mathbf{e}(x), \mathbf{h}(x))$ does not vary with the y coordinate since the permittivity is constant along this direction. We can find two independent set of eigenmodes with different polarizations. When the electric field is along the y direction, we say that they are TE modes, and then the eigenvalue problem becomes

$$\partial_x^2 e_y + \varepsilon(x) k_0^2 e_y = \beta^2 e_y. \tag{8.3.22}$$

When the magnetic field is along the y direction, we say that they are TM modes, and then the eigenvalue problem is recast into

$$\partial_x^2 h_y + \varepsilon(x) k_0^2 h_y - (\partial_x \ln \varepsilon(x))\, \partial_x h_y = \beta^2 h_y. \tag{8.3.23}$$

This classification is reasonable because they are decoupled with the components in the z direction [735, 757]. We can further manipulate Eq. (8.3.23) and obtain another form that makes the boundary conditions more transparent:

$$\varepsilon(x) \partial_x \frac{1}{\varepsilon(x)} \partial_x h_y + \varepsilon(x) k_0^2 h_y = \beta^2 h_y. \tag{8.3.24}$$

Now let us consider proper continuity conditions for the fields. Physically speaking, we necessitate the continuity of the field components that are tangential to the variations of ε that take place along the x axis. In the context of TE, these components are e_y and h_z. We obtain for TE polarization $h_z = \partial_x e_y / (\mathrm{i}\omega\mu_0)$ from Maxwell's equations, and as a result, we require the continuity of e_y and $\partial_x e_y$ in the TE case. The tangential components of TM polarization are h_y and e_z, and in this circumstance, we obtain $e_z = \mathrm{i}\partial_x h_y / (\omega\varepsilon_0\varepsilon)$, which leads to the necessity of the continuity of h_y and $\partial_x h_y / \varepsilon$ in the TM polarization.

Next, we will go over the strategy for solving Eqs. (8.3.22) and (8.3.24) on the basis of the PWE, which is utilized in the RCWA and FMM methods. In this part of the analysis, the field and the dielectric constant are expanded in plane-wave series. These techniques were initially developed for the purpose of researching periodic structures. In the following, we will consider a periodic geometry with a period of L_x, in which $\varepsilon(x + L_x) = \varepsilon(x)$ is satisfied.

Before moving on to examine the eigenvalue problem, we will first investigate a number of basic features of the PWE that will be required in the subsequent discussion. Let us take a look at two periodic functions in 1D, $f(x)$ and $g(x)$, both of which have a period of L_x. They can be expanded as [430, 742]

$$f(x) = \sum_{m=-\infty}^{m=\infty} F_m \mathrm{e}^{\mathrm{i}mKx} \tag{8.3.25}$$

and

$$g(x) = \sum_{n=-\infty}^{n=\infty} G_n \mathrm{e}^{\mathrm{i}nKx}, \tag{8.3.26}$$

where $K = 2\pi/L_x$, and m and n are the integers. These expansions are actually the Fourier series, and thus the method to be discussed is called the FMM. The product of the two functions, $p(x)$, can be obtained as

$$p(x) = f(x)g(x) = \sum_{m=-\infty}^{m=\infty} F_m e^{imKx} \sum_{n=-\infty}^{n=\infty} G_n e^{inKx} = \sum_{m=-\infty}^{m=\infty} \sum_{n=-\infty}^{n=\infty} F_{m-n} G_n e^{imKx}. \tag{8.3.27}$$

Its Fourier expansion series is

$$p(x) = \sum_{m=-\infty}^{m=\infty} P_m e^{imKx}. \tag{8.3.28}$$

By comparison, we can obtain the Fourier expansion coefficients as

$$P_m = \sum_{n=-\infty}^{n=\infty} F_{m-n} G_n. \tag{8.3.29}$$

This is, in fact, the convolution theorem [758]. The functions f, g, and p are now represented by their respective Fourier series that contain the expansion coefficients F_m, G_m, and P_m. Although these expansions are infinite series, in order to implement these expansions in practical numerical calculations, we need to truncate the summations in such a way that our Fourier series for, for example, the function f is given by the vector $\mathbf{F} = [F_{-M}; \ldots; F_M]$ of finite length $N = 2M + 1$. Now, on the basis of Eq. (8.3.29), we can see that the expansion coefficient vector $\mathbf{P}$ for the function $p(x)$ can be obtained using the Fourier series of f and g in the following way:

$$\mathbf{P} = \boldsymbol{\mathcal{F}} \mathbf{G}, \tag{8.3.30}$$

where we have defined the $N \times N$ Toeplitz matrix $\boldsymbol{\mathcal{F}}$ whose matrix elements are derived from the Fourier series $\mathbf{F}$ as $\mathcal{F}_{m,n} = F_{m-n}$, resulting in the expression

$$\boldsymbol{\mathcal{F}} = \begin{bmatrix} F_0 & F_{-1} & F_{-2} & F_{-3} & \cdots \\ F_1 & F_0 & F_{-1} & F_{-2} & \cdots \\ F_2 & F_1 & F_0 & F_{-1} & \cdots \\ \cdots & \cdots & \cdots & \cdots & \end{bmatrix}. \tag{8.3.31}$$

We further consider the Fourier series of the function $f^\dagger(x) = 1/f(x)$ as

$$f^\dagger(x) = \sum_{m=-\infty}^{m=\infty} F_m^\dagger e^{imKx}, \tag{8.3.32}$$

and we also denote the corresponding Toeplitz matrix as $\boldsymbol{\mathcal{F}}^\dagger$. Since $p = fg = \left(f^\dagger\right)^{-1} g$, we can therefore obtain the Fourier series for p as

$$\mathbf{P} = \left(\boldsymbol{\mathcal{F}}^\dagger\right)^{-1} \mathbf{G}. \tag{8.3.33}$$

The two matrix products, Eqs. (8.3.31) and (8.3.33), converge toward the same limit when $N \to \infty$. However, the convergence properties of the two products are

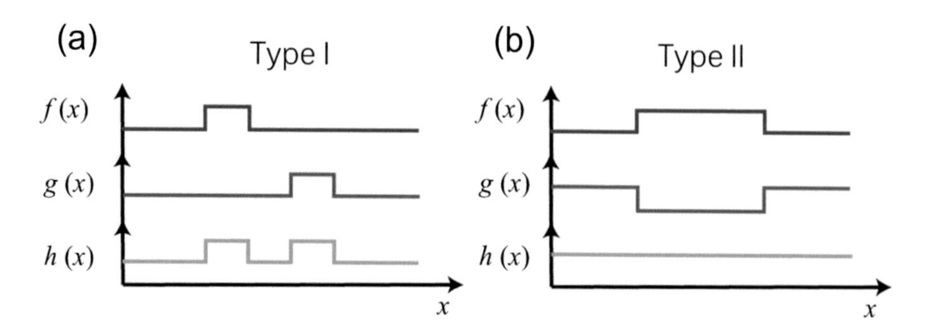

Figure 8.9 Functions $f(x)$ and $g(x)$ without (a) and with (b) complementary jump discontinuities. The discontinuity of $h(x)=f(x)g(x)$ in (b) is removable.

very different from one another with the increase of N. The speediest convergence can be achieved using either the direct rule (Eq. (8.3.31)) or the inverse rule (Eq. (8.3.33)), depending on the characteristics of the functions f and g. There are guidelines that can be used to select the appropriate product, which is known as Li's factorization rules [759].

Let us start by thinking about the discontinuous functions depicted in Fig. 8.9 so that we may get a better grasp on these rules. The selection of a matrix product is determined by the characteristics of the discontinuities that are inherent to the functions f and g. The locations of the discontinuities are referred to as jump points in this context. The functions f and g both have the jump points as depicted in Fig. 8.9(a); however, these points are located in different spatial positions. Therefore, the function $p=fg$ contains the jump points that match to the locations of the jump points found in the functions f and g. This scenario is dubbed type I. In Fig. 8.9(b), the type II scenario is presented. In this instance, the functions g and g are both discontinuous, but they demonstrate concurrent jump points. As a result, their product, p, is continuous. In this scenario, we state that the discontinuity in p is removable. We can thus summarize the factorization rules that should be as follows: (1) If the discontinuities are of type I, the fastest way to converge is to select the direct rule that is given by the matrix product, (Eq. (8.3.31)) and (2) if the discontinuities are of type II, the inverse rule of Eq. (8.3.33) provides the fastest way to converge. A more in-depth discussion of these rules can be found in [759].

Let us now go back to the eigenvalue problems represented by Eqs. (8.3.22) and (8.3.24) in order to find a solution to them via the PWE. We use periodic boundary conditions in such a way that $e_y(x+L_x)=e_y(x)e^{i\alpha}$ according to Bloch's theorem, where the parameter α can be arbitrarily chosen right from the beginning of the process. To begin, let us have a look at the eigenproblem for TE polarization, Eq. (8.3.22). In this equation, we can expand the mode profile $e_y(x)$ and the permittivity profile $\varepsilon(x)$ into PWEs:

$$e_y(x) = \sum_{j=-\infty}^{j=\infty} u_j \mathrm{e}^{\mathrm{i}(\mathrm{j}K+\alpha/L_x)x}, \tag{8.3.34}$$

$$\varepsilon(x) = \sum_{j=-\infty}^{j=\infty} \varepsilon_j \mathrm{e}^{\mathrm{ij}Kx}. \tag{8.3.35}$$

Since the profile of the electric field e_y in TE polarization is continuous as mentioned in the second paragraph of Section 8.3.3, the product εe_y is therefore of type I, and for this product, we should utilize the direct rule of the matrix product, Eq. (8.3.31). Therefore, Eq. (8.3.22) can be put out in a matrix form as

$$\left(\mathbf{K}^2 + k_0^2 \boldsymbol{\mathcal{E}}\right)\mathbf{u} = \beta^2 \mathbf{u}, \tag{8.3.36}$$

where $\mathbf{K}$ is the $N \times N$ diagonal matrix with diagonal elements given by $\mathrm{i}K$ $[-M;\ldots;M] + \mathrm{i}\alpha/L_x$, and $\boldsymbol{\mathcal{E}}$ is the Toeplitz matrix representing the expansion Eq. (8.3.35), where the matrix elements ε_j are computed as

$$\varepsilon_j = \frac{1}{L_x}\int_0^{L_x} \varepsilon(x)\mathrm{e}^{-\mathrm{ij}Kx}\mathrm{d}x, \tag{8.3.37}$$

which utilizes the periodicity of the permittivity profile. This equation can be evaluated analytically for simple profiles, while for more complicated geometries, a FFT algorithm is commonly used. Therefore, we have transformed Eq. (8.3.36) into a standard eigenvalue problem in the form of $\mathbf{Ou} = \lambda\mathbf{u}$.

When applied to the specific scenario of a uniform geometry with a dielectric constant equal to $\varepsilon(x) = \varepsilon$, the Toeplitz matrix $\boldsymbol{\mathcal{E}}$ is diagonal. If we numerate the eigenmodes m from $-M$ to M, then the solution e_{ym} to the eigenvalue problem for the uniform geometry is then essentially a single uncoupled plane wave with eigenvector coefficients $u_{m,j} = \delta_{mj}$ and a propagation constant $\beta^2 = \varepsilon k_0^2 - (mK + \alpha/L_x)^2$. After the e_{ym} profile has been computed, we are able to figure out the expansions for the magnetic fields, which are as follows:

$$h_{xm}(x) = \sum_{j=-M}^{j=M} v_{m,j} \mathrm{e}^{\mathrm{i}(\mathrm{j}K+\alpha/L_x)x} \tag{8.3.38}$$

and

$$h_{zm}(x) = \sum_{j=-M}^{j=M} w_{m,j} \mathrm{e}^{\mathrm{i}(\mathrm{j}K+\alpha/L_x)x}. \tag{8.3.39}$$

By using Faraday's law in Maxwell's equations, we find

$$h_x = -\frac{\beta}{\omega\mu_0} e_y \tag{8.3.40}$$

and

$$h_z = \frac{1}{\mathrm{i}\omega\mu_0}\partial_x e_y. \tag{8.3.41}$$

Therefore, the relations can be established between the Fourier expansion coefficients of the magnetic fields of the mth mode and those of the electric field

$$\mathbf{v}_m = -\frac{\beta_m}{\omega\mu_0}\mathbf{u}_m \tag{8.3.42}$$

and

$$\mathbf{w}_m = \frac{1}{\mathrm{i}\omega\mu_0}\mathbf{K}\mathbf{u}_m. \tag{8.3.43}$$

In a similar fashion, we can obtain the results for TM polarization, whose eigenvalue problem Eq. (8.3.24) can be written as

$$\varepsilon(x)\left(\partial_x \frac{1}{\varepsilon(x)}\partial_x + k_0^2\right) h_y = \beta^2 h_y. \tag{8.3.44}$$

The fields are expanded into the Fourier series as

$$h_{ym}(x) = \sum_{j=-M}^{j=M} v_{m,j}\mathrm{e}^{\mathrm{i}(\mathrm{j}K+\alpha/L_x)x}, \tag{8.3.45}$$

$$e_{xm}(x) = \sum_{j=-M}^{j=M} u_{m,j}\mathrm{e}^{\mathrm{i}(\mathrm{j}K+\alpha/L_x)x}, \tag{8.3.46}$$

$$e_{zm}(x) = \sum_{j=-M}^{j=M} w'_{m,j}\mathrm{e}^{\mathrm{i}(\mathrm{j}K+\alpha/L_x)x}. \tag{8.3.47}$$

As was previously mentioned, although $1/\varepsilon$ and $\partial_x h_y$ are both discontinuous, their product $1/\varepsilon(\partial_x h_y)$ is continuous, and therefore this is the type-II scenario. Therefore, the inverse rule, Eq. (8.3.33), is the one that should be used for this product. In addition, we need to take into account the product of ε and the parentheses times h_y on the LHS of Eq. (8.3.44). Although these functions are discontinuous, their product equals $\beta^2 h_y$ according to the eigenproblem equation, which is a continuous function. Hence, this product is also an example of type II. For the sake of dealing with this final product, we will define the Fourier series for $1/\varepsilon$ as

$$\frac{1}{\varepsilon(x)} = \sum_{j=-\infty}^{j=\infty} A_j \mathrm{e}^{\mathrm{ij}Kx}, \tag{8.3.48}$$

where

$$A_j = \frac{1}{L_x}\int_0^{L_x} \frac{1}{\varepsilon(x)}\mathrm{e}^{-\mathrm{ij}Kx}\mathrm{d}x \tag{8.3.49}$$

is represented by the Toeplitz matrix $\boldsymbol{\mathcal{A}}$. By utilizing the inverse product rule twice, we obtain the matrix form of the eigenvalue problem as

$$\boldsymbol{\mathcal{A}}^{-1}\left(\mathbf{K}\boldsymbol{\mathcal{E}}^{-1}\mathbf{K} + k_0^2\mathbf{I}\right)\mathbf{v} = \beta^2\mathbf{v}. \tag{8.3.50}$$

From the Eq. (8.3.50), the expansion coefficients of the magnetic field h_{ym} of the mode m can be solved, and by invoking Ampére's law, the electric field components can be calculated as

$$e_x = \frac{\beta}{\omega\varepsilon_0\varepsilon(x)} h_y, \tag{8.3.51}$$

and

$$e_z = \frac{\mathrm{i}}{\omega\varepsilon_0\varepsilon(x)} \partial_x h_y. \tag{8.3.52}$$

Inspecting the first equation reveals that there is an equivalence between the continuation of the magnetic field h_y component and the continuity of the product $e_x\varepsilon$. Therefore, when it comes to TM polarization, we can anticipate discontinuities in the electric field e_x component at sites where there are discontinuities in the permittivity profile. The expansion coefficient vectors of the electric field become

$$\mathbf{u}_m = \frac{\beta_m}{\omega\varepsilon_0} \boldsymbol{\mathcal{A}} \mathbf{v}_m, \tag{8.3.53}$$

$$\mathbf{w}'_m = \frac{\mathrm{i}}{\omega\varepsilon_0} \boldsymbol{\mathcal{E}}^{-1} \mathbf{K} \mathbf{v}_m. \tag{8.3.54}$$

Note that the direct rule is implemented in the first equation, and the inverse rule is used in the second equation.

8.3.4 The S-Matrix Theory

After determining the eigenmodes for layers that are homogeneous along the z axis, we will now pay attention to the two-layer structure again. We first expand the field on the eigenmodes in each layer as shown in Eq. (8.3.5) for an incoming field $\mathbf{E}_{\mathrm{inc},1}$ that only has one mode m, with expansion coefficients $a_{1j} = \delta_{jm}$. As shown in Fig. 8.10, the field travels in the forward direction in layer 1. However, when it reaches the layer interface, it is partially reflected, which results in the field traveling in the opposite direction in layer 1. Some of the incoming fields are also transmitted, and as a consequence, layer 2 now has a field that is traveling in the forward direction. The electric and magnetic fields in the two layers can thus be written as

$$\mathbf{E}_1(\mathbf{r}) = a'_{1m}\mathbf{e}^+_{1m}(\mathbf{r}_\perp)\,\mathrm{e}^{\mathrm{i}\beta_{1m}(z-z_{12})} + \sum_j^N b_{1j}\mathbf{e}^-_{1j}(\mathbf{r}_\perp)\,\mathrm{e}^{-\mathrm{i}\beta_{1j}(z-z_{12})}, \tag{8.3.55a}$$

$$\mathbf{E}_2(\mathbf{r}) = \sum_j^N a_{2j}\mathbf{e}^+_{2j}(\mathbf{r}_\perp)\,\mathrm{e}^{\mathrm{i}\beta_{2j}(z-z_{12})}, \tag{8.3.55b}$$

$$\mathbf{H}_1(\mathbf{r}) = a'_{1m}\mathbf{h}^+_{1m}(\mathbf{r}_\perp)\,\mathrm{e}^{\mathrm{i}\beta_{1m}(z-z_{12})} + \sum_j^N b_{1j}\mathbf{h}^-_{1j}(\mathbf{r}_\perp)\,\mathrm{e}^{-\mathrm{i}\beta_{1j}(z-z_{12})}, \tag{8.3.55c}$$

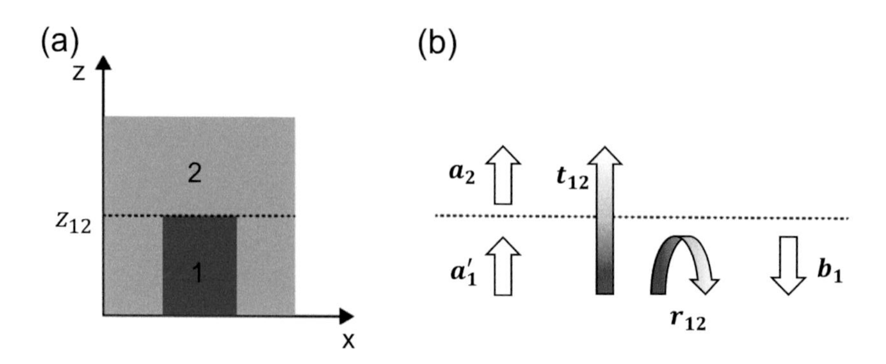

Figure 8.10 (a) A two-layer geometry. (b) The reflection and transmission of light at the interface.

$$\mathbf{H}_2(\mathbf{r}) = \sum_j^N a_{2j}\mathbf{h}_{2j}^{+}(\mathbf{r}_\perp)\,\mathrm{e}^{\mathrm{i}\beta_{2j}(z-z_{12})}, \tag{8.3.55d}$$

from which the EM fields are represented by $(\mathbf{E}_1, \mathbf{H}_1)$ for $z \leq z_{12}$ and by $(\mathbf{E}_2, \mathbf{H}_2)$ for $z \geq z_{12}$. Here $\mathbf{e}_{1m}^{+}$ represents the mth mode that is propagating in the positive z direction in layer 1 with a propagation constant β_{1m}, and similar meanings can be attributed to other mode vectors. By assumption, we set $a'_{1m} = 1$, while other coefficients b_{1j} and a_{2j} are unknown.

We will now define the reflection and transmission matrices r_{12} and t_{12}. Their matrix elements $r_{12;jm}$ and $t_{12;jm}$ are defined as $b_{1j} = r_{12;jm}a'_{1m}$ and $a_{2j} = t_{12;jm} \times a'_{1m}$, respectively. In order to obtain the coefficients of the reflection and transmission matrices, it is necessary to ensure that the boundary conditions are satisfied at the interface between layers 1 and 2. Specifically, it is necessary for the tangential components of the electric and magnetic fields to be continuous. By using Eqs. (8.3.55a) and (8.3.55d), we have

$$\mathbf{e}_{\perp 1m}(\mathbf{r}_\perp) + \sum_j^N \mathbf{e}_{\perp 1j}(\mathbf{r}_\perp)\,r_{12;jm} = \sum_j^N \mathbf{e}_{\perp 2j}(\mathbf{r}_\perp)\,t_{12;jm} \tag{8.3.56}$$

and

$$\mathbf{h}_{\perp 1m}(\mathbf{r}_\perp) - \sum_j^N \mathbf{h}_{\perp 1j}(\mathbf{r}_\perp)\,r_{12;jm} = \sum_j^N \mathbf{h}_{\perp 2j}(\mathbf{r}_\perp)\,t_{12;jm}, \tag{8.3.57}$$

where we should note a minus sign is given before the backward traveling magnetic field lateral components.

The method that is used to characterize the eigenmodes will now determine the procedure that is used to compute the reflection and transmission matrices. When dealing with eigenmodes that are defined through the use of the PWE, the lateral electric $\mathbf{e}_\perp$ and magnetic $\mathbf{h}_\perp$ fields can be readily represented through the use of their respective Fourier series vectors **u** and **v**. Therefore, above continuous boundary conditions (Eqs. (8.3.56) and (8.3.57)) can be recast into

$$\mathbf{u}_{1m}+\sum_{j}^{N}\mathbf{u}_{1j}r_{12;jm}=\sum_{j}^{N}\mathbf{u}_{2j}t_{12;jm}, \tag{8.3.58a}$$

$$\mathbf{v}_{1m}-\sum_{j}^{N}\mathbf{v}_{1j}r_{12;jm}=\sum_{j}^{N}\mathbf{v}_{2j}t_{12;jm}. \tag{8.3.58b}$$

By arranging these equations into the matrix forms, we have

$$\mathbf{U}_1+\mathbf{U}_1\mathbf{r}_{12}=\mathbf{U}_2\mathbf{t}_{12}, \tag{8.3.59a}$$

$$\mathbf{V}_1-\mathbf{V}_1\mathbf{r}_{12}=\mathbf{V}_2\mathbf{t}_{12}, \tag{8.3.59b}$$

where **U** and **V** are the matrices with columns m given by the Fourier series vector $\mathbf{u}_m$ and $\mathbf{v}_m$ for the eigenmode m. Through Eqs. (8.3.59a) and (8.3.59b), we can solve the transmission and reflection matrices as

$$\mathbf{t}_{12}=2\left(\mathbf{U}_1^{-1}\mathbf{U}_2+\mathbf{V}_1^{-1}\mathbf{V}_2\right)^{-1}, \tag{8.3.60a}$$

$$\mathbf{r}_{12}=\frac{1}{2}\left(\mathbf{U}_1^{-1}\mathbf{U}_2-\mathbf{V}_1^{-1}\mathbf{V}_2\right)\mathbf{t}_{12}. \tag{8.3.60b}$$

These obtained formulas are valid under the periodic boundary condition $\mathbf{E}(x+L_x)=\mathbf{E}(x)\mathrm{e}^{\mathrm{i}\alpha}$.

Let us now have a look at the three-layer geometry shown in Fig. 8.11. As before, we use light that is characterized by the mth eigenmode to illuminate the structure from below. The electric field in each layer is expressed as

$$\mathbf{E}_1(\mathbf{r})=a'_{1m}\mathbf{e}_{1m}^{+}(\mathbf{r}_\perp)\,\mathrm{e}^{\mathrm{i}\beta_{1m}(z-z_{12})}+\sum_{j}^{N}b_{1j}\mathbf{e}_{1j}^{-}(\mathbf{r}_\perp)\,\mathrm{e}^{-\mathrm{i}\beta_{1j}(z-z_{12})}, \tag{8.3.61a}$$

$$\mathbf{E}_2(\mathbf{r})=\sum_{j}^{N}a_{2j}\mathbf{e}_{2j}^{+}(\mathbf{r}_\perp)\,\mathrm{e}^{\mathrm{i}\beta_{2j}(z-z_{12})}+\sum_{j}^{N}b_{2j}\mathbf{e}_{2j}^{-}(\mathbf{r}_\perp)\,\mathrm{e}^{-\mathrm{i}\beta_{2j}(z-z_{23})}, \tag{8.3.61b}$$

$$\mathbf{E}_3(\mathbf{r})=\sum_{j}^{N}a_{3j}\mathbf{e}_{3j}^{+}(\mathbf{r}_\perp)\,\mathrm{e}^{\mathrm{i}\beta_{3j}(z-z_{23})}, \tag{8.3.61c}$$

respectively, and expressions for the magnetic fields in the three layers can be obtained similarly. New coefficients a'_{2j} and b'_{2j} for layer 2 can be defined as $a'_{2j}=p_{2j}a_{2j}$ and $b'_{2j}=p_{2j}b_{2j}$, where $p_{2j}=\mathrm{e}^{\mathrm{i}\beta_{2j}L_2}$ and $L_2=z_{23}-z_{12}$, as shown in Fig. 8.11(b). In contrast to the coefficients a_j and b_j, which have their backs against the interface, the coefficients a'_j and b'_j can be thought of as facing the interface.

Now let us have a look at how the three-layer geometry affects the reflection of light. Within layer 1 in Fig. 8.11(b), the illumination of mode m travels in the forward direction. It is partially reflected back into layer 1 at the interface, and this reflection is described by the matrix $\mathbf{r}_{12}$. However, some of the light is also allowed to pass through to the layer below. This transmitted part of the

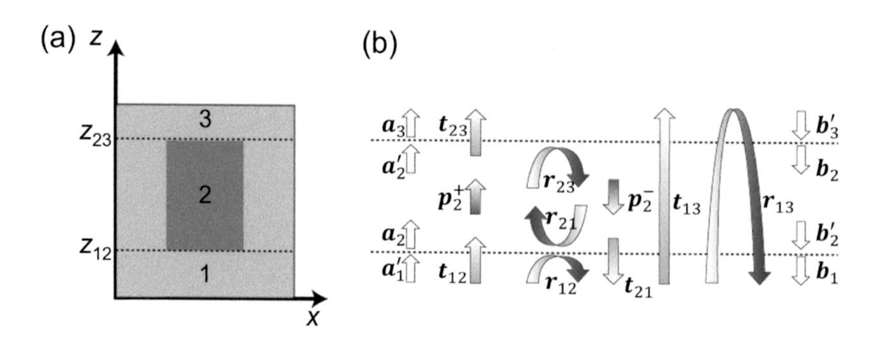

Figure 8.11 (a) A three-layer geometry. (b) The reflection and transmission at the two interfaces and the multiple scattering of light inside the central layer are illustrated.

light is further partially reflected as it propagates toward the interface between layers 2 and 3. The reflected light then travels in the opposite direction, back toward the first interface, and a portion of it is partially transmitted into layer 1. This trip may be represented by the subsequent multiplication of the matrices of individual processes $\mathbf{t}_{21}\mathbf{p}_2^-\mathbf{p}_2^+\mathbf{t}_{12}$, where $\mathbf{p}_2^+ = \mathbf{p}_2^- = \mathbf{p}_2$ is the propagation matrix for layer 2, which is a diagonal matrix consisting of the elements p_{2j}. At this point in the process, the $\mathbf{p}_2^\pm$ matrix does not depend on the direction $\pm$ of the propagation. Nevertheless, the $+$ and $-$ signs are still retained, in order to maintain compatibility with earlier parts. This trip, which consists of one round trip within layer 2, adds one more reflection to the one represented by $\mathbf{r}_{12}$. In layer 2, however, there are additional contributions from lights that have made two, three, and even more round trips. Given this, the scattering reflection matrix $\mathbf{r}_{13}$ that describes reflection at the interface between layers 1 and 2, taking into account numerous reflections within layer 2, may be written as

$$\begin{aligned}\mathbf{r}_{13} = &\mathbf{r}_{12} + \mathbf{t}_{21}\mathbf{p}_2^-\mathbf{r}_{23}\mathbf{p}_2^+\mathbf{t}_{12} + \mathbf{t}_{21}\mathbf{p}_2^-\mathbf{r}_{23}\mathbf{p}_2^+\left(\mathbf{r}_{21}\mathbf{p}_2^-\mathbf{r}_{23}\mathbf{p}_2^+\right)\mathbf{t}_{12}\\ &+ \mathbf{t}_{21}\mathbf{r}_{23}\mathbf{p}_2^+\left(\mathbf{r}_{21}\mathbf{p}_2^-\mathbf{r}_{23}\mathbf{p}_2^+\right)^2\mathbf{t}_{12} + \cdots,\end{aligned} \tag{8.3.62}$$

where the product $\mathbf{r}_{21}\mathbf{p}_2^-\mathbf{r}_{23}\mathbf{p}_2^+$ represents a round trip within layer 2. By employing the relation for a geometric series for matrices with absolute eigenvalues below unity, the summation over all orders of multiple round trips in layer 2 can be reduced, and we have

$$\mathbf{r}_{13} = \mathbf{r}_{12} + \mathbf{t}_{21}\mathbf{p}_2^-\mathbf{r}_{23}\mathbf{p}_2^+\left(\mathbf{I} - \mathbf{r}_{21}\mathbf{p}_2^-\mathbf{r}_{23}\mathbf{p}_2^+\right)^{-1}\mathbf{t}_{12}. \tag{8.3.63}$$

A similar process can be used to derive the scattering transmission matrix $\mathbf{t}_{13}$ by taking into account the multiple reflections in layer 2, which describes the transmission of light from layer 1 to layer 3, as well as the matrices $\mathbf{r}_{31}$ and

$\mathbf{t}_{31}$ corresponding to reflection and transmission of a backward traveling field in layer 3. These matrices read

$$\mathbf{t}_{13} = \mathbf{t}_{23}\mathbf{p}_2^+ \left(\mathbf{I} - \mathbf{r}_{21}\mathbf{p}_2^- \mathbf{r}_{23}\mathbf{p}_2^+\right)^{-1} \mathbf{t}_{12}, \tag{8.3.64}$$

$$\mathbf{r}_{31} = \mathbf{r}_{32} + \mathbf{t}_{23}\mathbf{p}_2^+ \left(\mathbf{I} - \mathbf{r}_{21}\mathbf{p}_2^- \mathbf{r}_{23}\mathbf{p}_2^+\right)^{-1} \mathbf{r}_{21}\mathbf{p}_2^- \mathbf{t}_{32}, \tag{8.3.65}$$

$$\mathbf{t}_{31} = \mathbf{t}_{21}\mathbf{p}_2^- \mathbf{t}_{32} + \mathbf{t}_{21}\mathbf{p}_2^- \mathbf{r}_{23}\mathbf{p}_2^+ \left(\mathbf{I} - \mathbf{r}_{21}\mathbf{p}_2^- \mathbf{r}_{23}\mathbf{p}_2^+\right)^{-1} \mathbf{r}_{21}\mathbf{p}_2^- \mathbf{t}_{32}. \tag{8.3.66}$$

The fields in layers 1 and 3 can be determined by using the relations $\mathbf{b}_1 = \mathbf{r}_{13}\mathbf{a}_1'$ and $\mathbf{a}_3 = \mathbf{t}_{13}\mathbf{a}_1'$, respectively. The expansion vectors $\mathbf{a}_2$ and $\mathbf{b}_2$ that describe the forward and backward traveling components of the field are used to calculate the field in the central layer 2 of the entire structure. The contributions to the vector $\mathbf{a}_2$ can be attributed to illumination from below and above, respectively, which are denoted by $\mathbf{t}_{12}\mathbf{a}_1'$ and $\mathbf{r}_{21}\mathbf{p}_2^- \mathbf{t}_{32}\mathbf{b}_3'$, respectively. Thus, its full expression is

$$\mathbf{a}_2 = \left(\mathbf{I} - \mathbf{r}_{21}\mathbf{p}_2^- \mathbf{r}_{23}\mathbf{p}_2^+\right)^{-1} \left(\mathbf{t}_{12}\mathbf{a}_1' + \mathbf{r}_{21}\mathbf{p}_2^- \mathbf{t}_{32}\mathbf{b}_3'\right), \tag{8.3.67}$$

in which the inverse matrix representing multiple round trips in layer 2 is also included. A similar expression for $\mathbf{b}_2$ can also be obtained:

$$\mathbf{b}_2 = \left(\mathbf{I} - \mathbf{r}_{23}\mathbf{p}_2^+ \mathbf{r}_{21}\mathbf{p}_2^-\right)^{-1} \left(\mathbf{r}_{23}\mathbf{p}_2^+ \mathbf{t}_{12}\mathbf{a}_1' + \mathbf{t}_{32}\mathbf{b}_3'\right). \tag{8.3.68}$$

Therefore, from above scattering matrices (Eqs. (8.3.63) to (8.3.66)) and the expansion vectors for the field (Eqs. (8.3.67) and (8.3.68)), the fields in the entire three-layer geometry can be completely determined.

Let us use a matrix notation for the above procedures for calculating the full S matrix connecting layer 1 and layer 3. The scattering matrix $\mathbf{S}_{pq}$ between the layers p and q can be defined as

$$\begin{bmatrix} \mathbf{a}_q \\ \mathbf{b}_p' \end{bmatrix} = \mathbf{S}_{pq} \begin{bmatrix} \mathbf{a}_p \\ \mathbf{b}_q' \end{bmatrix}, \tag{8.3.69}$$

where the thickness of the layer p is taken into consideration, but not the presence of adjoining layers below the layer p and above the layer q. By this definition, we can have the cascading relation for S matrices as

$$\mathbf{S}_{13} = \mathbf{S}_{12} \otimes \mathbf{S}_{23}. \tag{8.3.70}$$

When the scattering, reflection, and transmission matrices in the three-layer structure are calculated, we are able to recursively compute the field in a structure with any given number of layers based on the above recursive relation. Let us consider a four-layer structure as an instance. Following the steps given previously, we begin by computing the scattering matrices $\mathbf{r}_{13}$, $\mathbf{t}_{13}$, $\mathbf{r}_{31}$, and $\mathbf{t}_{31}$, respectively. We are therefore able to conceive the four-layer structure as a three-layer geometry. The scattering reflection matrix $\mathbf{r}_{14}$ is therefore given by a formula that is in a similar form to that of Eq. (8.3.63), in which the matrices $\mathbf{r}_{12}$, $\mathbf{t}_{12}$, are simply replaced by $\mathbf{r}_{13}$ and $\mathbf{t}_{13}$, respectively. More generally speaking, as soon as the reflection and transmission matrices $\mathbf{r}_{1,q}$, $\mathbf{t}_{1,q}$, $\mathbf{r}_{q,1}$, and $\mathbf{t}_{q,1}$ for the qth layer geometry are known, we are able to compute the matrices $\mathbf{r}_{1,q+1}$, $\mathbf{t}_{1,q+1}$, $\mathbf{r}_{q+1,1}$, and $\mathbf{t}_{q+1,1}$. This recursion formula links the fields in the first layer to the fields in the qth layer, which can be summarized as the following relation:

$$\mathbf{S}_{1,q+1} = \mathbf{S}_{1q} \otimes \mathbf{S}_{q,q+1}. \tag{8.3.71}$$

In a structure that has a total of n layers, we will also need to compute the reflection and transmission matrices $\mathbf{r}_{qn}$ and $\mathbf{t}_{qn}$, as well as the matrices $\mathbf{r}_{nq}$ and $\mathbf{t}_{nq}$ connecting the fields between the qth layer and the last nth layer. The recursive procedure is the same, with the exception that we begin by solving the three-layer geometry that consists of the layers $n-2$, $n-1$, and n, respectively. After obtaining the $\mathbf{r}_{n-2,n}$, $\mathbf{t}_{n-2,n}$, $\mathbf{r}_{n,n-2}$, and $\mathbf{t}_{n,n-2}$, we proceed recursively by conceptually reducing the four-layer structure consisting of layers $n-3$ to n to a three-layer geometry. Therefore, we can obtain the matrices $\mathbf{r}_{q-1,n}$, $\mathbf{t}_{q-1,n}$, $\mathbf{r}_{n,q-1}$, and $\mathbf{t}_{n,q-1}$ by knowing the matrices of $\mathbf{r}_{q,n}$, $\mathbf{t}_{q,n}$, $\mathbf{r}_{n,q}$, and $\mathbf{t}_{n,q}$, respectively. This recursive procedure can be represented as

$$\mathbf{S}_{q-1,n} = \mathbf{S}_{q-1,q} \otimes \mathbf{S}_{qn}. \tag{8.3.72}$$

In a general manner, we can summarize the recursive relations for all the scattering matrices used in the S-matrix formalism as follows:

$$\mathbf{r}_{1,q+1} = \mathbf{r}_{1q} + \mathbf{t}_{q1}\mathbf{p}_q^{-}\mathbf{r}_{q,q+1}\mathbf{p}_q^{+}\left(\mathbf{I} - \mathbf{r}_{q1}\mathbf{p}_q^{-}\mathbf{r}_{q,q+1}\mathbf{p}_q^{+}\right)^{-1}\mathbf{t}_{1q}, \tag{8.3.73}$$

$$\mathbf{t}_{1,q+1} = \mathbf{t}_{q,q+1}\mathbf{p}_q^{+}\left(\mathbf{I} - \mathbf{r}_{q1}\mathbf{p}_q^{-}\mathbf{r}_{q,q+1}\mathbf{p}_q^{+}\right)^{-1}\mathbf{t}_{1q}, \tag{8.3.74}$$

$$\mathbf{r}_{q+1,1} = \mathbf{r}_{q+1,q} + \mathbf{t}_{q,q+1}\mathbf{p}_q^{+}\left(\mathbf{I} - \mathbf{r}_{q1}\mathbf{p}_q^{-}\mathbf{r}_{q,q+1}\mathbf{p}_q^{+}\right)^{-1}\mathbf{r}_{q1}\mathbf{p}_q^{-}\mathbf{t}_{q+1,q}, \tag{8.3.75}$$

$$\mathbf{t}_{q+1,1} = \mathbf{t}_{q1}\mathbf{p}_q^{-}\mathbf{t}_{q+1,q} + \mathbf{t}_{q1}\mathbf{p}_q^{-}\mathbf{r}_{q,q+1}\mathbf{p}_q^{+}\left(\mathbf{I} - \mathbf{r}_{q1}\mathbf{p}_q^{-}\mathbf{r}_{q,q+1}\mathbf{p}_q^{+}\right)^{-1}\mathbf{r}_{q1}\mathbf{p}_q^{-}\mathbf{t}_{q+1,q}, \tag{8.3.76}$$

$$\mathbf{r}_{q-1,n} = \mathbf{r}_{q-1,q} + \mathbf{t}_{q,q-1}\mathbf{p}_q^{-}\mathbf{r}_{qn}\mathbf{p}_q^{+}\left(\mathbf{I} - \mathbf{r}_{q,q-1}\mathbf{p}_q^{-}\mathbf{r}_{qn}\mathbf{p}_q^{+}\right)^{-1}\mathbf{t}_{q-1,q}, \tag{8.3.77}$$

$$\mathbf{t}_{q-1,n} = \mathbf{t}_{qn}\mathbf{p}_q^{+}\left(\mathbf{I} - \mathbf{r}_{q,q-1}\mathbf{p}_q^{-}\mathbf{r}_{qn}\mathbf{p}_q^{+}\right)^{-1}\mathbf{t}_{q-1,q}, \tag{8.3.78}$$

$$\mathbf{r}_{n,q-1} = \mathbf{r}_{nq} + \mathbf{t}_{qn}\mathbf{p}_q^{+}\left(\mathbf{I} - \mathbf{r}_{q,q-1}\mathbf{p}_q^{-}\mathbf{r}_{qn}\mathbf{p}_q^{+}\right)^{-1}\mathbf{r}_{q,q-1}\mathbf{p}_q^{-}\mathbf{t}_{nq}, \tag{8.3.79}$$

$$\mathbf{t}_{n,q-1} = \mathbf{t}_{q,q-1}\mathbf{p}_q^{-}\mathbf{t}_{nq} + \mathbf{t}_{q,q-1}\mathbf{p}_q^{-}\mathbf{r}_{qn}p_q^{+}\left(\mathbf{I} - \mathbf{r}_{q,q-1}\mathbf{p}_q^{-}\mathbf{r}_{qn}\mathbf{p}_q^{+}\right)^{-1}\mathbf{r}_{q,q-1}\mathbf{p}_q^{-}\mathbf{t}_{nq}. \tag{8.3.80}$$

After iteratively using the above recursive formulas, the expansion vectors $\mathbf{a}_q$ and $\mathbf{b}_q$ of the field in the qth layer in an n-layer structure can be calculated as

$$\mathbf{a}_q = \left(\mathbf{I} - \mathbf{r}_{q1}\mathbf{p}_q^- \mathbf{r}_{qn}\mathbf{p}_q^+\right)^{-1} \left(\mathbf{t}_{1q}\mathbf{a}_1' + \mathbf{r}_{q1}\mathbf{p}_q^- \mathbf{t}_{nq}\mathbf{b}_n'\right),$$
$$\mathbf{b}_q = \left(\mathbf{I} - \mathbf{r}_{qn}\mathbf{p}_q^+ \mathbf{r}_{q1}\mathbf{p}_q^-\right)^{-1} \left(\mathbf{r}_{qn}\mathbf{p}_q^+ \mathbf{t}_{1q}\mathbf{a}_1' + \mathbf{t}_{nq}\mathbf{b}_n'\right). \quad (8.3.81)$$

As a consequence, the electric and magnetic fields in the qth layer can be simply obtained from the calculated expansion vectors as

$$\mathbf{E}_q(\mathbf{r}) = \sum_j^N \left[a_{qj} e^{i\beta_{qj}(z - z_{q-1,q})} \mathbf{e}_{qj}^+(\mathbf{r}_\perp) + b_{qj} e^{-i\beta_{qj}(z - z_{q,q+1})} \mathbf{e}_{qj}^-(\mathbf{r}_\perp)\right], \quad (8.3.82)$$

$$\mathbf{H}_q(\mathbf{r}) = \sum_j^N \left[a_{qj} e^{i\beta_{qj}(z - z_{q-1,q})} \mathbf{h}_{qj}^+(\mathbf{r}_\perp) + b_{qj} e^{-i\beta_{qj}(z - z_{q,q+1})} \mathbf{h}_{qj}^-(\mathbf{r}_\perp)\right]. \quad (8.3.83)$$

As a final remark, we provide a concise overview of generalizations of the theory to full 3D geometries. Two categories of geometries are taken into consideration. The matrix form of the eigenvalue problem (Eq. (8.3.4)) in Cartesian coordinates (x, y) is given as in the general case when there are no simplifying symmetries.

$$\begin{bmatrix} \nabla_\perp^2 + \varepsilon k_0^2 + \partial_x(\partial_x \ln\varepsilon) & \partial_x(\partial_y \ln\varepsilon) \\ \partial_y(\partial_x \ln\varepsilon) & \nabla_\perp^2 + \varepsilon k_0^2 + \partial_y(\partial_y \ln\varepsilon) \end{bmatrix} \begin{bmatrix} e_x \\ e_y \end{bmatrix} = \beta^2 \begin{bmatrix} e_x \\ e_y \end{bmatrix}. \quad (8.3.84)$$

When we hold the relative permittivity constant, we see that the nondiagonal terms of the operator matrix become null. Consequently, the two different components of the field are related to the varying values of the dielectric constant.

In most cases, the eigenvalue problem (Eq. (8.3.84)) does not have analytical solutions. Therefore, the PWE method should usually be utilized. When compared to the 2D problem, which involves an expansion of the field on N plane waves along the x axis, the 3D problem would require an expansion on a total of $2N^2$ plane waves in the (x, y) plane. Hence, the eigenvalue problem in 3D geometries becomes extremely difficult to solve numerically when N increases. For additional information regarding the numerical implementation of the 3D problem, one can refer to [747].

When the 3D geometry possesses rotational symmetry, by introducing cylindrical coordinates (r, ϕ), the eigenvalue problem can become a great deal easier to solve. In this case, the permittivity distribution is only a function of r, and we can describe the electric field as a sum of contributions from various angular quantum numbers in the cylindrical coordinates as

$$\mathbf{E}(\mathbf{r}) = \sum_l \mathbf{E}_l(r) e^{il\phi}, \quad (8.3.85)$$

where l represents the angular quantum number. The field contributions that have various angular quantum numbers l are decoupled in a structure that is rotationally symmetric. As a result, it is necessary to do computations for only one angular quantum number at a time in this structure. This brings the total number of effective dimensions down from 3 to 2, which enables precise 3D vectorial calculations to be performed using only a moderate amount of computation loads.

For a rotationally symmetric permittivity profile, the eigenvalue problem can be written as

$$\begin{bmatrix} \nabla_{\perp}^2 - \frac{1}{r^2} + \varepsilon k_0^2 + \partial_r (\partial_r \ln \varepsilon) & -\frac{2il}{r^2} \\ \frac{2il}{r^2} + \frac{il}{r} (\partial_r \ln \varepsilon) & \nabla_{\perp}^2 - \frac{1}{r^2} + \varepsilon k_0^2 \end{bmatrix} \begin{bmatrix} e_r \\ e_\phi \end{bmatrix} = \beta^2 \begin{bmatrix} e_r \\ e_\phi \end{bmatrix}, \quad (8.3.86)$$

where the notation ∂_r stands for $\partial/\partial r$. We make the observation that the e_r and e_ϕ components are coupled and that the equation must be solved for both components simultaneously, with the exception of the case in which the angular quantum number l is equal to zero. However, the azimuthal dependency on ϕ is handled analytically, and this eigenvalue problem only requires a determination of the fields along the r axis. This constitutes a significant advantage in comparison to the general 3D geometry. Note that this eigenvalue problem can be conveniently solved using the Fourier–Bessel expansion [760], which is in the same line of the PWE in Cartesian coordinates.

8.4 Computational Methods for Near-Field Thermal Radiation

In Section 8.3, we only consider systems that only interact with external illumination, typically in optics, photonics, and EMs. In this section, we further take thermal emission from the materials into account, as described in the framework of FE introduced in Chapter 7. Since in the far field, thermal emission can be coped with by invoking Kirchhoff's law in general circumstances and we can calculate absorptivity to represent emissivity, it is usually unnecessary to apply the FE and numerical EM techniques introduced in Section 8.3 [6]. Therefore, relevant numerical techniques for FE are developed with the intention to deal with NFRHT problems [761].

In light of the growing interest in nanoscale energy-harvesting devices as well as nanoscale manufacturing and sensing technologies, a computational method that takes into consideration all of the near-field modes of any particular problem is of the utmost relevance. Calculations for the near-field radiative transfer were first performed for many simple geometries, like two surfaces that were parallel to one another as discussed in Chapter 7. During the course of the 1990s and 2000s, a variety of computational approaches to investigations of NFRHT have been developed [6, 231, 761, 762].

On the basis of the generalization of the scattering approach, Bimonte [763] developed an exact method for computing the power of radiative heat transfer between two arbitrary nanostructured surfaces out of thermal equilibrium. Rodriguez et al. [397] presented a general numerical approach for calculating radiative heat transfer between arbitrary geometries and materials, in which white-noise sources are introduced into the evolution of Maxwell's equations in the FDTD method. This approach can be naturally extended to nonequilibrium situations because of the statistical independence of random currents in different objects. Didari and Mengüç also proposed a direct approach based on the FDTD method [722, 764, 765]. To calculate the nonequilibrium radiative heat transfer between a plate and compact objects of arbitrary shapes, McCauley et al. [766] developed a general numerical method on the basis of the scattering-theory formulations and the BEM, making the first accurate theoretical predictions for the total heat transfer and the spatial heat flux profile for 3D compact objects including corners or tips. Similarly, Rodriguez et al. [398, 721] presented a similar numerical approach to efficiently compute the radiative heat transfer between bodies of arbitrary shape, by developing a fluctuating-surface-current (FSC) approach [767] to nonequilibrium fluctuations based on the surface integral-equation (SIE) framework of EM scattering, which permits the direct application of the BEM solvers of classical numerical EMs. Polimeridis et al. [768] proposed a fluctuating volume–current formulation of EM fluctuations that deals with the RHT between arbitrarily shaped and inhomogeneous bodies. This approach exploited the techniques of the volume-integral equation (VIE) method, in which EM scattering is described in terms of volumetric, current unknowns throughout the bodies. Compared to the approach based on the FSC technique, this method can cope with more complex and inhomogeneous structures. Wen [396] proposed an approach to treat NFRHT problems between objects with arbitrary shapes and temperature distributions based on the Wiener chaos expansion (WCE). This approach first obtains the eigenmodes of randomly fluctuating thermal currents in the medium by means of WCE and calculates the net radiation flux by summing up the thermal radiation resulting from all of the different eigenmodes of randomly fluctuating thermal currents determined by solving the corresponding Maxwell's equations. By combining the scattering theory and the FMM (or rigorous coupled-wave analysis, RCWA), Liu and Zhang [769] investigated graphene-assisted NFRHT between gratings made of polar materials. The same approach has been used by many researchers to deal with NFRHT between materials with periodic nanostructures [718, 770–773]. On the basis of the coupled-dipole approximation for the VIE in the classical EM scattering problem, Edalatpour and Francoeur [400, 774–776] developed the so-called T-DDA method by introducing fluctuational dipoles, which can cope with the NFRHT between arbitrary 3D geometries.

We discussed several solution methods for the EM wave equations (without thermal emission or fluctuating currents) in Section 8.3. Among them, the FDTD method, as a full-wave solution, can be used for near-field thermal radiation problems as mentioned earlier by introducing white-noise sources. In this section, we will not discuss this kind of treatment in detail. By contrast, we present an introduction of the FSC approach, the T-DDA method and the WCE method, which are, to a greater extent, specific to NFRHT problems.

8.4.1 Wiener Chaos Expansion Method

From the FDT (Eq. (7.5.3)), the ensemble average of the fluctuating current densities is given by

$$\langle J_k(\mathbf{r}',\omega) J_n^*(\mathbf{r}'',\omega')\rangle = \frac{4}{\pi}\omega\varepsilon_0 \operatorname{Im}(\varepsilon(\omega))\delta_{mn}\delta(\mathbf{r}'-\mathbf{r}'')\Theta(\omega,T)\delta(\omega-\omega'). \quad (8.4.1)$$

We are aware of the fact that the spatially incoherent thermal current source described in Eq. (8.4.1) will result in the generation of EM waves that are spatially incoherent. In order to implement the WCE approach, the first thing that needs to be done is to formulate an expression for the fluctuating thermal current, $\mathbf{J}_{\text{thermal}}$, as a function of space and time [396]. Because $\mathbf{J}_{\text{thermal}}$ refers to the current that is caused by the random fluctuation of charges thermally generated in the material when its temperature is greater than 0 K, we have that

$$\mathbf{J}_{\text{thermal}} = \mathrm{d}W(\mathbf{r})V(t), \quad (8.4.2)$$

where $V(t)$ describes the time dependency of the thermal current and is equal to $J(\omega)$ in the steady state. And $\mathrm{d}W(\mathbf{r})$ depicts the independent spatial randomness of the thermal current at the position $\mathbf{r}$. To describe this randomness, we can use the WCE theorem [777, 778]. This is done by introducing a complete set of orthogonal basis function $\phi_{ijk}(\mathbf{r})$ and a set of independent random Gaussian variables ξ_i:

$$\xi_{ijk} = \int_0^{\mathbf{r}} \phi_{ijk}(\mathbf{r})\mathrm{d}W(\mathrm{r}), \quad i,j,k=1,2,\ldots, \quad (8.4.3)$$

whose inverse transform is expressed as

$$\mathrm{d}W(\mathbf{r}) = \sum_i\sum_j\sum_k \xi_{ijk}\phi_{ijk}(\mathbf{r}). \quad (8.4.4)$$

As a result, the thermal current is recast into [396]

$$\mathbf{J}_{\text{thermal}} = \left[J_\omega \sum_i\sum_j\sum_k \xi_{ijk}\phi_{ijk}(\mathbf{r})\right]\mathbf{I}, \quad (8.4.5)$$

with $\mathbf{I}$ being the identity matrix.

When the random thermal fluctuating current is decomposed into its eigenmodes $\phi_{ijk}(\mathbf{r})$ and then substituted into Eq. (7.5.2), the original stochastic PDE of FE is reduced to a set of deterministic PDEs. It is important to keep in mind that the contributions made by each eigenmode of the random thermal fluctuation current are independent due to the orthogonality of eigenmodes, as required by the property of the Gaussian random variables $\langle\xi_{ijk}\xi_{i'j'k'}\rangle = \delta_{ii'}\delta_{jj'}\delta_{kk'}$. As a result, the sum of the radiation intensities produced by each eigenmode of the random thermal fluctuation current then represents the total near-field thermal radiation intensity.

The WCE approach was utilized by Wen [396] in order to compute the NFRHT between two films separated by a vacuum gap. The first thing that has to be done in order to complete the computation is to determine the eigenmodes for random fluctuating currents within a thin film. Without loss of generality, we assume in the x and z directions, the film is semi-infinite, while in the y direction, the film has a thickness of D. In the y direction, a Fourier cosine series is used to expand the fluctuation current, and thus the eigenmode is given by [396, 731]

$$\phi_l(y)=\sqrt{\frac{2}{D}}\cos\left[\frac{l\pi y}{D}\right],\quad l=1,2,3,\ldots,\tag{8.4.6}$$

and

$$\phi_0(y)=\sqrt{\frac{1}{D}}.\tag{8.4.7}$$

The fluctuational current in the x and z directions can be expanded as orthonormal complex Fourier series. The eigenmode of the random thermal fluctuating current of the entire plate can be expressed as a product of the eigenmodes in the x, y, and z directions. When the length in these two directions, L, are taking to be infinite, namely, $L\rightarrow\infty$, we can obtain the product of the eigenmodes in the x, y and z directions as [396]

$$\phi_{lK_xK_z}=\frac{1}{\pi}\sqrt{\frac{2}{D}}\cos\left[\frac{l\pi y}{D}\right]\exp\left(\mathrm{i}K_x x\right)\exp\left(\mathrm{i}K_z z\right),\quad l=1,2,3,\ldots,\tag{8.4.8}$$

and

$$\phi_{0K_xK_z}=\frac{1}{\pi\sqrt{D}}\exp\left(\mathrm{i}K_x x\right)\exp\left(\mathrm{i}K_z z\right),\tag{8.4.9}$$

with $K_x=2\pi m/L$, $K_z=2\pi n/L$, and $m,n=\pm1,\pm2,\pm3,\ldots$. It is also noted that for the symmetry of this problem, we can define the in-plane wavenumber that considers the eigenmodes in the x and z directions as

$$\exp\left(\mathrm{i}K_x x\right)\exp\left(\mathrm{i}K_z z\right)=\exp\left(\mathrm{i}Kr\right),\tag{8.4.10}$$

with $K=\sqrt{K_x^2+K_y^2}$ and $r=K_x x/K+K_y y/K$. Therefore, we obtain the eigenmodes of the thermal fluctuating current for an infinite plate of finite thickness and constant temperature as

$$\phi_{lK}=\frac{1}{\pi}\sqrt{\frac{2}{D}}\cos\left[\frac{l\pi y}{D}\right]\exp\left(\mathrm{i}Kr\right),\quad l=1,2,3,\ldots,\tag{8.4.11}$$

and

$$\phi_{0K}=\frac{1}{\pi\sqrt{D}}\exp\left(\mathrm{i}Kr\right).\tag{8.4.12}$$

For these eigenmodes, the set of deterministic PDEs to be solved to obtain the total thermal radiation is

$$\begin{aligned}
&\nabla\cdot\mathbf{D}_{lK}=0,\\
&\nabla\cdot\mathbf{B}_{lK}=0,\\
&\nabla\times\mathbf{E}_{lK}=-\mathrm{j}\omega\mu\mathbf{H}_{lK},\\
&\nabla\times\mathbf{H}_{lK}=\mathrm{j}\omega\varepsilon\mathbf{E}_{lK}+J_{\omega}\phi_{lk}.\mathbf{I}.
\end{aligned}\tag{8.4.13}$$

From the EM fields produced by the eigenmodes, the time-averaged radiation intensity, given by the Poynting vector, due to each eigenmode can be calculated as [396]

$$\mathbf{S}_{lK}(x,y,z)=\frac{1}{2}\,\mathrm{Re}\left(\mathbf{E}_{lK}\times\mathbf{H}_{lK}^{*}\right).\tag{8.4.14}$$

The corresponding total thermal radiation intensity is obtained by integrating over all in-plane wavenumbers as

$$\mathbf{W}(\omega)=\int_{0}^{\infty}\left[2\pi\sum_{l=0}^{\infty}\mathbf{S}\left(\phi_{lK}\right)\right]K\mathrm{d}K.\tag{8.4.15}$$

The FDFD method [779, 780] can be used to solve Maxwell's equations for different eigenmodes of the random thermal fluctuating current as demonstrated in [396]. By using the WCE method, Wen [396] compared the numerically obtained results with that from the exact analytical solution based on Green's function, as shown in Fig. 8.12, in which the results for different polarizations for spectral near-field radiative heat flux between two thin gold plates (150 nm) separated by a 10-nm vacuum gap. It can be seen that a good agreement between the WCE method and analytical results is achieved. Recently, Li et al. [781] extended the WCE method to the structures with a permittivity or temperature inhomogeneity. In this approach, the inhomogeneous information of the structures is embedded into the modes of thermal fluctuating current via a spatially variant term, which improves the time efficiency when compared to traditional approaches that approximate inhomogeneous structures as the superposition of quasihomogeneous subregions.

8.4.2 Fluctuating-Surface-Current Formulation

The FSC formulation of thermal radiation was first proposed by Rodriguez et al. [398, 721]. This formulation is based on the SIE approach of classical EMs. The FSC approach produces compact trace formulations, which can be used for calculating the near-field heat transfer between two media, as well as the emission from isolated objects of arbitrary geometries. In a manner analogous to that of the scattering matrix method, the FSC approach makes use of the fact that the analytical EM solutions provided by Green's functions in homogeneous regions have been known. Then by matching the boundary conditions at the interfaces, one can obtain the final solution for NFRHT.

Nevertheless, despite the fact that both the FSC and the scattering matrix approach are stated in terms of a variety of matrices, there are a few significant

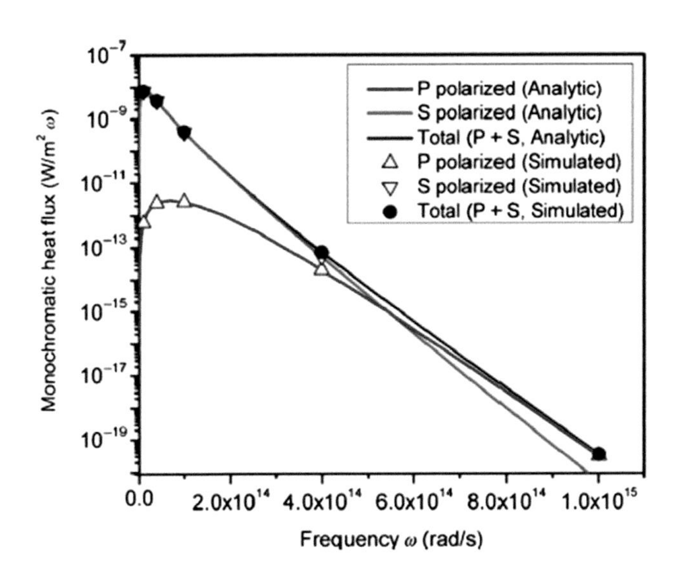

Figure 8.12 Comparison between WCE (simulated) and DGF (analytic) approaches for the spectral near-field flux between two thin gold plates (150 nm) separated by a 10-nm gap for different polarizations (s, p, and total). Used with permission of American Society of Mechanical Engineers (ASME), from [396]; permission conveyed through Copyright Clearance Center, Inc.

distinctions between the two. The scattering matrix approach works with a set of incoming and outgoing partial waves that are associated with the scattering matrices. Within the volume of the bodies, there are unknown fields. On the other hand, the FSC technique is phrased in terms of unknown electric and magnetic surface currents, and it does not track the waves that are coming in or going out. Since these currents are arbitrary and fictional vector fields and do not need to satisfy any wave equation, there is greater freedom in the choice of basis [721]. If appropriate spectral bases are chosen, the FSC trace formulation has the potential to produce semianalytical expressions for simple geometric configurations, in a manner analogous to Green's function method [7]. Furthermore, given that the currents are only present on the surfaces, it is possible to cut down on the number of unknowns that need to be solved by employing a numerical EM approach known as the BEM. In contrast to finite-element and finite-difference approaches, BEM typically produces dense matrices that call for intensive matrix operations, which in turn increase the computational time and complexity. As a result, the BEM is an effective method for analyzing systems that are not very complex, while it is not worthwhile for analyzing huge systems that need a great deal of computer resources. Rodriguez et al. have performed this method to solve NFRHT between two cylinders of different orientations, a cone over a circular plate, two cones [721], two rings, two circular plates [398], and a sphere or a cylinder over a perforated plate [397], etc.

It is known that the radiative heat transfer between two objects at temperature T_1 and T_2 is given by

$$Q_{12} = \int_0^{\infty} \mathrm{d}\omega \left[\Theta\left(\omega, T_1\right) - \Theta\left(\omega, T_1\right)\right] \Phi(\omega), \tag{8.4.16}$$

in which $\Phi(\omega)$ represents the ensemble averaged flux spectrum in medium 2 due to random currents generated in medium 1. It is given by [721]

$$\Phi = \frac{1}{2\pi} \operatorname{Tr}\left[\left(\operatorname{sym} G^1\right) W^{21*} \left(\operatorname{sym} G^2\right) W^{21}\right], \tag{8.4.17}$$

where the symbol W^{21} connects the equivalent currents at the interface of the medium 1 to the incident fields at the surface of medium 2, and the superscript * indicates a conjugate-transpose (adjoint) operation. G represents Green's function expansion. And sym $G = \frac{1}{2}\left(G + G^*\right)$ denotes the Hermitian part of G. According to [721],

$$W^{-1} = \begin{pmatrix} W^{11} & W^{12} \\ W^{21} & W^{22} \end{pmatrix}^{-1} = \begin{pmatrix} G^{0,11} & G^{0,12} \\ G^{0,21} & G^{0,22} \end{pmatrix} + \begin{pmatrix} G^1 & \\ & 0 \end{pmatrix} + \begin{pmatrix} 0 & \\ & G^2 \end{pmatrix}, \tag{8.4.18}$$

in which the first matrix on the RHS represents multibody interactions between basis functions on both objects, via waves propagating through the intervening medium 0 (namely, the vacuum gap in most cases), and the second and the third matrices indicate self-interactions via waves propagating waves within the two objects themselves.

In particular, for isolated plates, we only need to consider $G^{0,11}$ and G^1 in Eq. (8.4.18). The $G_\perp$ matrices, which correspond to purely electric surface currents (perpendicular polarization), are given by

$$G_\perp^{0,11} = \frac{1}{2} \begin{pmatrix} \frac{Z_0}{\gamma_0} & 1 \\ 1 & -\frac{\gamma_0}{Z_0} \end{pmatrix}, \tag{8.4.19a}$$

$$G_\perp^1 = \frac{1}{2} \begin{pmatrix} \frac{Z_1}{\gamma_1} & -1 \\ -1 & -\frac{\gamma_1}{Z_1} \end{pmatrix}, \tag{8.4.19b}$$

where $Z = \sqrt{\mu/\varepsilon}$ is the impedance and $\gamma_i = \sqrt{1 - (\mathbf{k}_\perp/k_i)^2}$ is the wavenumber in the z direction normalized to the wavevector k_i in medium i. The flux spectrum for the E (or the perpendicular) polarization is given by

$$\Phi_\perp = \frac{1}{4\pi} \operatorname{Tr}\left[\frac{\operatorname{Re}\left(\frac{\gamma_0}{z_0}\right) \operatorname{Re}\left(\frac{\gamma_1}{Z_1}\right)}{\left|\frac{\gamma_0}{Z_0} + \frac{\gamma_1}{Z_1}\right|^2}\right], \tag{8.4.20}$$

and the flux spectrum for the H (or the parallel) polarization is obtained as

$$\Phi_\parallel = \Phi_\perp \left(Z \to \frac{1}{Z}\right) \tag{8.4.21}$$

by simply conducting a replacement of the impedance. In Eq. (8.4.21), the $\mathrm{Tr}\,\Phi = \int \frac{\mathrm{d}^2\mathbf{k}_\perp}{(2\pi)^2}\Phi(\mathbf{k}_\perp)$ amounts to the integration over the parallel wavevector $\mathbf{k}_\perp$. Assuming the medium 0 is lossless, by substituting Eqs. (8.4.20) and (8.4.21) into Eq. (8.4.17), we can obtain the well-known formula for the emissivity [392]

$$\Phi(\omega) = \frac{1}{8\pi}\int_0^\omega \frac{\mathrm{d}^2\mathbf{k}_\perp}{(2\pi)^2}\sum_{\mathrm{p}=\{\perp,\parallel\}} \epsilon_\mathrm{p}(\mathbf{k}_\perp,\omega), \tag{8.4.22}$$

where $\epsilon_\mathrm{p} = \frac{1}{2}\left(1-|r_\mathrm{p}|^2\right)$ represents the directional emissivity of the plate for the polarization state of p, $\mathrm{p}=\{\perp,\parallel\}$, given by the Fresnel reflection coefficients as

$$r_\perp = \frac{\frac{\gamma_0}{Z_0}-\frac{\gamma_1}{Z_1}}{\frac{\gamma_0}{Z_0}+\frac{\gamma_1}{Z_1}}, \tag{8.4.23a}$$

$$r_\parallel = r_\perp\left(Z\to\frac{1}{Z}\right). \tag{8.4.23b}$$

Next, let us turn to the NFRHT between two plates. In this case, the self-interaction $F_\perp$ matrices are obtained as [721]

$$G_\perp^{0,ii} = \frac{1}{2}\begin{pmatrix}\frac{Z_0}{\gamma_0} & 1\\ 1 & -\frac{\gamma_0}{Z_0}\end{pmatrix},\quad G_\perp^{i} = \frac{1}{2}\begin{pmatrix}\frac{Z_i}{\gamma_i} & -1\\ -1 & -\frac{\gamma_i}{Z_i}\end{pmatrix}, \tag{8.4.24}$$

with $i=1,2$. For the interaction between the two plates, we can obtain the "translation" matrices as

$$G_\perp^{12} = G_\perp^{21} = \frac{1}{2}\begin{pmatrix}\frac{Z_0}{\gamma_0} & 1\\ 1 & \frac{\gamma_0}{Z_0}\end{pmatrix}\mathrm{e}^{\mathrm{i}k_0\gamma_0 d}. \tag{8.4.25}$$

From the above matrices, we can obtain the flux spectra for the two polarizations as

$$\Phi_\perp = \frac{1}{2\pi}\mathrm{Tr}\left[\frac{\left|\frac{\gamma_0}{z_0}\mathrm{e}^{2\mathrm{i}k\gamma_0 d}\right|^2}{|\rho_\perp|^2}\frac{\mathrm{Re}\left(\frac{\gamma_1}{Z_1}\right)\mathrm{Re}\left(\frac{\gamma_2}{Z_2}\right)}{\left|\frac{\gamma_0}{Z_0}+\frac{\gamma_\perp}{Z_1}\right|^2\left|\frac{\gamma_0}{Z_0}+\frac{\gamma_2}{Z_2}\right|^2}\right], \tag{8.4.26a}$$

$$\Phi_\parallel = \Phi_\perp\left(Z\to\frac{1}{Z}\right), \tag{8.4.26b}$$

where $\rho_p = \left|1-r_p^1 r_p^2 \mathrm{e}^{2\mathrm{i}k_0\gamma_0 d}\right|^2$, with r_p^q being the Fresnel reflection coefficient of plate q for the polarization state of p. Assuming a nonabsorbing medium, 0, the well-known formula describing the propagating and evanescent components of the flux spectra can be reached [392] (cf. Eqs. (7.5.33) and (7.5.34)):

$$\Phi_\mathrm{prop}(\omega) = \frac{1}{4\pi}\sum_p\int_0^\omega \frac{\mathrm{d}^2\mathbf{k}_\perp}{(2\pi)^2}\frac{\varepsilon_p^1\varepsilon_p^2}{\rho_p}, \tag{8.4.27a}$$

$$\Phi_\mathrm{evan}(\omega) = \frac{1}{4\pi}\sum_p\int_0^\omega \frac{\mathrm{d}^2\mathbf{k}_\perp}{(2\pi)^2}\left(\mathrm{Im}\,\mathbf{r}_p^1\right)\left(\mathrm{Im}\,\mathbf{r}_p^2\right)\frac{\mathrm{e}^{-2\,\mathrm{Im}(k_0\gamma_0)d}}{\rho_p}, \tag{8.4.27b}$$

where ϵ_p^q represents the emissivity of plate q for the polarization state of p.

8.4.3 Thermal Discrete-Dipole Approximation (T-DDA)

The T-DDA method has been presented by Edalatpour and Francoeur [400] for modeling NFRHT in 3D arbitrary geometries. The T-DDA is theoretically in the same line as the DDA as discussed in Chapter 7. The difference mainly lies in that, in the case of near-field heat transfer, the incident field is derived from thermal oscillations of dipoles as opposed to being caused by external illumination. A diagrammatic illustration of the problem under investigation is given in Fig. 8.13. There are a total of $m=1,2,\ldots,M$ objects that are at a temperature of T_m and are submerged in vacuum (medium 0), where they are exchange thermal radiation with one another. The emitters, marked by the symbol M_e, are composed of source points denoted by $\mathbf{r}'$, whereas the absorbers, denoted by M_a, are made up of points denoted by $\mathbf{r}$ where fields are calculated. The bodies have frequency-dependent dielectric functions ε_m and are in a state of thermodynamic equilibrium. The total electric field, denoted by $\mathbf{E}$, can be broken down into two parts, the incident field, denoted by $\mathbf{E}_\mathrm{inc}$, and the scattered field, denoted by $\mathbf{E}_\mathrm{sca}$, expressed as

$$\mathbf{E}(\mathbf{r},\omega)=\mathbf{E}_\mathrm{inc}(\mathbf{r},\omega)+\mathbf{E}_\mathrm{sca}(\mathbf{r},\omega). \tag{8.4.28}$$

It is noted the incident field is the field thermally generated by point sources. The magnetic field is determined from the electric field as

$$\nabla\times\mathbf{H}(\mathbf{r},\omega)=-\frac{\mathrm{i}}{\omega\mu_0}\nabla\times\nabla\times\mathbf{E}(\mathbf{r},\omega). \tag{8.4.29}$$

We can invoke the vector wave equation for the incident and scattered fields, which are given by

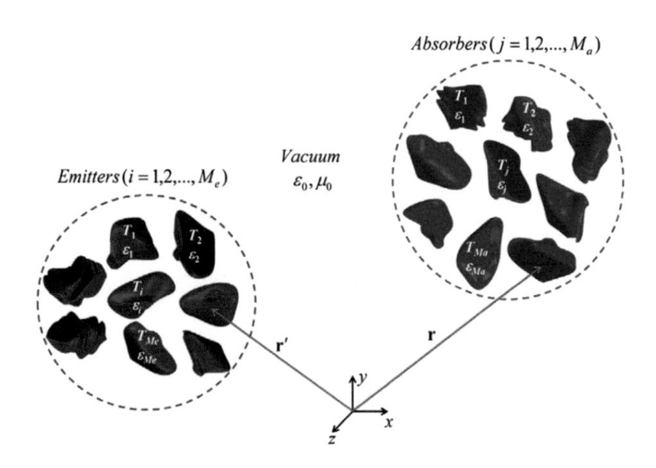

Figure 8.13 Schematic representation of the problem studied in [400] for T-DDA method. Reprinted from [400], Copyright 2014, with permission from Elsevier.

$$\nabla \times \nabla \times \mathbf{E}_{\text{inc}}(\mathbf{r},\omega) - k_0^2 \mathbf{E}_{\text{inc}}(\mathbf{r},\omega) = \mathrm{i}\omega\mu_0 \mathbf{J}^r(\mathbf{r},\omega), \tag{8.4.30a}$$

$$\nabla \times \nabla \times \mathbf{E}_{\text{sca}}(\mathbf{r},\omega) - k_0^2 \mathbf{E}_{\text{sca}}(\mathbf{r},\omega) = \left(k^2 - k_0^2\right) \mathbf{E}(\mathbf{r},\omega), \tag{8.4.30b}$$

where k_0 is the free-space wavevector and k is the wavevector inside the scatter. In Eq. (8.4.30b), the scattered field can be regarded as a field generated by an equivalent source function $\left(k^2 - k_0^2\right)\mathbf{E}$ due to permittivity difference, as frequently used in analytic wave theory. By using the DGF (Eq. (7.5.7)), we can obtain the formal solution for Eq. (8.4.30):

$$\mathbf{E}_{\text{inc}}(\mathbf{r},\omega) = \mathrm{i}\omega\mu_0 \int_{V_e} \mathbf{G}(\mathbf{r},\mathbf{r}',\omega) \cdot \mathbf{J}^r(\mathbf{r}',\omega)\,\mathrm{d}V', \tag{8.4.31a}$$

$$\mathbf{E}_{\text{sca}}(\mathbf{r},\omega) = \int_V \mathbf{G}(\mathbf{r},\mathbf{r}',\omega) \cdot \mathbf{E}(\mathbf{r}',\omega)\left(k^2 - k_0^2\right)\mathrm{d}V', \tag{8.4.31b}$$

where V_e is the volume of the emitting media, while V is the total volume of absorbing and emitting media. Note in Eq. (8.4.31a), the integration should be performed over the volume of the emitting media V_e, while in Eq. (8.4.31b), the integration is carried out over the total volume. The DGF in free space is expressed as [225]

$$\mathbf{G}(\mathbf{r},\mathbf{r}',\omega) = \frac{\mathrm{e}^{\mathrm{i}k_0R}}{4\pi R}\left[\left(1 - \frac{1}{(k_0R)^2} + \frac{\mathrm{i}}{k_0R}\right)\mathbf{I} - \left(1 - \frac{3}{(k_0R)^2} + \frac{3\mathrm{i}}{k_0R}\right)\hat{\mathbf{R}} \otimes \hat{\mathbf{R}}\right], \tag{8.4.32}$$

where $R = |\mathbf{r} - \mathbf{r}'|$ and $\hat{\mathbf{R}} = (\mathbf{r} - \mathbf{r}')/|\mathbf{r} - \mathbf{r}'|$.

Since the DGF has a singularity at $\mathbf{r} = \mathbf{r}'$, here we should apply the principal value method to avoid divergences. By assuming a spherical volume encircling the singularity, as done in Section 7.2.2, we have

$$\begin{aligned} & k_0^2 \int_V [\varepsilon(\mathbf{r}') - 1]\,\mathbf{G}(\mathbf{r},\mathbf{r}',\omega) \cdot \mathbf{E}(\mathbf{r}',\omega)\,\mathrm{d}V' \\ & = k_0^2\ \mathrm{PV} \int_V [\varepsilon(\mathbf{r}') - 1]\,\mathbf{G}(\mathbf{r},\mathbf{r}',\omega) \cdot \mathbf{E}(\mathbf{r}',\omega)\,\mathrm{d}V' - \frac{\varepsilon(\mathbf{r}) - 1}{3}\mathbf{E}(\mathbf{r},\omega), \end{aligned} \tag{8.4.33}$$

where PV implies the principal value. Hence, we derive the governing equation for the T-DDA method as [400]

$$\frac{\varepsilon(\mathbf{r}) + 2}{3}\mathbf{E}(\mathbf{r},\omega) - k_0^2\ \mathrm{PV} \int_V [\varepsilon(\mathbf{r}') - 1]\,\mathbf{G}(\mathbf{r},\mathbf{r}',\omega) \cdot \mathbf{E}(\mathbf{r}',\omega)\,\mathrm{d}V' = \mathbf{E}_{\text{inc}}(\mathbf{r},\omega). \tag{8.4.34}$$

In order to solve Eq. (8.4.34) numerically, it is necessary to discretize the emitter and absorber region into N_e and N_a subvolumes, respectively. The discretization of the space used for emitting and absorbing radiation should be carried out in such a way that the dimensions of each subvolume should be much smaller than the wavelength of radiation. When evaluated at the center $\mathbf{r}_i$ of a subvolume i, the discretized version of Eq. (8.4.34) becomes

$$\frac{\varepsilon_i+2}{3}\mathbf{E}_i - k_0^2\sum_{j=1}^{N}(\varepsilon_j-1)\left(\mathrm{PV}\int_{\Delta V_j}\mathbf{G}(\mathbf{r}_i,\mathbf{r}',\omega)\,\mathrm{d}V'\right)\cdot\mathbf{E}_j=\mathbf{E}_{\mathrm{inc},i}, \quad (8.4.35)$$

where i and j represent subvolumes with $i,j=1,2,\ldots,N$ and $N=N_e+N_a$, and ΔV_j denotes the volume of the subvolume. When $i\neq j$, the DGF has no singularity, and therefore the PV of the integral in Eq. (8.4.35) can be approximated as [400]

$$\mathrm{PV}\int_{\Delta V_j}\mathbf{G}(\mathbf{r}_i,\mathbf{r}',\omega)\,\mathrm{d}V'\approx\mathbf{G}_{ij}\Delta V_j, \quad i\neq j. \quad (8.4.36)$$

On the other hand, for $i=j$, the PV of the integral can be obtained as

$$\mathrm{PV}\int_{\Delta V_j}\mathbf{G}(\mathbf{r}_i,\mathbf{r}',\omega)\,\mathrm{d}V'\approx\frac{2}{3k_0^2}\left[\mathrm{e}^{\mathrm{i}k_0a_i}(1-\mathrm{i}k_0a_i)-1\right]\mathbf{I}, \quad i=j, \quad (8.4.37)$$

where $a_i=(3\Delta V_i/4\pi)^{1/3}$ denotes the effective radius of the subvolume. As a result, we can obtain the final discretized VIE:

$$\left[\frac{\varepsilon_i+2}{3}-\frac{2(\varepsilon_i-1)}{3}\left(\mathrm{e}^{\mathrm{i}k_0a_i}(1-\mathrm{i}k_0a_i)-1\right)\right]\mathbf{E}_i - k_0^2\sum_{\substack{j=1\\ j\neq i}}^{N}(\varepsilon_j-1)\Delta V_j\mathbf{G}_{ij}\cdot\mathbf{E}_j=\mathbf{E}_{\mathrm{inc}}, \quad i=1,2,\ldots,N. \quad (8.4.38)$$

Like the CDM, this equation describes a system of N vector equations in which the electric field in each subvolume $\mathbf{E}_i$ is unknown. Moreover, as the incident field is generated by random currents, the electric field $\mathbf{E}_i$ in each subvolume is also stochastic in nature. In particular, the incident field can be approximated as

$$\mathbf{E}_{\mathrm{inc},i}=\begin{cases}0, & i=1,2,\ldots,N_\mathrm{e},\\ \mathrm{i}\omega\mu_0\sum_{k=1}^{N_\mathrm{e}}\mathbf{G}_{ik}\mathbf{J}_k^r\Delta V_k, & i=N_\mathrm{e}+1,\ldots,N.\end{cases} \quad (8.4.39)$$

If it is assumed that each subvolume responds to external fields like an ED, then Eq. (8.4.38) can be expressed in terms of unknown dipole moments $\mathbf{p}_i$ rather than unknown electric fields $\mathbf{E}_i$. For doing so, we introduce the following relation to establish a connection between the two quantities:

$$\mathbf{E}_i=\frac{3}{\alpha_i^{\mathrm{CM}}(\varepsilon_i+2)}\mathbf{p}_i, \quad (8.4.40)$$

where $\alpha_i^{\mathrm{CM}}=3\varepsilon_0\frac{\varepsilon_i-1}{\varepsilon_i+2}\Delta V_i$ is the Clausius–Mosotti polarizability (cf. Eq. (7.2.18)).

We can also obtain the fluctuating current density similarly as

$$\mathbf{J}_k^r=\frac{-\mathrm{i}\omega}{\Delta V_k}\mathbf{p}_k^r, \quad (8.4.41)$$

where $\mathbf{p}_k^r$ represents a thermally fluctuating dipole moment describing thermal emission [225]. Inserting above formulas in terms of dipole moments, we can rewrite Eq. (8.4.38) as

$$\frac{1}{\alpha_i}\mathbf{p}_i - \frac{k_0^2}{\varepsilon_0}\mathbf{G}_{ij}\cdot\mathbf{p}_j = \mathbf{E}_{\mathrm{inc},i}, \quad i=1,2,\ldots,N, \tag{8.4.42}$$

where

$$\mathbf{E}_{\mathrm{inc},i} = \begin{cases} 0, & i=1,2,\ldots,N_{\mathrm{e}}, \\ \mu_0\omega^2\sum_{k=1}^{N_{\mathrm{e}}}\mathbf{G}_{ik}\mathbf{p}_k^r, & i=N_{\mathrm{e}}+1,\ldots,N. \end{cases} \tag{8.4.43}$$

The first term on the LHS of Eq. (8.4.42) represents the interaction of the i-dipole with itself, and the second term depicts the mutual interactions between all other dipoles denoted by j, with the i-dipole j. On the RHS of Eq. (8.4.42), the term represents the incident field impinging in the absorbing dipoles due to thermal emission from the emitting dipoles. In a matrix notation, this set of equations can be rewritten as [400]

$$\mathbf{A}\cdot\mathbf{P} = \boldsymbol{\mathcal{E}}_{\mathrm{inc}}, \tag{8.4.44}$$

in which $\mathbf{P}$ is the $3N$ stochastic column vector composed of the unknown dipole moments $\mathbf{p}_i$, $\boldsymbol{\mathcal{E}}_{\mathrm{inc}}$ represents the $3N$ stochastic column vector composed of the incident fields $\mathbf{E}_{\mathrm{inc},i}$, and $\mathbf{A}$ is the $3N\times 3N$ matrix characterizing the interactions between dipoles. Each 3×3 submatrix $\mathbf{A}_{ij}$ actually describes the interaction between the dipoles i and j, derived from the term containing the DGF as given in Eq. (8.4.42) . More specifically, Eq. (8.4.44) can be written as [400]

$$\begin{bmatrix} \mathbf{A}_{11} & \mathbf{A}_{12} & \cdots & \mathbf{A}_{1N} \\ \mathbf{A}_{21} & \mathbf{A}_{22} & \cdots & \mathbf{A}_{2N} \\ \vdots & \vdots & \ddots & \vdots \\ \mathbf{A}_{N1} & \mathbf{A}_{N2} & \cdots & \mathbf{A}_{NN} \end{bmatrix} \begin{bmatrix} \mathbf{p}_1 \\ \mathbf{p}_2 \\ \vdots \\ \mathbf{p}_N \end{bmatrix} = \begin{bmatrix} \mathbf{E}_{\mathrm{inc},1} \\ \mathbf{E}_{\mathrm{inc},2} \\ \vdots \\ \mathbf{E}_{\mathrm{inc},N} \end{bmatrix}. \tag{8.4.45}$$

For $i\neq j$, the submatrix $\mathbf{A}_{ij}$ can be analytically obtained as

$$\mathbf{A}_{ij} = C_{ij}\begin{bmatrix} \beta_{ij}+\gamma_{ij}\hat{r}_{ij,x}^2 & \gamma_{ij}\hat{\gamma}_{ij,x}\hat{r}_{ij,y} & \gamma_{ij}\hat{r}_{ij,x}\hat{r}_{ij,z} \\ \gamma_{ij}\hat{r}_{ij,}\hat{r}_{ij,x} & \beta_{ij}+\gamma_{ij}\hat{r}_{ij,y}^2 & \gamma_{ij}\hat{r}_{ij,y}\hat{r}_{ij,z} \\ \gamma_{ij}\hat{r}_{ij,}\hat{r}_{ij,x} & \gamma_{ij}\hat{\gamma}_{i,z}\hat{r}_{ij,y} & \beta_{ij}+\gamma_{ij}\hat{r}_{ij,z}^2 \end{bmatrix}, i\neq j, \tag{8.4.46}$$

where

$$\hat{r}_{ij,m} = \frac{r_{ij,m}}{r_{ij}}, m=x,y,z, \tag{8.4.47a}$$

$$C_{ij} = -\frac{k_0^2\mathrm{e}^{\mathrm{i}k_0 r_{ij}}}{4\pi\varepsilon_0 r_{ij}}, \tag{8.4.47b}$$

$$\beta_{ij} = \left[1-\frac{1}{(k_0 r_{ij})^2}+\frac{\mathrm{i}}{k_0 r_{ij}}\right], \tag{8.4.47c}$$

$$\gamma_{ij} = -\left[1-\frac{3}{(k_0 r_{ij})^2}+\frac{3\mathrm{i}}{k_0 r_{ij}}\right], \tag{8.4.47d}$$

where $r_{ij} = |\mathbf{r}_i - \mathbf{r}_j|$ is the distance between the ith and jth dipoles. When $i = j$, the submatrix is the self-interaction term, which is given by $\mathbf{A}_{ii} = \alpha_i^{-1}\mathbf{I}$.

From the dipole moments, the mean energy dissipated in the absorbers at a specific angular frequency ω is calculated as [400]

$$\langle Q_{\text{abs},\omega} \rangle = \frac{\omega}{2} \sum_{i=N_e+1}^{N} \left(\text{Im}\left[\left(\alpha_i^{-1} \right)^* \right] - \frac{2}{3} k_0^2 \right) \text{Tr}\left(\langle \mathbf{p}_i \otimes \mathbf{p}_i \rangle \right), \tag{8.4.48}$$

which is a standard result of the DDA except for the stochastic nature of the dipole moments in the present case [453]. It can be seen from this equation that the heat transfer flux can be derived from the trace of the dipole autocorrelation function. Hence, we do not need to solve the unknown dipole moments; the only thing we need to know is their autocorrelations.

From Eq. (8.4.44), we can formally obtain $\mathbf{P} = \mathbf{A}^{-1} \cdot \boldsymbol{\mathcal{E}}_{\text{inc.}}$. Let us define the correlation matrix of the dipole moment vector, which has a zero mean $\langle \mathbf{P} \rangle$, as [400]

$$\mathbf{R_{PP}} = \langle \mathbf{P} \otimes \mathbf{P} \rangle, \tag{8.4.49}$$

and thus $\mathbf{R_{PP}}$ should be a $3N \times 3N$ matrix composed of N^2 submatrices as

$$\mathbf{R_{PP}} = \begin{bmatrix} \mathbf{R}_{\mathbf{p}_1\mathbf{p}_1} & \mathbf{R}_{\mathbf{p}_1\mathbf{p}_2} & \cdots & \mathbf{R}_{\mathbf{p}_1\mathbf{p}_N} \\ \mathbf{R}_{\mathbf{p}_2\mathbf{p}_1} & \mathbf{R}_{\mathbf{p}_2\mathbf{p}_2} & \cdots & \mathbf{R}_{\mathbf{p}_2\mathbf{p}_N} \\ \vdots & \vdots & \ddots & \vdots \\ \mathbf{R}_{\mathbf{p}_N\mathbf{p}_1} & \mathbf{R}_{\mathbf{p}_N\mathbf{p}_2} & \cdots & \mathbf{R}_{\mathbf{p}_N\mathbf{p}_N} \end{bmatrix}, \tag{8.4.50}$$

where $\mathbf{R}_{\mathbf{p}_i\mathbf{p}_j} = \langle \mathbf{p}_i \otimes \mathbf{p}_j \rangle$ is the correlation matrix of the dipole moments $\mathbf{p}_i$ and $\mathbf{p}_j$. By this definition, Eq. (8.4.48) can be rewritten as

$$\langle Q_{\text{abs},\omega} \rangle = \frac{\omega}{2} \sum_{i=N_e+1}^{N} \left(\text{Im}\left[\left(\alpha_i^{-1} \right)^* \right] - \frac{2}{3} k_0^2 \right) \text{Tr}\left(\mathbf{R}_{\mathbf{p}_i\mathbf{p}_i} \right). \tag{8.4.51}$$

Therefore, by inserting $\mathbf{P} = \mathbf{A}^{-1} \cdot \boldsymbol{\mathcal{E}}_{\text{inc}}$ into Eq. (8.4.49), we can obtain

$$\mathbf{R}_{\text{PP}} = \mathbf{A}^{-1} \cdot \mathbf{R}_{\text{EE}} \cdot \left(\mathbf{A}^{-1} \right)^\dagger, \tag{8.4.52}$$

in which we have defined the correlation matrix of the incident field as

$$\mathbf{R}_{\text{EE}} = \begin{bmatrix} \mathbf{R}_{\mathbf{E}_1\mathbf{E}_1} & \mathbf{R}_{\mathbf{E}_1\mathbf{E}_2} & \cdots & \mathbf{R}_{\mathbf{E}_1\mathbf{E}_N} \\ \mathbf{R}_{\mathbf{E}_2\mathbf{E}_1} & \mathbf{R}_{\mathbf{E}_2\mathbf{E}_2} & \cdots & \mathbf{R}_{\mathbf{E}_2\mathbf{E}_N} \\ \vdots & \vdots & \ddots & \vdots \\ \mathbf{R}_{\mathbf{E}_N\mathbf{E}_1} & \mathbf{R}_{\mathbf{E}_N\mathbf{E}_2} & \cdots & \mathbf{R}_{\mathbf{E}_N\mathbf{E}_N}, \end{bmatrix}, \tag{8.4.53}$$

which is also a $3N \times 3N$ matrix with N^2 submatrices, and each submatrix is given by

$$\mathbf{R}_{\mathbf{E}_i\mathbf{E}_j} = \mu_0^2 \omega^4 \sum_{k=1}^{N_e} \sum_{n=1}^{N_e} \mathbf{G}_{ik} \cdot \langle \mathbf{p}_k^r \otimes \mathbf{p}_n^r \rangle \cdot \mathbf{G}_{jn}^\dagger, \quad i, j \geq N_e + 1. \tag{8.4.54}$$

The correlation functions of the random dipole moments due to thermal fluctuations can be determined from a modified version of the FDT as [400]

$$\langle \mathbf{p}_k^r \otimes \mathbf{p}_n^r \rangle = \frac{4\varepsilon_0 \operatorname{Im}\left(\alpha_k^{\mathrm{CM}}\right)}{\pi\omega} \Theta(\omega, T)\delta_{kn}\mathbf{I}. \tag{8.4.55}$$

By inserting Eq. (8.4.55) into Eq. (8.4.53), we reach the following expression:

$$\mathbf{R}_{\mathbf{E}_i\mathbf{E}_j} = \frac{4\varepsilon_0\mu_0^2\omega^3}{\pi} \sum_{k=1}^{N_\mathrm{e}} \mathbf{G}_{ik} \cdot \mathbf{G}_{jk}^\dagger \operatorname{Im}\left(\alpha_k^{\mathrm{CM}}\right), \quad i,j \geq N_\mathrm{e}+1. \tag{8.4.56}$$

As the DGF is connected to the interaction submatrix as

$$\mathbf{G}_{lk} = -\frac{\varepsilon_0}{k_0^2}\mathbf{A}_{lk}, \quad l \neq k, \tag{8.4.57}$$

we can substitute Eq. (8.4.57) into Eq. (8.4.56), which yields

$$\mathbf{R}_{\mathbf{E}_i\mathbf{E}_j} = \frac{4\varepsilon_0\Theta(\omega,T)}{\pi\omega} \sum_{k=1}^{N_\mathrm{e}} \mathbf{A}_{ik} \cdot \mathbf{A}_{jk}^\dagger \operatorname{Im}\left(\alpha_k^{\mathrm{CM}}\right), \quad i,j \geq N_\mathrm{e}+1. \tag{8.4.58}$$

From this equation, we can obtain the full correlation matrix $\mathbf{R}_{\mathbf{EE}}$, which is then substituted into Eq. (8.4.52) to obtain $\mathbf{R}_{\mathbf{PP}}$. By Eq. (8.4.51), the diagonal elements of the correlation matrix $\mathbf{R}_{\mathbf{PP}}$ are employed to compute the mean energy absorbed by the absorbers. Furthermore, through the reciprocity of the DGF, the power absorbed by the emitters due to the thermal emission from the absorbers can be straightforwardly obtained. As a consequence, from the above formulation, we can obtain the net radiative heat transfer between the emitters and absorbers.

In [400], Edalatpour and Francoeur employed T-DDA to calculate the spectral thermal conductance between two spheres of 200-nm diameter separated by a vacuum gap of 10 nm. The results were compared with the analytical solution derived in [394], as shown in Fig. 8.14 for three different permittivities of the spheres. When the dielectric function is close to unity (Fig. 8.14(a), with $\varepsilon = 1.2 + 0.1\mathrm{i}$), the T-DDA solution gets closer to the analytical results as the number of subvolumes gets higher, and the two results begin to converge at 280 subvolumes. When the real part of the permittivity grows, as shown in Fig. 8.14(b) for $\varepsilon = 2.5 + 0.1\mathrm{i}$, the T-DDA solution oscillates around the analytical solution as the number of subvolumes grows, and the best results are obtained for a total of 32 subvolumes in this circumstance. For an even larger permittivity of $\varepsilon = 7 + 0.1\mathrm{i}$, a similar oscillation between T-DDA and the analytical solution can be found, and the convergent solution can be reached for 136 subvolumes, as shown in Fig. 8.14(c).

The accuracy and convergence of the T-DDA were further investigated in [774] for the aforementioned case of two spheres that were separated from one another by a vacuum gap. They carried out a parametric study to determine the optimal values for the sphere diameter, gap size, and refractive index. For a fixed number of subvolumes, the T-DDA's accuracy becomes inferior along with the

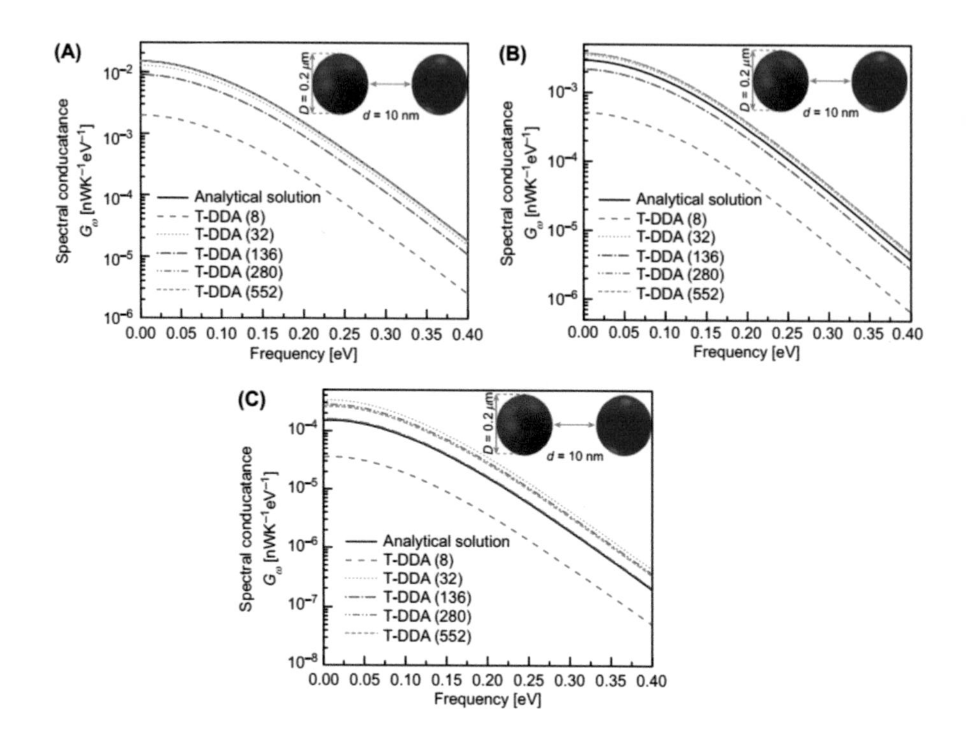

Figure 8.14 Comparison of the T-DDA method with analytical results for spheres for three different dielectric functions: (a) $\varepsilon = 1.2 + 0.1\mathrm{i}$, (b) $\varepsilon = 2.5 + 0.1\mathrm{i}$, and (c) $\varepsilon = 7 + 0.1\mathrm{i}$. The numbers in parentheses indicate the number of subvolumes used in the T-DDA analysis in each case [400]. Reprinted from [400], Copyright 2014, with permission from Elsevier.

increase in either the real or the imaginary part of the refractive index. This is because, for large refractive indices, the variation of the electric field and the DGF in the subvolumes becomes more significant, which amplifies the shape error. As the number of subvolumes increases, a converging trend can be observed. Edalatpour et al. [774] proposed a nonuniform discretization scheme to mitigate the large computational requirements by increasing the number of subvolumes, and the shape error induced by large sphere diameter-to-gap ratios, which also substantially speeds up the convergence. By using up to 82 712 subvolumes, they found that the scheme can lead to the errors of less than 5% in 74% of the cases investigated. They also showed that the T-DDA method is satisfactorily accurate in handling the cases in which surface polaritons dominate the NFRHT. Recently, the T-DDA method was used to treat the thermal NFI between the probe tips and the substrate [776], near-field thermal emission by periodic arrays of subwavelength elements [782], NFRHT between the finite objects of arbitrary shape that exhibit magneto-optical (MO) activity [783], super-Planckian far-field radiative heat transfer between subwavelength objects [784].

8.5 Summary

This chapter gives a brief introduction to various numerical methods that are commonly used to tackle MNTR problems. First, a general overview of numerical methods for solving Maxwell's equations is given. The two most widely used methods in MNTR studies are introduced in detail, including the FDTD method and the MM. The principles, boundary conditions, and implementation are described. Then comes a brief discussion on the simulation of EM wave propagation in disordered media, in order to delineate the connection of radiative transfer and EM wave theory. We further proceed to the specific algorithms dealing with near-field thermal radiation problems, in which fluctuating currents should be taken into account in Maxwell's equations. In particular, we introduce the frequently used WCE method, FSC method, and T-DDA, which can be readily used to deal with complex geometries.

9 Experimental Techniques for Micro/Nanoscale Thermal Radiation

Rapid progress has been made in the last two decades to measure thermal radiation at micro- and nanoscales. This chapter considers several important experimental techniques and methods. It first introduces the SNOM, starting from the principles and implementation of aperture SNOM. Then, the apertureless SNOM, which utilizes the tip or probe of an atomic force microscopy (AFM) to detect light–matter interactions, is discussed in detail. The apertureless and aperture SNOM technologies are the so-called active methods to imaging optical and radiative properties at the nanoscale, since an external light source should be used to excite optical responses. On the other hand, we also discuss the "passive" techniques that are an extension to apertureless SNOM, while the thermal radiation from the heated sample or tip is exploited as light sources, without the need of external sources. This set of methods include thermal radiation scanning tunneling microscopy (TRSTM), thermal infrared near-field spectroscopy (TINS), and scanning noise microscopy (SNoiM). The above methods are generally imaging techniques without consideration of the thermal radiation heat flux. So in the last section of this chapter, we discuss the various implementations of near-field heat transfer measurements.

9.1 Introduction

Scanning near-field optical microscopy, also known as SNOM or NSOM, is an important technique to image optical and thermal radiation signals at resolutions of nanoscales beyond the diffraction limit of conventional optical microscopy, that is, the resolution is restricted by diffraction to about half the wavelength, 0.205 μm for the visible light. Such a technology that combines the advantages of scanning probe microscopy with high lateral resolution and optical microscopy with excellent spectroscopic and temporal selectivity is not only able to resolve the optical and radiative modes in micro- and nanostructures and provide direct evidence of the mechanisms of light–matter interactions [785–787] but also offers a powerful method for comprehensively discerning a nano-object's shape, size, chemical composition [788], molecular structure [789], and dynamic properties [790], to name a few, due to its high spectral and spatial resolutions beyond the diffraction limit. Therefore, this technology can find a wide range of applications in microelectronics, materials sciences, chemistry, biology, medical sciences, and so on, and therefore it has been attracting great attention since its invention in the 1980s [791–797].

For conventional far-field optical microscopy and spectroscopy, the object can be regarded as being illuminated by a monochromatic plane wave from a simplified view. The light that is either transmitted or reflected by the object is then gathered by a lens and then imaged onto a detector, from which an optical image and corresponding reflectivity as well as transmissivity can be determined. In practice, the lens is positioned in the far field that is at least a few wavelengths λ of the illuminating light away from the surface of the object. High spatial frequencies that correspond to the details of the object generate Fourier components of the light field, which are in essence evanescent waves and therefore decay exponentially along the object normal [798]. As a result, it is impossible for the lens to gather these high-spatial-frequency components of the light field. This is the well-known Abbé diffraction limit, which can be expressed as $\Delta x = \lambda/(2\pi \mathrm{NA})$, with NA being the numerical aperture of the lens [225, 795]. Scanning confocal optical microscopy, in which a sharply focused spot of light substitutes the wide-field illumination, allows for a minor resolution increase [799]. Progress is still being made in this sector by employing nonstandard lighting and detector geometries, and occasionally combining these with nonlinear phenomena like multiphoton excitation [800].

From a historical point of view, early in 1928, an experimental approach was described by Irish physicist Synge [801], which would allow the optical resolution to reach into the nanoscale realm. As a means to create a tiny light source, he suggested hiding a powerful light source behind an ultrathin, opaque metal film with a hole of 100 nm in diameter. The pinpoint of light so produced can be utilized to highlight a small area of a thin biological slice. In order to ensure adequate local lighting, he stipulated that the metal film's aperture be positioned no more than the diameter of the aperture away from the biological slice, or less than 100 nm. A sensitive photodetector was to be used to record images as light passed through the slice. Synge's idea was revolutionary at the moment it was presented. However, he never made an effort to see his vision through. Due to the difficulty of taking samples for a planar screen, he abandoned this initial idea in favor of one he proposed in 1932 [802]. As an alternative to an aperture, he proposed using the image of a point source of light as an optical probe. An ellipsoidal mirror was planned to create the image, offering the biggest potential numerical aperture. Since near fields are not recoverable by any conventional imaging strategy, regardless of the numerical aperture, this second scheme would have never achieved SNOM-type resolution.

Nanometer-scale positioning technology based on piezoelectric materials became readily available in the early 1980s after the invention of the STM [803], and in 1984, Pohl, Denk, and Duerig at the IBM Zurich Research Laboratory demonstrated an optical microscope identical to Synge's suggested but forgotten method [804, 805]. Lewis and his team at Cornell University independently

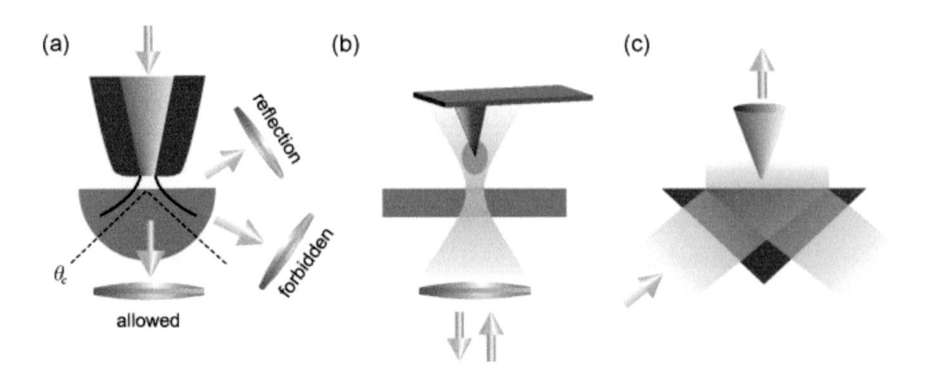

Figure 9.1 Different types of SNOM: (a) aperture SNOM with angular resolved detection, (b) apertureless configuration, and (c) scanning tunneling optical microscope. Reprinted from [795], with the permission of AIP Publishing.

suggested and implemented a similar approach [806–808]. Making a subwavelength optical aperture at the tip of a transparent probe that was then coated with metal was the major breakthrough. In addition, a feedback loop was added to ensure that the gap width between the fixed probe and the sample remained at a consistent few nanometers during the raster scan. At the same time, Fischer et al. [809–811] demonstrated that 100-nm circular apertures in a metal film as light sources were capable of achieving subdiffraction limit imaging, while no nanoscale positioning technology was utilized for the feedback control of sample–aperture distance.

Because of the potential to push optical microscopy beyond the diffraction limit, numerous experimental configurations have been developed that produce optical images with nanoscale resolution. The classical aperture SNOM configuration, as shown in Fig. 9.1(a), which depicts a probe with an aperture shining a light on a tiny section of the sample's surface [794, 796, 804, 806]. In the most generic setup for light detection, the sample is resting on a hemispherical substrate that collects all of the light emitted from the probe–sample interaction zone into the far field [812]. The fascinating apertureless NFO methods are depicted in Fig. 9.1(b) [813–821]. In this setup, an external far-field illumination creates a very constricted optical field at the tip of a precisely pointed probe. In this situation, the NFO signal is typically buried in a sea of far-field scattered radiation and must be retrieved. This kind of method is also dubbed "tip-enhanced" optical microscopy. The scattering-type SNOM (s-SNOM) is the most widely used mode of apertureless SNOM, which exploits the elastically scattered light from the tip–sample interaction. This technique has developed very rapidly in the recent decade that further extends the resolution limit of the aperture SNOM [822]. Nowadays, s-SNOM allows ultrabroadband optical (0.5–3 000 μm) nanoimaging, and nanospectroscopy with fine spatial (<10 nm), spectral ($<1\text{cm}^{-1}$), and temporal ($<$10 fs) resolution [821]. Direct detection of localized fields in the near field above a sample by uncoated dielectric tips is the basis of the third NFO microscopy setup. As shown in Fig. 9.1(c), a scanning tunneling optical microscope (STOM) [823], also known as a photon scanning tunneling microscope (PSTM) [824], is presented schematically. In a few experiments, the STOM yielded outcomes that were not attainable using other aperture-based NFO techniques [825, 826]. Note the difference between STOM or PSTM with the optical STM technology, which combines optical spectroscopy and STM probe, and the optical STM technology is of more interest for detecting electron tunnelling signal. [827].

9.2 Aperture SNOM

9.2.1 General Setup of Standard Aperture SNOM

Now let us focus on the aperture SNOM. Compared to a regular scanning confocal optical microscope with a typical numerical aperture of 1.4 [800, 828], aperture SNOM, as the most popular and well-developed NFO technique, has a resolution of 50–100 nm on a normal basis and can reach 10–30 nm under appropriate conditions, at least 5–10 times better [795]. Furthermore, the aperture SNOM method is used by the vast majority of commercially accessible devices, although s-SNOM is catching up in recent years in the market.

In Fig. 9.2, we see a schematic representation of a common aperture SNOM setup, as implemented in many laboratories and commercialized instruments. Laser light is coupled into an optical fiber terminating in an aperture probe. Before light can be coupled into the fiber, it must undergo polarization and spectrum filtering control. To get rough proximity to the sample, that is, into the range of displacement controlled by a feedback loop, the tip is attached to a mechanical support using adjustable screws. In addition, it is frequently essential to move the tip to different locations on the sample or to align it laterally onto the optical axis of the traditional microscope's objective.

A signal that strongly depends on the width of the gap between the tip and the sample is required for the accurate measurement and manipulation of its width. In the beginning, SNOMs used electron tunneling feedback [804, 805]. Most modern aperture SNOMs use a technique termed shear force feedback, which is similar to noncontact AFM [829, 830]. The fiber probe is excited by mechanical vibrations at one of its resonances parallel to the surface of the sample, with an amplitude below 1–5 nm in an ideal situation. A suitable displacement sensor is used to measure the amplitude and phase of this tiny fiber oscillation [829–831]. The resonance frequency is detuned with respect to the driving oscillator during the final approach ($\approx 0-20$ nm) due to the impact of shear forces, resulting in a decrease in the amplitude and the phase shift. It is not totally obvious where the influence of shear force comes from. Since the phase signal is unaffected by the

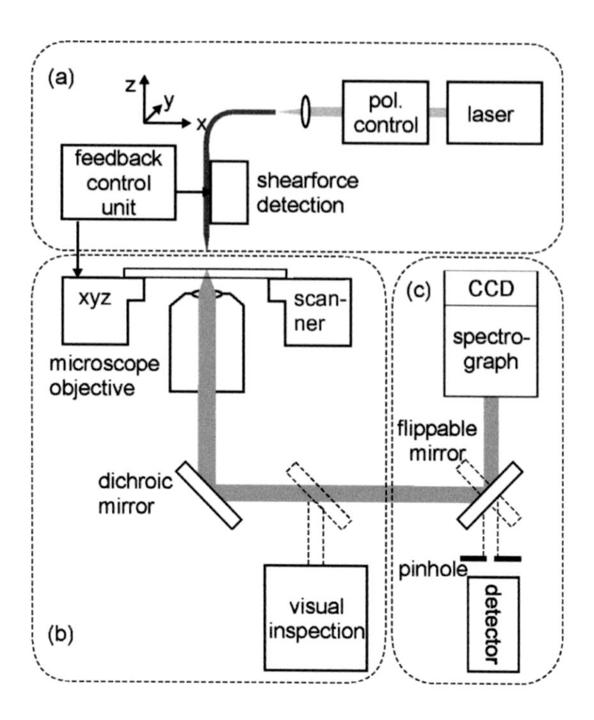

Figure 9.2 Standard aperture SNOM setup consisting of (a) an illumination unit, (b) collection and redistribution unit, and (c) a detection module. Reprinted from [795], with the permission of AIP Publishing.

loss of momentum in the resonance, it can respond more quickly to variations in the oscillation's state. As a result, fast shear force feedback systems benefit from either pure phase feedback [832] or a combination of amplitude and phase feedback [833]. In Fig. 9.2(a), it is shown how the light source, fiber, optical probe, shear force sensor, and mechanical support typically function in an SNOM system.

Aperture-emitted light locally interacts with the sample. The desired contrast mechanism in the sample will determine whether it is absorbed, phase-shifted, or locally excited fluorescence. In any scenario, high-efficiency light collection from the interaction zone is required to obtain usable signals. High NA (oil-immersion) microscope objectives or mirror systems are commonly utilized for this goal. When mounted on a plane-parallel substrate, an oil immersion objective has the particular advantage of capturing light that would otherwise be lost due to entire internal reflection [812, 834]. In phase- and amplitude-contrast images, this treatment can improve contrast and resolution [835]. Mirror-based detection occasionally makes use of hemispherical substrates offering a "solid immersion" to collect the otherwise lost signals of light [812, 834]. Then dichroic mirrors are used to split the collected light and send it to either the microscope's eyepiece or a detector (Fig. 9.2(c)). Unwanted parts of the spectrum can be filtered out using filters. A standard inverted optical microscope, like the one shown in Fig. 9.2(c), is useful for the collection, redistribution, and filtering of light.

9.2.2 Tip Fabrication

One of the most difficult parts of SNOM is getting the manufacture of the tips under control and repeatable. Despite advances in micro- and nanofabrication techniques, near-field tips are still not compatible with most AFMs and can only be utilized with specialized instruments [836, 837]. Therefore, many studies require custom-made probes to perform as expected in their intended context, although many types of high-quality probes are commercially available from a variety of suppliers [838]. For aperture SNOM, tips that are of good quality are characterized by a dense metal coating, a large cone angle, high brightness, a high threshold for optical damage, and a well-defined taper structure [795, 839, 840]. The most widely used type of aperture probe is based on optical fibers. In particular, fabricating the taper structure with a sharp apex and then coating the tip with a metal layer (usually aluminum) are the two primary steps in the process of making fiber-based optical probes. The former step is intended to achieve a high spatial resolution, and the latter step aims to reduce the leakage of light from side walls of the tip during illumination and collection.

The preparation of tapered optical fibers with a sharp tip and suitable cone angle can be done using two main methods. The first is the heating and pulling method, and the second is the chemical etching method. The heating and pulling technique involves first locally heating them with a CO_2 laser or a filament and then pulling them apart, as schematically shown in Fig. 9.3(b). The shapes of the obtained tips are highly sensitive to the parameters of the heating and pulling process, including temperature, time, and the size of the heated area [841, 842]. The glass surface on the taper is very smooth due to the pulling process, which is helpful to the quality of the coated metal layer during the evaporation process later on. In addition, a final fracture during pulling often results in flat facets at the apex, which makes it easier for an aperture to form during evaporation. Cone angles of large degrees of inclination are notoriously difficult to achieve using this method, and as a result, the transmission coefficient for a given aperture size is poor.

Since the introduction of Turner's etching technique [795, 843], chemical etching of optical fibers in hydrofluoric (HF) acid solutions has become commonplace. The procedure is to immerse the fiber in HF that is typically covered by a protective layer of a hydrophobic chemical solvent, such as toluene or p-xylene. Fiber tips are formed at the meniscus that occurs at the interface between the HF and the protective layer, as schematically shown in Fig. 9.3(a). The mechanism relies on the fact that the meniscus height depends on the diameter of the remaining fiber. Chemical etching paves the way for laboratory-scale reproducible mass production of optical probes, and it is simple and inexpensive to implement. The taper angle can be adjusted by changing the solvent used as the protective layer,

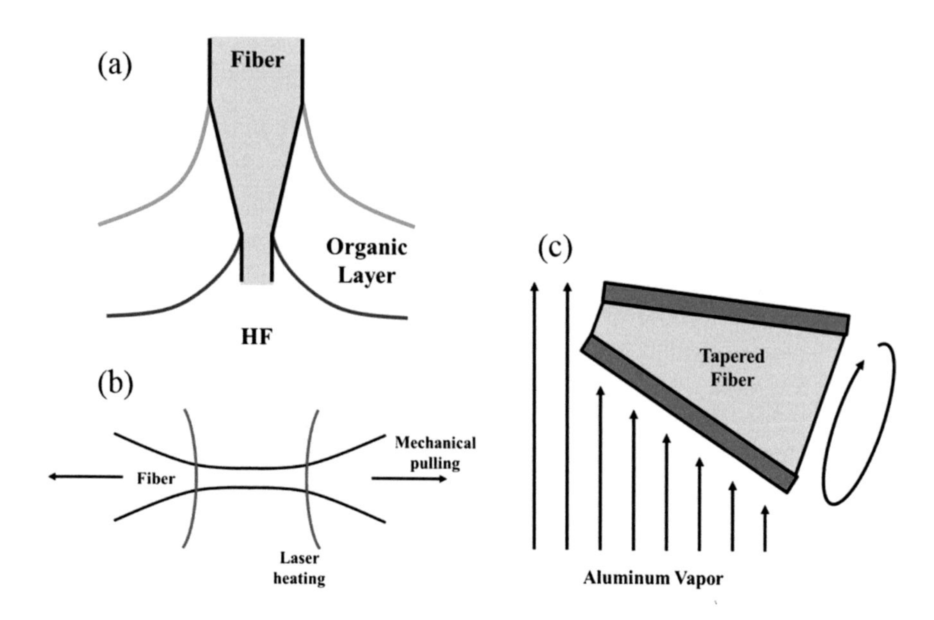

Figure 9.3 Typical fabrication techniques for tapered fiber aperture probes. The two common approaches to producing tapered fibers are (a) chemical etching and (b) heating and pulling. After the formation of the fiber tip, subsequent (c) metal deposition results in an aperture at the tip apex that provides light confinement. Evaporation geometry of the aluminum-coating process: Evaporation takes place under an angle slightly from behind while the tip is rotating. The deposition rate of metal at the apex is much smaller than that on the side walls. Reprinted from [822], with the permission of AIP Publishing.

which is a distinct benefit of Turner's approach. This allows for the fabrication of optical probes with a suitably high transmission coefficient [844, 845]. However, the microscopic roughness of the glass surface on the taper walls proved to be a key drawback of the meniscus etching method due to its sensitivity to environmental influences, which could lead to the formation of unwanted pinholes in the subsequently deposited metallic layer that interferes with the definition of the near-field aperture. This drawback was then solved by the tube-etching method, in which the etching process is conducted without striping off the polymer jacket of the fiber as a protection layer against environmental influences, resulting in significantly improved tip surface quality [839, 846]. Further improved methods on this basis have also been proposed in recent years [847–849].

Having finished with the taper at the fiber's tip, the aperture should be made. Typically, aluminum is evaporated into the sidewalls of the fiber taper surface in a configuration in which the metal deposition rate at the apex is much smaller than on the sides. This is done by pointing the probe tip away from the evaporation source and rotating it such that the sides are uniformly coated, as schematically shown in Fig. 9.3(c). The resulting aperture typically has a diameter of

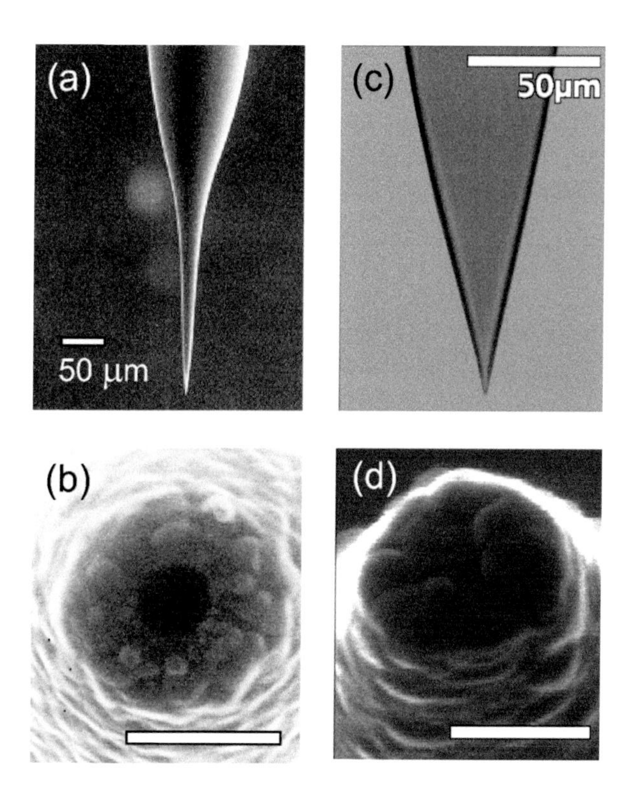

Figure 9.4 Aluminum-coated aperture probes prepared by pulling (a) and (b), and etching (c) and (d): (a) and (c) macroscopic shape, SEM, and optical image. (b) and (d) SEM close-up of the aperture region; the scale bar corresponds to 300 nm. Reprinted from [795], with the permission of AIP Publishing.

80–100 nm, and it is even possible to achieve a 20-nm diameter. Note that there are other constraints on aperture diameter besides evaporation conditions. For instance, the aperture diameter should be more than twice the skin depth of the metal. This is why aluminum is mostly used as the coating metal, since in the visible region, aluminum has the smallest skin depth and it is in theory the most suitable. Nevertheless, other metals such as silver, chromium, and platinum in conjunction with carbon have also been employed successfully.

In Fig. 9.4, we show typical images of fiber-based near-field probes. The macroscopic shape of a probe fabricated from the heating and pulling method before an aluminum coating is given in Fig. 9.4(a), while that of another probe prepared by the tube-etching method is presented in Fig. 9.4(c). It can be observed that the latter has a larger cone angle preferred for high light transmission efficiency. The SEM graphs of the tip apex regions of the two probes after aluminum coating are further shown in Figs. 9.4(b) and 9.4(d), respectively, in which the aperture can be observed for both tips. It can also be seen that the metal coating shows granular structures, which heavily depend on evaporation conditions [850].

Even if the metal layer does not block the aperture, the tip's protrusions at the end may still restrict how close it can get to the sample. To accurately clean the metallic protrusions at the tip, a focused ion beam (FIB) approach can be adopted recently, which can lead to a very flat and smooth apex of the tip [851, 852].

9.2.3 Applications of Aperture SNOM

A variety of applications can be achieved by the aperture SNOM, including near-field imaging of amplitude and phase of the light field, near-field fluorescence microscopy, near-field Raman spectroscopy, time-resolved near-field imaging, and so on [795, 822, 853]. In the MNTR community, the near-field characterization of amplitude and phase is of the most interest, from which an in-depth understanding of nanoscale light–matter interaction mechanism can be obtained by imaging the amplitude and phase in a wide range of micro- and nanostructures. A phase- or amplitude-contrast image of a sample can be obtained by directly collecting and recording light emitted from the probe–sample interaction zone using an appropriate photodetector, as discussed in Section 9.1. Theoretical studies of the tip–sample interaction mechanism and image formation in SNOM have received a lot of attention over the past several decades, and for a review, one can refer to [793, 854]. In this section, let us discuss several examples of the application of aperture SNOM to the characterization of photonic structures.

The first example is the observation of SPP. As mentioned earlier, plasmon polaritons are hybrid excitations of light and free electrons, which are able to confine the energy of radiation at a subwavelength scale. Wu et al. [855] demonstrated a visible-range phase-sensitive multiparameter heterodyne aperture SNOM that can measure both the optical phase and amplitude distributions in the near field [856]. As can be seen in Fig. 9.5(a), the system integrates a commercial SNOM with a homemade Mach–Zehnder interferometer. An aluminized aperture fiber probe, commercially available from NT-MDT (Russia), is used to collect data on the optical field at the sample's surface. The required field parameter data are present in the signals picked up by the photomultiplier. Mixing the two driving frequencies from the acousto-optic modulators directly yields a 20-kHz signal that is used as the lock-in reference. To retrieve the phase and amplitude, the signals would be demodulated at the beat frequency with a lock-in amplifier. To observe SPP in the near field, Wu et al. [855] fabricated 10 column grooves on a 150-nm-thick gold film using the FIB method, as shown in Fig. 9.5(b), in which the detailed parameters of the grooves are also shown. Polarized 633-nm He–Ne laser light is vertically focused on the back of the sample, that is, working in the collection mode, which can excite SPP waves on the sample surface.

The NFO amplitude and phase distributions of a single groove are shown in Figs. 9.5(c) and 9.5(d), respectively. The intensity distribution reveals very little about the propagation characteristics of the SPPs, whereas the phase distribution

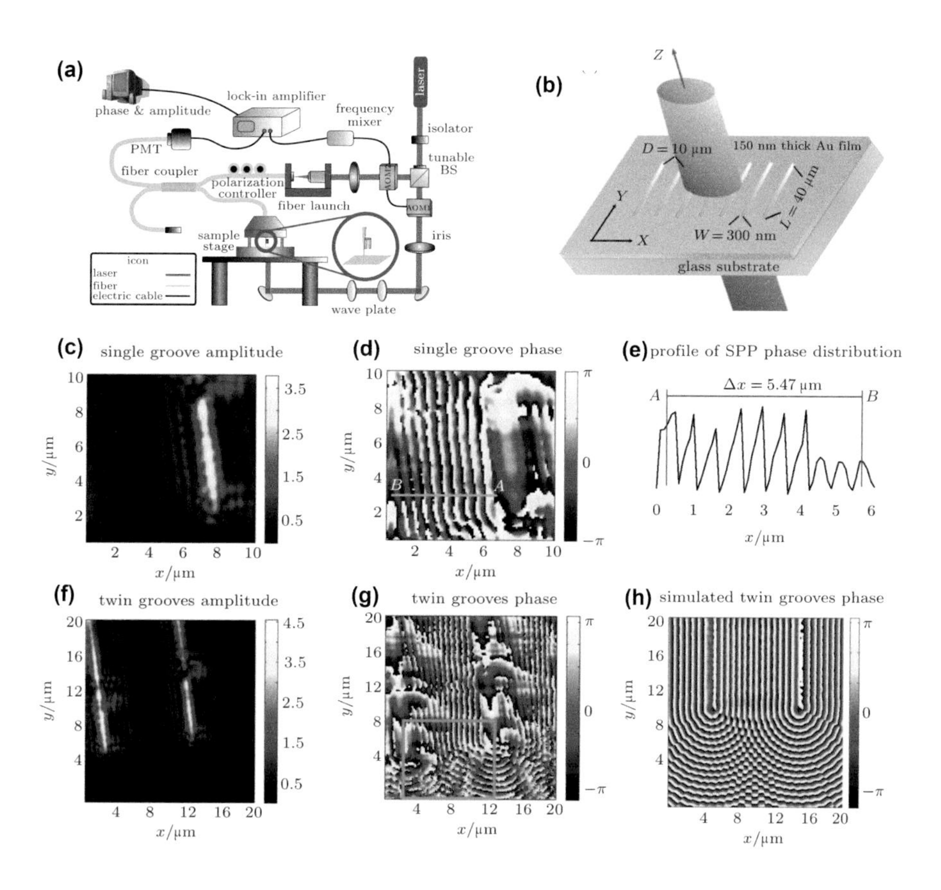

Figure 9.5 (a) Setup of the phase-sensitive SNOM. (b)Schematic of the groove sample. (c–h) Measured and simulated optical field distributions of SPP grooves. (c)–(e) single groove, (f)–(h) twin grooves. Reprinted figure from [855]. Copyright 2014 by the Chinese Physical B.

offers a great deal more information through its intense and quick fluctuations. Moreover, the SPP phase distribution associated with the segment AB (denoted in Fig. 9.5(d)) is displayed in Fig. 9.5(e). The average wavelength of the SPP wave across all nine peaks is $d = 5.47$ μm$/9 = 607$ nm, which agrees with the theoretical value of the SPP wavelength of 609 nm calculated from the standard dispersion relation. Furthermore, the optical amplitude and phase distributions of the twin grooves are shown in Figs. 9.5(f) and 9.5(g), respectively, with Fig. 9.5(h) displaying the simulated phase profile by means of the FDTD approach. Both measured and simulated results can demonstrate the propagation of SPP in a plane, which agree with each other well. It can also be noted that, unlike the phase distribution, the intensity or amplitude distribution cannot provide useful information about the propagation direction or interference pattern of SPPs. Furthermore, a square region is highlighted in Fig. 9.5(g), which clearly demonstrates an interference pattern formed by two SPP

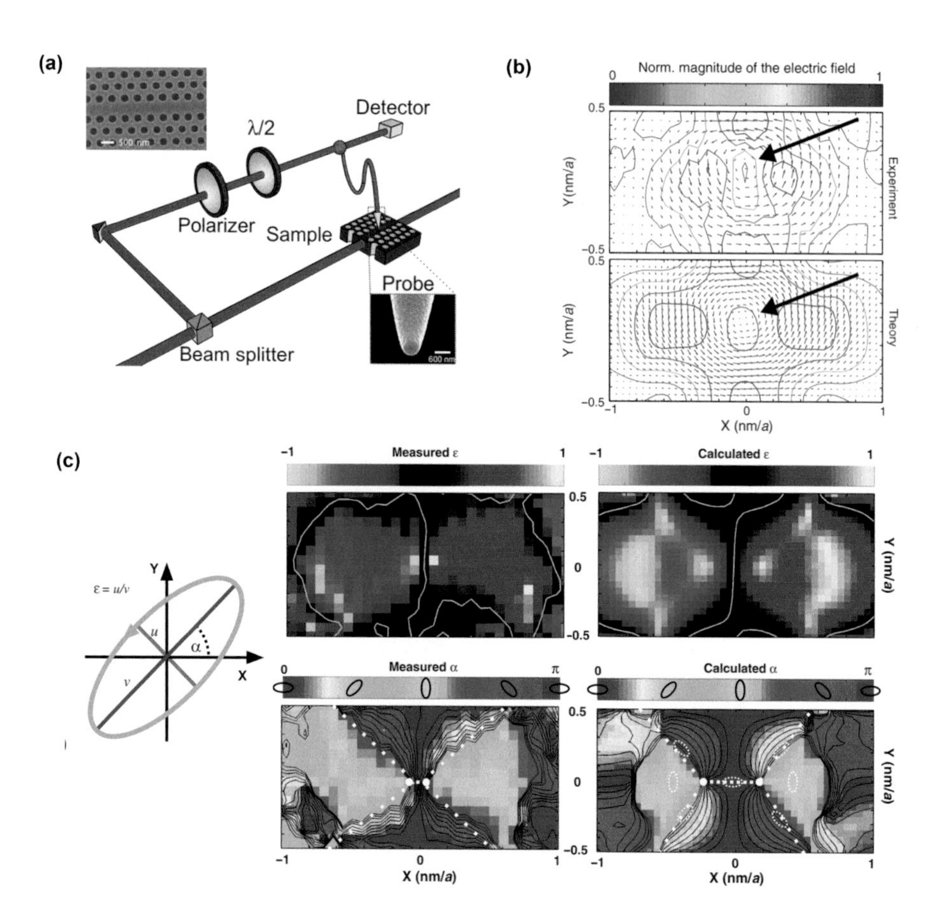

Figure 9.6 Full vectorial near-field characterization by aperture SNOM. (a) Schematic representation of the experimental setup of polarization- and phase-sensitive SNOM. (b) Experimentally and theoretically obtained 2D vector plots of the electric field. (c) Experimentally and theoretically obtained ε and α, whose definitions are schematically represented on the left. Reprinted figure with permission from [857]. Copyright 2009 by the American Physical Society.

waves. Therefore, by measuring the near-field amplitude and phase distribution, an improved comprehension of the interaction of light with SPP structures and the generation, localization, conversion, and propagation of SPP waves can be attained.

Full vectorial mapping of both electric and magnetic fields at nanoscale can be achieved using phase- and polarization-sensitive aperture SNOM scheme [857, 858]. For a comprehensive review on this subject, one can refer to [797]. Here we present an example of 2D PC waveguide (see Section 10.1.2 for an introduction of PCs). For instance, Burresi et al. [857] demonstrated a phase- and polarization-sensitive aperture SNOM and measured independently two in-plane electric-field components of light propagating in a 2D PC waveguide and the phase difference between the two components. They were therefore able to reconstitute the electric vector field distribution with subwavelength resolutions.

The experimental setup of Burresi et al. [857] is presented in Fig. 9.6(a). A diode laser at the wavelength of 1 463 nm with linear polarization is coupled into the PC waveguide, whose SEM image is shown in the upper-left corner. The sample consists of a silicon membrane with a 200-nm thickness, in which holes are etched in a periodic arrangement, with a lattice constant of $a = 450$ nm. The waveguide is formed by a single missing row of holes. The field above the sample is collected by the aperture of an aluminum-coated near-field probe, whose SEM image is given in the lower-right inset. The probe is kept 20 nm above the sample by means of the shear force feedback. The light collected by the probe is interferometrically mixed with a reference beam and then recorded using heterodyne detection in order to obtain the amplitude and phase information. Furthermore, when two orthogonal polarizations are present in the probe fiber, one can select either one by selecting the appropriate polarization for the reference branch. Hence, the near-field setup can identify the polarization state of the near-field.

They reconstructed the in-plane vector field of the electric field in a single unit cell of the PC waveguide by mapping the near-field amplitude and phase distribution of two well-chosen orthogonal polarizations. The electric-field vector plot, as derived experimentally and theoretically, is depicted in Fig. 9.6(b). The measured and computed magnitude of the electric field is depicted as contour lines. Since the out-of-plane component of the electric field is only noticeable close to the holes' borders and vanishes in the center, the electric field in the middle of the waveguide can be thought of as being entirely in-plane. It can be seen that both the theoretical predictions and the experiments are consistent with one another.

To get a better understanding of the polarization distribution of the light inside the waveguide, they calculated the ratio $\varepsilon = \mp u/v = \tan\{\arcsin[(\sin 2\psi) \sin \delta]/2\}$ and the orientation angle $\alpha = \{\arctan[(\tan 2\psi) \cos \delta]\}/2$ of the polarization ellipse, where v is the major semiaxis and u is the minor semiaxis (as depicted on the left-hand side of Fig. 9.6(c)). Since in these expressions, the angle $\psi = \arctan(|E_y| / |E_x|)$ is the amplitude ratio of the two electric-field components and the other angle $\delta = \delta_y - \delta_x$ indicates their phase difference, knowing the phase relation between the two field components is critical for deriving ε and α. The retrieved ε from the measured and calculated vector field distribution in Fig. 3 is displayed in Fig. 9.6(c). Right-handed polarization is represented by a negative value, while left-handed polarization is indicated by a positive one. The retrieved angle α from the measured and calculated field distribution is also shown in Fig. 9.6(c). The experimental and theoretical results show quite surprising agreement. Therefore, this work experimentally visualized the electric vector field of the light propagating through a PC waveguide, showing the power of the SNOM technology in studying nanoscale light–matter interactions.

9.3 Apertureless SNOM (Scattering-Type SNOM)

9.3.1 General Setup and Operation Principle of S-SNOM

In this section, we discuss the operating principles of the apertureless SNOM. The basic schematic of apertureless SNOM setup is already depicted in Fig. 9.1(b), in which an external far-field illumination creates a highly confined optical field at the tip of the probe that does not have an aperture. In this sense, the probe behaves like a rod antenna, although in this circumstance, the near-field signal is typically mixed with a strong signal of far-field scattered radiation and must be retrieved. The advantage is that its lateral resolution is mainly determined by the curvature radius of the tip, which is typically in the range of 120nm, regardless of the wavelength which can lie in the terahertz range, while the aperture SNOM, due to the waveguide cutoff effect, the aperture size in principle is limited to $\sim\lambda/10$ [817], not to mention the fact that there are no appropriate fibers to be used as probes in many wavelength ranges. As mentioned earlier, the s-SNOM is the most widely used mode of apertureless SNOM [225, 813–821, 859]. This technique has developed very rapidly in the past years, which can substantially raise the resolution limit of the aperture SNOM [822].

Intense work on the development of s-SNOM began in the late 1990s when the AFM technique had already reached a rather stable state of development. It was in 1996 that Lahrech et al. released one of the earliest works on s-SNOM [860]. They used s-SNOM to image a gold grating and gold surface and found excellent optical contrast at resolutions of below $\lambda/100$. Significant instrumentational development was made by Knoll et al. and Hillenbrand et al. [785, 816] in the late 1990s and the early 2 000s, laying the groundwork for contemporary s-SNOM. Modern s-SNOM, in contrast to aperture SNOM, typically employs a focused laser beam incident at an oblique angle to illuminate a metalized AFM probe. To improve scattering efficiency, a metal coating is used [861]. Si cantilevers, which are commercially available and cost-effective, are typically used as the basis for the metal-coated tips [862]. Figure 9.7 shows a schematic of how the scattering nanoscopic probe interacts with nanostructures, with micrographs of a real tip coated with Au. Nowadays, s-SNOM is capable of achieving ultrabroadband (0.5–3 000 μm) nanoimaging and nanospectroscopy with fine spatial (<10 nm), spectral ($<1\ \text{cm}^{-1}$), and temporal (<10 fs) resolutions [821].

Figure 9.8 depicts a typical configuration for the s-SNOM in the visible and infrared (IR) range. A beam splitter separates the incoming light into a reference beam and a probing beam. A focusing optical element, often a parabolic mirror or a lens, directs the probe beam to the AFM tip, where it is focused (within the diffraction limit). In the tip–sample system, light scatters elastically after interacting with the system. A detector collects the light that has been backscattered (or in certain circumstances forward scattered) along the same or a different path to the beam splitter. The backscattered light interferes with the reference beam at the detector, which is reflected from a flat mirror on a delay stage. This setup

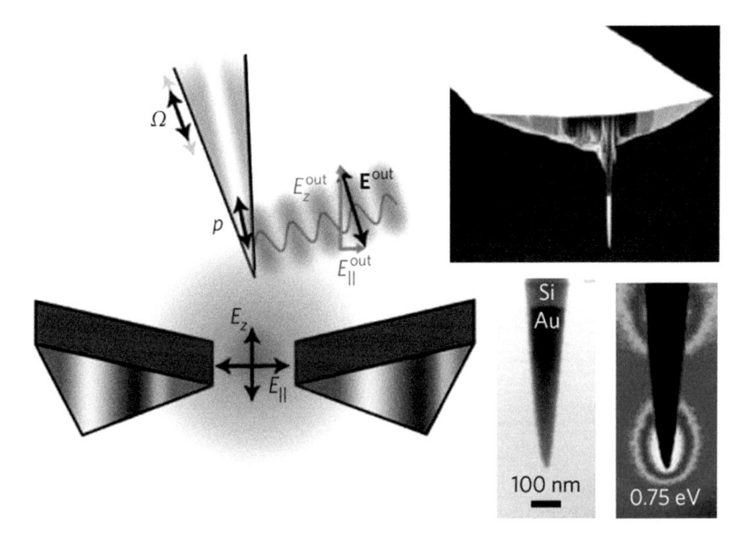

Figure 9.7 A scattering nanoscopic probe is brought into the near field of a structure. For a scattering NSOM, (a) shows the detected field is composed of both an in-plane component $E_{||}$ and an out-of-plane component $E_{z'}$ where their ratio and the orientation of $E_{||}$ are set by the direction at which $\mathbf{E}^{\text{out}}$ is detected. Reprinted from [797]. Copyright 2014 Springer Nature. The inset in (b) shows a typical scattering probe tip on a cantilever with a zoom-in on-the-probe tip (bottom-left inset), which in this case is resonant in the infrared. Reprinted with permission from [863]. Copyright 2013 American Chemical Society.

is, in fact, an asymmetric Michelson interferometer. Through this method, we are able to retrieve both the amplitude and phase information of the scattered light, opening the door to recovering local complex dielectric functions of the sample volume under the tip without relying on the K-K relations.

A theoretical topic of interest for s-SNOM is the enhancement of interaction efficiency between an external EM wave and a nano-object in the gap between the s-SNOM tip and the substrate, where the nano-object is a generalization of nanostructures on the sample surface. The simplest model assumes that the tip can be treated like a point dipole that is placed above a surface with a dielectric function ε_s [815]. However, this model predicts no field enhancement under the tip. A variety of articles have modeled a near-surface probing tip as a pyramid [864], hyperboloid [865], or spheroid [866–868]. To consider the field enhancement for a nano-object in the gap, a small gold sphere is placed between the tip and the surface in simulations as shown in [869]. In order to go beyond the electrostatic approximation used in analytic calculations, the excitation and contribution of the plasmonic waves traveling along the tip from its apex are taken into account in numerical simulations in which the tip is represented as a pyramid with a blunted tip apex [870]. Typical s-SNOM dependences are computed using these models, including the approach curve, which illustrates the signal's dependence

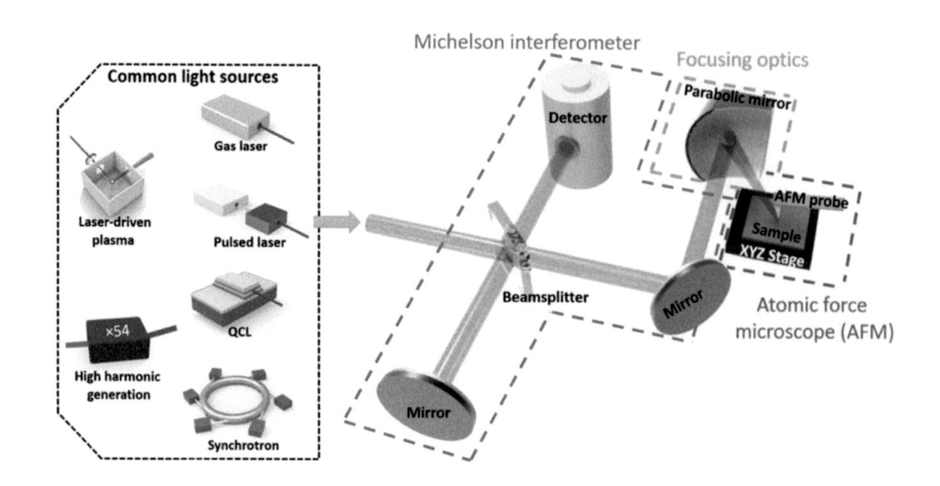

Figure 9.8 A typical s-SNOM setup with visible or IR light sources. For completeness, a THz CW source, with high-harmonic generation using a microwave, is included. Reprinted from [821]. Copyright 2019 WILEY-VCH Verlag GmbH Co. KGaA, Weinheim

on the tip–sample distance [869], and spectral and angular diagrams, which show how the signal is scattered by the tip when there is a surface present [870]. It is demonstrated that the angle-of-elevation diagram includes one or two directed lobes, depending on the tip and sample material. When using an s-SNOM, the best angle for radiation coupling to the tip is between 60° and 70° from the normal to the sample surface [864].

A variety of works theoretically [867, 871–873] and experimentally [865, 874] demonstrated that the field of an irradiating wave is significantly enhanced in the gap between the tip and the sample by a factor of 4–5 orders of magnitude. Fields stimulated around an ASNOM tip by external radiation in the presence of a sample surface and those radiated by an s-SNOM tip into the ambient space are theoretically considered in a variety of methods resulting in different results. Nevertheless, these methods lead to several common conclusions. First, the electric-field amplitude near the tip apex is several orders higher than the amplitude of the driving wave. Second, the tip–sample interaction is strongly enhanced at the distance of about the tip curvature radius (directed along the tip shaft). Third, the enhancement arises primarily for the polarization of the incident wave that is normal to the surface, while when illuminating with another polarization of light, the field enhancement is much weaker.

A widely used, simple but effective model assumes that the tip can be regarded as a sphere with the permittivity ε_t and a radius a, and its dipole polarizability under a homogeneous driving field can be calculated in the long wavelength limit since the tip apex is much smaller than the wavelength of radiation [875, 876]. The emission of a scattered EM field by the tip into space can be treated as radiation from an oscillating dipole of the sphere. The presence of the sample surface can be taken into account by setting an electrostatic image of the sphere when the retardation effect of wave propagation is ignored in the long wavelength limit. In this circumstance, the effective dipole polarizability of the tip with consideration of the contribution of its electrostatic image can be analytically obtained as

$$\alpha_{\text{eff}} = \frac{\alpha(1+\beta)}{1-\frac{\alpha\beta}{16\pi(a+z)^3}}, \tag{9.3.1}$$

where

$$\alpha = 4\pi a^3 \frac{\varepsilon_{\text{t}}-\varepsilon_{\text{i}}}{\varepsilon_{\text{t}}+2\varepsilon_{\text{i}}}, \quad \beta = \frac{\varepsilon_{\text{s}}-1}{\varepsilon_{\text{s}}+1}, \tag{9.3.2}$$

with ε_s being the dielectric function of the surface and ε_i that of the surrounding medium in which the tip and the sample is embedded, typically air. By using this effective dipole polarizability, the amplitude of the scattered wave at large distances, $E_{\text{sca}}(\infty)$, can be expressed as

$$E_{\text{scatt}}(\infty) \sim E_{\text{loc}}\alpha_{\text{eff}} = E_{\text{loc}}(\mathbf{r}_{\text{tip}})\,\alpha_{\text{eff}}(\mathbf{r}_{\text{tip}}, \varepsilon_{\text{s}}(\mathbf{r}_{\text{tip}})), \tag{9.3.3}$$

where E_{loc} is the local EM field that excites the dipole oscillation in the tip, and $\mathbf{r}_{\text{tip}}$ is the position of the tip with respect to the sample surface. This formula can be further expressed as

$$E_{\text{scatt}}(\infty) \sim E_{\text{loc}}(\boldsymbol{\rho}_{\text{tip}})\,\alpha_{\text{eff}}(z, \varepsilon_{\text{s}}(\boldsymbol{\rho}_{\text{tip}})), \tag{9.3.4}$$

where $\boldsymbol{\rho}_{\text{tip}}$ and z are the in-plane and vertical (namely, tip–sample distance) coordinates with respect to the sample surface.

Many works have calculated the elastic scattering of light by a tip in the presence of a surface [225, 866, 877–879]. One can calculate the scattering spectra of the probe tip by including the frequency dependence of the dielectric function of the sample. Results have been obtained for metals such as Au [816], Al, W [878]); nonresonant dielectrics such as quartz, PS and PMMA [816, 880], Si_3N_4, Si [878, 881]; and resonant dielectrics like SiC [817, 866]). Analytic calculations that account for the contribution of free carriers to the dielectric function of doped semiconductors show an excellent quantitative fit to the experimental data (maps of the s-SNOM amplitude and phase) [815, 817, 878, 882].

Accordingly, the s-SNOM can map the distribution of a local light field over the surface and a sample dielectric function. Nonetheless, it has been indicated that several parameters need to be modified for the theoretical simulations to fit the experimental findings (e.g., the gap between the tip and the surface has to be adjusted) [877]. Reconstructing the surface dielectric permeability from the s-SNOM signal is a challenging and uncertain inverse task.

Light scattered by the tip in the presence of a surface and subsequently collected by a photodetector as a signal has an infinitesimally small amplitude, due to the nanoscale small size of the tip. Optical homodyning [815, 884, 885] or heterodyning [883, 886] is a useful technical approach for detecting such a

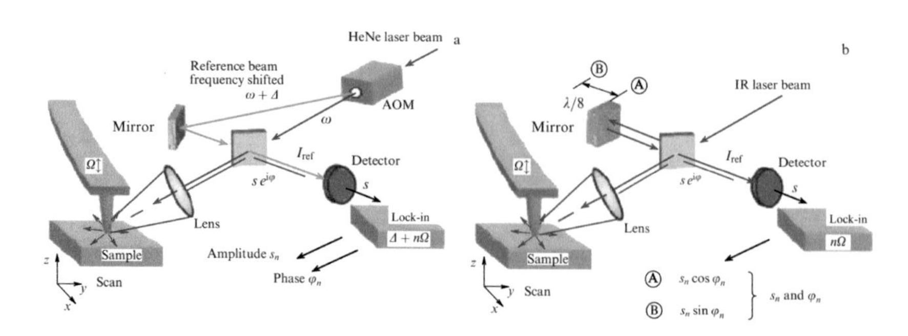

Figure 9.9 Optical (a) heterodyning and (b) homodyning of light scattered by an s-SNOM probe. Reprinted from [883]. Copyright 2003 The Royal Microscopical Society.

weak signal, as schematically shown in Fig. 9.9. Since the photodetector's signal depends on the intensity of the collected light, that is,

$$I_{\text{det}} \sim E_{\text{det}}^2 = (E_{\text{ref}} + E_{\text{sca}})^2, \tag{9.3.5}$$

which can be further recast into

$$I_{\text{det}} \sim (E_{\text{ref}} + E_{\text{sca}})^2 = E_{\text{ref}}^2 + 2E_{\text{ref}} E_{\text{sca}} + E_{\text{sca}}^2, \tag{9.3.6}$$

where the term E_{sca} represents the amplitude of the scattered light, and E_{ref} is the electric-field amplitude of the reference beam that is, in fact, constant. Hence, $2E_{\text{ref}} E_{\text{scatt}}$ is the leading term in changes of the detected signal. In the lack of a reference wave, the term E_{scatt}^2 would come into play. Since the scattered wave is vanishingly small, the detector's sensitivity can be increased by a factor of $E_{\text{ref}} / E_{\text{scatt}}$, or several orders of magnitude, due to the introduction of the reference beam. Note we have written these equations as if the field amplitudes are real numbers, although in practice they are complex quantities. Nevertheless, the same conclusion would be drawn. Given that near fields rapidly decay with distance, Eqs. (9.3.1)–(9.3.4) show that the effective tip polarizability relies on the tip–surface distance in a very nonlinear way.

In fact, it is possible to only retrieve the components of the optical signal that are uniquely induced by the NFO interaction between the tip and the surface, thanks to the nonlinear dependence of the optical signal on the tip–sample distance. Resonant cantilever vibration, employed in the tapping-mode feedback of AFM [887], is an example of a periodic modulation of the tip–sample distance that is common in s-SNOM systems. The complex amplitude of the light wave scattered by the tip is altered by the sinusoidal vibration of the tip. It should be noted that the only significant nonlinearity in the system is the nonlinearity of the NFO interaction between the tip and the sample at the separation distance. By eliminating the much stronger signals from scattering by the sample alone or by the tip alone, a near-field response can be obtained by recovering the higher harmonic components $I_{\text{det}}^{(n\Omega)}$ of the tip oscillation frequency Ω

(typically in the range 30–300 kHz) in the detector photocurrent, with n being an integer, $n \geq 2$ [876, 888, 889]. In a heterodyne system, which resembles the Mach–Zehnder interferometer (Fig. 9.9(a)), the light frequency in the reference beam is displaced by $\Delta = 80$ MHz relative to the light frequency ω in the laser beam being scattered by an s-SNOM tip [881]. The photodiode signal is retrieved using a lock-in detector set to the frequency of $\Delta + n\Omega$. It is possible to recast Eq. (9.3.4) in light of the homo- (hetero-)dyne recovery of the higher harmonic components as

$$I_{\text{det}}^{(n\Omega)}\left(\boldsymbol{\rho}_{\text{tip}}\right) \sim E_{\text{loc}}\left(\boldsymbol{\rho}_{\text{tip}}\right) \alpha_{\text{eff}}^{(n\Omega)}\left(\varepsilon_{\text{s}}\left(\boldsymbol{\rho}_{\text{tip}}\right)\right), \tag{9.3.7}$$

in which the averaged tip polarizability $\alpha_{\text{eff}}^{(n\Omega)}$ for high harmonics depends on the parameters of the vertical mechanical oscillation of the tip.

Initially, for s-SNOM, monochromatic illumination using the high power output (watts per wavenumber) and the narrow linewidth of continuous wave lasers (like CO_2 laser) to create high-quality images is widely implemented. In the meanwhile, broadband nanospectroscopy provides spectral data at individual points across a much wider range, usually from ≈ 200 to $\approx 5\,000$ cm^{-1} [821]. For investigating vibrational responses in materials, IR spectroscopy is crucial. With the advent of high-power tunable quantum cascade lasers and high repetition rate ultrafast fiber-based amplifiers [890–892], the practice of combining spectral measurements in s-SNOM rose to prominence in the late 2000s and the early 2010s [786, 788, 893–906]. Ultrabroadband sources from synchrotron radiation have also been utilized. FTIR spectroscopy methods are well established in far-field optics, allowing for regular IR spectroscopy with resolutions between 10 and 20 nm. The experimental setup of a broadband s-SNOM can be essentially different from that of a monochromatic s-SNOM depicted in Fig. 9.8. Broadband s-SNOM experiments involve scanning the reference mirror over a larger spatial range to generate an interferogram, that is, light intensity detected versus optical path difference. The near-field spectrum in complex numbers is found by performing a Fourier transform on the near-field interferogram. By contrast, CW s-SNOM measurements without an interferogram can only be performed with a tunable laser system by obtaining the results wavelength by wavelength.

9.3.2 Typical Applications

Since s-SNOM can reach a nanoscale resolution regardless of the working wavelength, it has become an extremely powerful microscopic characterization technology, especially in the IR and terahertz range. It is remarkable that the s-SNOM has given the first real-space observations of the exotic polariton waves in the mid-IR at nanoscale, which are important for applications in photonics and thermal radiation, including phonon polaritons in SiC [785], hyperbolic phonon polaritons hBN [907], graphene plasmons [908, 909], hybrid hyperbolic plasmon–phonon polaritons [898], and hyperbolic shear polaritons in certain monoclinic crystals [910], among others. Other applications include identifying nanoscale heterogeneities in materials [861] and biological samples [911], *in situ* characterization of chemical reactions [912], mapping the local carrier density in semiconductors and metals [878], studying the localized strain of phase-change materials [913], and identifying mode structure in nanophotonic structures and metamaterials [914]. In the following section, let us discuss several examples of the application of s-SNOM to the characterization of various micro/ nanostructures.

Visualization of Polariton Waves with S-SNOM

The first example is the near-field observation of graphene plasmon polariton (GPP) waves in the mid-IR. As the hybrid quasi-particles formed by the strong coupling of Dirac electrons and IR photons, GPPs offer great opportunities for exploring extreme light–matter interaction at the nanoscale. Since plasmonic dissipation in graphene is substantial, the propagation length of freestanding monolayer graphene is rather low. To tackle this problem and explore the fundamental limit of graphene plasmon lifetime, Ni et al. [915] proposed to investigate high-mobility encapsulated graphene at cryogenic temperatures. In this regime, the propagation length of plasmon polaritons is largely enhanced, with the only restrictions being the dielectric losses of the encapsulated layers, as well as a minor contribution from electron–phonon interactions. Basov et al. and McLeod et al. used s-SNOM to visualize the propagation and dissipation of GPPs at cryogenic temperatures, which allowed them to extract the propagation length and wavelength of GPPs directly [916, 917]. It was shown that at liquid-nitrogen temperatures, the intrinsic plasmonic propagation length of GPPs can exceed 10 μm or 50 times the wavelength of GPPs. Moreover, from the extracted results of propagation length and plasmonic wavelength, the response functions of the plasmonic medium, including the complex conductivity $\sigma(\omega) = \sigma' + i\sigma''$, can be obtained, from which information on both interactions and scattering processes in an electron liquid could be derived. Therefore, studying the near-field images of plasmon polaritons is of great value in understanding the fundamental electronic transport process in the media that support these polariton waves.

The sample studied by Ni et al. [915] is presented in Fig. 9.10(a), which is composed of a graphene monolayer encapsulated between two thin slabs of hBN. The graphene carrier density is tunable through a silicon back gate, as shown in Fig. 9.10(b). An s-SNOM that was customized to work at cryogenic temperatures is used to image GPPs at nanoscale, with its probe tip schematically shown in Fig. 9.10(c). A focused laser field at the IR frequency ω illuminates the metalized tip, which acts as an optical antenna that enhances the incident electric field at its apex. When the tip approaches closer to the sample, a concentrated field is created, which can excite polaritonic modes with a wavelength $\lambda_p(\omega)$ that is significantly shorter than the free-space wavelength of the incident infrared light [225, 918]. In near-field imaging, the signature of plasmon polaritons is periodic modulation (fringes) of the measured near-field signal as a function of tip location. These fringes have two different periodicities [919], λ_p

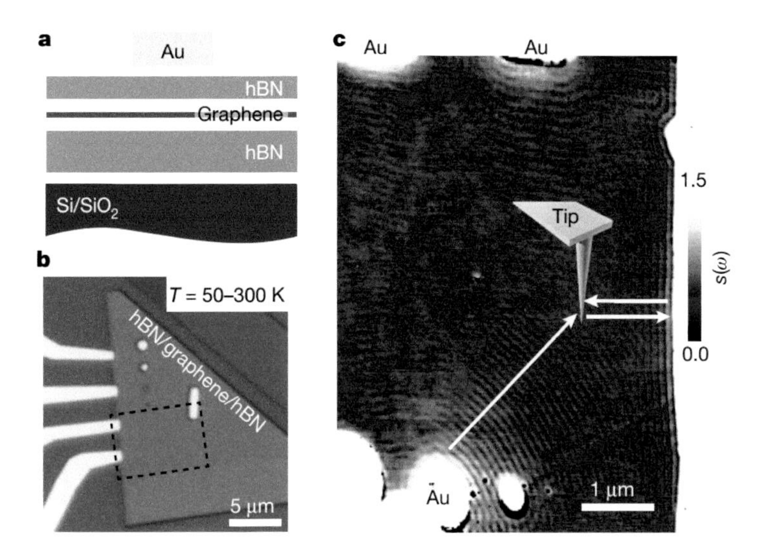

Figure 9.10 Nanoscale IR imaging of surface plasmons in Au/hBN/ graphene/hBN encapsulated structures at cryogenic temperature. (a) Sketch of the layered Au/hBN/ graphene/ hBN/SiO_2/Si heterostructure. (b) Optical image of the device. The black dashed rectangle marks the area shown in (c). (c) Nanoscale IR image of plasmonic interference fringes from the encapsulated graphene monolayer, expressed by the normalized scattering amplitude s acquired at a back-gate voltage of 97 V and a temperature of $T = 60$ K. The arrows represent the propagation direction of the plasmon waves. Reprinted from [915]. Copyright 2018, Springer Nature.

and $\lambda_p/2$, which can be observed from Figs. 9.10 and 9.11, respectively. In particular, the $\lambda_p/2$-period fringes are caused by plasmon polaritons that are emitted by the tip and complete a round trip between the tip and reflection from the sample edges. The λ_p-period fringes are generated from the interference between tip-launched plasmon polaritons and the evanescent component of the reflected plasmon polaritons [920]. The λ_p-period fringes can also be caused by plasmon polaritons launched by emitters other than the tip, which are present in the sample [919]. The emitter's dielectric polarizability, shape, and size all have a role in the efficiency with which plasmon polariton emission occurs. To be more precise, graphene edges are weak emitters, whereas long metallic objects mounted on graphene can serve as very efficient ones [921]. In this work, the gold contacts on top of the hBN-encapsulating layer function as fixed plasmonic antennas, as shown in Figs. 9.10 and 9.11.

In Fig. 9.10(c), a large-area (6 µm × 8 µm) image of plasmon polariton standing waves obtained at $T = 60$ K is given, where the incident laser wavelength is $\lambda_{IR} = 11.28$ µm, and a back-gate voltage of $V_g = 97$ V is applied. In this figure, the raw data of the amplitude s of the scattered field are shown, normalized by the amplitude detected at the gold contacts, whose optical response serves as a standard reference that is insensitive to changes in temperature and carrier

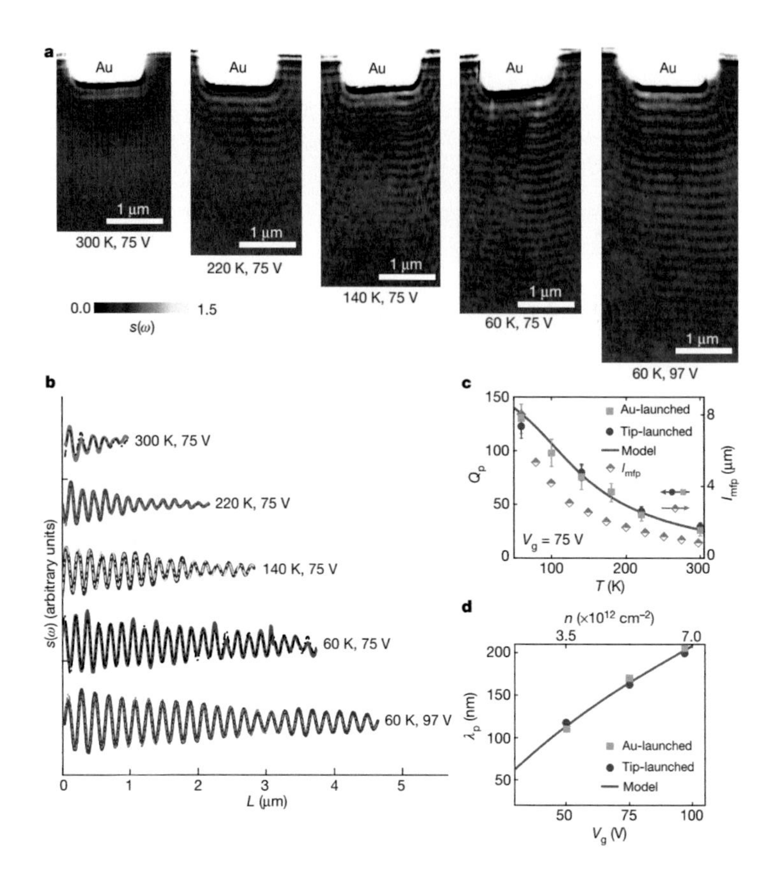

Figure 9.11 Temperature- and gate-dependent trends in surface plasmon propagation in graphene. (a) Nanoscale IR images of the normalized scattering amplitude s acquired at sequential sample temperatures and gate voltages. (b) Line profiles of plasmonic interference fringes propagating from left to right, as a function of the distance L from the gold launcher. (c) The temperature dependence of the quality factor Q_p for both λ_p- and $\lambda_\mathrm{p}/2$-period plasmon waves obtained from nanoscale IR images taken at $V_\mathrm{g} = 75$ V. The electronic mean free path $l_\mathrm{mfp}(T)$ is also plotted for comparison. (d) The voltage (V_g) and carrier density (n) dependence of the plasmon wavelength, obtained from nanoscale infrared data for both λ_p- and $\lambda_\mathrm{p}/2$-period plasmonic patterns. Reprinted with permission from [915]. Copyright 2018, Springer Nature.

density. It can be clearly observed that the entire field of view of this figure is filled with interference fringes of plasmonic polariton waves. On the right-hand side of the image, near a natural boundary of the graphene monolayer, the $\lambda_\mathrm{p}/2$-periodic fringes are more evident, whereas on the left-hand side of the image near the gold contacts, the λ_p-periodic fringes are more prominent. The plasmon polaritons shown in Fig. 9.10(c) are extremely confined, with $\lambda_\mathrm{IR}/\lambda_\mathrm{p} > 60$. Nevertheless, they can propagate over several micrometers, significantly exceeding the values at room temperature [919, 920].

As the temperature decreases, the plasmonic losses decrease dramatically, as demonstrated by the nanoscale infrared images and corresponding line profiles in Fig. 9.11. Detailed raster scans are taken near one of the gold antennas at $V_g = 75$ V and temperatures between 60 and 300 K, as depicted in Fig. 9.11(a). At room temperature, the plasmon polariton can only propagate for distances of less than 1 μm, while from Figs. 9.11(a) and (b), a consistent rise in the total travel distance and the number of discernible fringes is observed as the temperature is lowered. It can be estimated that at $T = 60$ K, the spatial extension of plasmonic oscillations is larger than 4 μm at $V_g = 75$ V and larger than 5 μm at $V_g = 97$ V. As a whole, these tendencies are consistent with the temperature dependency of the electronic mean free path, $l_{\rm mfp}$, in high-mobility devices manufactured using the same procedures [922]. Actually, the propagation range and $l_{\rm mfp}$ are comparable with each other, as presented in Fig. 9.11(c), which means that such real-space behavior is supported by ballistic electrons in graphene. The evolution of λ_p with the gate voltage (or carrier density) can be obtained from the near-field images, which is plotted in Fig. 9.11(d).

Imaging Resonant Mode Structure with S-SNOM

As the second example, we discuss the observation of optical anapole modes in certain dielectric resonators using s-SNOM. Due to the destructive far-field interference of concurrently generated toroidal dipole (TD) and ED modes, optical anapole modes exhibit strong near-field energy confinement and reduced radiative loss [923, 924]. A normally incident plane wave can excite the anapole in isolated dielectric particles, which is very desirable since it eliminates the need for a coupler as is required for conventional on-chip photonic devices. Therefore, low-loss anapole with profound near-field enhancement provide an excellent platform for realizing various intriguing physical phenomena and applications, such as the enhanced nonlinear effect [925–927], cloaking [928], high-Q cavities [929], and nanolasers [930].

Long-distance energy transfer is predicted via near-field coupling between neighboring anapoles [931], a phenomenon that could provide a viable alternative to conventional waveguides based on SPPs and line defects in PCs. It has been shown theoretically and experimentally by Huang et al. [914] that metachains made up of subwavelength silicon disks with air gaps that support anapole modes may efficiently and compactly transport EM energy over long distances. An air-gap structure was found to improve the energy confinement of the anapole within the isolated disks, which further enhances the transport efficiency of the chain. The s-SNOM was then exploited to experimentally assess the energy-transfer efficiency spectrum in the proposed metachain at mid-IR frequencies. Through the near-field signals, which reveal the anapole frequencies and portray the distributions of excited modes, it is possible to verify the connection between the anapole excitation and the high-efficiency energy transfer.

In the experiment, Huang et al. [914] fabricated the metachain composed of silicon nanodisks with geometric parameters $a = 2.375R, G = 0.5R$, where R is

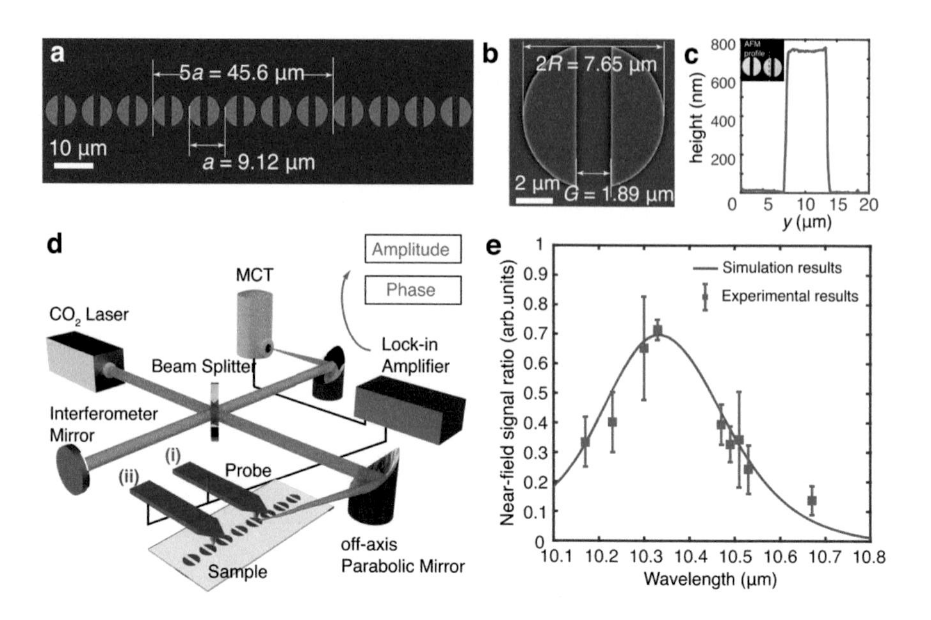

Figure 9.12 Near-field experimental descriptions of the metachain. (a, b) Overall and enlarged SEM images of the fabricated metachain. (c) Fabricated SD height extracted from a crossline (marked by the dashed red line in the inset) by AFM. (d) Illustration of the s-SNOM system for the near-field experimental verification of the metachain. (e) Simulated and experimental results of the near-field signal ratio, representing the energy-transfer performance. Note that the cyan error bars show the experimental standard errors. Reprinted with permission from [914]. Copyright 2021 American Chemical Society.

the radius of the nanodisk was designed to be $R = 3.85$ μm, as shown in the SEM images in Figs. 9.12(a–b). The substrate of the metachain was sapphire (Al_2O_3). The height of the nanodisk was measured by the AFM, as presented in Figs. 9.12(c). The s-SNOM (ANASYS NanoIR) used in this experiment is schematically shown in Fig. 9.12(d), where the relative position between the laser spot and the platinum-coated tip remains unchanged during raster scanning.

This allows us to compare the signal under two conditions in order to conduct an experimental study of the efficiency of the near-field energy transfer: (i) The laser spot overlaps with the tip, which is the default setup of the SNOM device, and (ii) the in-plane distance between the laser spot and the tip is tuned to be $\sim$37 μm along the metachain direction using a mercury cadmium telluride (MCT) detector and a built-in optimizer. The collected near-field signals of the metachain in scenarios (i) and (ii) are represented by $|Sg_{\mathrm{i}}|$ and $|Sg_{\mathrm{ii}}|$, respectively. Hence the experimental in-plane signal ratio, which can be understood as an indicator for the near-field energy-transfer efficiency, is given by $|Sg_{\mathrm{ii}}/Sg_{\mathrm{i}}|$. The results at different wavelengths are presented in Fig. 9.12(e) by blue solid squares

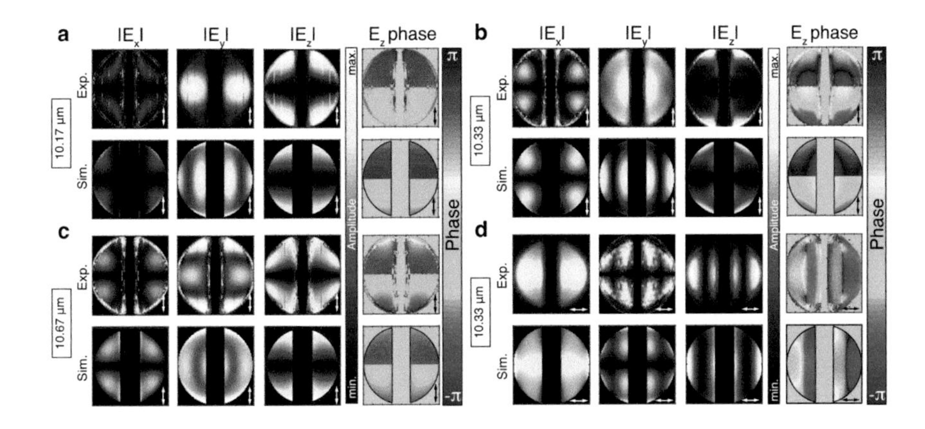

Figure 9.13 Experimental and simulated electric-field decompositions of the metachain. (a–c) Field decomposition of an SD in the metachain excited by the y-polarized laser at (a) 10.17, (b) 10.33, and (c) 10.67 μm. (d) Field decomposition of an isolated SD in the metachain excited by the x-polarized laser at 10.33 μm. Note that the arrows denote the polarization of the incident light. Reprinted with permission from [914]. Copyright 2021 American Chemical Society.

with the peak efficiency appearing at the anapole wavelength of ~10.33 μm, which agree well with simulation results.

Huang et al. [914] further obtained the near-field mode structure in the metachain at different wavelengths by means of s-SNOM. Since recorded raw near-field signals can be assumed to be the linear superposition of $\mathbf{E}_x$, $\mathbf{E}_y$, and $\mathbf{E}_z$, by exploiting the symmetry properties of the incident laser polarization and the periodicity, $\mathbf{E}_x, \mathbf{E}_y$, and $\mathbf{E}_z$ distributions can therefore be decomposed accordingly [932]. The measured and simulated near-field electric-field components at 10.17, 10.33, and 10.67 μm are shown in Fig. 9.13. The spatial distributions of amplitudes of $\mathbf{E}_x$, $\mathbf{E}_y$, and $\mathbf{E}_z$, and the phase of $\mathbf{E}_z$, when the chain is excited by the y-polarized source are given in Figs. 9.13(a–c), in which experimental and simulation results agree well. In Fig. 9.13(a) when the wavelength is shorter than the anapole wavelength, $\mathbf{E}_x$ is relatively weak due to the dominant magnetic quadruple mode. In Fig. 9.13(b), the $\mathbf{E}_x$ component gets enhanced, which indicates the appearance of electric vortexes induced by the TD mode at the anapole wavelength. Meanwhile, the existence of the anapole mode can be implied from the rapid change in the $\mathbf{E}_z$ phase in each quadrant of the nanodisk at 10.33 μm. Moreover, at this wavelength, the $\mathbf{E}_y$ field around the gap is more concentrated due to the interference of the ED and TD modes. The anapole-induced $\mathbf{E}_y$ enhancement around the gap becomes less significant at 10.67 μm due to the dominant ED mode at this wavelength. In Fig. 9.13(d), the near-field components excited by an x-polarized laser source at 10.33 μm are further presented. It is found that the electric-field distribution is distinct from those of the y-polarized cases since no anapole state is excited. Therefore, consistent with theoretical results, the s-SNOM system can clearly probe the near-field mode structure of the anapole state in the metachain, and unequivocally demonstrate the anapole state's essential role in the transfer of energy at subwavelength scales.

9.4 Passive SNOM Techniques for Thermal Radiation

The apertureless and aperture SNOM technologies in Section 9.3 are the so-called active methods used for imaging optical and radiative properties at nanoscale [933], since an external light source should be used to excite optical responses, by a focused laser beam [934], a blackbody thermal emitter [935, 936], or synchrotron IR radiations [861, 895, 902], among others. On the other hand, there are also a variety of *passive* techniques that are basically an extension of apertureless SNOM with the thermal radiation from the heated sample or tip being exploited as light sources, without the need for external sources. This set of methods includes TRSTM, TINS, and SNoiM, which will be discussed in what follows.

In 2003, Joulain et al. [704] theoretically proposed to use s-SNOM to directly measure the EM LDOS in the near field of the sample surface and then the thermal radiation energy. They showed that from EM LDOS at a vertical position z from the sample surface $\rho(\omega, z)$, that can be extracted from the signal of scattered waves, the EM energy density of thermal radiation emitted from the sample at temperature T_s can be calculated as

$$u(\omega, z, T_s) = \rho(\omega, z)\frac{\hbar\omega}{\exp(\hbar\omega/k_B T_s) - 1}, \tag{9.4.1}$$

where $\Theta(\omega, T) = \hbar\omega/[\exp(\hbar\omega/k_B T) - 1]$ is the mean energy of a Planck oscillator as noted earlier (cf. Eq. (7.5.10)). Therefore, they suggested that the probe tip is able to directly measure the thermal radiation energy and then EM LDOS at the nanoscale if the sample is heated [697, 787, 937]. If an oscillating tip is placed above a sample, the thermally emitted evanescent EM waves will be perturbed into propagating waves that will be detectable in the far field. This pure near-field signal can be extracted by demodulating the detector output at a higher harmonic of the tip oscillation frequency, as is done for conventional s-SNOM [936, 938].

The first type of passive s-SNOM for measuring thermal radiation LDOS at the nanoscale in the near field is also called TRSTM, developed by de Wilde et al. [697, 787, 937], with a schematic of the experimental setup shown in Fig. 9.14(a) [697]. In the experiment of de Wilde et al. [697], a dithering tungsten tip is brought close to a ($T = 170$ °C) SiC substrate on which Au disks (thickness of 80 nm) are patterned, as shown by the optical micrograph in Fig. 9.14(b), together with the AFM topography image (Fig. 9.14(c)). A full spectral range (6.5–11.5 μm) of an MCT detector is used to collect the far-field scattered waves, and the temperature is measured on the sample surface with a thermocouple.

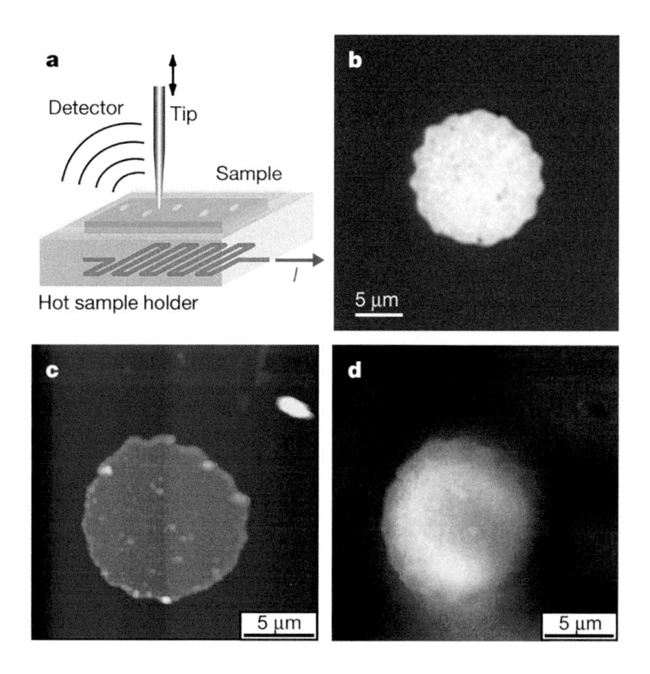

Figure 9.14 Thermal radiation scanning tunneling microscopy (TRSTM). (a) Experimental schematic where a cold tip oscillating at a frequency, Ω, scatters the thermal evanescent near-field toward a far-field detector. (b) Optical microscope image of a gold disk deposited on a SiC substrate. (c) AFM topography of the gold disk. (d) Tip-scattered thermal signal detected in the far-field by demodulating the detector signal at Ω (sample temperature of 170 K). (b)–(d) Reprinted from [697]. Copyright 2006, Springer Nature.

The measured image of near-field thermal radiation is presented in Fig. 9.14(d), in which an evident contrast between gold and SiC is observed in the thermal radiation image, with a higher signal on gold. It is possible to resolve the boundary between the two materials within 100 nm. When compared with what might be obtained using far-field IR microscopy, the resolution is significantly improved by two orders of magnitude.

Another approach, referred to as TINS [698, 699, 939], exploits a heated probe to generate near-field thermal radiation, where the tip can typically reach temperatures as high as 700 K. Similar to spectroscopy, this approach uses a Michelson interferometry configuration to spectrally resolve the tip-scattered light, as schematically shown in Fig. 9.15(a) [698, 699, 939]. The probe and sample stage are especially designed, whose temperatures can be controlled independently. By exploiting the motion of the delay arm (moving mirror), an interferogram is generated that depicts intensity versus delay distance. By implementing the FFT of the interferogram, the IR spectrum of the tip-scattered thermal radiation can be retrieved, which is then used to identify local spectral fingerprints at the nanoscale. In Fig. 9.15(b), Fourier transformed TINS spectra of the SiC

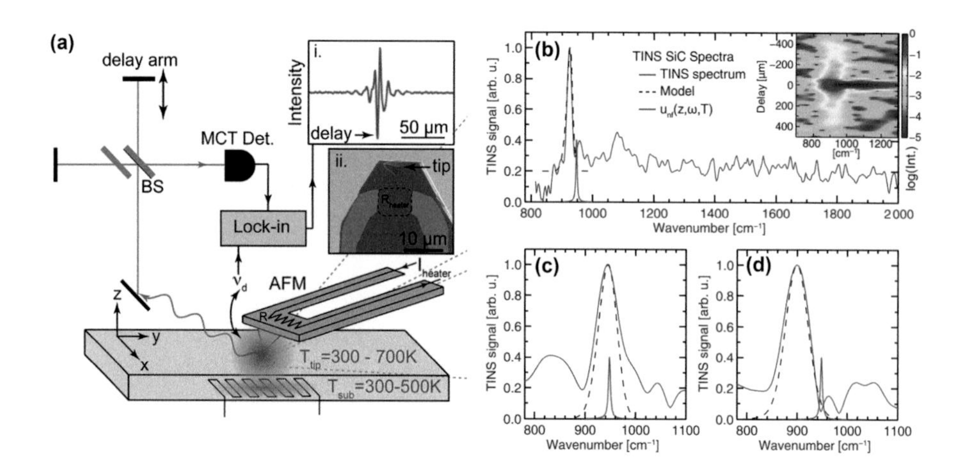

Figure 9.15 Thermal infrared near-field spectroscopy (TINS). (a) The schematic of TINS, in which the tip is heated to have a higher temperature than the sample. Inset (i) shows an interferogram containing spectral information, and inset (ii) shows the tip structure with an attached heater. Reprinted with permission from [698]. Copyright 2012 American Chemical Society. (b–d) Three separate NFRHT spectra of SiC scattered by three different Si tips. Reprinted figure with permission from [939]. Copyright 2014 by the American Physical Society.

sample is shown (with an inset of the interferogram), in obtaining which the tip was heated to 700 K [939]. There is a significantly enhanced emission peak at $\omega \sim 920$ cm^{-1}, indicating the contribution of SPhP resonance in SiC. We make note of the fact that the method is inherently broadband and that within the spectral range under investigation, there are no particular resonances related with the material or geometry of the Si tips. The theoretical near-field energy density $u_{\rm nf}(z=100\ \text{nm}, \omega, T=300\ \text{K})$ for SiC is also plotted in Fig. 9.15(b) for comparison, with a peak indicating the SPhP resonance. A pronounced spectral shift between the experimental and theoretical peaks is observed ($\sim$30 cm^{-1}). For the other two different Si probes, similar frequency shifts are also observed, as shown in Figs. 9.15(c–d), and in the latter case, the frequency shift can be as large as 50 cm^{-1}. These frequency shifts may be attributed to the collective effects of the tip–sample dipole coupling and an effective change in the dielectric environment due to the tip [939]. As a remark, a recent theoretical analysis raises the question of whether the detected signal is due to the evanescent field generated at the sample for the scenario in which the tip is heated [940].

The third approach, dubbed SNoiM, has been developed by Komiyama, Kajihara, Weng et al. [933, 943–947]. Different from previous passive technologies, this scheme does not rely on any external heating but uses an ultrasensitive detector to measure weak IR signals (see the schematic in Fig. 9.16(a)). The detector is an especially designed, charge-sensitive infrared phototransistor (CSIP) [943] that operates at 4.2 K and is sensitive to radiation at $\lambda \sim 14.1$ μm,

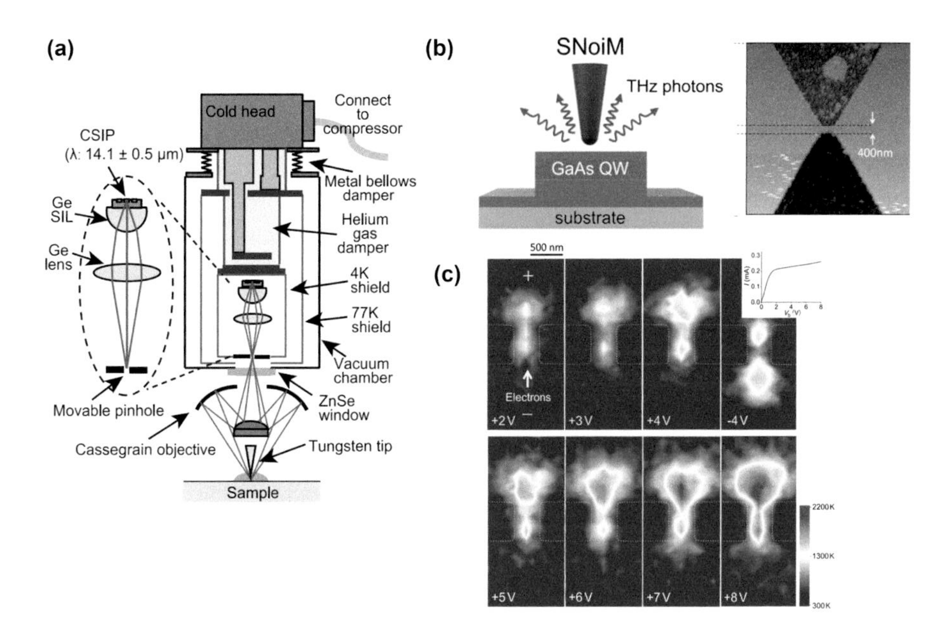

Figure 9.16 Scanning noise microscopy (SNoiM). (a) A schematic of the SNoiM. Reprinted from [941], with the permission of AIP Publishing. (b) An experimental setup of SNoiM to measure the fluctuating EM fields generated by local current fluctuations, or shot noise, induced by nonequilibrium electrons in the conducting channel of a GaAs/AlGaAs quantum well structure. The device's AFM graph is also shown. (c) SNoiM images of hot electron dissipation through the conducting channel at different bias voltages. Reprinted from [942] under the terms of the Creative Commons Attribution 4.0 International License.

which is by a factor of a few orders of magnitude more sensitive than a commercially available and widely used MCT detector in standard s-SNOM systems. The CSIP's high level of sensitivity makes it possible to take high-resolution thermal images by collecting tip-scattered thermal radiation. This can be done even when the tip is in the thermal equilibrium, with a sample and no external heating applied. When the sample is driven out from its state of thermal equilibrium, a significant distinction between active s-SNOM and SNoiM becomes apparent. Active s-SNOM is not sensitive to the heating of the sample since it maps the contrast of the sample's dielectric constant through the effective polarizability under external illumination as mentioned earlier. However, the tip-scattered signal in SNoiM not only contains the information on the dielectric constant of the sample through the EM-LDOS, but it is also impacted by the sample temperature through the mean energy of a Planck oscillator $\Theta(\omega, T_s)$ in Eq. (9.4.1). As a result, SNoiM is able to image a local hotspot within the joule-heated sample [933, 945]. Normalizing the SNoiM signal at a higher temperature T with that at room temperature T_0 $[\Theta(\omega, T_s)/\Theta(\omega, T_0)]$ removes the EM-LDOS and thus gives a temperature map of the sample.

As an example, SNoiM was utilized by Weng et al. [942] to visualize the hot electron distribution in a GaAs/AlGaAs heterostructure device for the purpose of understanding hot electron energy dissipation processes. The experimental setup is shown in Fig. 9.16(b), along with an AFM topographic image of the device, in which a narrow conducting channel can be seen. The images taken by the SNoiM at different bias voltages ranging from 2 to 8 V are presented in Fig. 9.16(c), where the image under an opposite bias of -4 V is also given. The local temperature rises to as high as $\sim$2000 K. It is also seen that the electron temperature profile exhibits a double-peaked structure, with the first hot spot within the channel close to the entrance and the second one, outside the channel, 100–250 nm away from the exit. Therefore, the SNoiM scheme provides a powerful platform for visualizing nanoscale carrier transport mechanisms and mapping local temperature distributions at nanodevices.

9.5 Near-Field Radiative Heat Transfer Experiments

For NFRHT, it is important to accurately measure the heat transfer rate or heat flux between two closely spaced objects. In this sense, near-field imaging techniques mentioned in Sections 9.2.3 and 9.3.2 that mainly provide the qualitative information of the dielectric constant contrast cannot meet this goal. In this section, we give an overview of the development of NFRHT measurement.

The experimental exploration of NFRHT already started in the 1960s. Tien et al. [406, 408] and Hargreaves [407] were the first to measure the energy flux of two parallel plates at cryogenic temperatures. In [408], copper disks were used as the emitter and receiver, with a diameter of 8.5 cm, and the gap spacing d can be varied from 10 μm to 2 mm using an external micrometer adjustment. Since the temperature was roughly 10 K, near-field effects were seen at $d < 200$ μm. The heat flux at $d = 10$ μm was more than three times the far-field value, while it was still much smaller than that between blackbodies. In 1969, Hargreaves [407] measured the NFRHT between two chromium-coated plates separated by a vacuum gap of roughly 6 to 1.5 μm near room temperature. The heat transfer rate was shown to be five times greater than the far-field value and also reached almost 40% of the heat transfer rate between two blackbodies.

In the 1990s, after the invention and rapid development of STM, Xu [948, 949] attempted to use this technique to approach the regime of NFRHT. In particular, for their experiments, Xu et al. [948, 949] employed an STM equipped with a heated indium needle with a flat tip surface having a diameter of 100 μm. To measure the temperature gradient, a thin-film thermocouple was evaporated onto a glass substrate, with a junction area of 160×160 μm^2. In 1999, Müller-Hirsch et al. [950] studied the heat transfer between a tungsten tip and a planar thermocouple on a substrate. It was challenging to quantitatively estimate the absolute heat flux between the tip and the substrate, despite the fact that the proximity effect was detected at distances as small as 10 nm. Measuring the temperatures of the tip and the substrate was also difficult [948–950].

To measure the temperature of the tip of an STM, in 2005, Kittel et al. [419] employed a platinum wire encased in a gold-coated glass micropipette to construct a thermal coupling junction, in which the thermoelectric voltage builds up between the inner platinum wire and the outer gold film. This thermal sensor was then integrated into the variable-temperature STM tip. NFRHT was measured between the warm tip kept at around 300 K and a plate, which was cooled to 100 K using liquid nitrogen. For comparison, the plate was made up of gallium nitride (GaN) or had a gold layer applied to it. In the experiment, thermal radiation heat transfer was measured at gap spacings between $d = 100$ and 1 nm. For a range of 10 nm $< d <$ 100 nm, the observed results are in agreement with those predicted by FE. Nonetheless, as $d <$ 10 nm, the measured heat transfer rate saturates and shows a different trend from the predicted one, which diverges as the distance approaches zero. This is then explained by the short-distance deficiency of macroscopic FE theory due to short-range correlations described by a correlation length L, which can be improved by using a phenomenological modification of the imaginary part of the dielectric function at short distances.

Starting from the above pioneering works, the experimental measurement of near-field thermal radiation heat transfer has made significant strides in recent years. [255, 383–385, 387, 388, 409–418, 421, 422, 951–962]. The measurements can be grouped into four broad classes based on their shown configurations: plate–plate, tip–plate, sphere–plate, and microfabricated suspended structures. Maintaining parallelism, determining gap distances, and precisely measuring temperatures and heat transfer rates are the primary challenges. Frequently, corrections must be made to account for heat conduction and far-field thermal radiation. Convection and gas conduction are usually quite minimal because all tests have been done in an evacuated chamber. Excellent reviews on this subject include [6, 231, 761, 963]. Now let us have a brief overview of some representative works on near-field heat transfer experiments.

In 2008, near-field heat transfer was observed by Hu et al. [412] between two parallel, optically flat glass (fused quartz) disks with diameters of 1.27 cm; PS spheres of 1-μm diameter were sparsely spaced between them to separate the two disks and create a vacuum gap. The microspheres were used to lessen the impact of the heat transfer by conduction by acting as spacers with poor thermal conductivity. The emitter can operate in the range of roughly 50 to 100 °C, controlled by a heating pad combined with a platinum resistance temperature sensor as the temperature controller. A thermocouple was used to monitor the temperature of the receiver, which was approximately 24 °C. A heat-flux sensor was placed between the receiver and the heat sink. The measured heat fluxes were in agreement with the FE expected values for a vacuum gap of $d = 1.6$ μm, demonstrating that the NFRHT exceeded that between two blackbodies by more than 35%. Recently, Lang et al. [951] used monodisperse SiO_2 nanospheres to shrink the gap spacing down to 150 nm, which was confirmed with interferometric measurements, and they achieved a heat flux that was over an order of magnitude greater than the blackbody limit between two 20-mm-diameter disks at room temperature.

Using a biomaterial AFM cantilever, Narayanaswamy et al. [409] were able to measure nanoscale radiative heat transfer between a sphere and a plate with a gap separation as small as 100 nm, as depicted in Fig. 9.17(a). Shen et al. [410], utilizing a better resolution piezoelectric controller, further refined the approach, allowing for measurements with a vacuum gap of as little as 30 nm. Microspheres made of silicon dioxide with radii of $a = 25$ and 50 μm were used. The substrate of the plate was either SiO_2, Si, or a gold-coated surface. At a gap spacing of $d = 30$ nm, it was shown that the heat transfer coefficient could be enhanced by approximately three orders of magnitude using a SiO_2-SiO_2 configuration that can support coupled SPhPs (Fig. 9.17(b)).

Because of the disparity between the thermal expansion coefficients of Au and SiN_x, the biomaterial cantilever, shown in Fig. 9.17(a), is able to monitor the temperature at the tip, where a SiO_2 microsphere is epoxied. The temperature of the sphere can be calculated by directing a focused laser beam at it and measuring the resulting deflection of the cantilever with a position sensitive detector (PSD). The cantilever is bent because the laser irradiation causes the tip (or sphere) to heat up. Some tens of degrees of temperature rise can be achieved at the tip. The cantilever's thermal conductance was measured by monitoring its bending in response to varying input powers. By moving the substrate with a piezoelectric motion controller and reducing the distance between the sphere and the substrate, it is then possible to quantify the cooling of the sphere caused by increased near-field radiation using the beam deflection signal. When the

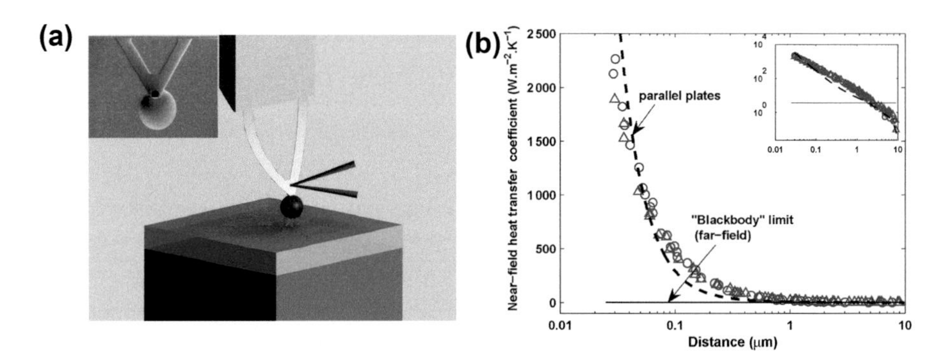

Figure 9.17 Experimental setup and measured results for sphere–plate configuration: an experimental setup; (b) measured radiative heat transfer coefficient, where the triangular marks are for radius $a = 25$ μm and circles are for $a = 50$ μm microspheres. The inset shown is the plot in the log–log scale. Reprinted with permission from [410]. Copyright 2009 American Chemical Society.

cantilever beam is positioned vertically, the bending of the cantilever causes very little vertical displacement of the sphere. It should be noticed that the base of the cantilever and the flat substrate are passively maintained at ambient temperature due to the minimal heating power. Hence, the additional bending of the cantilever can be used to quantify both the sphere temperature and the near-field radiative transfer [409, 410].

In Fig. 9.17(b), the obtained heat transfer coefficients between a SiO_2 sphere and a SiO_2 plate are shown against the far-field blackbody limit ($h_{r,BB} = 3.8$ W/m^2 K). The value of h_r is found to grow substantially as d is reduced, reaching a value of 2 230 W/m^2 K at $d = 30$ nm, which is around three orders of magnitude higher than the blackbody limit. According to AFM measurements [410], the surface roughness for these measurements is less than 4 nm. Later research conducted by Shi et al. [952] studied the near-field radiation heat transfer between a SiO_2 sphere and a silicon substrate of varying doping concentrations. They further investigated the NFRHT between a microsphere and a metamaterial of nickel wire arrays embedded in an anodic aluminum oxide (AAO) matrix [953]. The AAO matrix can be etched away in sections to reveal a 400-nm tall nanowire array. They discovered that the heat conductance is maximized in the protruded scenario but is much smaller in the nanowire array fully embedded in the AAO matrix, indicating that the protruded metal nanowires can act as a low-loss waveguide, which is capable of coupling high-k modes due to the hyperbolic dispersion.

In 2009, Rousseau et al. [411] devised an alternative approach for measuring nanoscale radiative heat transport between a sphere and a plate. Similar to the method used in [409], a biomaterial cantilever was vertically mounted with a microsphere attached to its tip near the surface. The deflection of the cantilever was tracked by measuring the interference pattern of the laser beam that was reflected off of it using an optical fiber system. A piezoelectric actuator was attached to the flat substrate, and it was heated to a few tens of degrees. Thermocouples were used to determine the temperatures of the substrate and the cantilever tip. By examining the thermal resistance network and calibrating the thermal conductance using far-field tests, the temperature of the microsphere can also be deduced. Note even at $d = 50$ nm, the radiative thermal resistance is still dominant, and therefore the temperature of the sphere is very near to that measured at the end of the cantilever. They used SiO_2 spheres of two different diameters of 40 and 22 μm, with surface roughnesses corresponding to 40 and 150 nm, respectively. Radiative heat transfer enhancement factors greater than 400 were measured for the 40-μm-diameter sphere when the distance d was decreased from 2 500 to 30 nm. Later, van Zwol et al. [954, 955] utilized this setup to measure near-field radiation transfer between a SiO_2 sphere and SiC, doped-Si, graphene-covered substrate, and a VO_2 (a metal–insulator phase transition material) plate. Therefore, it can be concluded that the near-field thermal radiation in the sphere–plate configuration has been measured quantitatively, and the results give experimental support for the FDT's prediction of the near-field radiative transfer rate at gap sizes as small as 30 nm [409–411, 952–955].

Although near-field heat transfer at submicron gaps has been measured using sphere–plate configurations, this geometry is not as perfect as the plate–plate geometry. An important drawback is that the employed microsphere's effective area is too little for practical use. There have been persistent efforts over the past decade to experimentally observe the near-field enhancement of radiative transfer for parallel-plate geometry with a relatively larger area (preferably larger than 1 mm^2) [255, 388, 413–415, 956–961].

In 2011, Ottens et al. [413] measured NFRHT between sapphire ($\alpha-Al_2O_3$) plates, each with a 50×50 mm^2 area and 5-mm thickness, from a large distance of 100 µm down to a near-field distance of ~2 µm. At the smallest gap spacings, they showed that the measured heat transfer coefficients near room temperature were larger than those between blackbodies in the far field. The emitter and receiver temperatures were measured using Si-diode thermometers. The flatness of the sapphire surfaces was shown to be 160 nm over a lateral displacement of 50 mm, making them optically smooth. Copper films of thickness 200 nm were applied to the four corners of the facing surfaces (each with an area of 1 mm^2) to create capacitor plates for gap separation measurements. The emitter's tip and tilt angular movements were adjusted using kinematic mirror mounts powered by stepper motors [413].

In 2012, a near-field radiation experiment at cryogenic temperatures was taken by Kralik et al. [956] between two parallel tungsten-coated 35-mm plates at separation distances ranging from 500 µm down to about 1 µm. In order to determine the gap size, a capacitance technique was used. To keep the emitter and receiver in perfect parallel alignment, they used a plane-parallel equalizer. When the cold plate temperature was between 10 and 40 K and the hot plate temperature was maintained at 5 K, higher at $d \approx 1$ µm, they found that the heat flux was enhanced by nearly four orders of magnitude over the corresponding far-field value (or two orders of magnitude over the blackbody limit).

In 2015, Ijiro and Yamada [957] measured NFRHT between two SiO_2 glass plates at room temperature, whose diameters were 25 mm. To keep the emitter and receiver surface aligned with a vacuum gap ranging from 100 to 1 µm, they employed a piezoelectric motor to drive a kinematic mount with a linear stage for both translational motion and tip/tilt angle adjustment. The gap size was determined by using an optic-fiber-coupled spectrometer. At ambient temperature, the radiative heat flux in the near-field is roughly twice that in the far-field scenario. In addition, they fabricated 5 µm $\times$ 5 µm $\times$5 µm microcavities on the glass substrate. Surprisingly, while the microcavities increased the far-field heat flow by 20%, they reduced the near-field heat flux. Near-field radiative transfer may be described by the proximity limit [769, 771], while the far-field enhancement may be explained by the improved emittance/absorptance due to guided modes or cavity resonance. However, when a layer of Au was deposited on the microcavities, the near-field radiative transfer was greatly improved in comparison to that between flat Au surfaces, as predicted by Guerout et al. [770].

In 2015, for the purpose of measuring the NFRHT between two large rectangular plates (19×8.6 mm^2) of fused quartz (SiO_2), Ito et al. [958] manufactured micropillars in a truncated square pyramid shape with thicknesses of 500, 1 000, and 2 000 nm. These spacers had 1 mm of lateral separation between the closest pillars and were etched into the SiO_2 surface. By compressing a spring with a very low spring constant, a modest force (about 1 N) was delivered to the plate. The emitter temperature was increased by 5, 10, 15, and 20 K above the 293 K receiver temperature. The heat flux to the thermoelectrically cooled receiver was measured using a flat heat-flux sensor. Theoretically, expected heat transfer coefficients were lower than the measured values, suggesting that heat conduction through the pad was contributing to the discrepancy; however, a quantitative study of this effect was not provided.

In the same year, Lim et al. [959] fabricated a 13.4×0.59-mm^2 strip film with a heater as the emitter and a 13.4×0.48-mm^2 strip film as the receiver (Fig. 9.18(a)). The strip materials were doped silicon, and the thickness of the strip film was 600 nm. The distance between the heating and receiving strips was controlled by modifying the usual load in an MEMS-based platform, and the gap was formed using patterned metal spacers, with an effective surface area for radiative heat transfer of around 6.4 mm^2 (Fig. 9.18(a)). The gap distance separating the

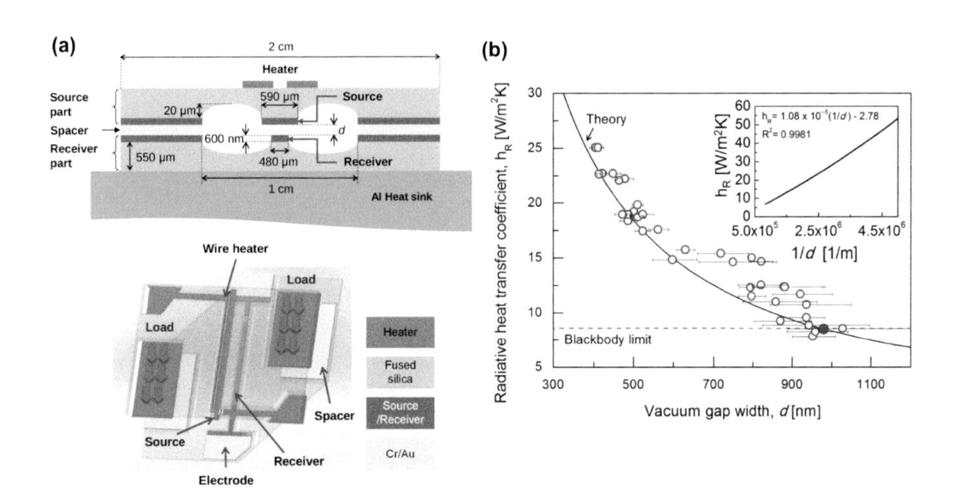

Figure 9.18 MEMS-based platform for measuring the near-field thermal radiation between doped-Si plates proposed in [959]. (a) (Top panel) Cross-sectional view of the MEMS-based platform. (Bottom panel) 3D schematic of the MEMS-based platform. (b) The measured radiative heat transfer coefficients. Reprinted figure with permission from [959]. Copyright 2015 by the American Physical Society.

emitter and the receiver was shown, by capacitance measurements, to vary from roughly 900 nm to as low as 400 nm. In particular, at $d = 400$ nm, a heat transfer coefficient 2.9 times the far-field blackbody limit can be experimentally achieved, as shown in Fig. 9.18(b).

In 2016, in order to separate large doped-Si plates (10 mm $\times$ 10 mm) with different vacuum gaps, Watjen et al. [960] manufactured SiO_2 micropillars as spacers, with the smallest vacuum gap down to 200 nm. They obtained near-field heat flux that was more than an order of magnitude higher than the blackbody limit. The sample consists of two doped-Si plates with a sparse array of SiO_2 micropillars. The micropillars were fabricated using photolithography on a silicon wafer and samples of different heights of micropillars ranging from 200 to 800 nm were manufactured. With the use of spring loading forces, the patterned plate was matched with a flat plate of the same size, and the two were forced together to keep the required gap spacing. The array of 1$-$μm-diameter SiO_2 pillars had a large span, that is, $S = 300, 400$, or 500 μm, and thus can maintain a reasonably low contribution to thermal conduction. It was shown that when S is greater than around 300 μm, thermal radiation will account for more than half of the heat flux. They further applied an FTIR spectrometer to measure the reflectance of the sample to determine the gap spacing before the heat transfer measurement. The gap spacing was modified by applying different forces during the FTIR and heat transfer measurements. The data analysis additionally accounted for the thermal resistances of the contacts. Results showed that the heat transfer rate for three different gap distances can agree well with the prediction of FE. It was expected that due to the stimulation of coupled SPPs in doped silicon samples, NFRHT was significantly amplified.

In 2016, a flexible Si membrane structure with a thickness of 20 μm was created by Bernardi et al. [414], which can support an Si emitter of size 5 mm $\times$ 5 mm and thickness 0.5 mm. The emitter can be heated and forced closer to the surface of the receiver with enough pressure. Without applying any external force, the vacuum gap distance was limited by SU-8 photoresist posts, each with a height of 3.5 μm and a diameter of 0.25 mm. To limit the size of the vacuum gap, 150-nm-tall SiO_2 stoppers were made and attached to the receiver plate below the emitter. For large temperature differences up to 120 K, the radiative heat flux between two intrinsic silicon plates was measured at gap distances ranging from 3 500 nm to 150 nm. A substantial enhancement over the blackbody limit by a factor of 8.4 was then obtained for the 150-nm-thick gap.

In 2018, four photoresist posts of varying heights (3 700, 1 400 or 430 nm) were utilized by Yang et al. [255] to generate a vacuum gap between the emitter and the receiver, with lateral dimensions of 20 mm $\times$ 20 mm. To prove that graphene surface plasmons increase radiative transfer, they employed this configuration to analyze the near-field radiative transfer between graphene-covered intrinsic silicon substrates separated by a vacuum gap as small as 430 nm, with

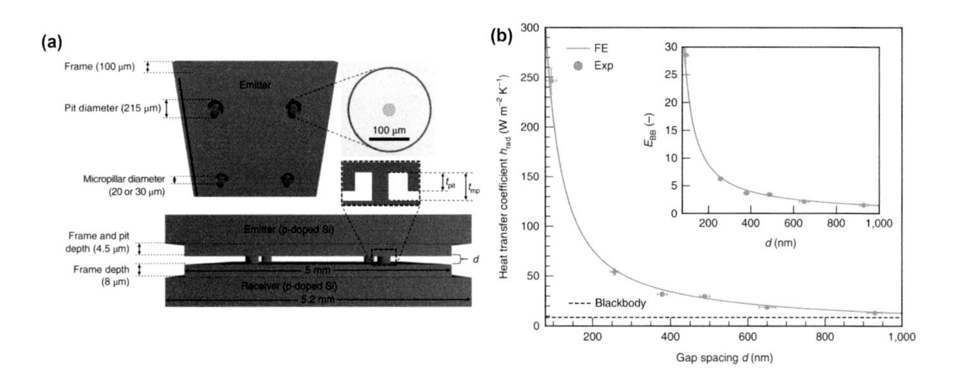

Figure 9.19 Near-field radiative transfer measurement by DeSutter et al. [388]. (a) Schematic of two doped-Si plates separated by SU-8 3005 micropillars. (b) Radiative heat transfer coefficient ($h_{\rm rad}$). The inset shows the enhancement with respect to the far-field blackbody limit ($E_{\rm BB}$), as a function of the vacuum gap spacing d. Reprinted from [388]. Copyright 2019, Springer Nature.

super-Planckian radiation with an efficiency 4.5 times greater than the blackbody limit reported. At the same year, Ghashami et al. [415] devised a nanopositioning platform with six degrees of freedom using piezometers, which can achieve 1-nm translational resolutions in all three axial directions and $1-\mu$rad rotational resolutions in each rotational direction. More than 40 times enhancement of the blackbody limit in the far field was observed for near-field thermal radiation transfer between crystalline quartz plates of a surface area of 5×5 mm^2 with a vacuum gap down to 200 nm at temperature differences up to 156 K.

In 2019, in order to reduce the amount of heat lost through conduction for plate–plate configuration with micropillars, DeSutter [388] built micropillars inside micrometer-deep trenches, and thus these pillars could be longer than the nominal gap size, as shown in Fig. 9.19(a). The emitter temperature was raised to around 100 K above the receiver temperature, which was held constant at 300 K. They achieved a gap distance between two doped-Si plates down to 110 nm, with a surface area of 5.2×5.2 mm^2, and experimentally obtained an enhancement in radiative transfer roughly 28.5 times that of the blackbody limit, as shown in Fig. 9.19(b). Recent innovations in the plate–plate structure show considerable potential for putting nanoscale heat radiation to use in the real world. Similarly, in 2020, Ying et al. [961] fabricated SU-8 polymer posts as spacers to measure the near-field radiative transfer between doped silicon plates with vacuum gaps from about 500 to 190 nm. The measured near-field radiative heat flux reaches a value of 7 260 W m^{-2} at the vacuum gap of $d=190\pm20$ nm with a temperature difference of 74.7 K, demonstrating an 11 times enhancement over the blackbody limit.

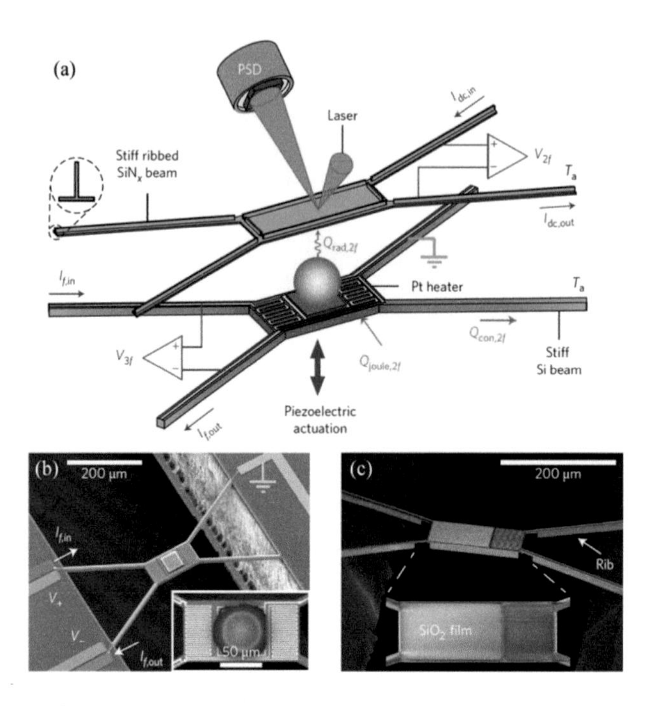

Figure 9.20 The nanopositioning platform to place two MEMS structures with nanoscale gaps in between. (a) Schematic drawing of the experimental setup; (b) SEM images of the suspended emitter with a sphere and heater/thermometer (inset); (c) SEM images of the receiver structure with a SiO_2 film and heater/thermometer (inset). Reprinted from [383]. Copyright 2015, Springer Nature.

An alternative method for measuring near-field heat radiation is to use MEMS to create suspended structures. In 2013, an MEMS device consisting of two suspended membrane islands of 77 μm × 77 μm was suspended with Pt heaters, and four long beams were built by Feng et al. [962]. The membrane was created using a beam of SiN that was sandwiched between two layers of SiO_2. They demonstrated that the near-field radiative transfer coefficient was around 10 times that of the far field, though still less than heat conduction through the beams, at a vacuum gap separation of 1 μm. Another MEMS device consisting of double nanobeams using an electrostatic actuator to control the gap separation was created by St-Gelais et al. [416, 417]. They first realized a 200-nm gap size [416], and then a nearly 40-nm gap was achieved later [417]. At $d \approx 42$ nm, they saw a near-field enhancement of over two orders of magnitude, with a temperature difference between the two beams of more than 200 K. The nanobeam had a length of 200 μm, a height of around 500 nm, and a width of 1–2 μm. The advantage is that this double-beam MEMS structure can be manufactured on-chip, and the gap spacing can be controlled by an applied voltage, despite having a relatively small effective area (of the order of 1 μm).

The transverse and rotational movements of one MEME structure relative to the other can be controlled by piezoelectric actuators and high-resolution stepper motors, as demonstrated by the Michigan group [383, 386, 418, 964]. This provides a versatile platform to measure and manipulate NFRHT at extremely small gap sizes. For an illustration of the MEMS structures built for implementing near-field measurements between a sphere and a plate, see Fig. 9.20 [383]. Both of the two MEME structures have integrated heaters and thermometers, and their side lengths are of the order of 100 μm. The mechanical contact was detected using a laser and a PSD. Using a SiO_2 sphere with a diameter of 53 μm, Song et al. [383] measured the near-field radiation in the sphere–plate configuration at distances from 10 μm down to 20 nm, as shown in Fig. 9.20. They have also studied the effect of varying the thickness of the flat SiO_2 layer from 50 nm to 3 μm. Based on theoretical predictions, they found that, in the near-field domain, the penetration depth of the photon depends heavily on the vacuum gap spacing [965]. To measure the near-field radiation between flat surfaces at separation distances down to near 30 nm, a 48 μm × 48 μm square mesa with a 20 μm height was then built [418]. They further demonstrated a near-field enhancement of almost 1 200-fold at the gap separation of 25 nm [386]. Using the same platform, a TPV device was constructed, with the output power 40 times higher than that in the far field [385].

Finally, let us remark that there has also been ongoing work to develop tip-based setups capable of sensing nanoscale heat radiation at ultrasmall distances, in order to examine the validity of FE and the transition from radiation to conduction. The ability to probe near-field heat transfer at vacuum gap spacings as small as 1 nm has been realized, but additional theoretical and experimental studies are still required to fully comprehend the dynamic interplay between phononic and photonic contributions [384, 405, 421, 422, 966, 967].

9.6 Summary

In this chapter, we have presented an overview of several important experimental techniques in MNTR. First, the aperture SNOM is introduced. We give a relatively detailed discussion on its principle and implementation. Then, we proceed to the apertureless SNOM that uses the tip or probe of an AFM to detect light–matter interactions. These are active methods for imaging optical and radiative properties at the nanoscale, which actually are already widely used in a variety of research fields like condensed matter physics, photonics, materials sciences, and chemistry. On the other hand, we also discussed those "passive" techniques, which is an extension of apertureless SNOM and directly measure the thermal radiation signals from the heated sample or tip, without applying external light sources like quantum cascade lasers. We discussed and compared three recently developed methods, including TRSTM, TINS, and SNoiM. These methods offer great opportunities for directly imaging thermal radiation at the nanoscale. We finally discuss the various implementations of measuring near-field heat transfer flux in plate–plate, sphere–plate, and tip–plate configurations.

10 Manipulation of Thermal Radiative Properties in Micro/Nanoscale

As introduced in previous chapters, the emitted radiation of macro objects is generally subjected to some fundamental constraints and shows typical broadband, incoherent, omnidirectional, and unpolarized characteristics as illustrated in Fig. 10.1(a). While, in the last decades, the unprecedented development of nanophotonics and nanofabrication techniques has challenged the conventional view of thermal radiation, opening a new era to engineer thermal radiation in more rich and original manners, the designed nanostructures, whose structural features are equivalent to the wavelength or even at the subwavelength scale, show dramatically different characteristics compared with conventional counterparts, leading to several exciting opportunities for extensive applications. For example, nanostructures could have coherent, narrowband, polarized, and directional thermal radiation properties (i.e., Fig. 10.1(b)). Especially, in some cases, radiation features would not follow classical laws of thermal radiation, exhibiting exotic behaviors such as super-Planckian radiation (Fig. 10.1(c)) and nonreciprocal thermal radiation (Fig. 10.1(d)). In this section, we will give a comprehensive overview of radiation control on micro/nanoscales, including the manipulation mechanism using different nanostructures, engineered radiation properties in several aspects, and some original phenomena beyond classical radiation laws.

10.1 Manipulation Mechanism of Micro/Nanothermal Radiation

In micro/nanoscale, the basic principle of thermal radiation control is to tailor light–matter interaction by using various nanostructures. During the last decades, a variety of tuning mechanisms have been proposed, providing many opportunities and yielding several novel discoveries in the area of micro/nanothermal radiation. Basically, we can roughly classify it into wavevectors- and phase-based strategies. Most of the works to control radiative properties are based on wavevector manipulation, such as enhancing absorption/emission by the surface plasmon, SPhP, and hyperbolic materials supporting large wavevectors. More recently, all-dielectric metamaterials have gained much attention in exciting multiple resonances and showing good potential in high-temperature thermal emitters instead of metal counterparts. The destructive or constructive interference (depending on the amplitude and phase information) between different

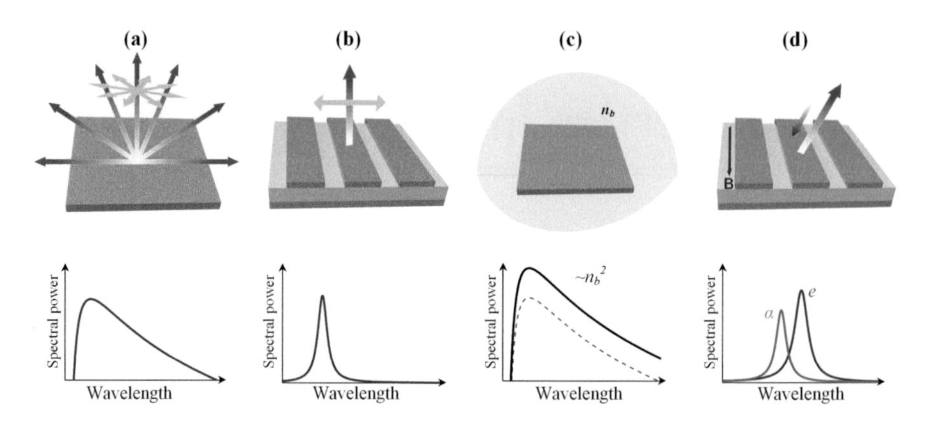

Figure 10.1 Overview of thermal radiation engineering in macro/nanoscale. (a) Conventional properties of thermal radiation: incoherent, unpolarized, omnidirectional, and broadband. (b)–(d) Nanophotonic structures could exhibit thermal radiation properties that are drastically different from conventional thermal emitters. (b) Nanophotonic structures could have control over the coherence, bandwidth, polarization, and directionality of thermal radiation. (c) Enhanced far-field thermal radiation by thermal extraction. The n_b is the optical index of the background. (d) Violation of Kirchhoff's law by breaking reciprocity.

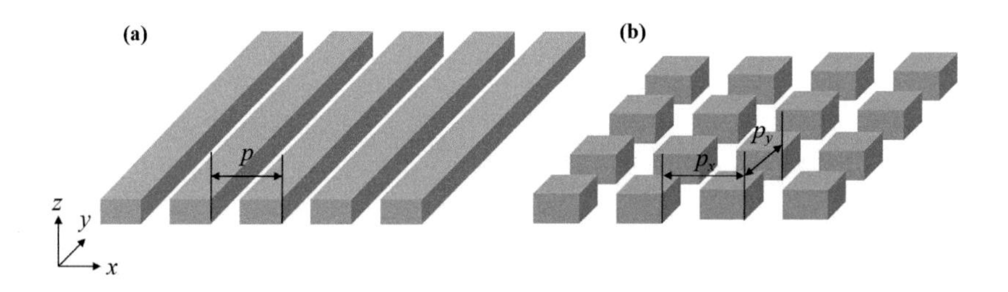

Figure 10.2 Illustration of general grating structures. (a) 1D grating structure with period along the x direction. (b) 2D grating structure with period along both the x and y directions.

modes also plays an important role in shaping radiative behaviors, which have not been extensively studied yet. To build a basic view of the development of micro/nanothermal emitters, instead, we would like to emphasize several methods for flexible radiation control based on the development history of thermal emitters/absorbers.

10.1.1 Grating structures

One of the earliest attempts to shape thermal radiation properties is based on grating structures. The geometrical profiles are given in Fig. 10.2. As for 1D grating structures (Fig. 10.2(a)), the direction of period (p) is along the x axis, and there are two periods in the xOy plane for two directions, denoted as p_x and

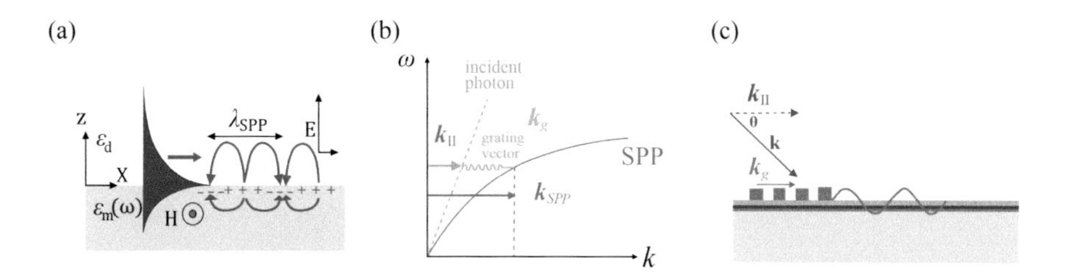

Figure 10.3 The role of gratings in wavevector compensation theory. (a) Surface waves along the interface of the dielectric and metal interface under TM polarization. (b) The principle of momentum matching when adding a periodic wavevector $k_g = m\frac{2\pi}{p}$.

p_y, respectively. The radiative response of 1D gratings will be more sensitive to incident polarization.

The pioneering theoretical works can date back to the end of the last decade, that is, [351, 968]. In theory, one of the essential roles of grating structures is attributed to providing momentum compensation from the periods, which contributes to converting surface waves to propagating waves that can be used for engineering thermal radiation. To get a clear view of this problem, we would like to briefly introduce the generation of surface waves. The most simple configuration to sustain surface waves is that of a metal slab as shown in Fig. 10.3. The upper plane can be a dielectric material with permittivity $\varepsilon_d > 0$, that is, air. The metal materials are usually described by the Drude model defined as $\varepsilon_m(\omega) = 1 - \frac{\omega_p^2}{\omega(\omega + \mathrm{i}\Gamma)}$, in which ω_p and Γ are the bulk plasma frequency and the damping rate, respectively. When considering TM polarization (i.e., (k_z, E_x, H_y) in Fig. 10.2(a)), the dispersion relation of SPPs (surface waves supported in the metal–dielectric interface) propagating at the interface between two materials is expressed by

$$\beta = k_0\sqrt{\frac{\varepsilon_d\varepsilon_m}{\varepsilon_d + \varepsilon_m}}. \tag{10.1.1}$$

The k_0 is the incident wavevector in a vacuum. It is obvious that, in the case of $\omega < \omega_p$, Eq. (10.1.1) remains satisfying $\beta > k_0$ (below the light line as illustrated in Fig. 10.3(b)), leading to the evanescent decay on both sides of the interface. Similarly, polar materials such as SiC and GaP, whose permittivity shows Lorentz–Drude lineshape, can also support surface waves, namely SPhPs. Therefore, one of the effective ways to excite surface waves is to provide wavevector compensation by using grating structures. According to Bloch's theory, the 1D grating, for example, can provide an extra wavevector $k_g = n_x\frac{2\pi}{p_x}$, where n_x

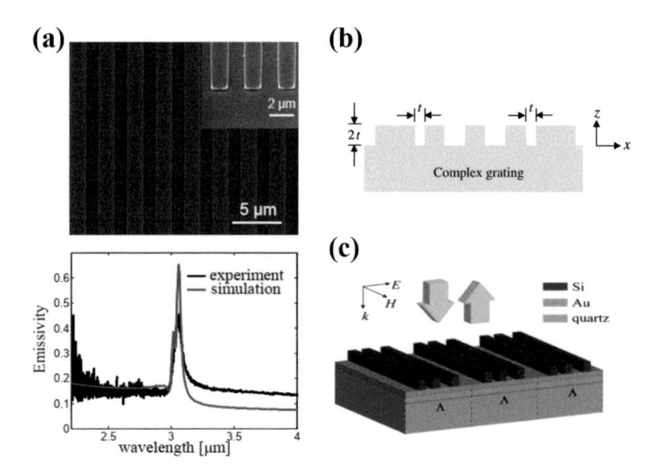

Figure 10.4 Thermal radiation control using 1D gratings. (a) SEM images of TiN gratings and compared spectral emissivity at $\theta = 0°$ and 540 °C for a TiN 1D grating covering with a thin Si_3N_4 film sitting on the sapphire substrate. Reprinted from [969]. (b) The schematic of a complex grating. Reprinted from [970], Copyright 2007, with permission from Elsevier. (c) Schematic view of dual- and narrowband absorbers with a complex unit. Reprinted from [971].

is an integer denoting the diffraction order. So, wavevector compensation takes place whenever the condition

$$\beta = k_0 \sin\theta \pm k_g \tag{10.1.2}$$

is fulfilled, where θ is the incident angle. Likewise, for 2D scenarios, the condition is changed to

$$\beta = \sqrt{\left(k_x + n_x \frac{2\pi}{p_x}\right)^2 + \left(k_y + n_y \frac{2\pi}{p_y}\right)}, \tag{10.1.3}$$

in which n_y denotes the diffraction order along the y direction. The $k_x = k_0 \sin\theta$ and $k_y = k_0 \cos\theta$ are the x and y tangential components of incident light, respectively.

Based on the aforementioned theory, there are several works focusing on exciting surface waves (SPPs and/or SPhPs) to tailor thermal radiation properties. Here, we would like to mention a series of representative works. One typical experimental work has been demonstrated by Greffet et al. [335], in which the 1D SiC grating was used to excite SPhPs. In detail, by tuning the grating's period, directional emission control can be achieved at a certain angle. It can also be regarded as a pioneering work for exploring thermal radiation with coherent and directional properties. Similarly, Liu et al. [969] fabricated a thermal emitter composed of a dielectric grating sitting on a TiN substrate, as shown in Fig. 10.4(a). The designed emitters show well-collimated emission at ~3 μm in

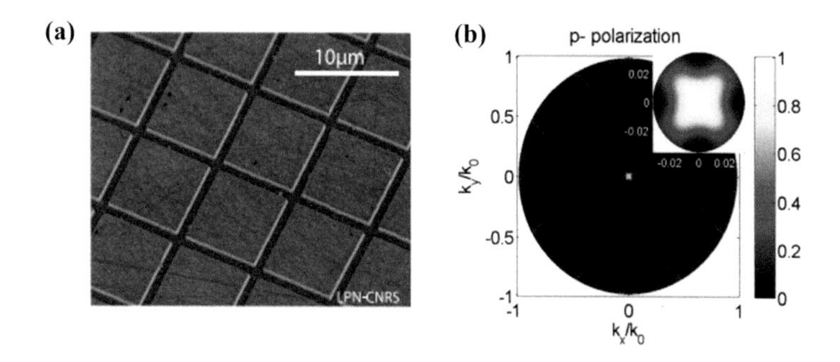

Figure 10.5 Thermal radiation control using 2D gratings. (a) Scanning electron microscope image of the gratings with 2D squares. (b) Polar representation of the calculated emissivity under p polarization. Reprinted figure with permission from [687]. Copyright 2012 by the American Physical Society.

Fig. 10.4(b) at a high temperature (~540 °C), owing to the excitation of SPPs. In addition, there are some works to combine more than one period in 1D grating structures, that is, binary gratings. The combination of different periodical units provides an extra degree of freedom to engineer emission properties. For example, Chen and Zhang's [970] work showed that the complex grating (Fig. 10.4(b)) enabled us to support wider emission peaks, which would be promising in TPV. Recently, an asymmetric grating backed with gold film (Fig. 10.4(c)) has been proposed to realize dual-band absorption [971], which consisted of three spaced dielectric strips in one unit. Coupling SPPs and Fabry–Perot (FP) resonance between different strips, the proposed asymmetric grating can support two perfect emission peaks at mid-infrared regions.

Two-dimensional gratings have also been extensively studied as perfect thermal emitters or absorbers. A seminal work was conducted by Heinzel et al. [972], and they proposed a 2D-grating patterned tungsten structure. The emission frequencies can be tailored with the dispersion relation of SPPs, which also exhibited strong angle-dependent properties. Another experimental work done by Arnold et al. [687] also demonstrated that, a 2D cross-slit SiC grating (see Fig. 10.5(a)) can achieve a high emissivity in both polarizations at Restrahlen region, thanks to the excitation of SPhPs. In particular, the p-polarized emission showed highly directional in Fig. 10.5(b).

On the other hand, apart from exciting surface waves, the designed grating structures can also support confined cavity modes. One of the pioneering works investigated thermal emission in 1D deep Silicon gratings [351]. Normal spectral emissivity measurements in the mid-infrared region were carried out for 45-μm deep, near square-wave gratings made of heavily phosphorus-doped silicon at 400° for both s and p polarizations. The deep gratings show superior thermal emission to the purely smooth surface of doped silicon at the same waveband. And the results also revealed that the resonance emission was attributed to the generation of standing waves in air slots. Maruyama et al. [343] proposed a 2D Cr-coated Si surface, of which the emission wavelength is approximately equal to the dimension of cavity parameters. Similarly, on the basis of the cavity resonances, Kohiyama et al. [973] compared the influence of open or closed microcavities in narrowband thermal emission manipulation. Results indicated that the closed one showed superior performance with intense and isotropic emission over a wide angle.

Furthermore, due to their relative easiness of fabrication, recently, a bunch of works were carried out on grating structures to further improve the emission performance and achieve multifunctionalities combined with other novel materials [974], for example, graphene, hBN, or other 2D materials. While, in these scenarios, the periods are generally seen as the characteristic lengths being comparable with the working wavelength, which will inevitably limit the tuning capability toward the subwavelength or even deep subwavelength scales.

10.1.2 Photonic Crystal

Artificial passbands and forbidden bands (i.e., PBGs) can be engineered by the arrangements and periodicity of PCs, which can also serve as an effective tool to tailor thermal emission. In theory, the basic generation mechanism of PBGs originated from the destructive interference of scattering waves induced by period, which are described by Bragg's diffraction [975]. It indicates that the scale of designed structures is generally comparable to the working wavelength. Figure 10.6 shows the basic types of PCs according to the dimensions of structures, including 1D, 2D, and 3D PCs. The different colors in the figures denote different kinds of materials. The edge states in PCs, guided modes, or localized states in the appearance of defects, also provide several possibilities of tailoring thermal radiation in various aspects.

The 1D PCs possess alternating multilayers, which are relatively easy to be fabricated. One of the earliest works to use 1D PCs for thermal radiation control was that by Narayanaswamy and Chen [691]. They developed an analytical method combining the modified DGF and the FDT to derive the radiative properties of the 1D PC consisting of two alternating metallic layered structures, providing theoretical guidance to shape thermal radiation. The selective emission can be obtained in the infrared and visible range by adjusting the period. Further, they extended the 1D PC-based emitters for TPVs applications as shown in Fig. 10.7(a) [691]. Similarly, by considering different materials, such as dielectric or polar materials, Lee et al. [336] proposed a 1D PC structure sitting on the SiC substrate. The unit of a PC is a sandwich-like structure, in which the middle layer is adjacent to two different layers with the same dielectric material. In this case, the role of a 1D PC is to provide extra wavevectors to excite SPhPs at different angles and wavelengths. But it is obvious that the 1D PC is sensitive to the polarization of the incident light.

Two-dimensional PCs also show a great potential in engineering thermal radiation. Actually, 2D PCs can also be regarded as 2D gratings, both of which are

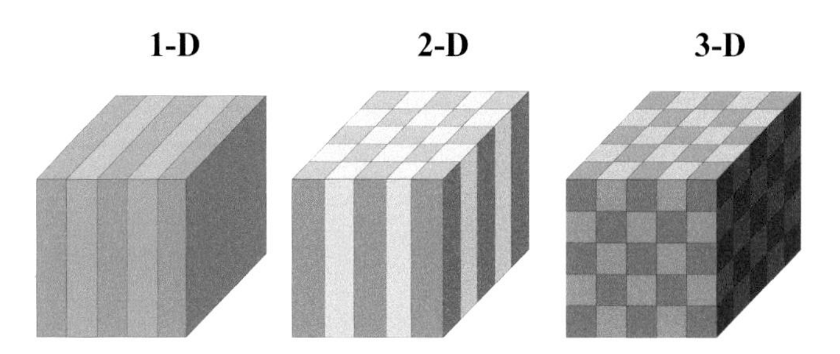

Figure 10.6 Simple examples of 1D, 2D, and 3D PCs. The different colors represent materials with different dielectric constants. The defining feature of a PC is the periodicity of the dielectric material along one or more axes.

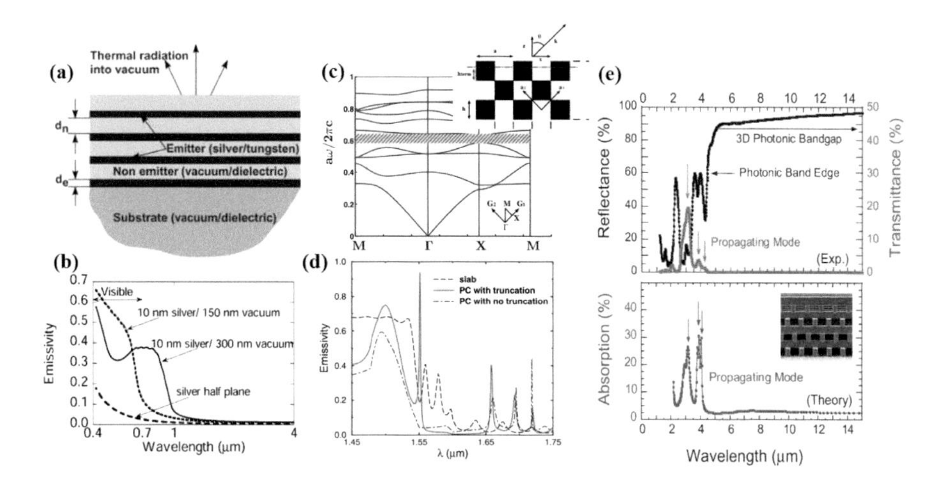

Figure 10.7 Thermal emitters based on PCs. (a) Schematic of 1D PC with the alternative dielectric and metal layers. (b) Compared emission spectra for emitters with different thicknesses of the dielectric layer. (a)–(b) Reprinted figure with permission from [691]. Copyright 2004 by the American Physical Society. (c) Band structure of a 2D PC as shown in the top-right inset. (d) Compared emission spectra for 2D PC slabs with and without truncation. (c)–(d) Reprinted figure with permission from [372]. Copyright 2006 by the American Physical Society. (e) Theoretical and experimental results of spectral reflection and absorption for the proposed 3D PC are shown in the inset. Reprinted with permission from [976] © The Optical Society.

composed of wavelength-scale periodic arrays or cavities. Therefore, for 2D PCs, one can also consider exciting surface waves or resonant cavity modes to shape thermal radiation properties. In 2006, Chan et al. [977] discussed the role of the periodicity of a 2D photonic slab in influencing waveguide cutoffs, waveguide resonances, diffraction peaks, and surface plasmon modes as well. Figure 10.7(c) shows the band structure of a 2D infinite PC in which the black areas denote the Ge material, and the white areas depict air. In this case, the large emissivity below $\lambda = 1.55$ μm is due to the loss of the Ge material. Several emission peaks over 1.55 μm are ascribed to FP resonances. For the infinite PC without truncation, high emission is forbidden because of the appearance of a PBG. By contrast, there is a sharp emission peak within the gap due to the resonant excitation of leaky surface waves in PC with truncation. Besides, several works have also paid attention to the use of 2D PCs to achieve broadband emission. One of the interesting works was by Niu et al. [978], in which they designed alternating cavities filled with different materials (HfO_2 or TiO_2) embedded in the tungsten film. In this case, broadband absorption covering visible to near-infrared regions can be obtained. They further proposed methods to fabricate the structures as SEM images.

Furthermore, more comprehensive control of photonic band structures can be achieved by thermal emitters based on 3D PCs. As a typical 3D PC, woodpile structure thermal emitters were extensively studied by Fleming et al. [976, 979, 980]. The large difference between 3D PCs from others is that they can effectively suppress light transport in all directions at a certain frequency due to the appearance of a complete bandgap. A representative 3D PC thermal emitter made of the tungsten rod [976] is presented in Fig. 10.7(e), in which narrowband emission can be obtained over a wide temperature (475–850 K). High thermal emission peaks occur in the vicinity of the photonic band edge because of the high DOS. To overcome the fabrication difficulty of woodpile structure, Han et al. [981] theoretically proposed that metallic inverse opals, a type of 3D PCs that can be fabricated using the much simpler self-assembling method, may exhibit similar or even better thermal emission performance. Arpin et al. [982] fabricated this kind of 3D PC thermal made of tungsten using self-assembled templates, which demonstrated a spectral selective thermal emissivity suitable for solar TPVs at high operating temperatures. These 3D PC thermal emitters have great potential in realizing high-efficiency TPVs and flexibly tailoring emission profiles.

However, being similar to gratings, the above-mentioned PC-based strategies only ensure the manipulation of EM behaviors of thermal radiation when the characteristic length is of the order of the working wavelength, limiting the tunability and locking up many novel light–matter interaction phenomena.

10.1.3 Metamaterials and Metasurfaces

Over the past decades, we have witnessed unprecedented development in metamaterials and metasurfaces that utilize engineered nanostructures at subwavelength scales to tailor radiation properties in more flexible manners. According to the different tuning mechanisms, we would like to separate this part into several small subsections to show their superior performance in thermal emission engineering.

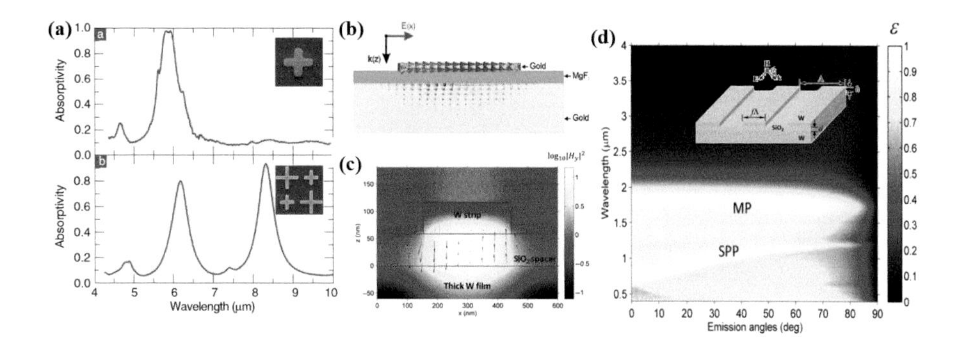

Figure 10.8 MIM-based thermal emitters. (a) Experimental results of the absorptivity of the single-band (top) and dual-band (bottom) absorbers. Reprinted figure with permission from [337]. Copyright 2011 by the American Physical Society. (b) Electric currents distributions in MIM structures. Reprinted with permission from [983]. Copyright 2010 American Chemical Society. (c) EM field distribution of magnetic polaritons in the dielectric layer. (d) Angle-resolved emissivity spectra of the W-SiO_2-W structure shown in the inset. (c)–(d) Reprinted from [984], with the permission of AIP Publishing.

Metal–dielectric–metal Metamaterials

The metal–dielectric–metal (MIM) metamaterials have gained much attention in manipulating thermal emission properties, owing to rich resonant features and advanced fabrication techniques as well. The MIM nanostructures are generally composed of basic three layers, in which two metallic layers are separated by a dielectric interlayer. In particular, the upper metal layer is generally tailored with different shapes as shown in Fig. 10.8(a). The essential roles of MIM metamaterials can be attributed to their superior ability to excite SPPs, magnetic polaritons, or mutual coupling with other resonant modes.

A seminal work of MIM-based thermal emitters was demonstrated by Liu et al. [337], in which a cross-shaped Au resonator was designed sitting on a silicon layer backed with a gold substrate. Based on the impedance matching principle, single- and dual-band perfect absorbers were achieved as presented in Fig. 10.8(a) with different metallic resonant units. In this case, the physical mechanism is attributed to the generation of antiparallel electric currents at both the top and bottom metallic layers with localized surface plasmon resonances [983] as shown in Fig. 10.8(b). Such properties have already shown as promising prospects in infrared sensing [983, 985, 986]. Besides, Fig. 10.8(c) shows a typical EM distribution in general MIM structures, known as magnetic polaritons, which also contributes to enhancing spectra emissivity. Furthermore, by coupling MIM metamaterials with other functional materials like 2D materials, that is, graphene or phase-change materials, the MIM-based nanostructures will show much more flexibility and superior performance, which will be discussed in Section 10.2.

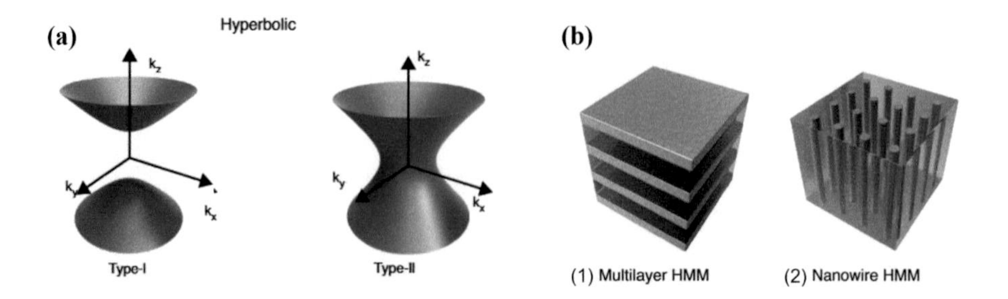

Figure 10.9 (a) Effect frequency counter of TM waves in type I (left) and type II (right) hyperbolic metamaterials. (b) HMMs fabricated in multilayer (left) and nanowire (right) structures. Reproduced with permission from [987]. Copyright 2022, Nature Publishing Group.

Metamaterials with Topological Transitions: Hyperbolic Metamaterials and Epsilon-Near-Zero Metasurfaces

Other notable categories of metamaterials for efficient thermal emission engineering include hyperbolic metamaterials and epsilon-near-zero (ENZ) metamaterials. Here, we would like to combine these two types of structures in one subsection for the similar tuning mechanism on the basis of topological transition around resonant frequencies.

Hyperbolic Metamaterials

Hyperbolic metamaterials can be considered as anisotropic materials showing an extremely anisotropic dielectric tensor, in which one of the principal components of relative permittivity tensor has an opposite sign compared with the other two elements, as

$$\widehat{\varepsilon_s} = \begin{pmatrix} \varepsilon_{xx} & 0 & 0 \\ 0 & \varepsilon_{yy} & 0 \\ 0 & 0 & \varepsilon_{zz} \end{pmatrix}, \tag{10.1.4}$$

in which the in-plane elements $\varepsilon_{xx} = \varepsilon_{yy} = \varepsilon_{\perp}$ and the out-of-plane element $\varepsilon_{zz} = \varepsilon_{\parallel}$, and $\varepsilon_{\parallel} \cdot \varepsilon_{\perp} < 0$. The subscripts $\perp$ and $\parallel$ denote the elements parallel and perpendicular to the anisotropy axis, respectively. Then, derived from Maxwell's equations, as for TM waves for example, the properties of hyperbolic metamaterials can be understood by viewing the equifrequency surface that is given by

$$\frac{k_x^2 + k_y^2}{\varepsilon_{\parallel}} + \frac{k_z^2}{\varepsilon_{\perp}} = k_0^2, \tag{10.1.5}$$

in which k_x, k_y, and k_z are the wavevectors in the x, y, and z directions, respectively. The $k_0 = 2\pi/\lambda$ is the wavevector in free space. Equation (10.1.5) shows a hyperboloid when $\varepsilon_{\parallel} \cdot \varepsilon_{\perp} < 0$, thus leading to an unbounded frequency surface that is totally different from the closed one of the isotropic media. By contrast,

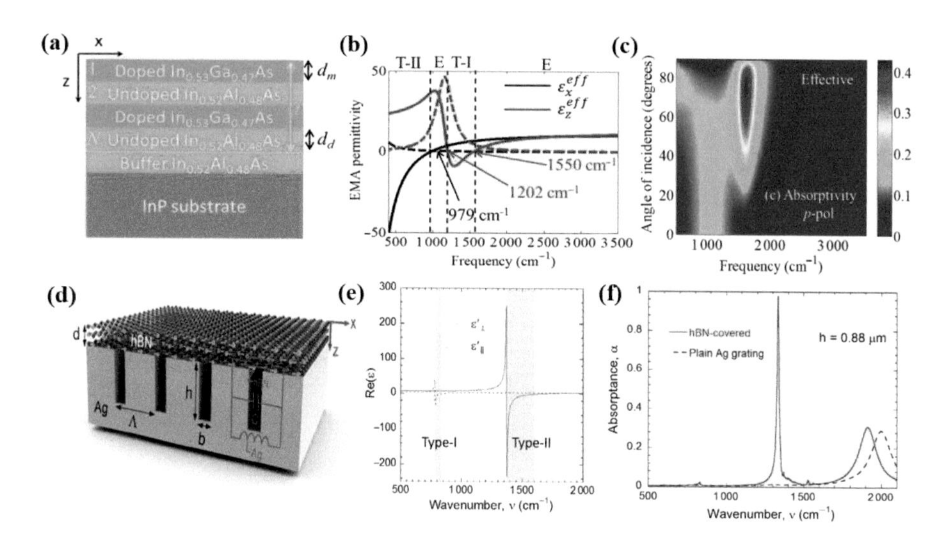

Figure 10.10 Hyperbolic materials-based thermal emitters. (a) The semiconductor hyperbolic metamaterial is made of by alternating 50 pairs of 10-nm-thick doped $In_{0.53}Ga_{0.47}As$ and 8-nm-thick undoped $Al_{0.48}In_{0.52}As$ layers. The superlattice is on top of a 200-nm-thick buffer $Al_{0.48}In_{0.52}As$ layer and a 0.65-mm-thick InP substrate. (b) Effective permittivity functions of the homogeneous SHM computed using a local anisotropic effective medium model. (c) Angle-resolved absorption spectra under p waves. (a)–(c) Reprinted from [990]. Copyright 2016, Nature Publishing Group. (d) Schematic of the hBN/metal grating hybrid structure. (e) Real part of the hBN material with natural hyperbolic properties. (f) Absorptance spectra of plain Ag gratings and the hBN-covered Ag gratings. (d)–(f) Reprinted from [991], Copyright 2017, with permission from Elsevier.

there are two types of hyperbolic surfaces as shown in Fig. 10.9: (left) $\varepsilon_{\|} < 0$ and $\varepsilon_{\perp} > 0$; (right) $\varepsilon_{\|} > 0$ and $\varepsilon_{\perp} < 0$. It is worth noting that, as for these types of metamaterials, they can only support the propagation of high-k modes, and there is a cutoff frequency for small wavevectors as shown in Fig. 10.9(a). As reported, there are a few natural materials showing hyperbolic dispersion. In the area of thermal radiation control, hBN is one of the most popular materials, which shows two types of hyperbolic dispersion in the mid-infrared region. Using artificial nanostructures like multilayer or nanowires as shown in Fig. 10.9(b), it is reliable to design hyperbolic metamaterials at desirable wavelengths. A more detailed introduction can be found in several review papers [988, 989].

Because of the excitation ability of large wavevectors of hyperbolic metamaterials, they have gained much attention in near-field thermal emission control, which will be discussed later in Section 10.2. Similar to far-field thermal properties, hyperbolic materials also show outstanding performance. Campione et al. [990] studied the emission properties of semiconductor hyperbolic metamaterials composed of alternating 50 pairs of 10-nm-thick $In_{0.53}Ga_{0.47}As$ and 18-nm-thick undoped $Al_{0.48}In_{0.52}As$ layers as presented in Figs. 10.10(a)–(c). Then, two types

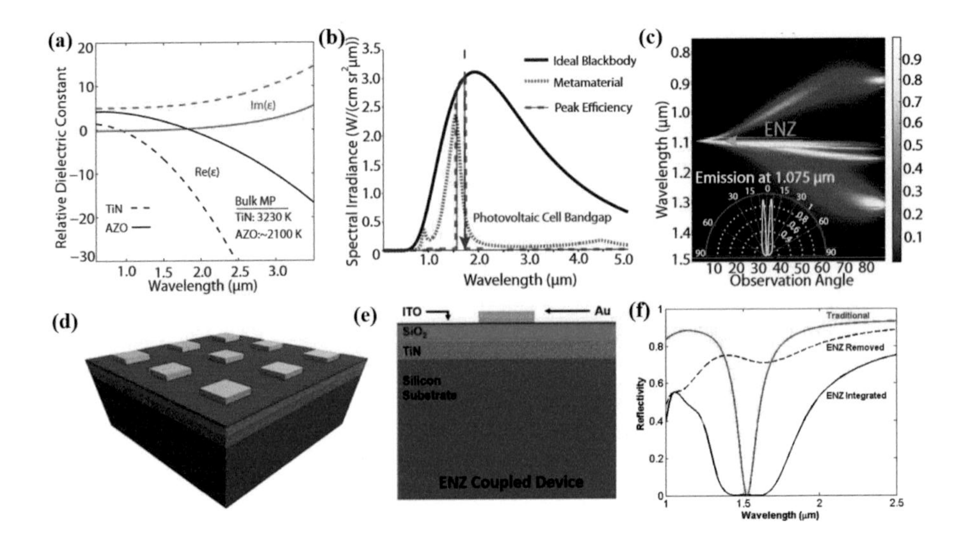

Figure 10.11 ENZ materials-based thermal emitters. (a) Drude models of relative permittivity of TiN and AZO used in [349]. (b) Comparison of emission spectra between the titanium nitride metamaterial, an ideal blackbody, and an emitter that maximizes the efficiency of energy conversion at 1 500. (c) Angular and spectral emission of the proposed metamaterial emitters composed of a host matrix of aluminum oxide (Al_2O_3) embedded with 15-nm diameter silver nanowires in a 115 nm^2 unit cell. Reprinted with permission from [349]. Copyright 2012, Optical Society of America. (d) Schematic of the nanostructures consisting of (e) a TiN metallic thick film, a silicon dioxide dielectric layer, an ITO nanofilm, and a patterned periodic array of metal squares. (f) Results of reflectivity of structures with (black) and without an ENZ layer. Reprinted from [993]. Copyright 2018 American Chemical Society.

of hyperbolic metamaterials have been observed based on the effective medium theory. Besides, some natural materials like hBN also show hyperbolic properties as presented in Fig. 10.10(d). The appearance of the hBN layer on Ag grating contributes to introducing a narrow absorption peak around the type II Reststrahlen band. Combined with natural hyperbolic materials hBN, Kan et al. [992] proposed a compact hBN/metal metasurface as presented in Fig. 10.10(b). When introducing hBN/metal multilayer into sawtooth gratings, it is possible to excite hyperbolic phonon–plasmon polaritons in the Reststrahlen band and surface phonon–plasmon polaritons out of the Reststrahlen band of hBN, thus resulting in a near-perfect absorption broadband.

Epsilon-Near-Zero Metamaterials

Another category of metamaterials that looks promising for thermal emission control is ENZ metamaterials [994] whose effective permittivity approaches zero at a certain frequency. In this scenario, the large-field enhancement can be reliable based on the boundary condition, showing as $E_i = \frac{\varepsilon_j}{\varepsilon_i} E_j \to \infty$ when $\varepsilon_i \to 0$

[995]. On the basis of this principle, Molesky et al. [349] proposed effective ENZ metamaterials similar to Fig. 10.9(b) as high-performance emitters for TPV applications. By tuning the filling ratio of the metallic or dielectric components, the parallel component $\varepsilon_{\|}$ of the effective permittivity of multilayer ENZ metamaterials can approach zero, thus leading to omnidirectional, polarization-insensitive emitters as shown in Figs. 10.11(a)–(c). In the following, the ENZ behavior has been explored in several systems. Dyachenko et al. [348] proposed a W/HfO_2 multilayered metamaterials for thermal emitters design. The angle-insensitive and wavelength-selective near-perfect thermal emission can be attributed to the ENZ property associated with the optical transition of effective permittivity of the nanostructures. The topological transition indicates a specific transition at a certain frequency the equifrequency surface transfers from an ellipsoid to a hyperboloid, where the ENZ point appears for the parallel effective permittivity. Such behaviors are quite similar to the hyperbolic metamaterials introduced above. The ENZ properties near topological transition contribute to enhancing absorption or emission, and the hyperbolicity provides an extra degree of freedom to tune the polarization or directionality.

On the other hand, there are several materials possessing ENZ behavior, including metallic materials described by the Drude model at their plasma frequencies, polar materials described by the Lorentz–Drude model at longitudinal optical phonons frequencies, and semiconductor materials such as InAs and InP. Since the plasma frequency for most metals is located at a very short wavelength, that is not attractive for emission control, while in the area of manipulation thermal radiation properties, the doped semiconductors [994], such as ITO or ZTO, show excellent performance to broaden the absorption band. For example, by introducing an extremely thin ITO layer into conventional MIM structures, numerical and experimental results show that the absorption peak around 1.5 μm can be broadened [993]. Similar performances are also proved using Au–HfO_2–ITO nanostructures with electric tuning. Note that the ENZ-based metamaterials show large flexibility, for which their ENZ wavelength can be tuned by different levels of doping of aforementioned semiconductors or changing different filling ratios of components in composite multilayer or nanowire metamaterials.

Dielectric Metasurfaces

Recently, we have witnessed an unprecedented development of dielectric metasurfaces for high efficiency compared with metallic counterparts. In terms of the basic definition of metasurfaces, in fact, they are regarded as a new branch of metamaterials in two dimensions, which are generally composed of planar and subwavelength meta-atoms (which can be resonant or not) in a periodic or aperiodic arrangement [376, 439, 999]. The EM response of the whole configuration can be artificially tailored by engineering an individual resonant meta-atom, including the shape, size, and composite material, and also the spatial and orientational arrangements of these meta-atoms. In this section, we will overview

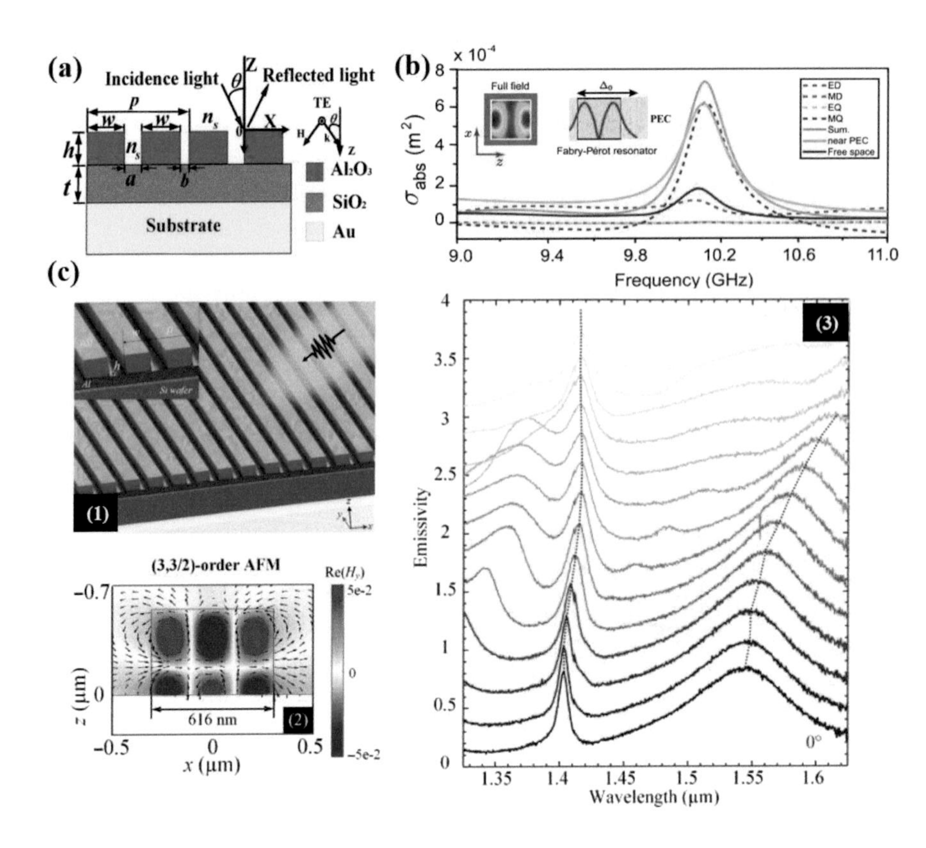

Figure 10.12 Metal-backed dielectric thermal emitters. (a) Schematic of dielectric gratings proposed in [996]. Reprinted from [996], Copyright 2016, with permission from Elsevier. (b) Absorption cross section and its approximated multipole decomposition of the stand-alone ceramic cube in free space or placed 3 mm above the PEC sheet. Reprinted figure with permission from [997]. Copyright 2019 by the American Physical Society. (c) Nanograting consists of periodical a-Si nanobars and a metal-like substrate (1). The magnetic field patterns of high-order AFM resonance (2). Measured spectral emissivity at different angles with a step of 2.5°. Reprinted with permission from [998]. Copyright 2021 American Chemical Society.

the recent typical works for tuning absorption or emission properties on the basis of the resonant performance of dielectric meta-atoms.

A general category of dielectric-based absorbers/emitters is summarized in Fig. 10.12. Even though they are similar to some of the nanostructures introduced before, like gratings, the working principles are totally different. In these cases, the metal substrate is still needed to ensure a perfect reflection. For example, in Fig. 10.12(a), the role of the dielectric grating is to excite cavity resonances [996], which leads to an ultra–narrowband absorption with a extremely large quality factor ($\sim 10^4$). Similarly, Fig. 10.12(b) shows another feasible method to introduce a dielectric cavity between a reflector substrate and dielectric meta-atoms, where there is a typical standing-wave pattern as illustrated in the inset. At this

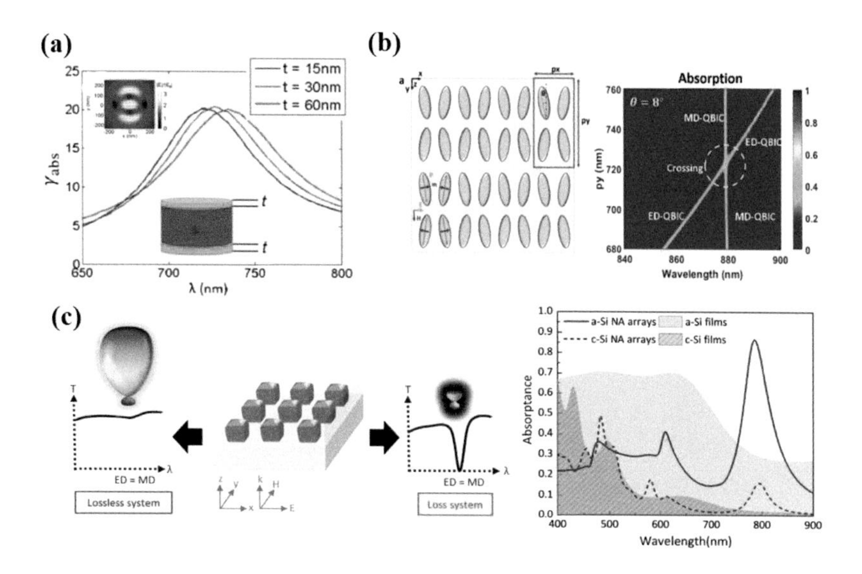

Figure 10.13 All-dielectric thermal emitters. (a) The effect of absorption rate enhancement change on three representative cases with different thicknesses t in the inset (Si nanodisk capped with thin ITO films). Reprinted from [1000]. Copyright 2016, Optical Society of America. (b) Schematic of a high-Q silicon metasurface, along with the spectral absorption changing with the period at the y direction. Reprinted with permission from [1001]. Copyright 2020 American Chemical Society. (c) Left: Schematic diagram of a silicon narrowband absorber. Directional scattering from a single nanoparticle and near-unit transmission from the nanoparticle array take place in the effectively lossless structure. By contrast, a lossy structure causes a dip in the transmittance. Right: Comparison of the absorptance spectra of a-Si NA arrays (blue solid line), c-Si NA arrays (blue dashed line), a-Si films (light gray area), and c-Si films (gray area). Reprinted with permission from [1002]. Copyright 2018 American Chemical Society.

resonant frequency, the absorption cross section σ_{abs} are enhanced [997]. Differently, subwavelength meta-atoms also support a series of high-order multipolar resonances. In Fig. 10.12(c), Liu et al. [998] proposed a new mechanism based on the excitation of higher order antiferromagnetic resonances to realize high-Q wavelength-selective thermal emitters. The antiferromagnetic resonances are originated from hybrid modes of magnetic resonances and high-order FP resonances in an individual meta-atom, showing several antiparallel magnetic dipoles as presented in field patterns as shown in Fig. 10.12(c). With the experimentally measured emission spectra are also given, the achieved narrowband emission can be protected over a wide range of angles.

There are other novel works using dielectric metasurfaces for emission engineering without the need of a metal substrate. Here, we would like to introduce some interesting works as presented in Fig. 10.13. Inspired by the superior light confinement ability of nonradiative anapole modes that can totally confine energy in the near field, Wang et al. [1000] employed a-Si nanodisk covered with ITO materials on two sides, as shown in Fig. 10.13(a), which enables a large enhancement of absorption cross section. The two ITO layers work as a good reflector and open a loss channel because of a nonzero imaginary part of permittivity. In addition, more recently, Tian et al. [1001] introduced the concept of bound states in the continuum (BICs) in designing wavelength-sensitive thermal absorbers. The physics origin of quasi-BIC modes appears, thanks to a strong coupling between leaky modes in optically guided nanostructures or dielectric metasurfaces, with energies embedded in the continuous spectrum of radiating waves. BICs have been predicted in quantum mechanics for a long time but have not been experimentally verified. While optical systems provide a feasible platform to achieve this beautiful phenomenon with a very large Q factor obtained in lossless cases, in a thermal radiation system, the loss is of particular importance. Tian et al. [999] numerically studied an all-dielectric metasurface supporting dipole quasi-BIC resonance with engineered optical loss. Perfect and ultra-narrowband absorption has been realized by properly tuning the geometrical parameters of designed meta-atoms or periods. But it should be noted that this behavior will be deteriorative if there is a moderate intrinsic loss for dielectric materials. On the other hand, by taking advantage of the intrinsic of a-Si in the visible range, Yang et al. [1002] compared the performance of dielectric metasurfaces composed of c-Si and a-Si nanocubes. Under the second Kerker condition in which the excited electric and magnetic dipoles are oscillated out of phase, high absorption can be observed for the high loss in a-Si metasurfaces as illustrated in Fig. 10.13(c).

All-dielectric metasurfaces are good candidates for compact thermal absorbers/emitters when further combining them with integrated nanodevices. Besides, the melting points of several dielectric materials are ideally high, which could be applicable to several high-temperature working situations. This is an emerging area that has not been fully explored but shows great potential.

10.1.4 Aperiodic or Disordered Nanostructures

So far, most of the mentioned works are based on periodic structures in either one or two dimensions. Recently, there are also some efforts devoted to aperiodic or disordered media, which are comparatively easier to fabricate scalably in a large area and become less insensitive to fabrication errors. The disordered or random media provide a platform to enhance the light–matter interaction, thus leading to improved absorption in lossy cases. Here, we would like to briefly introduce some typical disorder designs as shown in Figs. 10.14 and 10.15.

To enhance the total absorption of absorbers in solar cells, scientists have proposed nanohole-based structures to enhance absorption in the visible range. For example, Vynck et al. [1007] proposed patterned disordered nanoholes in a thin silicon film. The broad absorption spectra, as compared with a bare thin film, show great potential in light harvesting. Further, Fang et al. [1003] studied

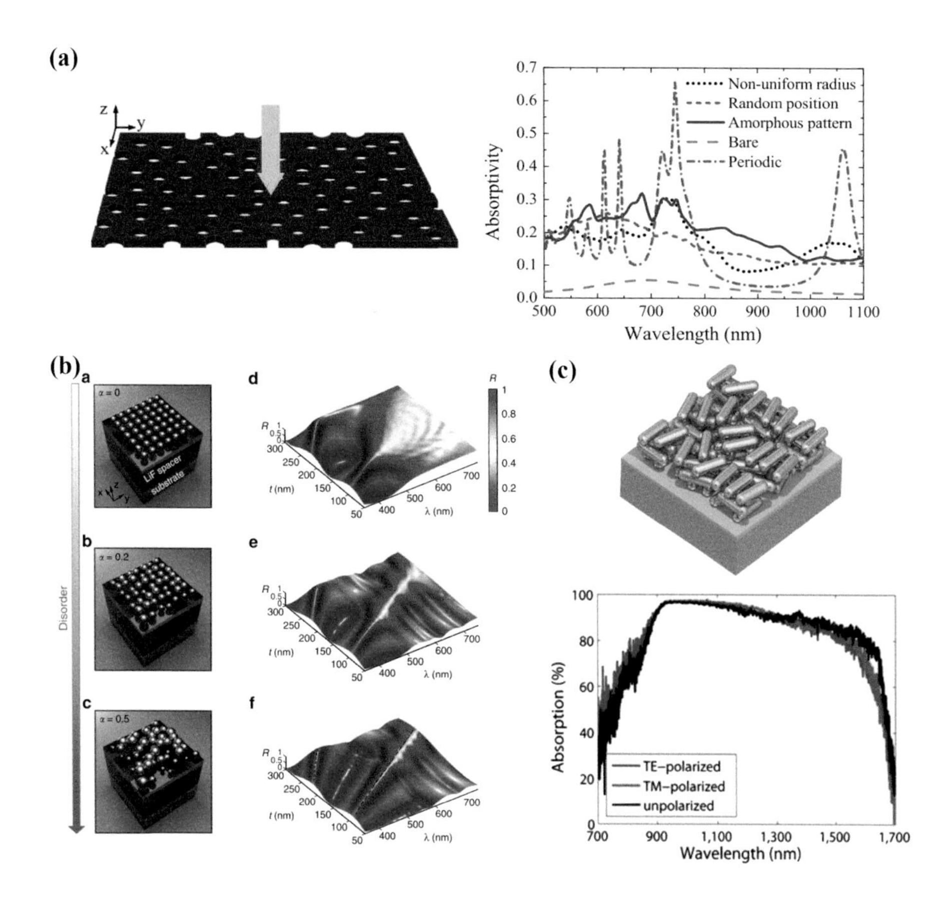

Figure 10.14 Disordered-driven broadband absorption performance. (a) The schematic of silicon thin film etched with disordered nanoholes (left), along with absorptivity spectra of thin films with different types of nanohole patterns [1003]. Reprinted from [1003], Copyright 2015, with permission from Elsevier. (b) The reflection spectrum (right-hand column) changes with the different degrees of disorder. Reprinted from [1004]. Copyright 2020, Nature Publishing Group. (c) Morphology of the nanostructure based on multilayer gold nanorods, along with absorption spectra for different polarizations of the light: unpolarized (black curve), TE-polarized (blue curve), and TM-polarized (red curve) light. Reprinted with permission from [1005]. © 2013 Optical Society of America.

the thin silicon film etched with nanoholes in different distributions, including periodic structures, amorphous patterns, random patterns, or nanoholes with different radii. Results indicated that amorphous patterns should be superior to enhance absorption over a broader band as shown in Fig. 10.14(a). Another type of disorder design, to realize broadband absorption, randomly distributes plasmonic nanoparticles on the surface or embedded in the PCs, as shown in Figs. 10.14(b)–(c). The random plasmonic nanoparticles can support multiple localized surface plasmonic resonances, contributing to enhancing the light–matter

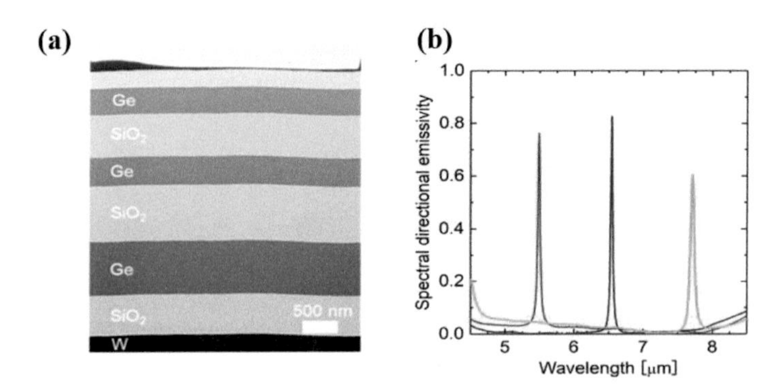

Figure 10.15 Aperiodic-induced narrowband absorption performance. (a) Cross-sectional TEM images of the fabricated aperiodic multilayered sample. (b) Measured spectral directional emissivity at different targeted wavelengths. Reprinted from [1006]. Copyright 2019, American Chemical Society.

interaction and improving light absorption. Figure 10.14(c) shows the changes in reflection spectra as a function of spacer thickness and the degree of disorder. The large degree of disorder contributes to trapping more light within the structures with reduced reflection. Besides, the thickness of space will also change the position of reflection bands. Similarly, by randomly distributing the Au nanorods on a gold film spaced by a dielectric layer, polarization-insensitive broadband absorption has been achieved over a broad near-infrared region [1005]. In the same manner, some works placed nanoparticles within the nanoholes in PCs [1008, 1009] to realize the perfect absorption of solar spectrum, which can be applied in vapor steam generation.

Despite promising applications in broadband absorption, aperiodic nanostructures have also been applied to design narrowband thermal emitters. One of the seminal works is the aperiodic multilayered metamaterial emitter proposed through the Bayesian optimization method, which is a design algorithm based on machine learning [1006, 1010] as shown in Fig. 10.15. As for a given absorption spectra (either quite narrow or moderately broad peaks), the optimal configuration of multilayers can be efficiently identified from over several billion candidates consisting of aperiodic multilayered metamaterials. The measured absorption spectra agree well with the simulated ones.

10.1.5 Reconfigurable Control Based on Active Materials

For most of the aforementioned manipulation methods, the designed radiative functions are usually fixed at a certain wavelength, angle, or polarization, which cannot be flexibly tuned after fabrication. In order to dynamically control thermal radiation and achieve multiple functions, several strategies have been proposed, including but not limited to temperature modulation and time-dependent manipulation using active materials, such as vanadium dioxide (VO_2),

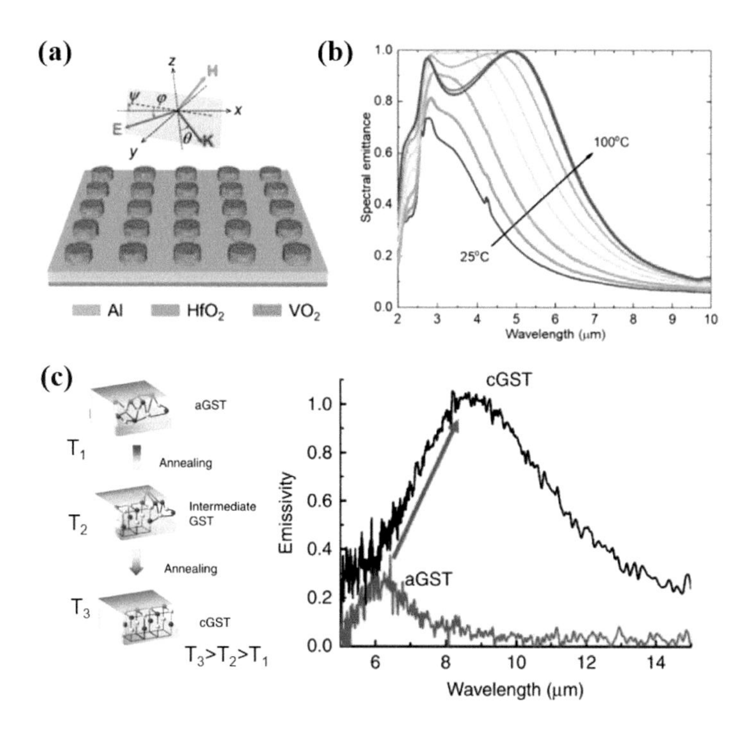

Figure 10.16 Temperature-dependent thermal emission control. (a) Schematic of the directional measurement of VO_2-HfO_2-Al metamaterials. (b) Results of spectral emittance at different temperatures under normal incidence. (a)–(b) Reprinted with permission from [1011]. Copyright 2020 American Chemical Society. (c) Illustration of the thermal camouflage with different background temperatures. (d) The measured emissivity at the normal angle for aGST-Au and cGST-Au devices. Reprinted from [1012]. Copyright 2018, Nature Publishing Group.

$Ge_2Sb_2Te_5$(GST), liquid crystal, or 2D materials. In this section, we would introduce some typical dynamical tuning techniques and emphasize the corresponding pros and cons.

Thermal radiation is directly related to temperature. Despite changing spectral emission power distribution based on Planck's law, there are some methods that can dynamically tailor the emissivity of nanostructures with temperature by employing temperature-dependent materials. One of the most popular temperature-dependent materials for thermal emission is VO_2 that could experience a reversible phase transition from an insulating state to a metallic state when its temperature exceeds about 68 °C [1014]. The phase change makes the optical index of the two states different, leading to different emission performances. As exampled in Fig. 10.16(a), Long et al. [1011] designed an MIM metasurface in which the upper layer metallic meta-atoms are VO_2 disks. When increasing the temperature from 25 °C to 100 °C, the spectral emittance changes from a narrow band to a broader band as presented in Fig. 10.16(b). Results indicate that

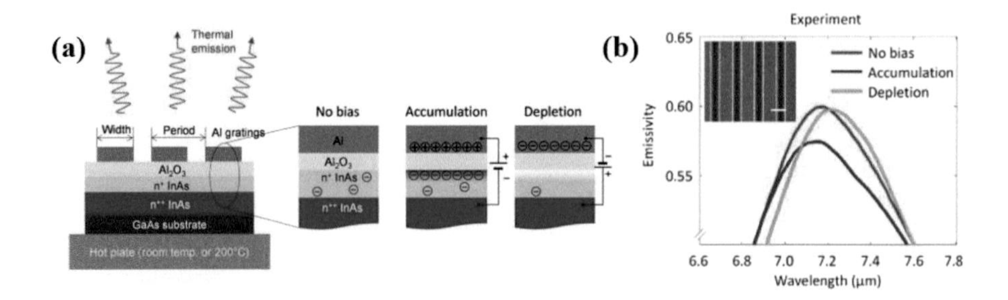

Figure 10.17 Electric gate control of thermal emission. (a) Device schematic of active metasurface composed of InAs multilayers with different carrier densities, with (b) measured emissivity spectrum for no bias (red curve), depletion (green curve), and accumulation (blue curve) on the right panel. Reprinted from [1013]. Copyright 2018, American Association of the Advancement of Science.

insulting VO_2 metamaterials are capable of suppressing emission, while metallic VO_2 metamaterials enable enhanced emission at a high temperature. In particular, by carefully fabricating tungsten-doped VO_2 where the tungsten fraction is judiciously graded across a thickness less than skin depth, the emissivity can be engineered to change with temperature as the inverse of T^4, which means that the thermal radiation of the whole nanomaterials will not increase with the fourth power of absolute temperature known as the Stefan–Boltzmann law. Tang et al. [1015] experimentally investigated such intriguing phenomena and further extended the application for thermal camouflage. In addition, GST materials are also promising in reconfigurable thermal emission engineering. Similarly, it can also undergo a phase transition between the amorphous phase and crystal phase by controlling the annealing temperature. More specifically, the deposited GST film at room temperature usually shows an amorphous state with a disordered atomic distribution, while it will be altered to a well-organized distribution by annealing it at 200 °C. In particular, when choosing different annealing times, it is possible to obtain different intermediate phases [1012], showing flexible tunability. Figure 10.16(c) shows the measured spectral emissivity of the GST thin film with two different states. Obviously, the crystal GST thin film shows superior emission properties covering broad mid-infrared regions. Further, by considering MIM structures composed of GST materials, it enables us to shape spectral emissivity with desirable emission peaks [1016]. Even though thermal tuning is promising and attractive in several energy-based applications, the tuning speed is slow.

On the other hand, electric modulation based on 2D materials, varactor/PIN diodes, doped semiconductors, and liquid crystal is a promising platform to improve modulation speed. Benefiting from the excellent thermal, electric, and mechanical properties of graphene, Luo et al. [1017] designed hBN encapsulated graphene devices with a photonic cavity consisting of SiO_2 and hBN dielectrics. The intensity of thermal emission can be actively tuned by changing the applied

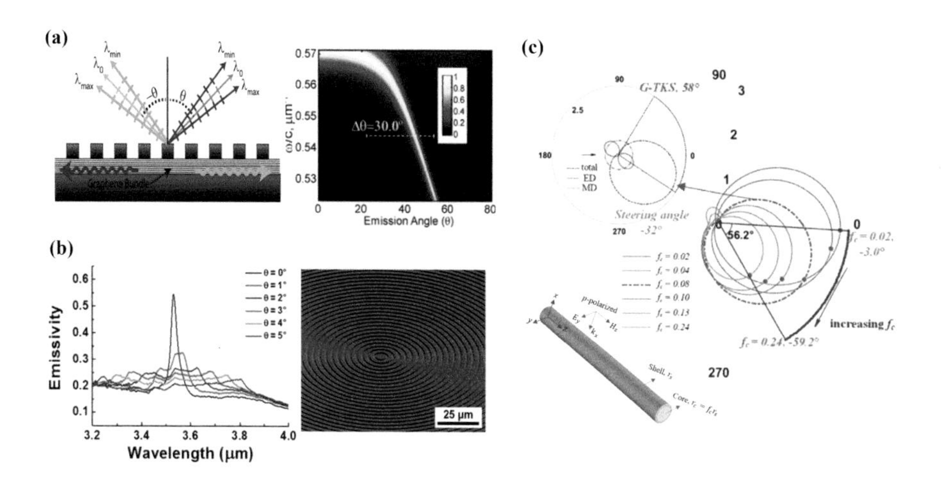

Figure 10.18 Angular control of thermal radiation. (a) Measured thermal-emission spectra at various incident angles, with the SEM image of the fabricated sample on the right. Reprinted figure with permission from [1020]. Copyright 2016 by the American Physical Society. (b) Schematic geometry of the SiC slab with the graphene bundle and dielectric gratings of periodicity, along with the emissivity spectrum of SiC gratings on top of the SiC layer at a fixed period [1020]. Reprinted with permission from [370]. Copyright 2016, ACS Publications (https://pubs.acs.org/doi/10.1021/acsphotonics.6b00022). (c) Arbitrary beam steering function using a dielectric–Weyl semimetal cylinder with different filling ratio [1021]. Reprinted from [1021], Copyright 2021, with permission from Elsevier.

bias voltage. Similarly, Fig. 10.17 demonstrates an electrically tunable thermal emission, proposed by Park et al. [1013], by adding active InAs layers into $Al/Al_2O_3/Al$ metasurfaces. The emission peak (around 7.3 μm) can be shifted to short or long wavelength via varying the charge density in the low-doped (n^+) InAs layer. UV plus can also be used to achieve the fast modulation of thermal emission with nanosecond tuning speed. Xiao et al. [1018] proposed a method using 200-fs laser plus at 515 nm to pump unpatterned silicon and gallium arsenide. The response of the free carrier dynamics in the sample contributes to the nanosecond-scale modulation of thermal emissivity in the mid-infrared region. Similarly, Coppens et al. [1019] designed a larger, infrared metamaterial composed of the ITO film, ZnO film, and gold resonators. The external stimuli, UV illumination, will generate free carriers in the photosensitive ZnO spacer layer, leading to the flexible tuning of spectral emissivity.

10.2 Overview of Thermal Radiation Control in Micro/Nanoscale

Section 10.1 has elaborately reviewed different mechanisms to engineer thermal radiation in micro/nanoscales, which has challenged conventional views of thermal radiation that are generally constrained by, for example, Planck's law and Kirchhoff's law. It is known that thermal radiation possesses spectral, angular, and polarized features, even though conventional radiators are typically broadband, incoherent, unpolarized, and omnidirectional. By taking advantage of artificial nanostructures, in this section, we would like to give a brief overview of nanophotonic control of thermal radiation from the aspects of thermal properties mentioned above in Section 10.1, including spectral/angular/polarization features and some intriguing phenomena beyond some classical laws of thermal radiation, such as near-field thermal radiation, thermal extraction, and nonreciprocal thermal radiation.

10.2.1 Spectral/Angular/Polarization Control of Thermal Radiation

In contrast to the conventional radiation properties, in this section, we would like to briefly introduce some typical works, focusing on engineering thermal radiation in spectral, angular, and polarization aspects. Noting that, some unique features, being different from conventional emitters/absorbers, can occur in one nanostructure, like narrowband polarization-insensitive thermal emitters. Herein, we would like to emphasize on one of the most inspiring characteristics.

As for directional radiation control, one of the seminar works was proposed by Greffet et al. [335]. By exciting SPhPs in SiC grating, a highly directional coherent radiation pattern can be realized. Likewise, Park et al. [370] designed a bull's eye-like patterns (Fig. 10.18(a)) composed of tungsten and molybdenum supporting SPPs. Then, the measured thermal emission was spectrally narrow (tens of nanometers) and highly directional (2° angular divergence). The directional emission can also be tailored by making use of diffraction light. In Fig. 10.18(b), Inampudi and Mosallaei [1020] proposed a method to mitigate the angular dispersion (change of emission angle with frequency) by inserting bundled graphene sheets into the substrate and SiC gratings. This method achieved a substantially lower angular dispersion of 16° in a relatively broad wavelength range of 11–12 μm. Distinctly, despite of exciting surface waves, there are some methods to tailor directional radiation using Mie scatterers by exciting several EM modes like electric/magnetic dipoles or quadrupole [1022, 1023]. For example, Liu and Zhao [1021] proposed a core-shell cylinder composed of dielectric and MO materials. The MO materials contribute to exciting nondegenerate EDs with a collective rotated dipolar pattern. Such rotated radiation patterns further interfere with a magnetic dipole, providing a versatile platform to realize arbitrary beam steering as shown in Fig. 10.18(c).

By taking advantage of several resonant phenomena, it is achievable to tailor the emission peaks at a desirable wavelength or waveband, which is promising in several applications, such as TPV, solar cells, radiative cooling, and thermal camouflage. Here we would like to show some representative works, and more details about their specific applications are presented in Section 10.2.2. Figure 10.19(a) shows the measured reflection and emission spectra of multilayer structures supporting metal–optical Tamm states [1024]. The measured Q factor can reach over

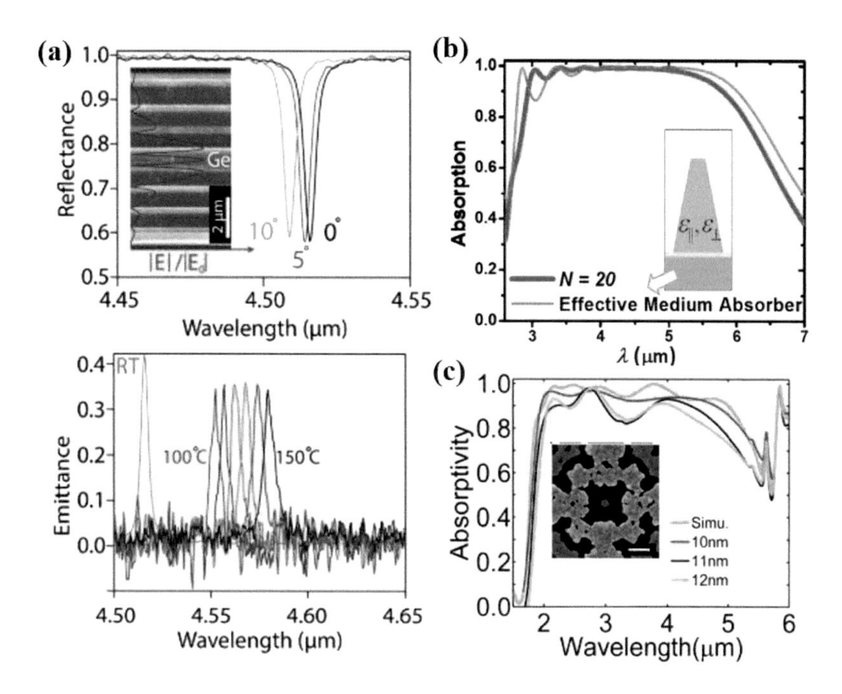

Figure 10.19 Spectral control of thermal radiation. (a) Measured reflectance spectra of the fabricated hybrid metal–OTS structure for three different incident angles (0, 5, and 10°.) under TE-polarized incident light, along with the emission spectrum of the fabricated hybrid metal–OTS structure at different temperatures. Reprinted with permission from [1024]. Copyright 2020 American Chemical Society. (b) Absorption spectra for the sawtooth AMM absorber with the number of periods $N = 20$ (thick line) and the effective homogeneous sawtooth structure that is shown in the inset (thin line) [353]. Reprinted with permission from [353]. Copyright 2012 American Chemical Society. (c) Simulation and measurements for Au-based MMA under unpolarized illumination at normal incidence with different Au thicknesses. The inset shows the top view of nanostructures. Reprinted with permission from [1025]. Copyright 2014 American Chemical Society.

750 around 4.5 μm. Dual-band emitters were also achievable in Fig. 10.19(b). By arranging several resonant scatterers with different geometric parameters, the emission waveband can be broader. Besides, the sawtoothed anisotropic structure is also a kind of metamaterial to achieve broadband performance as shown in Fig. 10.19(c). Such a design works well at a very wide range of angles as well. It can be explained that the sawtooth with different width accounts for the continuous absorption spectrum. Similarly, ultrabroadband absorption can also be realized by considering complex nanostructures as illustrated in Fig. 10.19(d) [1025], in which the measured angular absorption spectra are also polarization-insensitive.

Even though many efforts have been dedicated to spectral and angular manipulation of thermal radiation, polarization responses are also of importance.

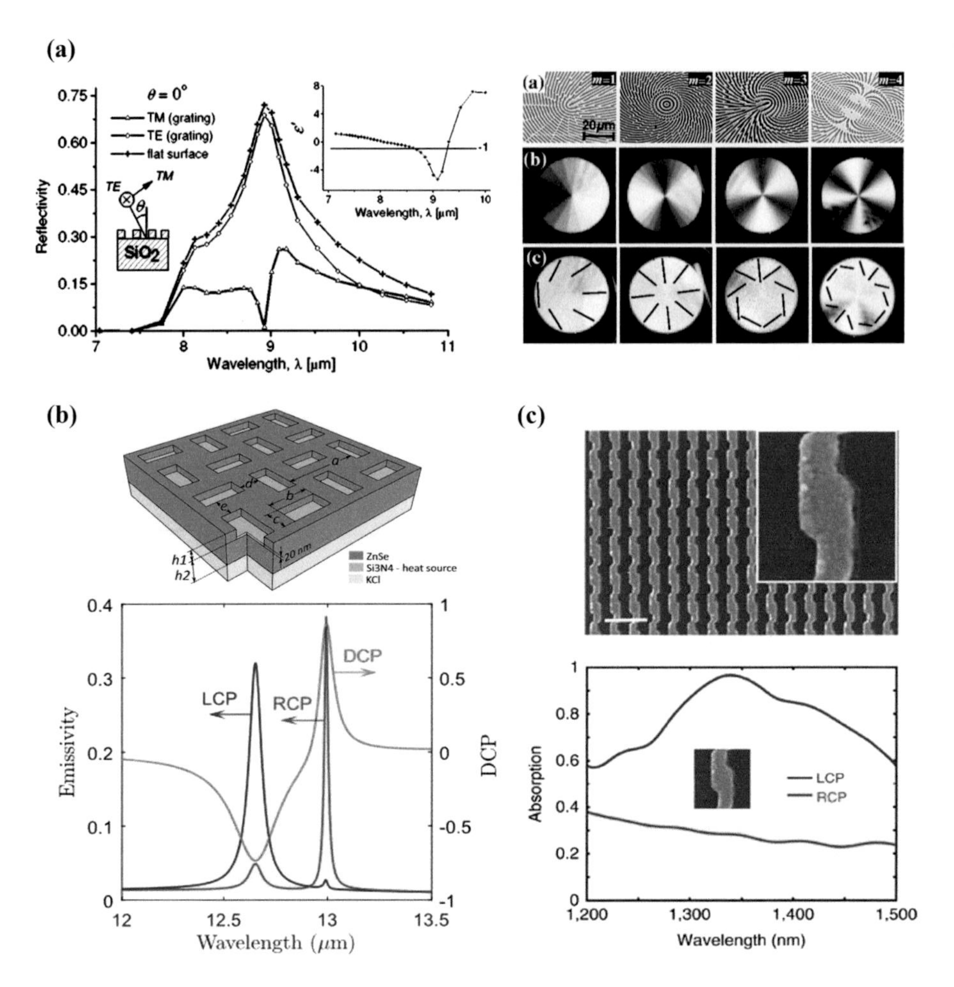

Figure 10.20 Polarization control of thermal radiation. (a) Different polarized responses of reflectivity for the TE and TM waves (left), along with the space-variant polarization manipulation of thermal emission (right). Reprinted from [1026], with the permission of AIP Publishing. (b) Proposed schematic of chiral metasurface, along with LCP (blue) and RCP (red) emissivity spectra. The green line denotes DCP. Reprinted figure with permission from [1027]. Copyright 2018 by the American Physical Society. (c) SEM images of the left-handed chiral metamaterials, along with experimentally measured optical absorption spectra under LCP (blue) and RCP (red) illumination [1028]. Reprinted from [1028]. Copyright 2015, Nature Publishing Group.

Generally, conventional thermal emitters are unpolarized. But the radiative behaviors of nanostructures become different for different polarized lights. Figure 10.20 shows some polarized-sensitive thermal emitters. As for a nanowire antenna, the optical responses to TM and TE waves are quite different [362]. Strong polarized thermal radiation can also be realized using grating, PCs, resonant cavities, and so on. By using an aperiodic array of antennae, thermal radiation with more complex polarization properties can be observed. For example, Fig. 10.20(a) shows the results of polarization manipulation using aperiodic SiO_2 gratings with different orientations. Besides, Dyakov et al. [1027] proposed that a slab ZnSe waveguide etched with chiral microstructures of fourfold the rotational symmetry can thermally emit circularly polarized radiation in a narrowband, as presented in Fig. 10.20(b). Because of the circularly polarized eigenmodes supported by such metamaterial, it is seen that the degree of circular polarization (DCP) can be as high as 0.87 at 13 μm for RCP and for LCP, the DCP reached −0.73, and the narrowband emission arises from the Fano resonance nature of the quasi-guided modes. The broad circularly polarized responses can also be obtained by using chiral plasmonic metamaterials [1028] in Fig. 10.20(c), which can be further utilized in detectors.

10.2.2 Manipulation of Near-Field Thermal Radiation

The parallel-plate configuration allows us to illustrate not only the impact of evanescent waves in the near-field regime but also the importance of the choice of materials. There are two main classes of materials when it comes to NFRHT, namely, metals and dielectrics. As an example of the NFRHT for these two types of materials, two parallel plates are made of Au and SiO_2, respectively. For Au, the NFRHT rate is dominated by TE evanescent waves that originate from eddy currents inside the Au plates. This typically leads to a saturation of the heat-transfer coefficient for small gaps. On the contrary, in the SiO_2 case, NFRHT is dominated by TM evanescent waves that can be shown to stem from SPhPs [706].

In principle, the previously discussed plate–plate configuration is ideally suited to experimentally investigating NFRHT because some of the largest enhancements in this regime are expected to occur in this setting. However, this configuration is difficult to realize in practice because it is very complicated to achieve and maintain good parallelism between macroscopic plates at nanometer separations. In recent years, several groups have developed novel techniques to explore the plate–plate configuration in the near-field regime. Some of those experiments have made use of macroscopic planar surfaces [388, 413, 415, 1029], while others are based on microscopic plates ($50 \times 50\,\mu m^2$) [387, 416–418]. With the help of microdevices to facilitate the parallelization of the systems, it has become possible to explore gaps as small as 30 nm [1030].

Distinguished from metals and dielectrics, NFRHT between nanostructures exhibits different properties. A central idea for NFRHT of nanostructures or anisotropic materials is to make better use of surface EM modes. Two-dimensional materials are revolutionizing materials science they also hold promise in the field of NFRHT. In particular, graphene has attracted much attention, as it can support SPPs that can contribute to NFRHT in spite of graphene's ultrasmall (one-atom) thickness [1030, 1031]. For example, graphene-based composite structures significantly enhance the NFRHT [718]. To further analyze the underlying

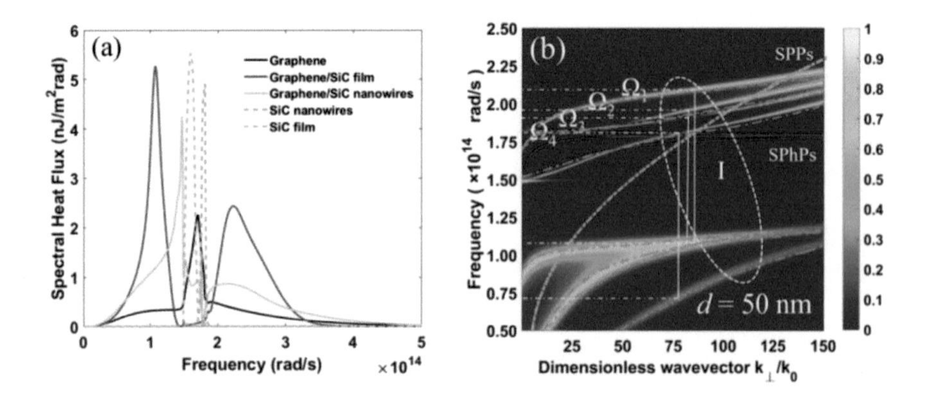

Figure 10.21 (a) Spectral heat flux of various structures. Contours of energy transmission coefficients for (b) graphene/SiC film composite structures with a gap distance of 50 nm. Reprinted figure with permission from [718]. Copyright 2019 by the American Physical Society.

mechanisms, energy transmission coefficients, and near-field dispersion relations are obtained. The dispersion relations of composite nanostructures are substantially different from those of isolated graphene, SiC films, and SiC nanowire arrays due to the strong coupling effects among surface polaritonic modes. Four pairs of strongly coupled polaritonic modes are identified with considerable Rabi frequencies in graphene/SiC film composite structures that greatly broaden the spectral peak, as illustrated in Fig. 10.21.

What makes these surface modes so attractive compared to SPhPs in polar dielectrics is the possibility of modulating them electronically [1032], which can be achieved by controlling graphene's chemical potential by means of a nearby gate electrode.

NFRHT between dissimilar materials is important for near-field applications, such as near-field photonic thermal diode and nanoscale thermal imaging. It is demonstrated that monolayer graphene leads to a significant enhancement in the NFRHT between dissimilar materials [1033]. The presence of graphene is not only able to permit a large variation and amplification of NFRHT through its chemical potential but also to fully compensate for the mismatch between the resonance frequencies of the two dissimilar materials, as illustrated in Fig. 10.22.

In general, several theoretical studies have shown that coating structures with graphene sheets may lead to a substantial increase in NFRHT [1034, 1035]. Furthermore, the role of graphene in NFRHT has been theoretically studied in a wide variety of hybrid structures [718, 1036, 1037].

Also inspired by nanophotonic concepts, NFRHT between periodically patterned systems has been intensively investigated from a theoretical point of view, both in one grating and in PCs and periodic metasurfaces. Again, the goal of such nanostructure is to tune the spectral heat transfer and enhance NFRHT. The key idea, in this case, is to use nanostructure to create new surface modes, whose

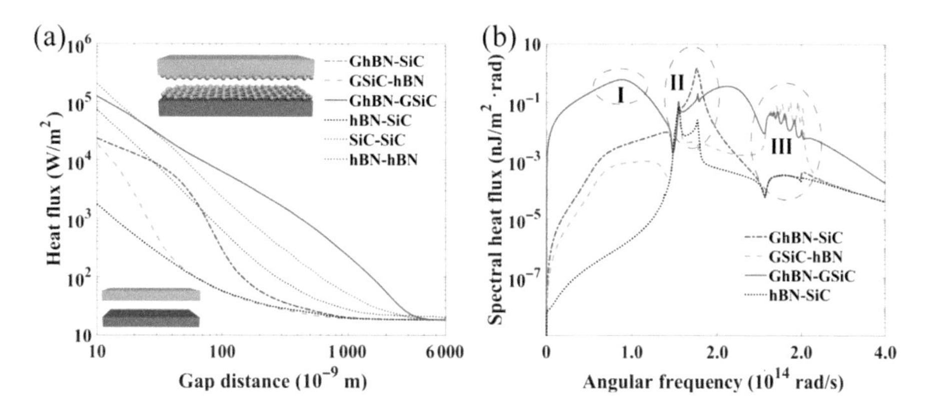

Figure 10.22 (a) Comparison of the NFRHT as a function of gap distance. The temperatures of the emitter and receiver are set at 310 K and 290 K, respectively. (b) Spectral heat flux of various configurations with a vacuum gap distance of 20 nm. The thicknesses of hBN and SiC are 50 nm, and the chemical potential of graphene is 0.3 eV. Reprinted from [1033], Copyright 2022, with permission from Elsevier.

frequencies can be adjusted by tuning the length scales of these periodic systems so that their surface modes can be thermally populated at the desired working temperature. The reported results have demonstrated the possibility of enhancing NFRHT over the corresponding planar bulk materials. However, NFRHT in these periodically patterned metallic structures continues to be smaller than that observed in simple planar polar dielectrics.

To regulate and control the spectral of NFRHT of parallel plates, particularly, to obtain high-quality quasi-monochromatic NFRHT, several physical mechanisms have been studied [1038]. Polaritons originating from plasmons or optical phonons have been demonstrated, while the peak of NFRHT is fixed near the EM resonant frequency of the material [1029]. It has been demonstrated that the quasi-monochromatic NFRHT can be achieved by a hyperbolic metamaterial layer [1039]. What's more, an adaptive hybrid Bayesian optimization algorithm has been developed to design the high-quality quasi-monochromatic or multipeak NFRHT [1040]. This is a new idea to use machine learning methods to tune the spectral of NFRHT [1041–1043].

In order to further improve the NFRHT, three-body or many-body configurations have been applied. The three-body effects in NFRHT were discussed by Messina et al. [1044]. They considered a system made of three parallel slabs, as shown in the inset of Fig. 10.23(a). The intermediate slab, of thickness δ, is placed at a distance d from the external slabs, which are assumed to have infinite thickness. This configuration is compared to the standard two-body scenario, shown in the inset of Fig. 10.23(a), where the intermediate slab is removed and d is now the distance between the external slabs. More recently patterned intermediate media [1045], 2D atomic systems [1046], and hyperbolic media [1047] have also been considered to enhance the transfers. The use of such three-body

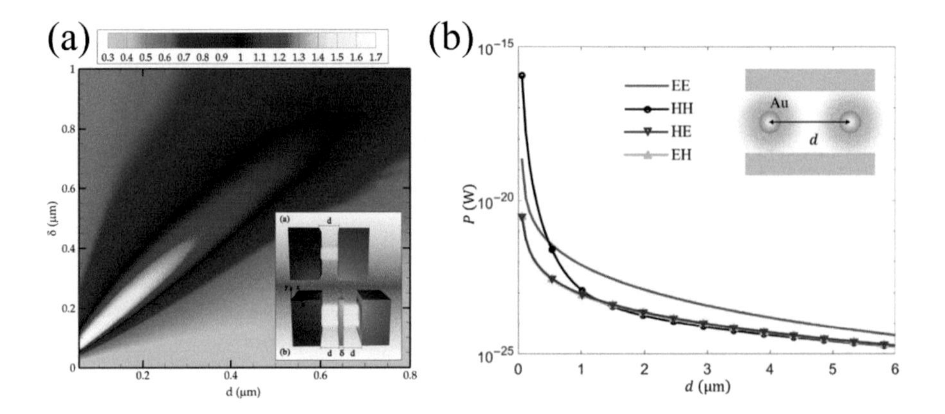

Figure 10.23 (a) Heat-flux amplification in a three-body configuration compared with a two-body configuration. Reprinted figure with permission from [1049]. Copyright 2012 by the American Physical Society. (b) The EE, HH, HE, and EH terms of the radiative heat transfer between two Au nanoparticles in case c as a function of the gap distance. Reprinted from [1050], Copyright 2022, with permission from Elsevier.

control of heat flux was proposed to design many-body heat engines [1048], with thermodynamic performances being better than their two-body counterpart and the thermal analog of the transistor [694] driven by photons.

The theoretical model can be proposed to investigate the NFRHT between two or more SiC/Au nanoparticles inside a cavity configuration composed of two semi-infinite (SiC) plates [1050]. The cavity configuration can effectively enhance the NFRHT between two nanoparticles. Compared with the configuration of two isolated nanoparticles, the maximum amplification of the cavity configuration goes beyond seven orders of magnitude for SiC nanoparticles. This amplification effect can be attributed to a strong coupling between the nanoparticle SPhPs and the planar SPhPs. Besides, the electric contribution to the NFRHT can exceed the magnetic contribution for Au nanoparticles, as illustrated in Fig. 10.23(b).

10.2.3 Nonreciprocal Thermal Radiation

The design principle of thermal emitters is always based on Kirchhoff's law, which indicates that the emissivity e at a given wavelength and angle is equal to the corresponding absorptivity α showing as $e(\lambda,\theta,\phi)=\alpha(\lambda,\theta,\phi)$, while under this rule, the systems like solar cells will possess intrinsic loss, since the same portion of solar absorbers received will be emitted to the sun at the same time. So, violating this detailed balance is of great necessity. In fact, Kirchhoff's law, which is not required by the second law of thermodynamics, intrinsically originates from the Lorentz reciprocity of Maxwell's equations [1051]. Therefore, an effective way to break this detailed balance is to break the reciprocity. Actually, the concept of nonreciprocal thermal radiation has been initially proposed in

1998, but the real demonstration has been realized recently by Zhu and Fan's work [1052] in 2014. In detail, this problem can be understood by studying the energy balance of a thermal emitter interacting with two blackbodies A and B as depicted in Fig. 10.24 for both reciprocal and nonreciprocal cases under thermal equilibrium. In theory, the emitter will radiate part of energy toward A and B, as described by e_A and e_B, respectively. At the same time, it will absorb part of the emission from A and B, as described by α_A and α_B, respectively. Then, as required by the second law of thermodynamics, the net energy flow in the whole system tends to be zero, which does not rely on the condition of either reciprocal or nonreciprocal. In detail, the emission from A is either absorbed by the emitter α_A or reflected blackbody B denoted by $r_{A\to B}$; thus, we have

$$\alpha_A + r_{A\to B} = 1. \tag{10.2.1}$$

At the same time, blackbody A will receive emission from both the emitter and part of the emission from blackbody B, given as

$$e_A + r_{B\to A} = 1. \tag{10.2.2}$$

Then, by combing Eqs. (10.2.1) and (10.2.2) and considering a similar situation for emitter B, one can obtain

$$e_A - \alpha_A = r_{A\to B} - r_{B\to A} = \alpha_B - e_B. \tag{10.2.3}$$

Apparently, as for generally reciprocal cases $r_{A\to B} = r_{B\to A}$ is consistent with Kirchhoff's law, while for nonreciprocal cases, the condition of $r_{A\to B} \neq r_{B\to A}$ results in $e_{A,B} \neq \alpha_{A,B}$.

The violation of this detailed balance will indeed offer a new strategy to engineer radiation but also shows promising prospects in several energy-related applications such as solar cells, radiative cooling, and TPV. Nevertheless, up to now, there is little attention focused on this promising area. In 2014, Zhu and Fan [1052] first designed a 1D PC-patterned InAs structure as shown in Fig. 10.24(b). The results of α and e are no longer identical when a large $\mathbf{B} = 3$ T is applied, and their difference has been maximized up to unity at an AOI 61.28° in the mid-infrared region by exciting guided mode resonances. Further, to enhance the MO effect and reduce the magnitude of $\mathbf{B}$, Zhao et al. [1053] recently proposed composite nanostructures in which a SiC grating was added over a thick InAs film. As a result, Kirchhoff's law can be largely violated at 65° under a small $\mathbf{B} = 0.3$ T as presented in Fig. 10.24.

More recently, a new class of topological materials named Weyl semimetal [1058] have gained much attention in nonreciprocal optics. Due to its unique topologically nontrivial electronic state and inherent time-symmetry breaking known as the anomaly in the Hall effect [1059, 1060], the nonreciprocal light transport can be realized without the magnetic field [1061]. Zhao et al. [1054] first demonstrated the capability of realizing nonreciprocal emission considering grating-patterned Weyl semimetal like Eu_2IrO_7 [1062] at AOI = 80°. Later on,

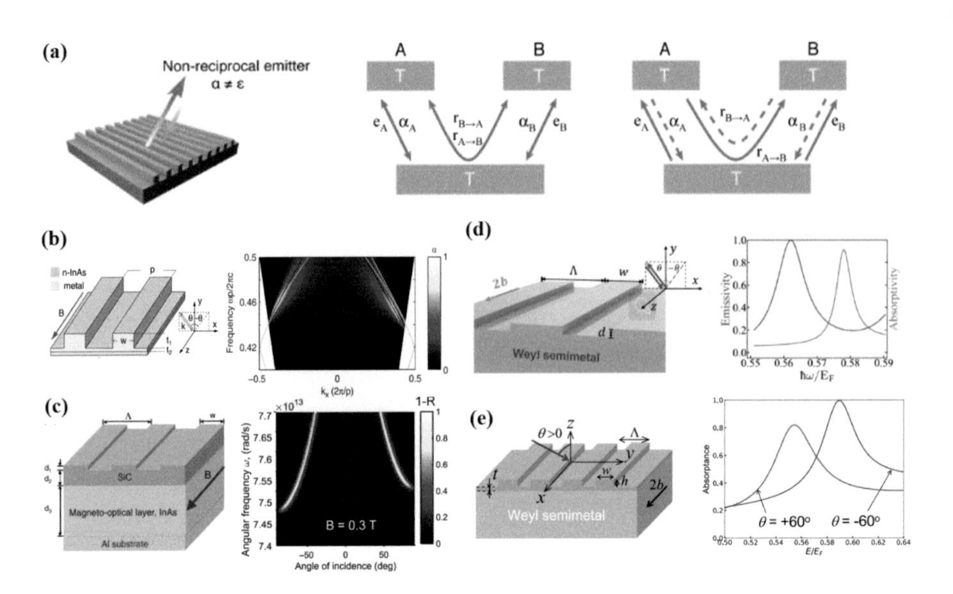

Figure 10.24 Nonreciprocal thermal radiation: breaking Kirchhoff's law of thermal radiation. (a) Schematic of nonreciprocal thermal radiation ($e \neq \alpha$). Reprinted from [232]. Copyright 2018, Optical Society of America. (b) Aschematic of an MO PC structure composed of the InAs grating (left), along with the asymmetric absorption spectrum as functions of the angle and frequency [1052]. Reprinted figure with permission from [1052]. Copyright 2014 by the American Physical Society. (c) The structure consists of (from top to bottom) a SiC grating, an MO film made of InAs, and an Al substrate proposed in [1053], with asymmetric absorption spectra on the right. (d) The Weyl-semimetal PC in [1054] (left) and absorption/emission spectra at a specific angle (right). Reprinted with permission from [1054]. Copyright 2020 American Chemical Society. (e) An optical grating structure consisting of a low-loss dielectric on top of a semi-infinite magnetic Weyl semimetal [1055]. The right-hand panel shows the asymmetric absorption and emission spectra at a fixed angle. Reprinted figure with permission from [1055]. Copyright 2020 by the American Physical Society.

the similar properties studied by Pajovic et al. [1063] have also been obtained under the incidence of 60° in composite structures, where the dielectric grating was added on a fictitious ferromagnetic Weyl semimetal. Despite novel properties without external stimulus, such intrinsic nonreciprocity appears only if the energy of the Weyl point is close to the Fermi level, which relies much on the temperature change [1064]. Especially, for the most of available Wely semimetals, extremely low working temperature is strictly required, hindering its further applications in practice.

Even though we have witnessed the rapid development of nonreciprocal thermal radiation, there is little experimental realization of asymmetric absorption performance. Until recently, Shayegan et al. [1056] first demonstrated a nonreciprocal thermal emitter with dielectric gratings atop a thick MO InAs wafer as shown in Fig. 10.25(a). By taking advantage of exciting asymmetric guided

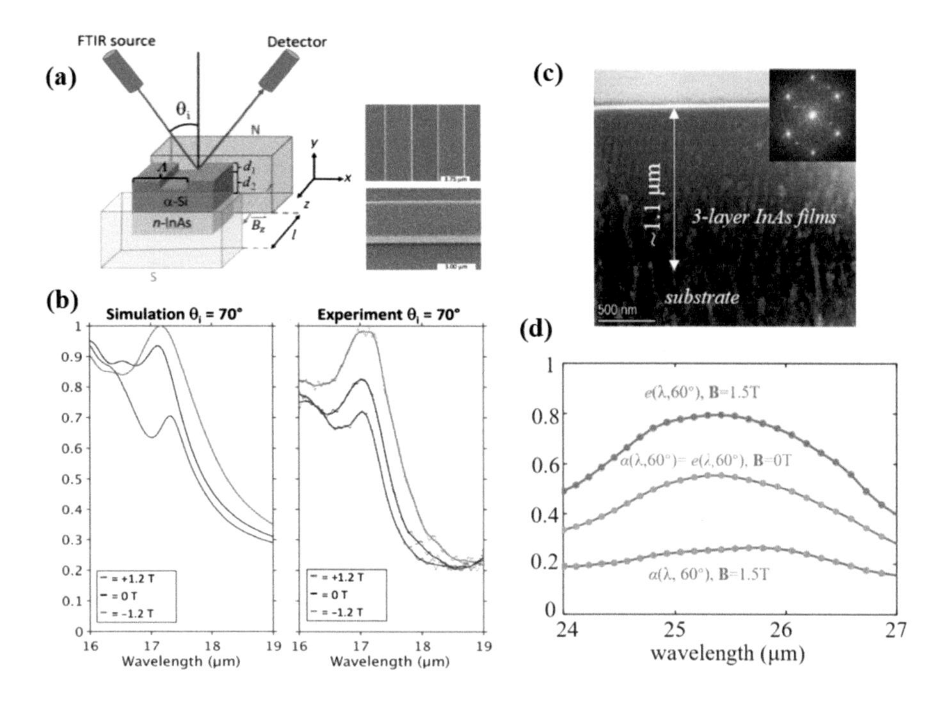

Figure 10.25 Experimental realization of nonreciprocal thermal radiation. (a) Schematic of the measurement scheme for measuring nonreciprocal reflectivity. (b) Numerical and experimental results of nonreciprocal absorption under a $B = 1.2$ T magnetic field. Reprinted from [1056]. Copyright 2022, American Association of the Advancement of Science. (c) TEM image of three-layer InAs films with gradient doping concentrations, along with (d) the measured absorption and emission spectra at a moderate $B = 1.5$ T. (c)–(d) Reprinted from [1057].

resonances, it is possible to observe nonreciprocal absorption spectra at the certain wavelength and angle. The maximum difference between $\alpha(-\theta)$ and $\alpha(\theta)$ can be 0.3 at $\mathbf{B} = 1.2$ T. While the above nonreciprocal spectra can only be observed at the fixed waveband and limited to a narrow band. Differently, Liu et al. [1057] proposed multilayered structures with gradient doping concentrations in each layer (Fig. 10.25(c)–(d)), which ensure to experimentally realize broadband nonreciprocal absorption. The presented experimental evidence provides new possibilities for renovating next-generation energy devices to break the current limit of energy harvesting and conversion.

In summary, though the existing work has presented some heuristic exploration in violating Kirchhoff's law of thermal radiation, it should admit that all the proposed strategies are based on the same mechanism by exciting guided resonances in grating configurations. In these cases, several inherent requirements, like very large angle conditions (>60°) for satisfying phase matching, cannot be ignored, which inevitably limits the manipulation flexibility and tunable range. Besides, all the studied structures usually possess large thicknesses to ensure high emission or absorption, making it challenging for cost-efficient fabrication and photonic integration. On the other hand, most of the reported works are theoretical analysis, experimental demonstration, and practical analysis have seldom been discussed. All in all, nonreciprocal thermal emission control and its practical realization are still an open challenge.

10.3 Summary

With the unprecedented development of nanofabrication and nanophotonics, we have been witnessing great processes in engineering thermal radiation on micro/nanoscales. In this chapter, an overview of micro/nanostructures to control far- or near-field thermal emission has been elaborated from two aspects. First, the tuning mechanisms of different structures are presented to reveal the deep physics of emission engineering, including grating, photonics, metamaterials, and metasurfaces, which provide versatile platforms to shape the absorption and emission properties in various aspects, that is, angular, spectral, and polarized responses. This chapter is not intended to be an exhaustive review, and it is also not possible to make it exhaustive because there are so many excellent works in the last several decades trying to tailor thermal emission using different micro/nanostructures and materials based on different design principles.

11 Applications of Micro/Nanoscale Thermal Radiation

As introduced in Chapter 10, the light–matter interaction between thermal EM waves and micro/nanostructures has provided new platforms to shape angular, spectral, and polarized properties of thermal radiation and has reopened several new cutting-edge areas, including thermal extraction, near-field thermal radiation and nonreciprocal thermal emitters. Especially, as emphasized before, metamaterials and metasurfaces have empowered and renovated several advanced technologies and applications in energy harvesting, conversion, thermal management, lighting, and so on. In this chapter, some typical metamaterials-based thermal devices will be highlighted, including thermal management devices, such as daytime radiative cooling and thermal camouflage; energy devices like TPV; and photothermal conversion devices.

11.1 Metamaterials-Based Thermal Management: Daytime Radiative Cooling

According to the classical thermal radiation theory, all materials at a temperature above absolute 0 K can emit and absorb thermal radiation continuously. This intrinsic property of any terrestrial object gives it a chance to realize temperature control of objects via the adjustment of the emissive and absorptive properties. The cold universe, at a temperature of about 3 K, is an ideal huge natural heat sink for the Earth and the terrestrial objects on it. Coincidently, the atmosphere of the Earth is intrinsically transparent to thermal radiation in some specific spectrum regions. This phenomenon is called the atmospheric transparency windows, including 8–13 μm (the first atmospheric window) and 16–24 μm (the second atmospheric window). Within the atmospheric transparency window, the thermal radiation emitted by terrestrial objects can transmit to the cold universe.

11.1.1 Fundamental Principles of Daytime Radiative Cooling

According to the heat transfer theory, the steady-state energy balance equation of a radiative cooler can be expressed as [1065]

$$P_{\mathrm{cool}}(T) = P_{\mathrm{rad}}(T) - P_{\mathrm{sun}} - P_{\mathrm{atm}}(T_{\mathrm{atm}}) - P_{\mathrm{non\text{-}rad}}, \tag{11.1.1}$$

where T_{atm} is the ambient temperature and $P_{cool}(T)$ is the net cooling power of the RC determined by four parts of energy emitted or received by the RC at a temperature of T, which are the thermal radiative emission of RC, the absorption of atmospheric radiation, the absorption of solar irradiation, and the nonradiative heat transfer, respectively.

At present, to quantify the radiative cooling performance, two parameters can be used: one is the lowest temperature when the net radiative cooling power is zero ($P_{cool} = 0$), and the other is the net radiative cooling power when the temperature of RC is the same as ambient temperature ($T = T_{atm}$).

11.1.2 Recent Progress of Daytime Radiative Cooling

Unlike nocturnal radiative coolers, the requirement of optical properties for daytime radiative cooling cannot be easily fulfilled by natural materials. The development of artificial material science provides an opportunity to manipulate the optical properties of artificial metamaterials through their material and structural properties, which makes daytime radiative cooling possible.

Before 2014, daytime radiative cooling was proven to be feasible in theory. Rephaeli et al. [1066] first reported a theoretical design of daytime radiative cooler on the basis of the combination of patterned surface and multilayer, as shown in Fig. 11.1. The simulated results showed that this RC can achieve a net cooling power above 100 W m^{-2} at ambient temperature.

In 2014, Raman et al. [1065] first realized daytime radiative cooling experimentally by a photonic structure. The multilayer film is composed of seven alternating layers of HfO_2 and SiO_2 with various thicknesses stacked on top of 200-nm Ag. With the assist of the opaque silver reflector and HfO_2/SiO_2 layers with high emittance within the atmospheric window, about 97% of incident solar irradiance can be reflected. Meanwhile, the multilayer film exhibits a selective high emissivity in the atmospheric window for strong thermal radiation. the outdoor experiment results showed that 5 °C temperature drops below ambient can be achieved under direct sunshine. Meanwhile, a net cooling power of approximately 40.1 W m^{-2} was realized at ambient temperature even under a parasitic cooling loss process. This study proved the feasibility of daytime radiative cooling and started academic research on daytime radiative cooling. Since then, various works have shown that artificial photonic structures can also be seen as metamaterials and can realize daytime radiative cooling [1066–1068]. These photonic structures can be widely applied to the cooling of electronic/optoelectronic devices and thermal devices, such as TPV, concentrating photovoltaics (CPV), and concentrated solar energy devices.

However, though these photonic structures have excellent cooling performance, it is still difficult to scale them up cost-effectively and meet the large area fabrication requirements. Later on, some daytime radiative coolers based on inorganic/organic disordered media were demonstrated, which can fulfill the large area fabrication requirements. In 2017, Zhai et al. [295] developed a

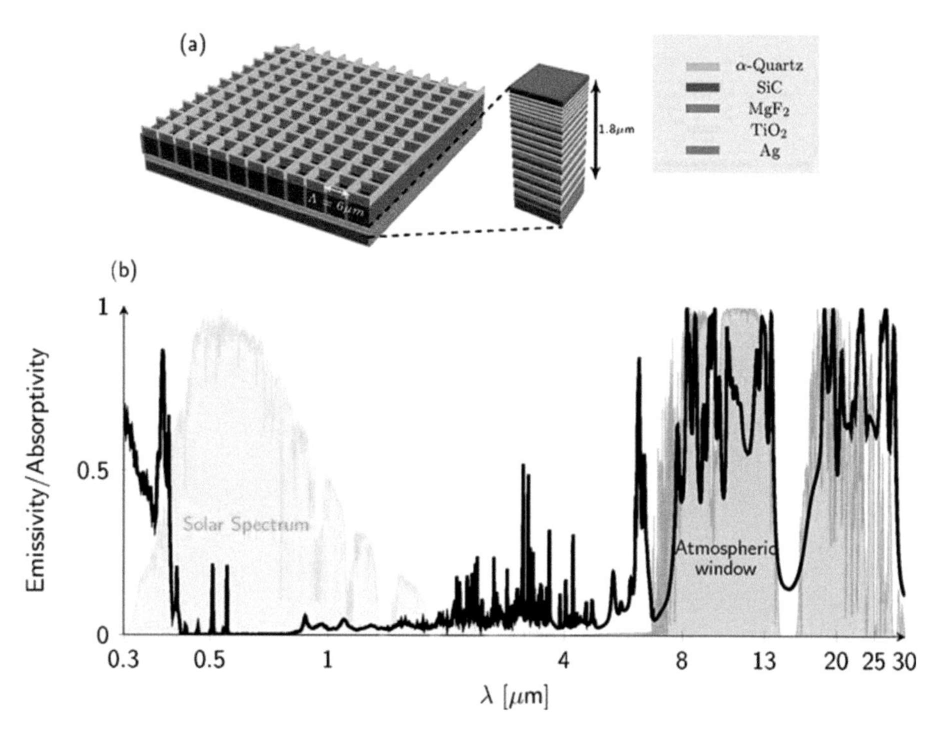

Figure 11.1 (a) Optimized daytime radiative cooler design that consists of two thermally emitting PC layers comprised of SiC and quartz, below which lies a broadband solar reflector. (b) Emissivity of the optimized daytime radiative cooler shown in (a) at normal incidence (black) with the scaled AM1.5 solar spectrum (yellow) and AT (blue) plotted for reference. Reprinted with permission from [1066]. Copyright 2013 American Chemical Society.

polymer-based metamaterial daytime radiative cooler. The radiative cooler consists of polymethylpentene as a matrix material, a 200-nm bottom Ag layer as a solar irradiance reflector, and SiO_2 microspheres with high emissivity in the atmospheric window due to the phonon-enhanced Frohlich resonances. The RC showed a net cooling power of 93 W m^{-2} under direct sunshine in the noontime. Moreover, the hybrid metamaterial can be fabricated by high throughput, economical roll-to-roll manufacturing. In the same year, Bao et al. [1069] proposed a double-layer daytime radiative cooler composed of one layer containing rutile TiO_2 particles as a solar irradiance reflector and the other layer containing SiC/SiO_2 nanoparticles as an emitter due to the SPhP induced by the nanoparticles in the atmospheric window, as shown in Fig. 11.2(a). The theoretical temperature drops of the double-layer RC can achieve 17 °C below ambient at night and 5 °C below ambient under direct solar radiation (AM 1.5), whereas the on-site measurement showed that the RC can cool the covered Al foil for 8 °C at daytime and 5 °C at nighttime, and cool the covered black surface for more than 30 °C at daytime, as shown in Fig. 11.2(b). The RC they designed can also

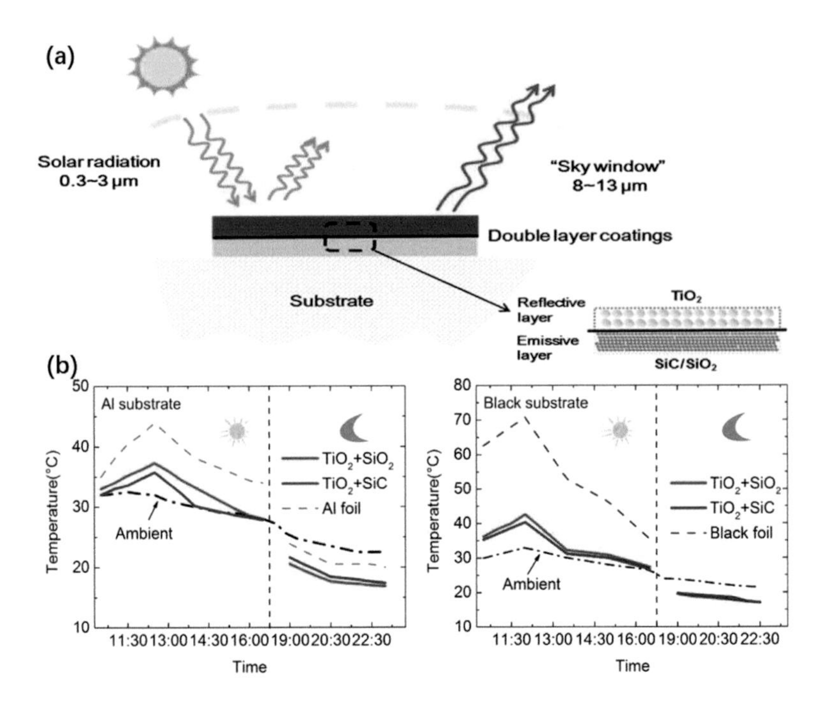

Figure 11.2 (a) Schematic of a double-layer coating design for efficient radiative cooling. (b) Measured emissive spectra of different nanoparticles from the ultraviolet to the infrared: The coatings are deposited on an Al foil (left) and the coatings are deposited on a black substrate (right). Reprinted from [1069], Copyright 2017, with permission from Elsevier.

achieve relatively low-cost and scalable production of coatings for radiative cooling purposes. These works enlightened researchers to realize a largely scalable radiative cooler based on polymer materials and randomly distributed inorganic nanoparticles. The large area fabrication of these hybrid materials makes radiative cooling have an opportunity to extend energy saving for buildings, which is a place with desirable application for daytime radiative cooling.

The existence of the bottom metal layer still makes the large area fabrication difficult and costly. And the ultraviolet absorption of TiO_2 nanoparticles makes an insufficient solar irradiance reflectance that will reduce the cooling performance of RC. As a result, the design of materials that can achieve both the metal-free substrate and the sufficient high sunlight reflection has become a research hotspot. In 2018, Mandal et al. [1070] demonstrated a novel porous polymeric radiative cooling coating on the basis of poly(vinylidene fluoride-co-hexafluoropropene) ([P(VdF-HFP)$_{HP}$]) that is a common polymer electrolyte material. The porous [P(VdF-HFP)$_{HP}$] coating can strongly reflect the sunlight without a metal reflective layer due to the high optical backscattering of the porous in the solar spectrum, which makes the solar reflectance achieve 0.96 ± 0.03. Meanwhile, [P(VdF-HFP)$_{HP}$] has intrinsic high emittance in the atmospheric window due to the functional bonds it has. The experiment showed that this coating can realize subambient temperature drops of 6 °C and cooling powers of 96 W m^{-2} under solar intensities of 890 and 750 W m^{-2}. Since then, the porous polymer has become the ideal material to realize high reflectance in the solar spectrum and high emissivity in the atmospheric window or the whole infrared region, simultaneously. Until now, novel porous polymer materials for daytime radiative cooling are presented constantly. The excellent ductility, low-cost, scalable production, and intrinsic high emissivity in the atmospheric window or the whole infrared region make the porous polymer more practical.

Polymer is a type of material with excellent multifunctional properties. The most of daytime radiative coolers based on polymer, multifunctional radiative coolers were proposed and designed in recent two years. In 2019, Li et al. [1071] designed a multifunctional passive radiative cooling material based on wood. The engineered wood is composed of multiscale cellulose fibers or fiber bundles, which can be severe as disordered scattering elements for sunlight reflectance. Meanwhile, cellulose has strong MIR emission for radiative cooling. As a result, the wood can realize daytime radiative cooling experimentally. Moreover, the mechanical strength of engineered wood is more than eight times that of natural wood. This multifunctional, scalable cooling-wood material holds promise for future energy-efficient and sustainable building applications.

In 2021, Zeng et al. [1072] reported a large-scale woven metafabrics with high emissivity (0.945) in the atmospheric window and high reflectivity (0.924) in the solar spectrum, which can be attributed to the hierarchical morphology design of the randomly dispersed scatterers throughout the metafabric. These metal fabrics can exhibit desirable mechanical strength, waterproofness, and breathability for commercial clothing while maintaining efficient radiative cooling ability. Their practical application tests purposed that a human body covered by the metal fabric could be cooled 4.8 °C lower than the one covered by the commercial cotton fabric. This work incorporates passive radiative cooling structures into personal thermal management technologies.

11.1.3 The Design Principles of a Radiative Cooler

In this section, we will introduce the design principles and optimization methods of radiative coolers according to two evaluation indicators given in Section 11.2.1, respectively.

The Design of RC in the MIR Region

For radiative cooling, the MIR radiative properties of an RC directly determine its performance. Meanwhile, under different evaluation indicators, the requirement of MIR spectra is also different. In this section, we will give a detailed introduction of the design principles of the emissivity spectra in an MIR region for RC. Two terms in Eq. (11.1.1) are determined by the MIR emissivity spectra of RCs, which are $P_{\mathrm{rad}}(T)$ and $P_{\mathrm{atm}}(T_{\mathrm{atm}})$.

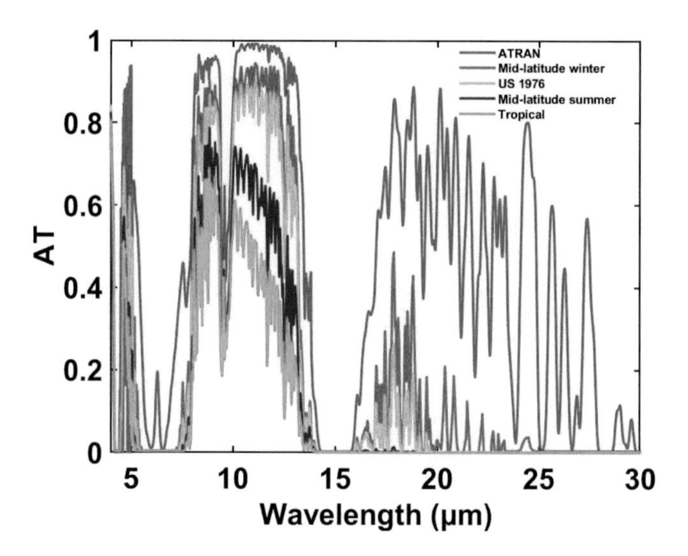

Figure 11.3 AT under five typical climatic conditions.

The radiative power emitted per area of a radiative cooler is

$$P_{\text{rad}}(T) = \int_0^{\infty} \int_0^{2\pi} \int_0^{\pi/2} \sum \varepsilon(\lambda,\theta) I_{bb}(T,\lambda) \sin\theta \cos\theta \mathrm{d}\theta \mathrm{d}\phi, \tag{11.1.2}$$

where $I_{bb}(T,\lambda) = \frac{(2hc^2)}{\lambda^5} \frac{1}{(e(hc(\lambda k_B T)) - 1)}$ is the spectral blackbody radiance of RC at temperature of T, and $\varepsilon(\lambda,\theta)$ is the spectral directional emissivity of the RC.

The absorbed power of RC due to the incident atmospheric thermal radiation $P_{\text{atm}}(T_{\text{atm}})$ is

$$P_{\text{atm}}(T) = \int_0^{\infty} \int_0^{2\pi} \int_0^{\pi/2} \sum \varepsilon_{\text{atm}}(\lambda,\theta) I_{bb}(T_{\text{atm}},\lambda) \sin\theta \cos\theta \mathrm{d}\theta \mathrm{d}\phi, \tag{11.1.3}$$

where $I_{bb}(T_{\text{atm}},\lambda)$ is the spectral blackbody radiance of atmosphere at T_{atm}. $\varepsilon_{\text{atm}}(\lambda,\theta)$ is the spectral directional emissivity of atmosphere, which can be commonly calculated by $\varepsilon_{\text{atm}}(\lambda,\theta) = 1 - t(\lambda,0)^{\wedge(1/\cos\theta)}$. $t(\lambda,0)$ is the atmospheric transmittance (AT) in the zenith direction. In this equation, $t(\lambda,0)$ is the AT in the zenith direction, which is influenced by the climatic conditions. Figure 11.3 provides the AT under five typical climatic conditions. With the increase of relative humidity and atmospheric temperature, the value of $t(\lambda,0)$ drops significantly. Then the $\varepsilon_{\text{atm}}(\lambda,\theta)$ will correspondingly increase and lead to a larger atmospheric thermal radiation P_{am}, which will weaken the achievable net radiative cooling power and temperature drops.

Therefore, the value of $P_{\text{rad}}(T) - P_{\text{atm}}(T_{\text{atm}})$ is determined by the ambient temperature (T_{atm}), the AT, and the MIR emissivity spectra of RC. For different evaluation indicators, the requirements for MIR spectra are different.

Obtaining Large Net Radiative Cooling Power at $T = T_{\text{atm}}$

To realize a considerable net radiative cooling power $P_{net} = P_{\text{rad}}(T_{\text{atm}}) - P_{\text{atm}}(T_{\text{atm}})$, the radiation emitted out from RC (P_{rad}) should be maximized and the power absorbed by RC should be minimized according to the theoretical calculation principles shown above in Section 11.1.3.

According to Planck's blackbody radiation law and Wien's displacement law, the irradiance and the radiative emission of terrestrial objects are mainly in the infrared region. As a result, it is necessary to ensure a sufficient spectral emissivity in the infrared region for radiative cooling. However, according to Kirchhoff's law, high spectral emissivity also means high spectral absorptivity, which leads to an increase in the absorption of atmospheric thermal radiation $P_{\text{atm}}(T_{\text{atm}})$. At $T = T_{\text{atm}}$, since the $\varepsilon_{\text{atm}}(\lambda, \theta) = 1 - t(\lambda, 0)^{1/\cos\theta}$ is always less than 1, the value of integrand in Eq. (11.1.2) is always greater than the integrand in Eq. (11.1.3) at all wavelengths in the MIR region (4–25 μm). Under this condition, a near-blackbody broadband emitter with a near-unity emissivity in all infrared wavelengths is more suitable for the realization of large net radiative cooling power. Therefore, the net radiative cooling power can be made large enough by selecting or designing materials with MIR emissivity as close as possible to the blackbody.

Generally, the broadband MIR near-unity emissivity can be easily realized by some macro-sized bulk materials with an MIR attenuation index in a suitable range ($k < 1$), which can avoid the impedance mismatch. Among the existing materials, organic materials, mainly polymers, can greatly meet this requirement. For organic materials, the high emissivity property is mainly caused by the vibrational resonance of functional bonds. Different functional bonds have different wavelengths of vibrational resonance. For example, C–O–C (1 260–1 110 cm^{-1}), C–OH (1 239–1 030 cm^{-1}), –CF3 (1 148 cm^{-1}), Si–O–Si (1 100 cm^{-1}), C–H (1 260 cm^{-1} of methyl and 900 cm^{-1} of C=CH_2 and monosubstituted alkenes) and C–O (1 250–1 300 cm^{-1} of carboxylic acids and 1 100–1 300 cm^{-1} of esters) have bond vibrations only in the atmospheric window (8–13 μm). By contrast, -CHO (2 810–2 710 cm^{-1}), C=C (1 880–1 785 cm^{-1}), C=O (1 825–1 725 cm^{-1}) and C-S (weak vibrational peaks at 600–650 cm^{-1}), as well as S–S (below 500 cm^{-1}) have bond vibrations outside the atmospheric window. Moreover, the MIR attenuation index arising from these vibrational resonances of functional bonds is mostly within the suitable range ($k < 1$). Here, to exhibit what optical properties of the organic materials are suitable for the realization of the broadband emitter, the refractive index of two common polymeric materials utilized for broadband RC are provided in Fig. 11.4.

For the polydimethylsiloxane (PDMS) shown in Fig. 11.4(a), the attenuation index κ appears at the wavelength about 7 μm and two attenuation index peaks at 9 and 12 μm, which is conducive to realize the high emissivity in the atmospheric window of 8–13 μm. Moreover, the attenuation index κ is in a suitable range ($0 < \kappa < 0.73$). Since the excellent properties the PDMS has, various works

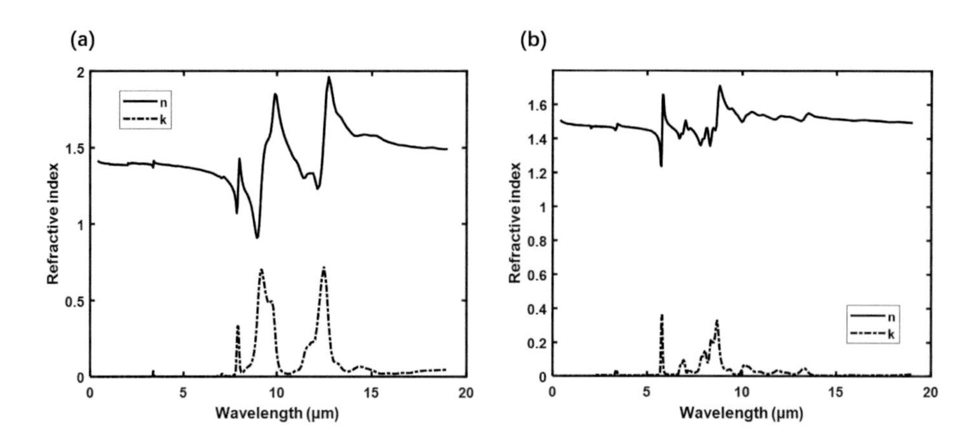

Figure 11.4 Refractive indices of (a) PDMS and (b) PMMA.

based on this material were proposed: For example, Yang et al. [1073] developed a functional radiative cooling bilayer consisting of PDMS film and highly scattering polyethylene (PE) aerogel film inspired by the fur of a polar bear. The experiment showed that 5–6 °C temperature drop can be achieved at noontime in the summer of Beijing, China. Jeon et al. [1074] proposed a multifunctional radiative cooler chosen PDMS as the radiative polymer matrix. The outdoor experiment showed that the devices can achieve 7.4 °C and 6.0 °C temperature drop below ambient.

For the polymethyl methacrylate (PMMA) shown in Fig. 11.4(b), the C=C band, C=O band, and C–O band make this polymer have attenuation index κ in the whole IR region with a suitable range ($0 < \kappa < 0.4$), which makes it have broadband emission property as a bulk matrix material. Some daytime radiative cooler coating for energy-saving buildings adopted this polymer as matrix material due to its low-cost and good coating ability. For example, Wang et al. [1075] reported a hierarchically porous array PMMA (PMM_{AHPA}) film with a close-packed micropore array combined with random nanopores which can exhibit a high reflectance (0.95) in the solar spectrum and broadband high emittance (0.98) in infrared spectrum for daytime radiative cooling. The experiment result showed that the below average ambient temperature of the PMM_{AHPA} film is 8.2 °C during the night and 6.0 °C during midday in the autumn of Shanghai. China. Li et al. [1076] demonstrated a sub-ambient daytime radiative cooling paint composed of acrylic as a matrix for sky window emissivity and $CaCO_3$ with a high particle concentration of 0.6 and a broad size distribution mainly for achieving high sunlight reflection. The benefits of cost-effective and convenient paint form make this $CaCO_3$-acrylic radiative cooling paint have the potential to be utilized in practice.

In addition to the organic materials, some works designed photonic structures to obtain broadband emission in IR region. For instance, Li et al. [1077] have demonstrated 1D photonic films consisting of alternating layers of $Al_2O_3/SiN/TiO_2/SiN$ with aperiodic arrangement of thickness ($N = 11$) and a single SiO_2 on the top for antireflection. The photonic films can lower the temperature of solar cells by over 5.7 K. Some metamaterials are also purposed for broadband emission. Wu et al. [1078] and Kong et al. [1079] have adopted the moth-eye structure to achieve broadband emission in IR region. However, the fabrication of these RCs is complex and costly, restricting them from large-area fabrication at present.

Obtaining Large Temperature Drops at $P_{\text{cool}} = 0$

Since the AT $t(\lambda, 0)$ is near-zero out of the atmospheric windows, the spectral emissivity of the atmosphere is near-unity in the wavelengths outside the atmospheric windows according to $\varepsilon_{\text{atm}}(\lambda, \theta) = 1 - t(\lambda, 0)^{1/\cos\theta}$. Therefore, when the temperature of RC is lower than T_{atm}, the value of the integrand in Eq. (11.1.3) is always larger than the integrand in Eq. (11.1.2) in the wavelengths outside the atmospheric windows, which will impair the achievable temperature drops at $P_{\text{cool}} = 0$. Therefore, to obtain large temperature drops at $P_{\text{cool}} = 0$, a selective emitter (SE) with high emissivity in the atmospheric windows is more suitable. An ideal SE for radiative cooling has near-unity spectral emissivity in the wavelengths of atmospheric window (8–13 μm), and near-zero emissivity in other wavelengths, minimizing the absorbed atmospheric radiation.

To achieve this design goal, some researchers demonstrated novel polymer-based radiative coolers. For example, poly(vinylidene fluoride-co-hexafluoropropene) ([P(VdF-HFP)$_{\text{HP}}$]) are utilized as the material of an SE due to the –CF3 band it has by Mandal et al. shown above. The RC can realize selective high emission in the wavelengths of 8–18 μm. And some Similar polymers with the –CF3 band, such as polyvinylidene fluoride (PVDF), and polytetrafluoroethylene (PTEF) are applied to the MIR emissive matrix of RC as well. Li et al. chose and designed polyethylene oxide (PEO) as an ideal polymer that can present desirable selective absorption bands overlapping the atmospheric window since it only contains C–C, C–O, and C–H bonds for emission. As expected, the PEO film only has high broadband emissivity in the wavelengths of 7–12 μm, closer to the ideal SE for radiative cooling than the ([P(VdF-HFP)$_{\text{HP}}$]) film proposed by Mandal et al. [1070]. Banik et al. [1080] used silicon oxycarbonitride (SiCNO) emitter coating consisting of Si–CH3, Si–O–Si, Si–N–Si, Si–C bonds for thermal emission, which can also realize desirable selective emission in 8–13 μm atmospheric window and achieve a temperature reduction of 6.8 °C below ambient temperature.

There are also many SEs based on multilayered inorganic thin films. Inorganic materials including silicon-based materials, metallic oxides, carbon-based materials, and metallic acid salt have been selected to design the SE. For these materials, the thermal emission is mainly caused by the interaction between photon and optical phonon. Generally, the inorganic SE for radiative cooling can be realized based on these materials via some optimization methods. Numerous studies have designed different SEs based on these materials. For instance, the multilayered RC demonstrated by Raman et al. [1065] are designed based on the SiO_2/HfO_2 layers with submicron thickness and needle optimization method. Chae et al. [1067] used particle swarm optimization to perform optimization of RC composed of Al_2O_3, Si_3N_4, and SiO_2 on a silver substrate. The optimized RC has a low solar and a high mid-IR emissivity of 0.91 in the wavelength region of 8–13 μm. However, these RCs cannot achieve selective high emission only within the atmospheric window of 8–13 μm, which is probably attributed to the broad attenuation index peak of HfO_2 and Al_2O_3 extending to wavelengths outside the atmospheric window. In this view, the SiO_2, Si_3N_4, and SiC are more suitable for the realization due to their narrow attenuation index peak within the atmospheric window (8–13 μm). For instance, Chen et al. [1081] purposed a highly selective RC matched to the atmospheric transparency window, consisting of Si_3N_4/Si layers and Al reflector. This SE can achieve a maximal temperature reduction of 42 °C during daytime compared to the ambient when the parasitic heat losses are minimized. Mira et al. [1082] designed highly SEs on the basis of Si_3N_4, SiC, and SiO_2 via memetic algorithm, which can provide a cooling temperature of 49.8 and 44 K with a cooling power of 100.72 and 112.27 W m^{-2} at equilibrium.

Except for 1D organic/inorganic materials and thin films, some metamaterials are also purposed for the design of SE. For example, the first theoretical daytime radiative cooler demonstrated by Rephaeli et al. [1066] used a combination of a two-layer 2D patterned surface and a chirped multilayer. Hossain et al. [1083] also used a specially patterned surface consisting of an array of symmetrically shaped conical metamaterial pillars, each comprising multilayers of Al and Ge. This metamaterial has a near-ideal emission in the 8–13 μm wavelength range. Meanwhile, when the inorganic materials mentioned above are made into nanoparticles, the induced SPhP leading to a strong and selective optical absorption can be utilized to design a highly SE. Liu et al. [1084] reported a reconfigurable photonic structure composed of nanoparticle-embedded PDMS thin film. The SiC, Si_3N_4, and BN nanoparticles inside the RC have separate extinction coefficient peaks from 8 to 13 μm (SiC: 12.8 μm; Si_3N_4: 8.5 and 12.5 μm; and BN: 7.09 and 12.45 μm), resulting in highly selective thermal emission under the excitation of SPhP in these wavelengths.

At present, most of the research on SE mainly focuses on spectral design. However, some studies have shown that some designed angular SEs can achieve larger temperature drops than the omnidirectional ideal SE. At higher angles, measured from the zenith, the atmosphere's emissivity increases as it becomes thicker. At these high incidence angles, the value of the integrand in Eq. (11.1.3) is larger than the integrand in Eq. (11.1.2) when $T < T_{\mathrm{atm}}$, impairing the achievable largest temperature drops. Therefore, reducing the emissivity to 0 at these high incidence angles will help to further reduce the atmospheric radiation absorption and achieve larger temperature drops. Jeon et al. [1085] have quantitatively illustrated that significant net radiative absorption at high zenith angles

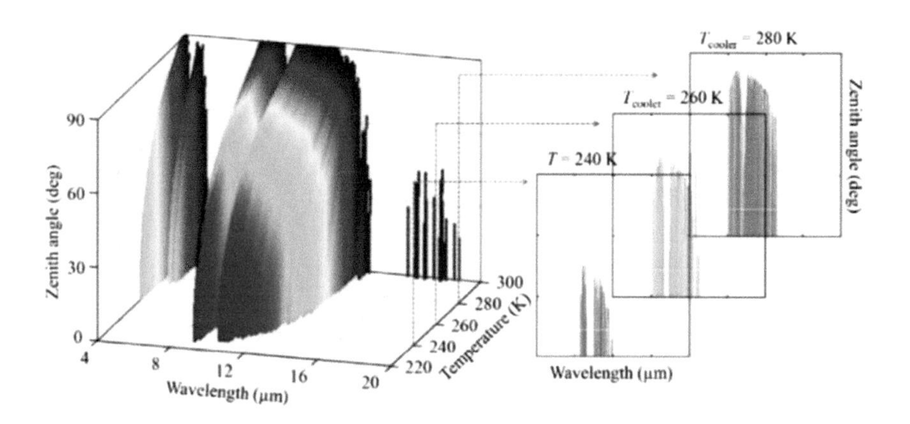

Figure 11.5 Ideal spectral directional emissivity depending on the cooler's temperature, where the colored region presents unit emissivity. Reprinted from [1085].

limits the performance of such isotropic emitters. Then, they demonstrated that simply cutting off components corresponding to high angles can substantially improve the cooling performance of commonly used isotropic emitter designs. The ideal spectral directional emissivity depending on the RC's temperature T is given in Fig. 11.5. With the increase of the desirable temperature drops, the cutoff incidence angle correspondingly decreases. Chamoli et al. [1086] also investigated how angular selective thermal emission can improve radiative cooling under various conditions by considering the effect of angular and spectral selectivity on the cooling performance in ideal conditions, environments with high humidity and low atmospheric transparency, and with parasitic heating. Based on this analysis, they designed a 10 layer-SE, which can achieve a temperature drop of 40 °C below ambient temperatures under experimentally realizable conditions.

The Comparison of Two Ideal Emitters for Radiative Cooling

To clearly show and compare the performance of two ideal RC, we calculate the temperature drops and net cooling power of the ideal SE emitter and ideal blackbody emitter under the identical AT (Mid-latitude winter calculated by MODTRAN model) and ambient temperature ($T_{\mathrm{atm}} = 20$ °C), as shown in Fig. 11.6. The ideal blackbody emitter has larger net radiative cooling power (100 W m^{-2}) than the ideal SE (90 W m^{-2}) at $T = T_{\mathrm{atm}}$. Then, as the RC temperature decreases, the P_{net} of the ideal blackbody emitter decreases faster than the ideal SE due to its larger P_{atm}($P_{atm-BB} = 268$ W m^{-2}, while $P_{atm-SE} = 42$ W m^{-2}). Therefore, the ideal SE can achieve much larger temperature drops (60 °C) than the ideal blackbody emitter (20 °C). Moreover, the high absorption of ideal blackbody emitter in the wavelengths other than the atmospheric window can also bring an impairment of the radiative cooling performance when the RC is in

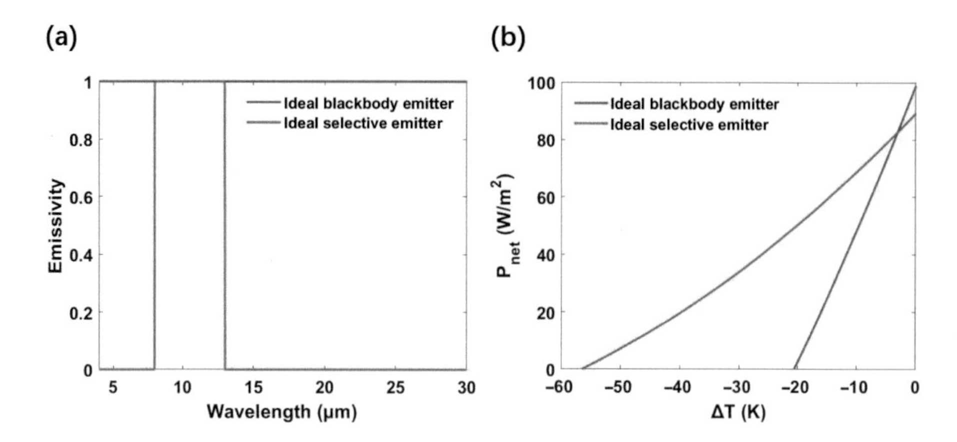

Figure 11.6 (a) Conventional ideal emitter for radiative cooling. (b) The radiative cooling performance of two ideal emitters

a complex environment receiving thermal radiation from surrounding structures. Fan and Li [1087] also provided a detailed comparison of the radiative cooling performance between the ideal blackbody emitter, ideal SE, and optimized angular SE. The optimized angular SE can achieve a temperature drop of 100 °C, much higher than the conventional ideal SE and blackbody emitter.

The Influence of Nonradiative Heat Transfer

In practical applications, the nonradiative heat losses caused by convection and heat conduction are not negligible. The nonradiative absorbed power $P_{\text{non-rad}}$ due to convection and conduction can be calculated by

$$P_{\text{non-rad}} = h_c(T_{\text{atm}} - T), \tag{11.1.4}$$

where h_c is the nonradiative heat transfer coefficient containing the effect of convection and conduction. The value of h_c can be calculated by the wind speed u_a as [1088]

$$h_c = 2.8 + 3.0u_a, \tag{11.1.5}$$

where h_c can also be calculated by the lumped capacitance method [1089], and some specific values calculated by previous works, such as 6.9 W m^{-2} based on the energy-balance principle. Generally, the value range of h_c is from 0 W m^{-2} (ideal situation without nonradiative heat transfer) to 15 W m^{-2}.

The value of nonradiative heat transfer will directly affect the performance of radiative cooling and the comparison between the two ideal emitters shown above. With the increase of the nonradiative heat transfer coefficient, the achievable temperature drops of two emitters will significantly decrease, impairing the radiative cooling performance. Meanwhile, the SE is more affected by the nonradiative heat transfer due to its smaller P_{rad} than the ideal BB emitter. Therefore,

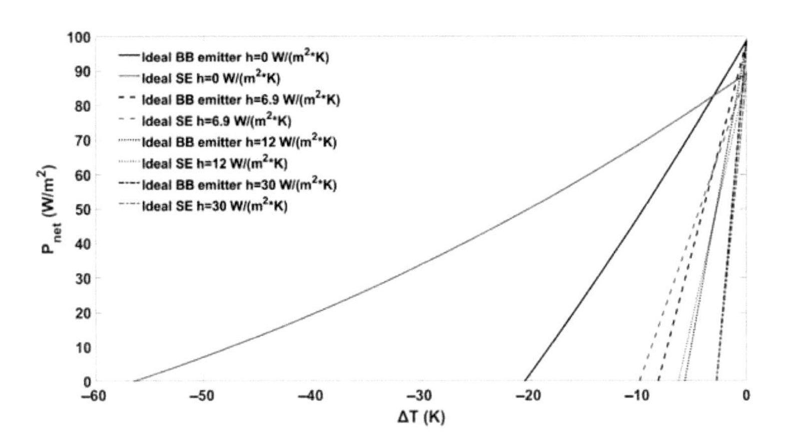

Figure 11.7 The cooling performance of two ideal RCs under different nonradiative heat transfer coefficients.

the difference between the temperature drops of two ideal emitters will significantly decrease, as shown in Fig. 11.7.

To avoid the influence of nonradiative heat transfer and accurately estimate the radiative cooling performance by the measured temperature drops, researchers usually use materials with high transmittance in the VIS and IR region (such as H/LDPE film, pigmented PE foils, and ZnSe glass) as wind cover to seal the RC in a chamber during the outdoor experiments, which can limit the nonradiative heat transfer coefficient in the suitable range.

The Design of RC in the Solar Wavelengths

For RC, the difference between daytime radiative cooling and nocturnal radiative cooling is mainly due to solar radiation absorption (P_{sun}). Obviously, compared to nocturnal radiative cooling, daytime radiative cooling is more difficult to realize due to the absorption of solar radiation. The sun at a temperature of 5 800 K has a large irradiance up to 1 000 W m^{-2}, which is about an order of magnitude larger than the irradiance of terrestrial objects at a temperature of 300 K. The large power-density mismatch between the solar irradiation and the thermal radiation of terrestrial objects clearly requires a daytime radiative cooler to have a very low solar absorptivity, which means a requirement of high reflectance in the solar spectrum region due to the opaque of the most RC. The absorbed power of RC due to the incident solar radiation (P_{sun}) can be calculated by

$$P_{\text{sun}} = \int_0^{\infty} \varepsilon(\lambda, \theta_{\text{sun}}) I_{AM1.5}(\lambda) \mathrm{d}\lambda, \tag{11.1.6}$$

where $I_{AM1.5}(\lambda)$ is the solar radiance power, θ_{sun} is the angle at which RC faces the sun, and $\varepsilon(\lambda, \theta_{\text{sun}})$ is the spectral emissivity of the RC at the angle of $\theta_{\text{sun}} = 0°$. Commonly, $\varepsilon(\lambda, \theta_{\text{sun}})$ is represented by the spectral emissivity of the RC under the normal incident $\varepsilon(\lambda, 0)$, where $\theta_{\text{sun}} = 0°$. Actually, the term of

P_{sun} is used only for the energy balance of the daytime radiative cooler. For the nocturnal radiative cooler, P_{sun} can be easily considered as 0.

Typically, the high reflectance of sunlight can be realized by the bottom metal layer or strong scattering nanoparticles and pores. Metal such as Ag, Al and Cu are the most common materials for the Back-reflector. Due to the high reflectance in the whole spectrum region, a metal back-reflector is the ideal natural material to fulfill the requirement of high sunlight reflectance for daytime radiative cooling. Metal back-reflector is suitable for combining with all the materials for MIR emission shown above. By adding a metal back-reflector to the emitter, daytime radiative cooling can be easily realized. However, the metal back-reflector is too costly and not suitable for large-area fabrication, meanwhile, the high reflectivity in the whole spectrum region restricts the structural design (the metal layer can only be placed at the bottom).

Strong scattering nanoparticles, such as TiO_2 nanoparticles can be applied to fulfill the requirement of high sunlight reflectance as the substitute for metal back-reflector. TiO_2 has a wide bandgap, stable chemical properties, high refractive index and being transparent to most infrared radiation. The submicron anatase and rutile TiO_2 particles could be solar reflector to scatter sunlight efficiently, both under independent scattering and dependent scattering. Bao and Huang utilized submicron rutile TiO_2 particles-embedded layers for sunlight reflectance, one is densely packed (dependent scattering) and the other is with a volume fraction at a small value 5 % (independent scattering). Both these two layers have excellent reflective performance in the solar spectrum region and realize daytime radiative cooling theoretically. However, the ultraviolet absorption of TiO_2 nanoparticles make an insufficient solar irradiance reflectance which will reduce the cooling performance of RC. As a result, some other inorganic nanoparticles are applied to realize strong scattering in the solar spectrum under the dependent scattering, such as $CaCO_3$, $BaSO_4$, and $Mg_{11}(HPO3)_8(OH)_6$, which are transparent in the solar spectrum region.

Except for nanoparticles, the high reflectance of sunlight can also be realized by the nanopores or micropores, which can strongly scatter the light in the solar spectrum. The strong-scattering property of pores make the polymers have ability to realize MIR emission and sunlight reflection simultaneously by themselves. Since the novel porous polymeric radiative cooling coating based on poly(vinylidene fluoride-co-hexafluoropropene) ($[P(VdF\text{-}HFP)_{HP}]$) proposed by Mandal et al., various works were exhibited to the design of porous polymers. For example, Zhou et al. presented a nanoporous polymer matrix composite (PMC) which enable fabrication by commercially polymer processing techniques such as molding, extrusion, and 3D printing. The PMC is composed of nanoporous polyethylene as a polymer matrix and randomly disturbed SiO_2 particles as the radiative emitter in the atmospheric window. The experiment showed that a temperature drop of 6.1 °C can be realized a solar irradiance of 747 W m^{-2} on a sunny day in Champaign, Illinois. Up to now, the design of porous polymers for RC is still a hotspot in the domain of daytime radiative cooling

Energy-efficient Buildings

Building energy consumption accounts for approximately 40% of the total energy consumption of the world, where a large amount of energy is used for indoor thermal management via conventional HVAC systems. As a result, the passive radiative cooling method that cools objects without additional energy input can make a difference in establishing energy-efficient buildings. Wang et al. [1090] proposed a photonic radiative cooling system for office buildings and estimated the corresponding energy savings. Simulation results showed that the electricity saving is between 45% and 68% relative to conventional cooling methods. Goldstein et al. [1091] experimentally demonstrated a diurnal nonevaporative fluid radiative cooling system that passively achieves 5 ° below ambient air temperature. Smith et al. [1092] proposed a similar idea of diurnal radiative cooling and air-conditioning integration.

Thermal Device Cooling

Some thermal devices, such as PV, TPV and concentrating PV systems suffer from the excessive thermal load which will limit the energy conversion efficiency. The passive radiative cooling method can be applied to resolve this problem. Lu et al. [1093] developed a universal routine to improve the radiative cooling ability of silicon solar cells by adding ultra-broadband versatile textures on the cells. The average emissivity of the modified solar cell within 8–13 μm is improved above 0.96 from spectral testing and the PV efficiency is also increased by 3.13% relative to the commercial glass encapsulated PV module. Wang et al. [1094] utilized the passive radiative cooler for the CPV system cooling. Their experiment results showed that by applying a cheap, passive, and lightweight radiative cooler on top of the traditional cooling approaches, 36 °C temperature drop of the solar cell operating temperature under a heat load of 6.1 W can be achieved, which can lead to a 31% increase of open-circuit voltage for GaSb cell and 4 to 15 times predicted lifetime extension. Cho et al. [1095] developed a cost-effective, wafer-scale, on-chip radiative cooler for the cooling of concentrated solar energy devices. This radiative cooler can exhibit high emissivity via photon-tunneling-mediated optical resonance. The experimental results showed that the cooler can decrease 31 °C of the temperature of a Si wafer, which indicates an improvement of the semiconductor solar-cell efficiency by more than 5%.

Water Harvesting

From the second half of the last century to the beginning of this century, researchers have been exploring the use of radiative cooling to condense the water vapor in the atmosphere to obtain liquid water. With the development of RC in recent years, water harvesting of RC has been demonstrated in some works. Zhou et al. [1096] presented a comparative experiment using high-performance RC for water vapor condensation, the results showed that the water harvesting device based on the RC they designed has larger water production than commercial materials. Li et al. [1097] utilized radiative cooling from solar panels for nighttime water harvesting applications. With the assistance of nighttime radiative cooling, the solar panels can realize water generation and self-cleaning. Moreover, the collected water can also be used for other applications including agrophotovoltaic and evaporative cooling of solar panels during the day and can be extended to other solar energy harvesting systems.

Personal Thermal Management

Personal thermal management by passive radiative cooling and heating is an emerging topic in the engineering field and has huge potential to considerably reduce fossil energy consumption. For example, radiative cooling can be applied to the design of smart cloths, Cai et al. [1098] designed a novel textile based on nanoporous metalized polyethylene to meet the spectral selective demand for passive heating, which can enable more than 7 °C skin temperature reduction compared with that of normal textile. Hsu et al. [1099] designed a dual-mode textile combining radiative cooling and heating for the human body based on the contribution of single-mode radiative cooling and heating via a bilayer radiator embedded inside a nanoPE layer.

Radiative cooling can be utilized in the thermal management of wearable devices, Lee et al. first presented a colored radiative cooler composed of MIM structure as colorant structure and SiO_2/Si_3N_4 bilayer as an SE. By depositing on an Al foil and encapsulated by PDMS film, the colored radiative cooler can be applied to the electronics of wearable devices. Kang et al. [1100] presented a novel tissue oximeter integrating nano/microvoids polymer (NMVP) and a self-produced patch-type tissue oximeter (PTO). The experimental results showed that the novel PTO can realize sub-ambient daytime radiative cooling and provide reliable measurement in tissue oxygenation. Moreover, compared to the commercial PTO, the novel PTO can eliminate the thermal interference outdoors, which can provide more accurate results for measurement.

Light Generation

Passive power generation is a typical potential application of radiative cooling. In 2019, light generation was realized by nocturnal radiative cooling. Raman et al. [1101] designed a thermal system that can use the temperature difference between the cold side cooling by passive radiative cooling and the hold side heated by the surrounding air of thermoelectric generators to generate electricity at night. The experiment showed that 25 mW m^{-2} of power generation can be realized, which is sufficient for an LED.

11.1.4 Discussion and Prospects

Although there have been great advances in the study of radiative cooling in recent years, some limitations also exist, which prevent it being directly applicable to production and life. Firstly, the performance of daytime radiative cooling is restricted by environmental conditions, including solar irradiance and atmospheric conditions. Actually, the abundant atmosphere water vapor can significantly absorb infrared (IR) radiation and close the transparent window; thus, radiative cooling performance is largely limited in humid places. As a result, daytime radiative cooling is more suitable in dry regions such as North America, the Middle East, and Northwestern China. Secondly, the net cooling power of daytime radiative cooling is lower compared to the active cooling device at present. The maximum net cooling power of most RC is only about 60–80 W m^{-2} which makes the radiative cooling not able to up to the enormous demand for energy. Thirdly, the fabrication of daytime radiative coolers is still complex and costly, especially the RC based on photonic structures and metamaterials. The difficulties in processing make it difficult for daytime radiative cooling to be applied in large-area fabrication at present. Moreover, most of the radiative coolers at present are single-functional and with the color of white or silver due to the high reflectance of visible light, restricting the possible applications and aesthetics of daytime radiative cooling.

To break the current regional limitations of daytime radiative cooling, ultra-high solar reflectance radiative coolers should be designed and fabricated for cooling in humid places. The candidate materials such as porous polymers, inorganic particles-based paintable coating, and even metal-reflector-based multilayers can be extensively investigated to fulfill the requirement. Moreover, due to the excellent multifunctional property the polymers have, it is possible to solve the single-functional drawback of daytime radiative cooling at present. Multifunctional daytime radiative coolers, not just based on polymeric materials, should be extensively investigated in the future. In addition, to fulfill the aesthetic requirement of daytime radiative cooling, novel photonic structures and materials that can delicately balance the tradeoff between coloration and cooling performance should be explored.

11.2 Metamaterials-Based Thermal Management: Thermal Camouflage

In nature, many animals possess the ability to camouflage by blending themselves perfectly into the surrounding environment, such as a chameleon, squid, frog, and so on. Such natural phenomena are instructive to develop camouflage technologies, showing a wide application in several areas, including military weapons, anti-counterfeiting, thermal management, etc. Thermal camouflage, in particular, has prospered over the past decades owing to brilliant properties of metamaterials and metasurfaces. In general, there are two basic strategies to realize thermal camouflage: one is to change the local temperature, and the other is to control the local emissivity of protected objects. So, thermal camouflage can be categorized into conductive and radiative thermal camouflage, respectively. Usually, the realization of conduction-typed camouflage is on the basis of transformation thermotics. In this part, we will mainly focus on the recent progress of radiative camouflage.

11.2.1 Basic Principles

Radiative thermal camouflage is designed to prevent detection from IR cameras. When the background temperature is below the object temperature, an effective way to realize camouflage is to reduce the emissivity of objects to decrease thermal radiation energy. According to the Stefan–Boltzmann law, the radiation power is proportional to the fourth power of temperature, which means that there is a large portion of energy even with a small temperature rise, destroying camouflage performance. Besides, reducing emissivity will increase the corresponding reflection, which can also be detected from reflected signals. Then, wavelength-selective thermal emitters based on various metamaterials have provided a feasible platform to solve these problems.

Figure 11.8 shows the atmosphere transmittance curves in the blue region, where there are two wavebands with high transmittance: mid-wavelength ranges ($3-5$ μm) and long-wavelength ranges ($8-14$ μm). These two regions are well known as atmosphere windows, in which most of IR detectors work within the waveband. As introduced above, the objective of thermal camouflage is to keep the thermal balance of the background and objects, which suggests the ideal emission spectra match with atmosphere transmittance curves. According to the temperature difference between the background and objects, there are also two basic rules to design the emissivity of the camouflaged object. When the

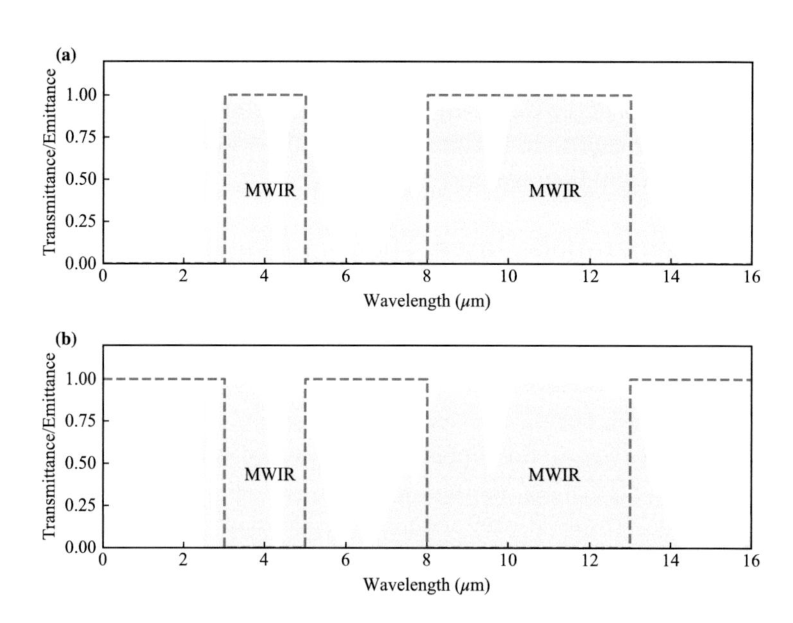

Figure 11.8 Ideal emissivity spectrum for thermal camouflage at (a) high and (b) low background temperatures. Shaded regions denote the atmosphere transmittance.

background temperature is lower than the objects, in this case, the thermal radiation of objects should be reduced to blend into the background. It means that the emissivity should be small enough among two atmosphere windows as shown in Fig. 11.8(a). At the same time, the emissivity outside the two ranges can be high, resulting in cooling the surface for better radiative heat dissipation. By contrast, if the background temperature is high, the condition changes. In this regard, within the two atmosphere windows, the emissivity of the objects should be high as shown in Fig. 10.10(b), and the emissivity of other wavebands can be low to reduce the heat dissipation. Such types of wavelength SEs can be realized by artificial nanostructures, like metamaterials or metasurfaces.

11.2.2 Radiative Thermal Camouflage

Based on the concepts mentioned above, many efforts have been dedicated to emissivity-engineered thermal camouflage. One seminal work is by Mohammad et al. [1102]. As shown in Figs. 10.11(a)–(b), the authors proposed and designed silicon nanowires and Ag nanoparticles on a flexible polyimide substrate to achieve wideband thermal camouflage. The fabricated samples can absorb or scatter a broad band of infrared light covering 2.5–15.5 μm, thus enabling to hide infrared signals toward the detectors. Figure 10.11(b) (bottom left) gives the compared measured reflection and transmission spectra, which indicates high absorption performance of nanowire structures possessing better light-trapping ability. The two right-hand panels in Figs. 10.11(a)–(b) show the thermal camouflage performance of the designed sample, where the temperature of positions covered with camouflage emitter is nearly the same as the background, showing excellent camouflage properties. Recently, thanks to the rapid development of MXene, Li et al. [1103] demonstrated high-temperature indoor/outdoor thermal camouflage using a series of ultrathin $Ti_3C_2T_x$ MXene films whose thickness can be as small as 1 μm. This new type of 2D transition metal carbide/nitrides has attracted tremendous interest owing to their excellent properties in several aspects, such as adjustable chemical properties, ultrahigh electric conductivity, better infrared thermal radiation performance, and superior mechanical strength [1104, 1105]. As shown in Fig. 11.9(b), the fabricated MXene films show better thermal camouflage properties compared with other commonly-used materials like Graphene, stainless steel and so on. The emissivity of the MXene films is quite low over the wide infrared waveband (< 20%, 7–14 μm). The authors have done a series of experiments working at a wide temperature (from below 10 °C to over 500 °C), in which the designed MXene shows superior high-temperature indoor/outdoor thermal camouflage performance, long-term fire stability, disguised Joule heating capability and high EM interference shielding efficiency. Especially, due to the excellent flexibility of MXene, the ultrathin MXene films can be cut into different shapes as illustrated in Fig. 11.9(b), and the MXene

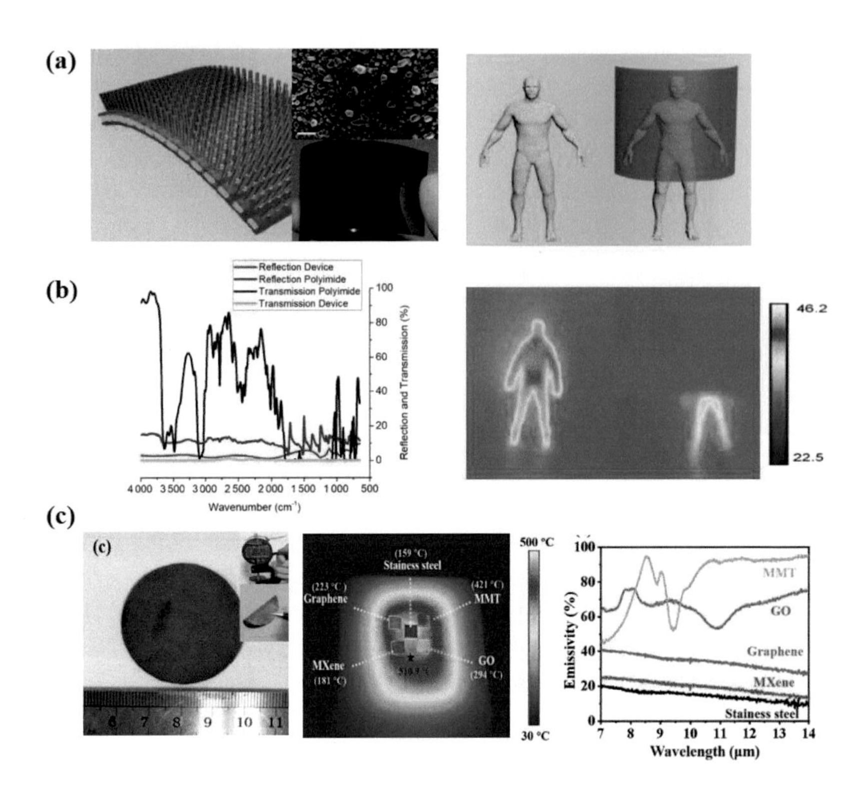

Figure 11.9 (a) Schematics of a stealth sheet with absorbers, thermal insulation, and micro-emitters (left) proposed in [1102]. The human model with and without the stealth sheet is given on the right. (b) A comparison of transmission and reflection spectra on the flexible polyimide substrate (left), and the IR image is given on the right. (a) and (b) are reproduced with permission from [1102]. Copyright 2018 Wiley-VCH GmbH, Weinheim. (c) Digital images of the as-obtained MXene film (left), along with the IR image of compared samples including 45-μm-thick MXene, 45-μm-thick graphene, 30-μm-thick stainless steel, 10-μm-thick graphene oxide and 35-μm-thick montmorillonite films. The right-hand panel shows the measured mid-IR emissivity. Reproduced with permission from [1103]. Copyright 2021 Wiley-VCH GmbH, Weinheim.

dispersion inks can also be used to write unique images. The finds of this work show great potential of MXene materials for flexible thermal camouflage, which is promising in several areas like infrared stealth, security protection, counter-surveillance, etc.

Noting that the proposed structures for thermal camouflage discussed above are all broadband within one single waveband. As introduced in Section 11.2.1, multi-spectral camouflage should be considered, which is challenging but more promising in real situations. To this end, Kim et al. [1106] proposed hierarchical

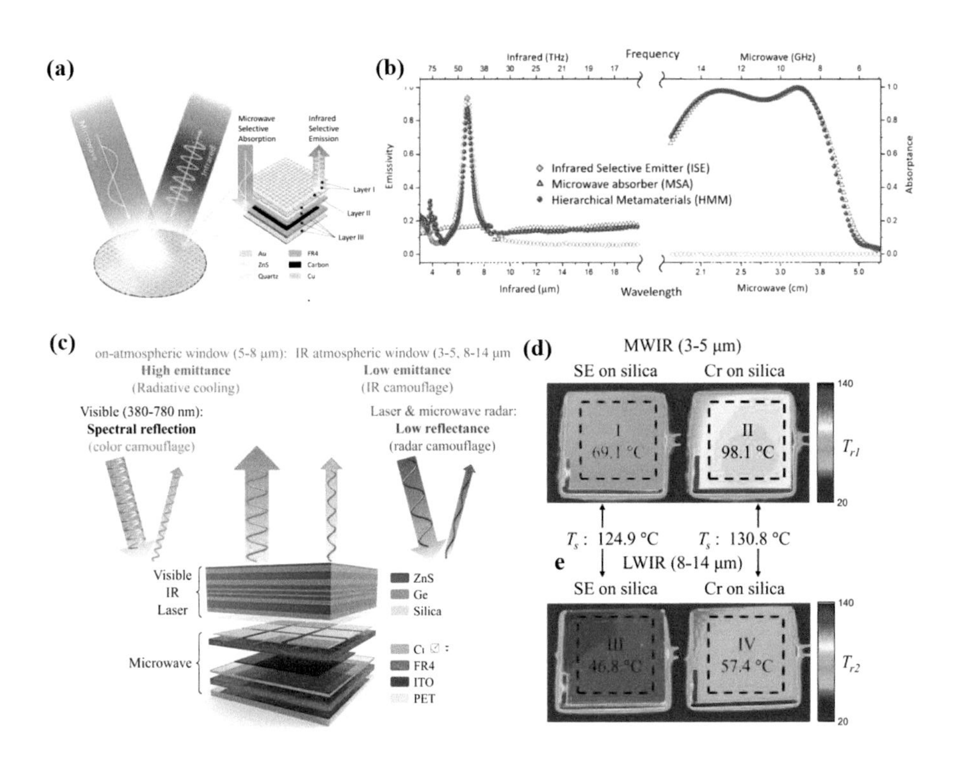

Figure 11.10 (a) Schematic showing incoming microwave and outgoing IR radiation as well as the structure and composition of the hierarchical metamaterials that incorporate the IR SE and microwave selective layers. (b) Measured emissivity over the IR spectrum and absorptance over the microwave spectrum. (a) and (b) are reproduced with permission from [1106]. Copyright 2019 Wiley-VCH GmbH, Weinheim. (c) Schematic of multi-spectral camouflage for different SEs and the composition of proposed metasurfaces. (d) Compared thermal images at different wavebands. (c) and (d) are reproduced from [1107]. Copyright 2021, Nature Publishing Group.

metamaterials composed of several layers to realize camouflage functions in both infrared and microwave bands as illustrated in Fig. 11.10(a). The micro signals can be absorbed by the microwave-selective absorber first, and the absorbed energy is then dissipated into heat emitted through the infrared selective emitters within an atmospheric window (5–8 μm). The obtained emission spectra are also shown here, in which infrared and microwave emitters work well as expected. Recently, Zhu et al. [1107] also proposed multispectral camouflage working within an ultrabroad band, covering infrared (3–5 μm, 8–14 μm), visible, lasers (1.55 and 10.6 μm) and microwave (8–12 GHz) band as shown in Fig. 11.10(b). In detail, the multispectral camouflage is composed of a ZnS/Ge multilayer as wavelength SEs, and a Cu-ITO-Cu metasurface for microwave absorption. In addition, the designed camouflage also shows excellent radiative cooling performance.

11.2.3 Active Thermal Camouflage

As discussed above, thermal camouflage has gained much attention and a series of new materials and nanostructures have been used to improve the working performance. Notably, the aforementioned design is usually working at targeted conditions, whose properties cannot be tuned once fabrication is finished. Therefore, several dynamical methods have also been proposed to actively control thermal camouflage with broad applications, including thermal control using phase-changing materials, optical, electric, or chemical modulations, and even considering bio-inspired designs, etc. In this section we will name a few seminal works from different aspects.

Phase-changing Materials

Phase-change materials exhibiting different optical properties at different phases provide a versatile platform to actively control thermal emission as well as application of thermal camouflage. For example, as for typical phase-changing materials like germanium-antimony-tellurides (GST) and vanadium oxide (VO_2), their optical index will change significantly when phase transition happens.

As for commonly used GST materials, such alloys undergo the reversible phase transition between crystalline phase to amorphous phase as if the temperature of GST is raised above the crystallization and melting point, which could be triggered by thermal [1012], optical [1108, 1109] or electrical plus [1110, 1111]. As introduced before, the imaginary part of amorphous GST (aGST) is negligible, working like dielectric materials. Differently, the magnitude of imaginary permittivity of crystallization GST (cGST) will increase associated with non-negligible absorption in the mid-infrared region. Such properties allow to tailor thermal camouflage performance dynamically. Besides, the phase change is nonvolatile, and no external power is required to maintain the modified properties. For instance, Qu et al. [1012] fabricated a 350-nm GST film sitting on Au substrate, and continuous thermal camouflage was actively realized for background temperatures ranging from 30 °C to 50 °C. Generally, the as-deposited GST shows an amorphous phase with a disordered atomic distribution, then the atomic patterns could be re-organized by annealing the material at 200 °C, that is, on a hot plate. The annealing time also plays a role in phase changing, and 60 s is adequate to obtain cGST in this case. The near-perfect emissivity can be obtained in the mid-infrared region when the phase of GST changes to crystallization states. The authors further explored the camouflage performance when choosing different background temperatures and different annealing times. The samples keep the temperature above the environment (T_0). When the background temperature is 30 °C, the object is hardly distinguishable from the background because the GST is at initial amorphous states with low emissivity over the mid-infrared region. While, after 60s-annealing process, the emissivity of the object becomes

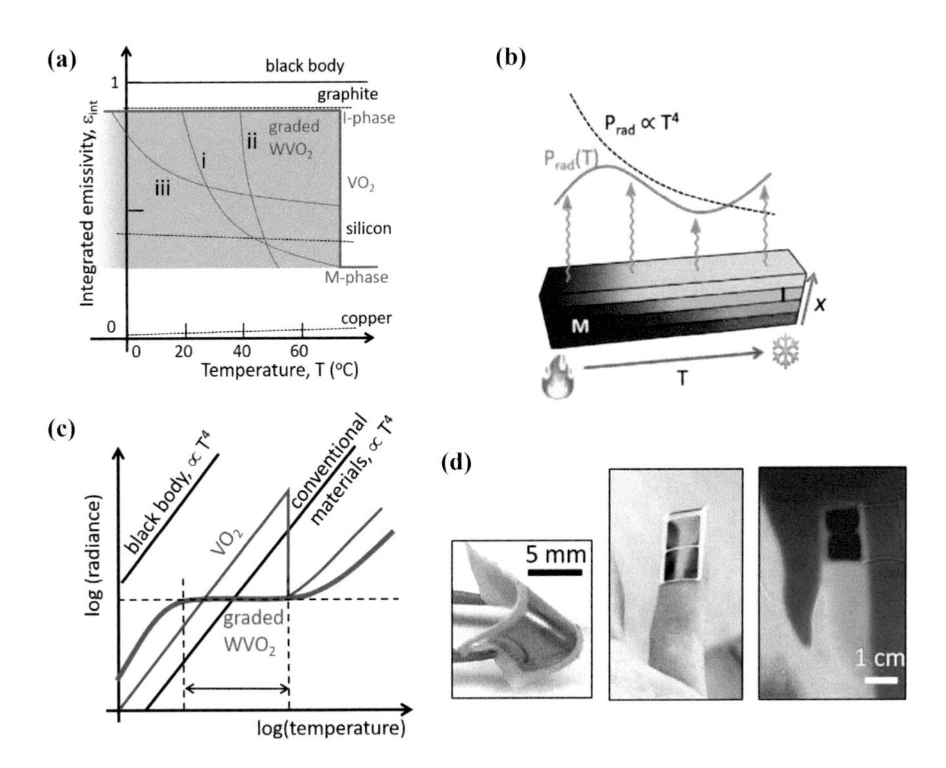

Figure 11.11 VO_2-based thermal camouflage. (a) Schematic illustrating realization of the nearly arbitrary temperature dependence of integrated emissivity (ε_{int}) within the shaded area using graded W-doped VO_2, represented by the three arbitrarily designed emissivity curves denoted by i, ii, and iii. The behavior of typical high, moderate, and low-ε_{int} materials and of VO_2 is also shown for comparison. (b) Schematic showing that by rational design of the W doping profile (x along thickness direction) of $W_xV_{1-x}O_2$, emissivity can be programmed to regulate thermal radiation (P_{rad}) for distinctly different behavior from the Stefan–Boltzmann T^4 law. (c) Schematic illustrating constant thermal radiance over a wide range of temperature from the camouflage, in stark contrast to the Stefan–Boltzmann T^4 law from conventional materials. (d) Optical image of a graded WVO_2 film transferred onto a PE tape showing high mechanical flexibility. The visible color of the sample is related to the WVO_2 thickness and lighting on the surface. The fingertip is hidden from the IR camera when covered with camouflage. Reproduced with permission from [1015]. Copyright 2020 Wiley-VCH GmbH, Weinheim.

high, thus it can emit power to infrared cameras and match with the background temperature.

Another popular phase-changing material used in thermal camouflage is VO_2. Differently, the phase transition happens between metal and insulator at the critical temperature of 67 °C [1112]. The metal-insulator transition can also be triggered by heat, light, electricity, magnet, and so on. One of the important works was done by Tang et al. [1015]. Apart from thermal modulation, they proposed a W-doped VO_2 ($W_xV_{1-x}O_2$) platform, in which the phase transition can be realized at any temperature ranging between 0 °C and 67 °C depending on the fraction x of W. Generally, it is known that thermal radiation from a blackbody increases with the fourth power of absolute temperature (Stefan–Boltzmann law). Actually, this law is only valid for materials whose emissivity is insensitive to the temperature. As for VO_2, its emissivity is moderately high at low temperatures owing to the insulator states, then there is an abrupt drop after the transition temperature (63 °C) since the VO_2 changes to metal states. In [1015], the authors designed multilayer $W_xV_{1-x}O_2$ films, where x varies continuously across the thickness direction as shown in Fig. 11.11. In this case, the emissivity is directly proportion to T^{-4}, thereby the emission power is not dependent on temperature [Figs. 11.11(b)–(c)], being a perfect candidate for thermal camouflage. Fig. 11.11(d) shows the performance of designed WVO_2 camouflage on PE substrate with high mechanical flexibility. The measured IR image shows (right) that the temperature of the area covered by WVO_2 is similar to that of the background environment as expected. Besides, when gradually changing the object's temperature between 15 °C and 70 °C, the camouflage properties are still preserved and insensitive to temperature, which is ascribed to $\varepsilon_{int} \sim T^{-4}$.

Electric Modulation

The aforementioned thermally tunable methods are limited by the modulation speed, thereby electric modulation [1013, 1115] has gradually gained much attention. By controlling the externally applied voltage, it is feasible to modify the optical index of electric-controlled materials like InAs, and graphene, allowing to tune emissivity of the systems. Fig. 11.12 shows some typical examples of tunable thermal camouflages based on electric modulation. For example, Salihogulu et al. [1113] reported an active thermal camouflage design, that can realize real-time electric control of thermal emission as illustrated in Figs. 10.24(a)–(b). The active thermal emitter is composed of a multilayer-graphene electrode, a porous polyethylene membrane, and a Au-electrode coated on heat resistive nylon. Initially, the spectral emissivity is considerably high over the mid-infrared region. When the voltage is applied, i.e., 3V, the value of emissivity drops significantly, for which the IR image cannot capture the thermal information as compared in Fig. 11.12(b). Notably, several thermal camouflages (low-e devices) can only work in conditions where the temperature is higher than the background. To solve this problem, Hong et al. [1114] demonstrated a wearable and adaptive infrared camouflage device, consisting of flexible thermoelectric materials covered with a high-e polymer layer. Such a design can well work for a broad range of background temperatures. The surface temperature of thermoelectric materials can be controlled to realize temperature matching with the background and protect the objects. As shown in the second row of Fig. 11.12(c), the camouflage prop-

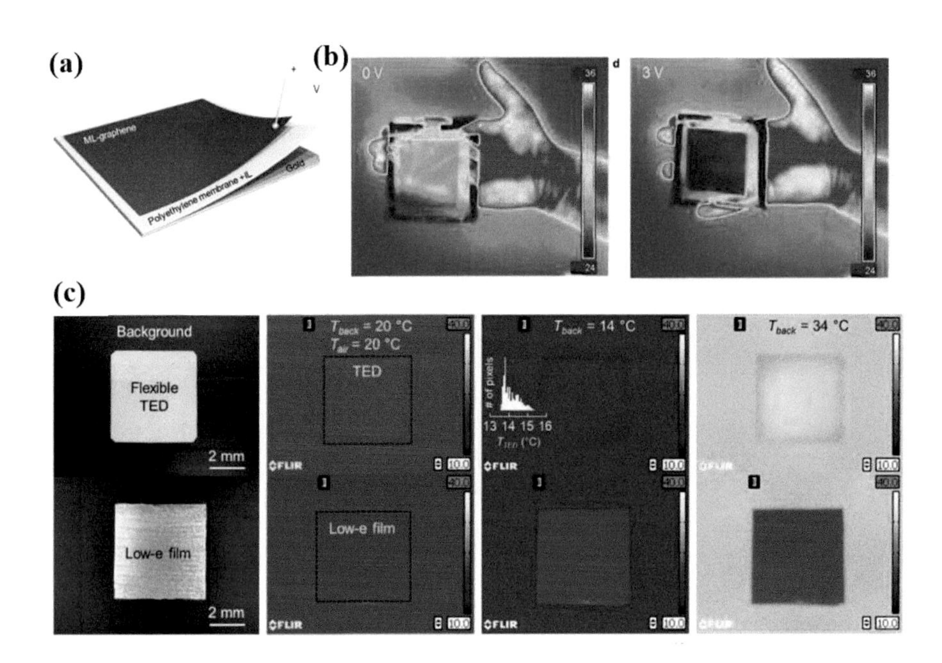

Figure 11.12 Electric modulation of thermal camouflage. (a) Schematic drawing of the active thermal surface consisting of a multilayer-graphene electrode, a porous polyethylene membrane soaked with a RTIL, and a back gold-electrode coated on heat resistive nylon. (b) Thermal camera images of the device were placed on the author's hand under the voltage bias of 0 and 3 V, respectively. (a) and (b) are reprinted with permission from [1113]. Copyright 2018 American Chemical Society. (c) Photographs of the TED (top) and low-e film (bottom) on the measurement setup. The second row of IR images of the TED and low-e film with the same background (Tback) and ambient temperature (Tair) of 20 °C. Both devices are maintained at 28 °C and showed the camouflage effect. The third row of IR images of the TED and low-e film at T_{back} = 14 °C and 34 °C. Reproduced with permission from [1114]. Copyright 2020 Wiley-VCH GmbH, Weinheim.

erties are visible for both low-*e* and designed thermoelectric device (TED) when the background temperature is the same as the air temperature. However, when the background temperature is higher or lower than the air temperature, the object covered with low-*e* materials will be perceived, while the designed TED can still protect the object by showing superior thermal camouflage performance.

Chemical Modulation

Chemical modulation is also an attractive method in shaping radiative spectra. Especially for some metal materials like Mg [1117], the controllable chemical reactions make it possible to actively tune the optical properties of designed structures. For example, Li et al. [1116] designed nanoscopic platinum (Pt) film-based reversible silver (Ag) electrodeposition devices as shown in Fig. 11.13(a).

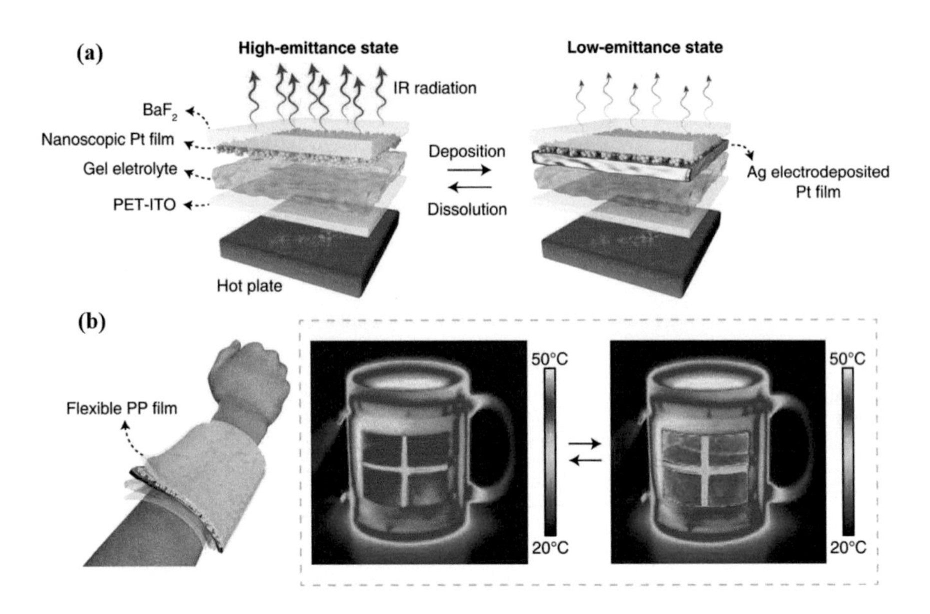

Figure 11.13 Chemical modulation of thermal camouflage. (a) Schematics of a nanoscopic Pt film-based RSE device before (left) and after (right) electrodeposition. (b) Schematic of a flexible surface-based adaptive device (left). Plots within dashed box: LWIR images of the flexible PP-based device on a curved cup surface before (left) and after (right) electrodepositing for 25 s. Reproduced from [1116]. Copyright 2020, American Association for the Advancement of Science.

The spectra emissivity can be controlled by the electrodeposition process of Ag. Initially, before the electrodeposition of Ag (left), the devices have high emissivity but low reflectivity. By contrast, the low-emittance state is available when Ag is gradually deposited on the Pt film. Specially, the Ag film could be dissolved or deposited reversibly, enabling dynamically control emissivity of the devices. In order to fabricate flexible surface-based adaptive camouflage devices, the authors used polypropylene (PP) film and thin Pt film as the top electrode as shown on the left-hand side of Fig. 11.13(b). The PP-based devices are attached to a cup filled with water at the temperature of 50 °C. In Fig. 11.13(c) (plots in dashed box), controllable camouflage performance could be visible after electrodeposition, where the area covered with the PP-based camouflage device shows similar temperature profiles with the background.

11.2.4 Discussion and Prospects

Recently, we have witnessed the flourishing development of thermal camouflage technologies using different types of materials or designs. Especially, several works have paid much attention to actively controlling working performance. For practical considerations, adaptive and soft thermal camouflage may become more attractive. Soft materials with desirable thermal emission characteristics could cover the object with an arbitrary shape. On the other hand, multispectral camouflage is also important in the next thermal camouflage. Even though there are some works that have designed camouflage working at multiple wavebands, there are still several challenges in design, fabrication, and increasingly advanced detection technologies.

11.3 Metamaterials-Based Energy Nanodevices: Thermophotovoltaics

TPVs is a thermoelectric conversion technology with great development potential, which can utilize various forms of energy including fossil energy, solar energy, nuclear energy, and industrial waste heat, and theoretically can realize efficient energy conversion and utilization. Early in 1956, a TPV system based on the silicon photocell is proposed by Kolm [1118], which is supposed to have an output power of 1 W. However, due to the intrinsic disadvantages of silicon cells and germanium cells, the development of TPV was greatly limited in the last century. From the end of the last century to this century, with the rapid development of micro- and nanofabrication and the advent of semiconductor PV cells with low bandgap group III–V elements, researchers began to explore and try the application of technology in military, transportation, aerospace, and industrial production and other fields.

Compared to other technology, TPV has some intrinsic advantages: (1) high energy output density; (2) the energy that can be utilized as the heat source is various; (3) low noise, the energy conversion part of the whole system has no moving parts, which makes this system less prone to failure and easier to maintain than a traditional power system; (4) compared to solar PV, the system is easier to achieve continuous power supply and is less affected by external environmental conditions.

At present, TPV systems still face a major challenge in that the energy conversion efficiency is low, which should be solved by the development and optimization of the major components.

11.3.1 Fundamental Principles of Thermophotovoltaics

A typical TPV system includes a heat source, emitter, optical filter, TPV cells, and auxiliary devices. The energy conversion principle of the TPV system is shown in Fig. 11.14 [1119]. The heat source of the TPV system can be diverse, solar energy, fossil energy, nuclear energy, industrial waste heat, and biomass energy can be utilized as input heat to the TPV system. In the TPV system, the input heat from the heat source mentioned above is absorbed by the emitter. Then the thermal radiation of the emitter is partially absorbed by TPV cell to generate electrical energy. The energy conversion efficiency of TPV system η_{TPV}

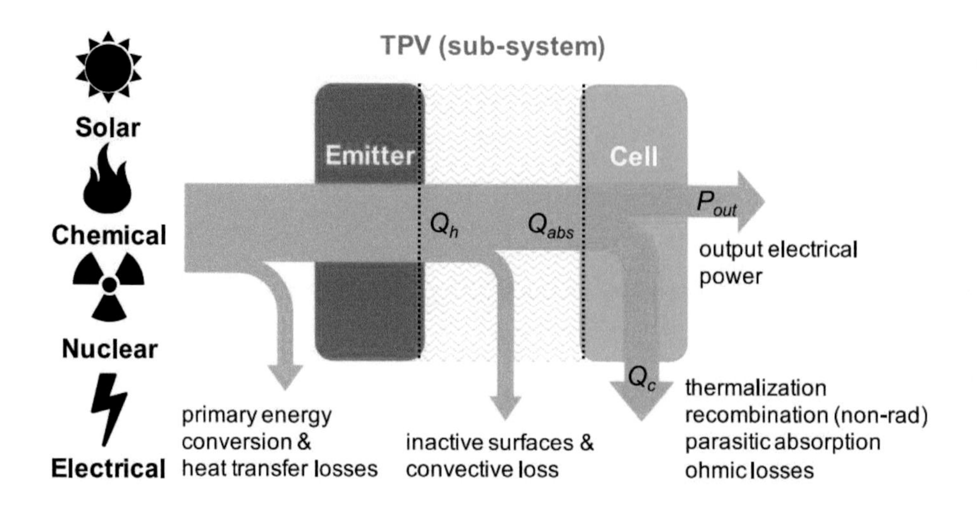

Figure 11.14 Energy transport and conversion in a TPV generator. Reprinted from [1119], Copyright 2020, with permission from Elsevier.

can be expressed as

$$\eta_{\mathrm{TPV}} = \frac{P_{\mathrm{out}}}{Q_{\mathrm{th}}}, \tag{11.3.1}$$

where Q_{th} is the net energy flow from the emitter and P_{out} is electrical power generated by the TPV cell. Actually, for a TPV cell having a bandgap of E_g, only the high-energy (in-band) photons ($E > E_g$) can be absorbed. The remaining unconverted photons will cause out-of-band (OOB) absorption of the TPV cell, which will seriously restrict the energy conversion efficiency of the TPV cell and the whole system.

In the practical systems, except for OOB absorption, some other energy loss will also restrict the energy conversion efficiency. As shown in Fig. 11.14, the important loss pathways include the emission and absorption of OOB photons, the thermalization of in-band photons, electron–hole pair recombination, Ohmic losses along the conduction pathway, and parasitic heat losses to the surroundings. The efficiency of TPV-based energy systems is largely dependent on the energy losses, and these losses can be attributed to the optical properties of the emitter and the cell, electrical properties of the cell, system configuration, the degree of matching between the parts, etc.

In order to evaluate the performance of each part and the suitability of the parts in the TPV systems, the energy conversion efficiency of TPV systems can be expressed as

$$\eta_{\mathrm{TPV}} = (\mathrm{SE} \cdot \mathrm{IQE}) \cdot \mathrm{VF} \cdot \mathrm{EF} \cdot \eta_{\mathrm{cavity}}. \tag{11.3.2}$$

In this equation, SE is the spectral efficiency, IQE is the internal quantum efficiency, and the SE $\cdot$ IQE, which describes the fitness of emitters for cells, can be expressed as

$$\mathrm{SE}\cdot\mathrm{IQE}=\frac{J_{\mathrm{sc}}\cdot V_g}{Q_{\mathrm{abs}}}=\frac{E_g\int_{E_g}^{\infty}\varepsilon_{\mathrm{eff}}(E)\cdot\mathrm{IQE}\cdot b(E,T_h)\mathrm{d}E}{\int_0^{\infty}E\cdot b(E,T_h)\mathrm{d}E}, \quad (11.3.3)$$

where $b(E,T_h)=\frac{2\pi E^2}{c^2h^2[\exp(\frac{E}{k_BT})-1]}$ is the spectral photon flux of a blackbody (c is the speed of light in vacuum, h is Planck's constant, and k_B is Boltzmann's constant). $\varepsilon_{\mathrm{eff}}(E)$ is the effective emissivity of an emitter–cell pair: $\varepsilon_{\mathrm{eff}}(E)=\frac{\varepsilon_e(E)\varepsilon_c(E)}{\varepsilon_e(E)+\varepsilon_c(E)-\varepsilon_e(E)\varepsilon_c(E)}$, where ε_e and ε_c are the emissivity of the emitter and that of the TPV cell, respectively. E_g is the bandgap of the TPV cells.

The voltage factor VF is a measure of bandgap utilization, described as the ratio of the open-circuit voltage V_{oc} to the bandgap voltage V_g:

$$\mathrm{VF}=\frac{V_{\mathrm{oc}}}{V_g}. \quad (11.3.4)$$

The fill factor FF is the ratio of the generated power P_{out} to the product of the short-circuit current density J_{sc} and the open-circuit voltage V_{oc}:

$$\mathrm{FF}=\frac{P_{\mathrm{out}}}{J_{\mathrm{sc}}\cdot V_{\mathrm{oc}}}. \quad (11.3.5)$$

The cavity efficiency η_{cavity} can be expressed as

$$\eta_{\mathrm{cavity}}=\frac{Q_{\mathrm{abs}}}{Q_h}. \quad (11.3.6)$$

The value of η_{cavity} is closely related to the view factor between the emitter and the cell and the parasitic convective loss.

As a result, to ensure the energy conversion efficiency of the TPV system, spectral manipulation is commonly utilized to restrict the absorption of the low-energy photon by the TPV cell. Except for OOB absorption, the loss pathways of the TPV cells include nonradiative recombination, thermalization, and Ohmic losses.

11.3.2 Major Components

TPV Photocells

A photocell is an essential part of the TPV system, which is directly related to energy conversion efficiency. There are some factors determining the thermal-electrical energy conversion inside the photocell.

The most important factor is the matching between the cell and the emitter, which can be evaluated from Eq. (11.3.3). It requires that the cell should possess high IQE at the in-band wavelength range. Besides, the bandgap, E_g, of the cell should be as small as possible, and the in-band emissivity/absorptivity of the cell should be close to 1 to make full use of thermal radiation from the emitter. Moreover, effective cooling of photocells is essential so as to avoid a decrease in cell efficiency.

Table 11.1 TPV cell performance.

Materials	E_g (Ev)	λ_G (μm)	J_{SC} (A cm^{-2})	V_{OC} (V)	FF	V_F
InGaSb	0.56	2.2	3	0.27	–	0.48
$InGa_{0.53}As_{0.47}$	0.74	1.7	1	0.47	0.64	0.63
$InGa_{0.69}As_{0.31}$	0.60	2.1	2.26	0.36	0.67	0.59
InGaAsSb	0.53	2.4	1	0.3	–	0.57
GaInAsPSb	0.34	3.6	0.29	0.03	0.33	0.08
InAs	0.32	3.9	0.89	0.06	0.37	0.19

The bandgap (E_g), cutoff wavelength (λ_C), short-circuit current (J_{sc}), open-circuit voltage (V_{oc}), fill factor (FF), and voltage factor (VF) are reported in [1120].

So far, there are a series of photocells that can be applied in different situations:

Silicon (Si) (with a bandgap energy of about 1.1 eV), which is widely utilized in the solar PV system, can also be utilized in TPV systems, although its bandgap is usually deemed to be high. It needs near-perfect spectral management to achieve acceptable efficiency. However, its advantage includes the inexpensive cost compared with low bandgap cells, commercial availability in large quantities, and nontoxic fabrication, making it still vital for the TPV system.

Germanium (Ge) with a bandgap energy of 0.66 eV is another photovoltaic material that has been investigated for applications in TPV technology. Germanium is a better match for TPV applications than silicon due to the lower bandgap and larger cutoff wavelength up to 1.88 μm. As a result, Ge can be suitable for applications in TPV devices, where the emitter temperature is usually less than 1800 K. However, the efficiency of Ge cells is much lower than that of Si and GaSb photocells due to the high effective electron mass of Ge, which results in low open-circuit voltage. Ge cells are more expensive than Si cells but much cheaper than III–V low bandgap cells such as GaSb and GaInAsSb. As a result, this material has a certain application value similar to the TPV photocell.

GaSb (with a relatively low bandgap energy of 0.72 eV, J_{SC} of 3 A cm^{-2}, V_{OC} of 0.41V, and VF of 0.57) has the ability to operate under elevated irradiation densities, which makes it suitable for use in TPV systems. Actually, the GaSb cells were one of the main reasons for the renewed interest in TPV technology since the early 1990s. The suitable bandgap and EQE, which can match the spectral radiant intensity distribution of objects with temperatures from 1000 to 1800 K and the stability under elevated irradiation densities make this material become one of the most commonly used materials of the TPV photocell. However, the GaSb has the disadvantages of being toxic and costly.

In addition to GaSb, other III–V photovoltaic converters, such as InGaAsSb, InAsSbP, InAsSb, InGaSb, GaInAsPSb, InAs, and InAsSbP, can be utilized as the materials of the TPV photocell. The efficiencies of these materials can be comparable to or higher than those of the GaSb cells. The TPV cell performances for these materials are listed in Table 11.1. However, these materials also suffer from the disadvantage of being costly and toxic similar to the GaSb cell.

Heat Source

For the TPV systems, the temperature of the heat source is typically between 1 000 and 2 000 K. The input energy of the hole system is driven by the radiative heat of the heat source, which can convert the combustion [1121, 1122], radioisotopes [1123–1125], high-temperature industrial processes [1126], and even the solar energy [1127, 1128]. Compared with the solar PV system, the TPV system can take energy from a wider range of heat sources.

Emitters

An emitter is another essential part of the TPV system that is directly related to the energy conversion efficiency and electrical output power density. For high-temperature heat sources, according to Planck's blackbody radiation law, only about a little percent of its radiative power is at photon energies above the bandgap energy of PV. For example, a blackbody emitter at $T = 1800$K has only about 6% radiative power at photon energies above the bandgap energy of silicon photocells. Therefore, the energy conversion efficiency of a TPV system will be greatly limited if the blackbody emitter is adopted without any other components, such as an optical filter [1129] or a back surface reflector (BSR) [1130].

To improve the energy conversion efficiency of TPV systems, a straightforward way is to design an SE that can restrict the radiative power emitted by the heat source to be photons with in-band energies. This will be introduced in the "TPV-Driven Emitters" section.

11.3.3 TPV-Driven Emitters

As mentioned the "Emitters" section, the input energy of the TPV system is determined by the temperature and emissive property of the emitter. For the selection of emitters, several requirements need to be considered. Good thermal stability in a selected atmosphere is important for the long-term operation of the TPV system. It is also important to note that the emitters need a high emissivity at the in-band wavelength range to realize high electrical output power density. A low emissivity at the OOB wavelength range may be further needed when attempting to improve the energy conversion efficiency of TPV systems from the emitter side. Considering wide applications in future, the availability and economics of the emitters should be taken into consideration.

According to the emissivity spectrum of the emitters, they can be classified into two categories: broadband and SEs. In the early years of TPV research, broadband-emitting emitters were chosen, including SiC, graphite, and silicon nitride (Si_3N_4). Such types of emitters can exhibit high emissivity of about 0.9 and high mechanical and thermostructural stability at high temperatures.

However, although the broadband emitters can provide sufficient radiative power at high temperature as a blackbody, the property of a photocell that only the energy of in-band photons, whose energy is higher than the intrinsic

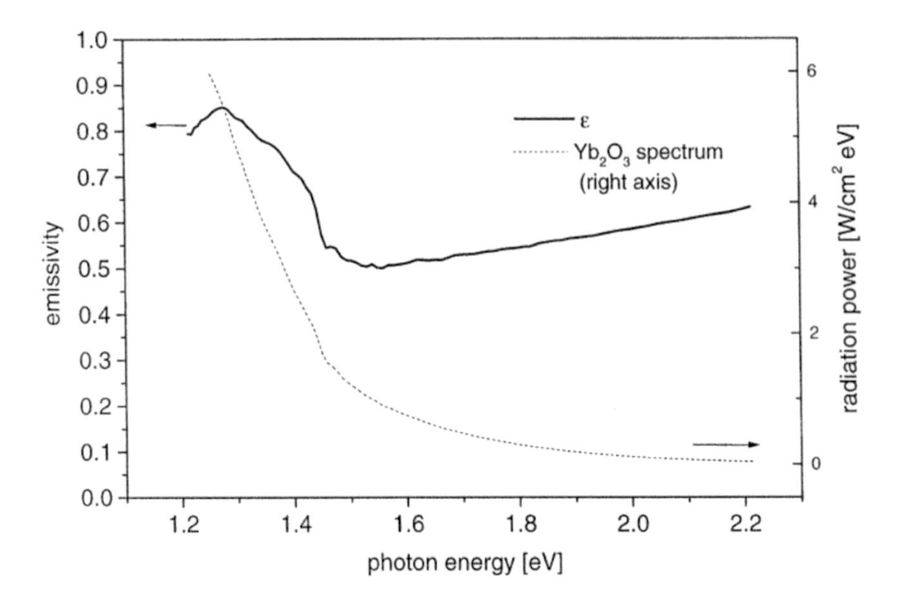

Figure 11.15 Spectral emissivity and radiative power of rare-earth-based emitters: Emissivity (solid line) and radiative power (dashed line) of a Yb_2O_3 mantle emitter measured with the flash-assisted multiwavelength pyrometry method. Reprinted from [1131], Copyright 2002, with permission from Elsevier.

bandgap of a photocell, can be transformed to electric energy and utilized makes a waste of OOB photon energy, which can have an adverse impact on its performance. Due to this property of the photocell, the emissivity of the emitter should be carefully designed to provide utilizable energy to the photocell to the maximum extent. These emitters are called SEs. To attain a suitable spectral match between emitters and given photocells, there are two general approaches as follows: (1) material design including doping, coating, and mixing; and (2) physical structure design including the PBG material design.

Rare-Earth-Based Emitters

In the early years of TPV research, researchers chose SE and high–melting-point materials from nature. Rare-earth-based material is a kind of natural material with selective emissivity over narrow bandwidths in the infrared region. As a result, these materials are expectedly investigated as the SE of the TPV system.

Aiming at the early TPV system equipped with Si or Ge photocells, rare earth oxides, including ytterbia (Yb_2O_3) and erbium oxide (Er_2O_3), were selected as the SE. Yb_2O_3 is specifically suited to match with silicon photocells (Fig. 11.15), which has a peak of emissivity at about 1.27 eV and a high melting point of about 2 400 °C. Er_2O_3, a rare earth oxide with an emission peak at 0.805 eV and a quite high melting point of about 2 400 °C, can match with the Ge photocell (with a bandgap energy of 0.66 eV) and the GaSb photocell (with a bandgap energy of 0.72 eV) that were developed later.

Moreover, rare earth elements can also be doped into the matrix of an appropriate ceramic (e.g., alumina and SiC) to achieve exotic selective emission properties. By this means, the mechanical and thermostructural properties can also be improved. Similarly, rare earth oxides can be mixed with some structuring materials, such as quartz and alumina. For example, the researchers at the Paul Scherrer Institute have made Yb_2O_3- and Er_2O_3-coated foam ceramics from the broadband emitters of Al_2O_3 and SiC. The coating of SiC or Al_2O_3 substrates was accomplished by plasma spraying. Expect for these common substrates, intermetallic molybdenum disilicide ($MoSi_2$) has been utilized as the material of the substrate for the TPV emitter. The excellent oxidation resistance in oxygen-containing atmospheres at high temperatures of up to 2 400 °C and the good material pairing with both ytterbia and erbia coatings make $MoSi_2$ a promising material of the substrate for the TPV emitter.

Nano/Microstructure Selective Emitters

In recent years, with the development of micro/nanotechnology and further understanding of the fundamental principles of nanophotonics, the physical phenomena occurring at nanoscale, such as surface plasmon/phonon polaritons (SPPs and SPhPs), PBG effects, and optical microcavity effects, can be utilized to manipulate the spectral properties of SEs. The SEs can be divided into two categories as antireflection coating-based SEs and SEs with periodic micro/nanostructures.

Antireflection Coating-Based Selective Emitter

The antireflection coating-based SE usually consists of a metal layer for radiative emission and an antireflection coating for the realization of optical selectivity. Among all the types of metals. Transition element materials are good candidates for the TPV emitter. When a transition element atom is located in a lattice with an asymmetric crystal field, the degeneracy of d orbitals is removed. As a result, the symmetry properties of the crystal field do influence the emission characteristics of a transition-element-based material, which provides an opportunity for the design of a selective TPV emitter. Among all the transition elements, tungsten (W) is well suited to spectrally match with the GaSb photocells with a bandgap of 0.72 eV (corresponding to a wavelength of 1.73 μm). Due to the advantages of owning an exceptionally low absorption coefficient in the infrared at wavelengths greater than 2 μm, and a very low evaporation rate at high temperatures, tungsten has been investigated extensively in the development of TPV emitters and has gradually become the mainstream material of the selective TPV emitter. As a result, the AR/W structure is a common type of the antireflection coating-based SE. For example, Fraas et al. [1132] demonstrated a refractory-metal-coated emitters comprising tungsten (W) as metal owing to low emittance at long wavelengths and a refractory oxide with a refractive index of 2.0 that can control the spectral emissivity within the wavelength range higher than λ_c by tuning its thickness. A spectral efficiency of 75% and a GaSb cell power density of 1.5 W cm^{-2} can be achieved under 1 555 K. Moreover, the thicknesses of the two layers can be tuned to achieve enhanced radiation in the appropriate wavelength region corresponding to the bandgap of a given photocell. They also have proposed a dual-layer-coated SiC emitter composed of an inner layer of Ta_2O_5 serving as an antireflection coating for the NIR radiation, and an outer layer of the intermetallic $TaSi_2$ that selectively emits the NIR.

Under the advantages of simple fabrication and large-area fabrication ability, the antireflection coating-based SE has been utilized in the applied TPV system [1133].

Selective Emitters with Periodic Micro/Nanostructures

The research on the SEs with periodic micro/nanostructures is mainly based on the research of 2D periodic SEs, including 2D gratings consisting of periodic cylinders or rectangular columns and a 2D PC SE consisting of cylindrical or rectangular holes. The most common absorptive material of the SE is also tungsten due to its advantages shown in the "Antireflection Coating-Based Selective Emitter" section. Currently, most works of the TPV SE design are based on 2D periodic structures and tungsten as an absorptive material.

Tungsten-based SEs: For the investigation of tungsten-based 2D periodic SEs, early in 1999, Heinzel et al. [1134] demonstrated a 2D tungsten grating structure as an SE. By adjusting the period and depth of gratings, the manipulation of the resonant peak frequencies can be realized. Reactive ion etching and chemical etching were utilized to fabricate the corresponding sample. Then in 2000–2005, Hitoshi Sai et al. [1135–1138] used RCWA and FDTD to calculate the spectral emissivity of the metal tungsten gratings structure. Two different resonant peaks were found in the spectral emissivity of the structure, one of which is caused by SPPs and affected by the period and angle of the tungsten gratings, and the other is caused by the optical microcavity effects and mainly affected by the depth of the grating. This group also designed the 2D surface-relief gratings with a period of 1.0–0.2 mm composed of rectangular microcavities and realized fabrication by atomic beam etching, as shown in Fig. 11.16. The results showed that the microstructured W radiators behave as good selective radiators, with both high efficiency and high power density.

And then, in 2008, Celanovic et al. [1139] presented systemic theory, design, fabrication, and optical characterization of 2D W PC as selective thermal emitters. They fabricated 2D W-PhC by the atomic beam etching. They also presented simple design rules and a largely scalable microfabrication process that is not restricted to tungsten but can be extended to platinum, silver, molybdenum, etc. The state-of-art technology of TPV SEs by Celanovic et al. [1137] has became the basis of TPV technology developed at the Institute for Soldier Nanotechnologies. In 2010, Chen et al. [1140] investigated the 2D grating structure (convex) and PC structure (concave) SE by the RCWA method on the basis of predecessors. They optimized the geometric dimensions of the microstructure based on the numerical simulation. The results showed that the spectral emittance of optimized structures is above 0.9 at wavelengths between 1.0 and

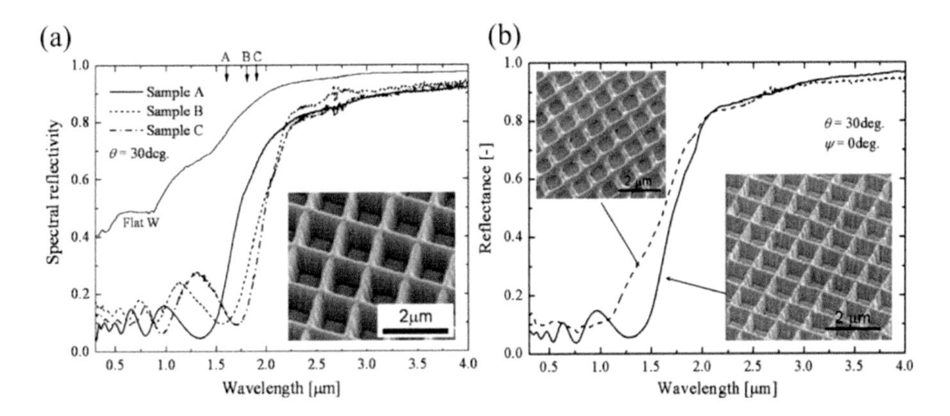

Figure 11.16 Reflectance spectra of typical tungsten-based 2D periodic SEs. (a) The comparison between the spectral reflectivity of the W gratings and flat W at the incident angle of 30°. Reprinted from [1138], with the permission of AIP Publishing. (b) Reflectance spectra of the W gratings after heating at 1400 K for 1 h in a reductive atmosphere. The single crystalline grating keeps its spectral feature (solid line) and microstructure as shown in the inset. But the spectral property of the polycrystalline sample becomes worse than before heating (dashed line). Reprinted from [1137], with the permission of AIP Publishing.

2.0 μm but below 0.2 at wavelengths between 2.0 and 4.0 μm. In addition to the work of Chen et al. [1140], Araghchini et al. [1141] and Yeng et al. [360] did some research on the development of W-based 2D SEs.

In addition to the 2D periodic SE, some researchers also put a lot of effort into the investigation of SEs with 1D and 3D periodic structures. For example, to tackle the limitation of incomplete bandgaps in 1D and 2D PCs, 3D PCs of W have been developed for the TPV SEs by Gee et al. [1142], Fleming et al. [979, 980], and Nagpal et al. [1143]. To realize easy and large-area fabrication, Zhang et al. [1010] designed an aperiodic 1D multilayer structures (Fig. 11.17) for the TPV SE by machine learning (Bayesian optimization). The optimized structure has a high spectral efficiency of 82.16%, which is further proved by the experiment.

Other material-based SEs: TPV SEs made from materials other than tungsten have also been investigated. For example, tantalum is a candidate for the absorptive material of the TPV emitter due to its high melting point (3290 K) and the small coefficient of thermal expansion it has. Some highly selective PC emitters fabricated from tantalum have been reported [1145, 1146]. The thermal emitters exhibited high spectral selectivity, enhanced emissivity, and sharp cutoff between the high and low emissivity regions. A combination of pre-annealing treatment and conformal deposition of HfO_2 resulted in thermally stable structures.

In summary, great progress has been made in the design and fabrication of nano/microstructure SEs. However, some challenges still exist. First of all, these structures are usually so elaborate that they are hard to be used in long-term

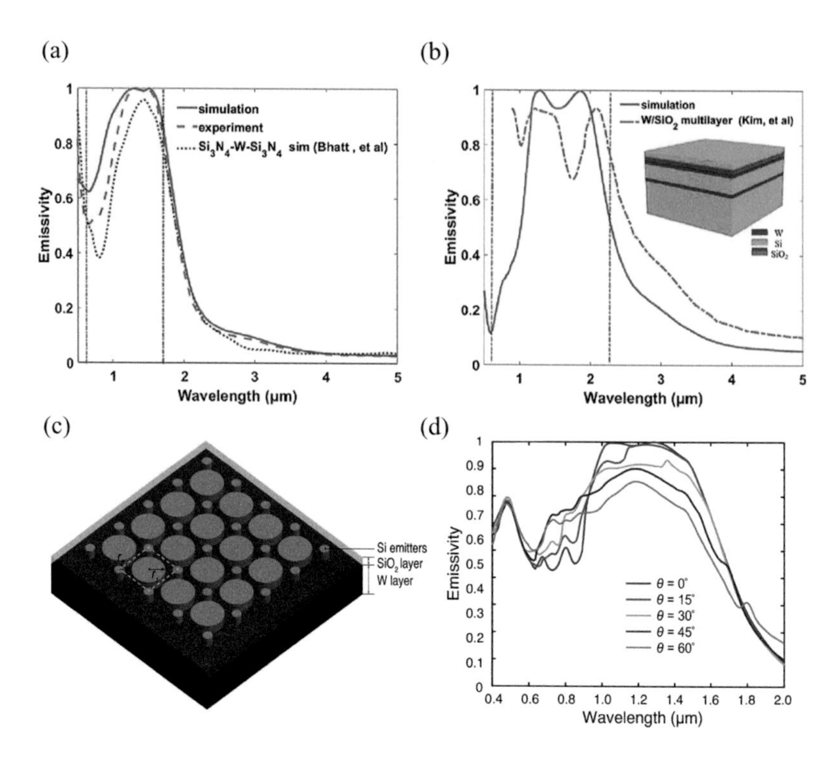

Figure 11.17 (a) Fabricated sample of the optimal selective TPV emitter. (b) Cross-sectional SEM image of the fabricated sample for the optimal structure. (a)–(b) Reprinted with permission from [1010]. Copyright 2021 American Chemical Society. (c) Simulated emissivity of the optimal structure, the experimentally measured emissivity of the fabricated sample, and simulation results. (d) A simulated emissivity of the optimal structure. (c)–(d) Reprinted from [1144], Copyright 2022, with permission from Elsevier.

high-temperature operating conditions and stand thermal shock. Second, most of the emissivity spectra of the SEs are measured indirectly at room temperature. However, material emissivity at high temperatures, especially above 1 000 K, can be larger than that at room temperature, which means that the SEs cannot show expected spectral management performance under working conditions in a TPV system.

11.3.4 Far-Field Thermophotovoltaics

Far-Field TPV Devices

With the development of TPVs, practical systems have emerged naturally. This section will introduce the applied and commercialized TPV systems, all of which belong to far-field TPV.

The world's first commercial TPV system called Midnight Sun^{TM} was designed and fabricated by Fraas et al. [1147, 1148] working for JX Crystals Inc.,

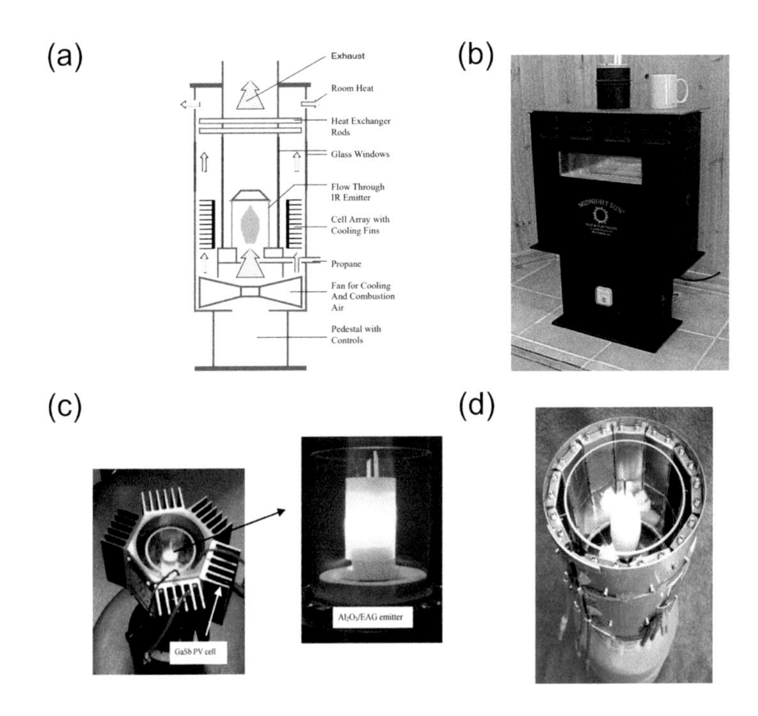

Figure 11.18 (a)–(b) The JX Crystals Midnight Sun™ TPV stove shown here was the first commercial TPV product. Reprinted from [1148], with the permission of AIP Publishing. (c) General view of the single-burner GaSb-based commercial TPV device. Reprinted from [1149], Copyright 2005, with permission from Elsevier. (d) TPV prototype system with a system efficiency of 4%. The Yb_2O_3 mantle emitter, the quartz tube for the photocell protection and the cylindrical photocell module are clearly visible. Reprinted from [1150], Copyright 2013, with permission from Elsevier.

Issaquah, WA, USA, which takes GaSb as photocell and realizes the production and practical utilization of TPVs. This TPV system takes a propane-based combustor as a heat source whose temperature can be tuned from 1 200 to 1 400 °C and utilizes air circulation fans for system cooling. It operates with a propane fuel burn rate of 7.5 kW and produces an output power of about 100 W. The device diagram of this TPV system is shown in Fig. 11.18.

Fraas et al. [1133] also designed a GaSb-based TPV furnace generator for the home as distributed combined heat and power systems, which can provide both heat and electric power supply. This system can produce an output power of about 1.5 kW and an electrical power density of 1 $W\,cm^{-2}$ that is 100 times higher than the typical solar cell under the input burn rate of 12.2 kW.

In addition to the Fraas group from JX Crystals Inc., Narihito et al. [1149] designed a GaSb-based commercial TPV device. That device used Al_2O_3/EAG as an SE, which has an emissivity peak at 1.5 μm. Moreover, the EAG SE has a higher energy conversion efficiency than the Er_2O_3 emitter. The structure of this TPV system is shown in Fig. 11.18.

Although the GaSb photocell has a higher energy conversion efficiency than the silicon photocell, the disadvantages of being expensive and toxic limit its application. At present, the most widely used and commercially available photovoltaic cell on the market is still silicon photocell. For the Si-based TPV system, some researchers also provided the corresponding system design. For example, Durisch et al. [1151] proposed a Si-based prototype TPV system and grid-interfacing device. In this device, a water layer between the emitter and photocells was utilized to protect the photocells against overheating by absorbing the nonconvertible emitter radiation. Then, in 2003, they adopted the Si photocells with a higher energy conversion efficiency of 21.1% (under solar irradiance in standard test conditions) and improved their active cell area. By changing the shape of the emitter from a sphere to a cylinder, the Si photocells could attain more uniform radiative energy, which produced an output power of 51 W and the highest system energy conversion efficiency of 2.4% as a Si-based TPV system at that time. Afterward, in 2013, they utilized Si photocells with an energy conversion efficiency of 23% for the TPV device, which led to a system energy conversion efficiency of 3.96% (Fig. 11.18).

At present, with spectral management and photocell optimization, the energy conversion efficiency, the ratio of electrical output-to-thermal radiation energy, of the TPV system can surpass 20%, while the system efficiency is still hard to exceed 10%. To further improve the system efficiency, some requirements are inevitable, including efficient conversion from primary energy to thermal radiation, an efficient cell-cooling system, and achieving a homogeneous flux density of thermal radiation incident on different cells.

Leading TPV Emitter–Cell Pairs

In a TPV system, the emitter–cell pair is the key part. Over the past few decades, energy conversion efficiency has improved dramatically with the development of TPV photocells and emitters. Except for improving the cell performance, spectral management is a vital method to realize an efficient emitter–cell pair. Spectral management includes utilizing the aforementioned SE and adding a rear reflector (also called the BSR) on the back of the cell. Table 11.2 compares the working efficiency of some typical TPV emitter–cell pairs.

The pairs with BSR show leading efficiency in the experiment, although SE is proposed earlier. BSR reflects photons with energy lower than the cell bandgap back to the emitter and recycles their energy, improving the energy conversion efficiency. For the most widely used and commercially available silicon TPV cells, applying BSR made its pairwise energy conversion efficiency ($\eta_{\text{pairwise}} = P_{\text{out}}/Q_{\text{abs}}$) achieve 29% in 1978 [1153].

Due to the bandgap of the silicon cell, to realize this high efficiency, the emitter's temperature should be as high as 2300 K. For Ge and GaSb cells with a narrower bandgap, this method is also valid, and their requirement on emitter temperature is lower than that of the silicon cell. Fernández et al. [1154] demonstrated a microstructured W emitter/Ge cell TPV system. With the help of BSR, the TPV emitter–cell pairs can achieve a high η_{pairwise} of 16.5% assuming

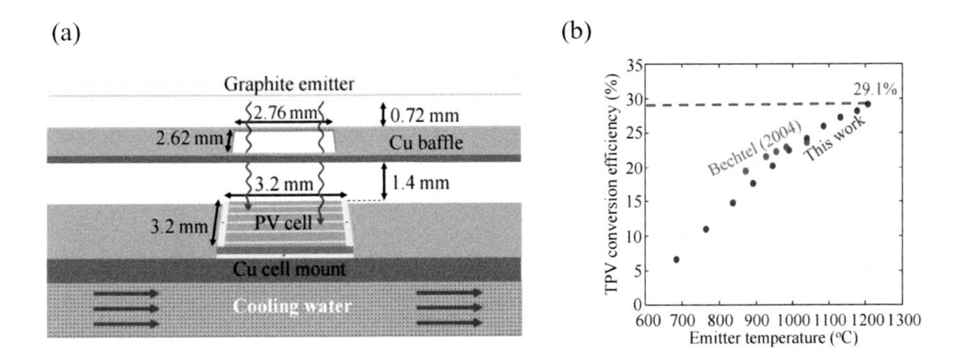

Figure 11.19 (a) Schematic illustration of the graphite/$In_{0.55}Ga_{0.45}As$ TPV system. A planar graphite emitter is Joule-heated by a large electric current. Beneath the emitter, a copper baffle is used to limit the area of the photovoltaic cell exposed to illumination. The InGaAs cell is mounted on a copper base, with water at 20 °C flowing through the copper mount. (b) TPV efficiency (corresponding to η_{pairwise}) at different emitter temperatures. The maximum efficiency is 29.1% at an emitter temperature of 1 207 °C. Reprinted from [1152]. Copyright 2019, National Academy of Science.

that the microstructured W emitter is at 1 373 K. Fraas et al. [1155] working for JX Crystals Inc. designed and fabricated a 500-W water-cooled TPV generator with the AR-coated W foil emitter/GaSb cells; this TPV system had η_{pairwise} of 21.5% working at the emitter temperature of 1 548 K.

In recent years, with the development of other III–V photovoltaic converters, the narrow bandgap cell with satisfying performance is proposed, and the pairwise efficiency is further improved. In 2019, Omair et al. [1152] proposed a $In_{0.55}Ga_{0.45}As$ TPV cell paired with a 1 480 K graphite emitter, as shown in Fig. 11.19(a). They used BSR to boost the voltage and reused infrared thermal photons, which greatly enhanced the energy conversion efficiency of the system. That system could achieve a high η_{pairwise} of 29.1% at 1 480 K (Fig. 11.19(b)). It is worth noting that Fan et al. [1130] further improved the performance of BSR in 2020. They demonstrated the near-perfect reflection of low-energy photons by embedding a layer of air bridge between the thin-film $In_{0.53}Ga_{0.47}As$ cell and the Au film. The air-bridge reflector can realize 99% OOB photons reflection and then suppress the OOB absorption. This system can realize a η_{pairwise} of 32.0% under the illumination of a SiC globar at 1 455 K.

Multi-junction cells, multiple p–n junctions made of different semiconductor materials, in a TPV system can utilize thermal radiation energy in a wider wavelength range and improve the pairwise efficiency. Recently, LaPotin et al. [1156] demonstrated the TPV systems with a W halogen bulb emitter–multi-junction PV cells pairs. The pairwise energy conversion efficiency of up to 40% can be achieved, as shown in Fig. 11.20.

The improvement of TPV pairwise efficiency is arresting. It is worth noting that these efficiencies depend on testing conditions, including but not limited

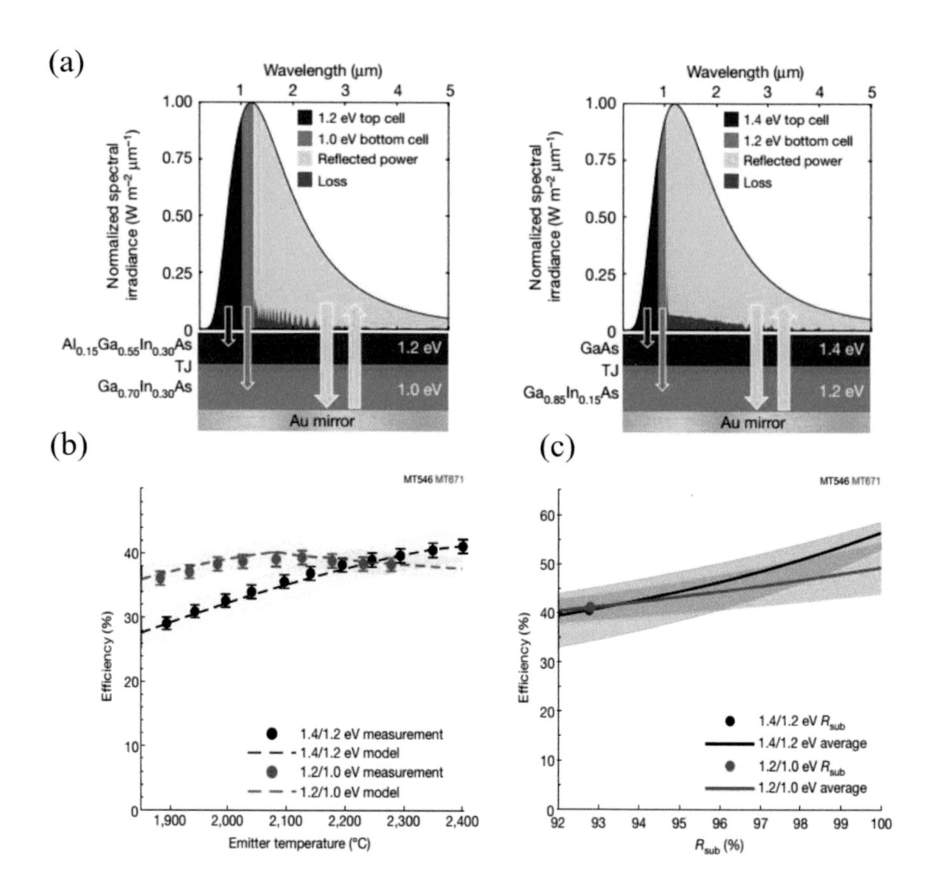

Figure 11.20 (a) The 1.2/1.0 eV (left) and 1.4/1.2 eV (right) tandems that were fabricated and characterized in [1156], and a representative spectrum shape at the average emitter temperature (2 150 °C blackbody) indicating the spectral bands that can be converted to electricity by the top and bottom junction of a TPV cell. A gold mirror on the back of the cell reflects approximately 93% of the below bandgap photons, allowing this energy to be recycled. T_j represents the tunnel junction. (b) TPV efficiency is measured at different emitter temperatures ranging from approximately 1900 and 2400 °C. (c) Predicted efficiency (η_{pairwise}) of the 1.4/1.2 eV and 1.2/1.0 eV tandems as the weighted sub-bandgap reflectance (R_{sub}) is extrapolated assuming a W emitter with AR = 1, VF = 1 and a 25 ° cell temperature. Reprinted from [1156]. Copyright 2020, Nature Publishing Group.

to the emitter temperature, the cell temperature, the view factor between the emitter and the cell, and the spectrum of incident thermal radiation. Great efforts are still required to achieve pairwise efficiency in a real TPV system.

11.3.5 Prospects and Challenges

Since 1956, TPVs have grown immensely over the past few decades. For the far-field TPV, the energy conversion efficiency has improved dramatically with the

Table 11.2 Leading TPV emitter–cell pairs.

Cell	Emitter	T_h(K)	η_{pairwise} (%)	Reference
Si	W (presumed blackbody)	2 300	29	[1153]
Ge	Microstructured W	1 373	16.5	[1154]
GaSb	W foil with ARC	1 548	21.5	[1155]
$In_0.55Ga_{0.45}As$	Graphite	1 480	29.1	[1152]
$In_0.53Ga_{0.47}As$	SiC emitter	1 455	32.0	[1130]
$GaAs/In_0.53Ga_{0.47}As$	W halogen bulb emitter	2 673	41.1	[1156]
$Al_0.15Ga_{0.55}In_{0.3}As/Ga_{0.7}In_0.3As$	W halogen bulb emitter	2 400	39.3	[1156]

development of TPV photocells and emitters. In particular, some recent works have demonstrated that the BSR fabricated at the back of TPV photocells can significantly enhance TPV efficiency. Currently, a pairwise energy conversion efficiency of up to 40% has been achieved. These works bring TPVs closer to realizing their practical application in providing power. However, the far-field TPVs are also faced with some challenges, such as the insufficient thermal stability of emitters at high temperatures and the tension between cost per power and Ohmic losses associated with high current densities. Moreover, a consensus regarding efficient testing and reporting of TPV systems needs to be reached in this field. For the near-field TPV, most of the work is still in the laboratory stage. More efforts are needed to realize its application in the future.

11.4 Metamaterials-Based Energy Nanodevices: Photothermal Conversion Devices

Solar energy has become one of the most important renewable energy sources for several decades, which plays an irreplaceable role in photothermal, photoelectric, and photocatalysis applications. Especially, in recent years, solar-steam generation has gained much attention due to the rapid development of advanced nanofabrication technologies and imperious demands for fresh water in several countries and regions. Extracting fresh water from seawater, wastewater or, even atmospheric water has become a critical challenge. Recently, several research groups have proposed some technologies to design solar-steam-generation devices to obtain fresh water, in which various materials, structures, and devices are designed. In this section, we would like to give a brief introduction to the development of this area.

11.4.1 Basic Introduction and Design Principle

The key process in solar-steam generation is to convert solar energy directly into heat for water evaporation even below the boiling temperature. Therefore, one of the key issues is to improve the absorption of sunlight. Figure 11.21 shows the solar spectrum covering a broad range of the ultraviolet, visible, and NIR regions. As for the standard AM1.5, the corresponding portions of power intensity are denoted in Fig. 11.21. A desirable solar absorber should exhibit broadband near-perfect absorptivity over the whole solar spectrum.

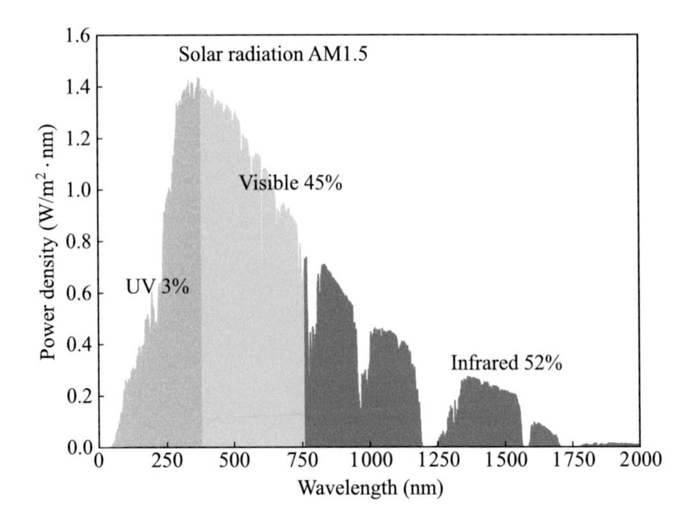

Figure 11.21 Power density distribution of solar spectrum.

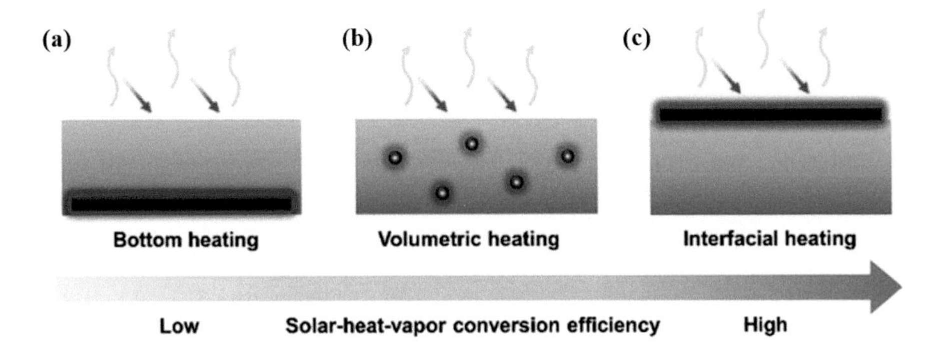

Figure 11.22 Schematic illustration of different strategies for solar-steam generation. Used with permission of Royal Society of Chemistry, from [1157]; permission conveyed through Copyright Clearance Center, Inc.

Another important factor is thermal management when heating the water, and better heat localization ensures high solar-thermal conversion efficiency. Previously, the bottom heating and volumetric heating methods have been used as shown in Figs. 11.22(a) and (b). One of the seminal works of volumetric solar-thermal conversion was done by Neumann et al. [1158]. As shown in Fig. 11.22(a) by employing the absorbing metal and carbon nanoparticles dispersed in water, the authors experimentally demonstrated the direct steam generation, in which 80% absorbed sunlight could convert into water vapor. Later, Chen's group [1159] also proposed a nanofluid-based solar receiver for steam generation. In Fig. 11.23(b), the designed nanofluid possesses graphitized carbon black,

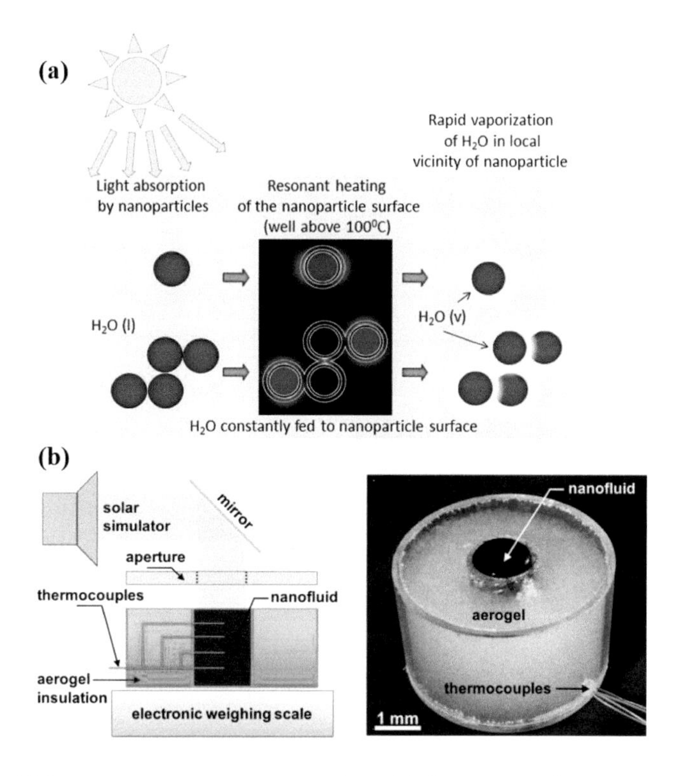

Figure 11.23 Volumetric solar-steam generation. (a) Schematic of nanoparticle-enabled solar-steam generation. Reprinted with permission from [1158]. Copyright 2013 American Chemical Society. (b) Left: Schematic of solar vapor generation device. Right: Image of the nanofluid container shows the aerogel insulation, black nanofluid, and thermocouple feed through. Reprinted from [1159], Copyright 2015, with permission from Elsevier.

carbon black, and graphene. In this regard, the measured total solar-thermal conversion efficiency can be up to 57%. Notably, such strategies will inevitably heat the bulk water and introduce thermal loss into the surrounding water, lowering the total conversion efficiency. To mitigate this problem, the interfacial solar-steam-generation method was proposed as shown in Fig. 11.22(c). In this regard, the solar-absorption and water evaporation both occur at the interface. Many efforts have been paid on the interfacial heating strategy combined with several superior solar absorbers, which will be discussed in detail in Section 11.4.2.

On the other hand, continuous water supply and steam escape are also important in solar-steam-generation designs. The weight of the interfacial absorber should be lower to float on the water surface, and the hydrophobic property should also be considered to maintain an adequate water supply for evaporation. Accordingly, in some review papers [1157], the authors divided methods of water supply into three types, as shown in Fig. 11.24, including 1D, 2D, and 3D

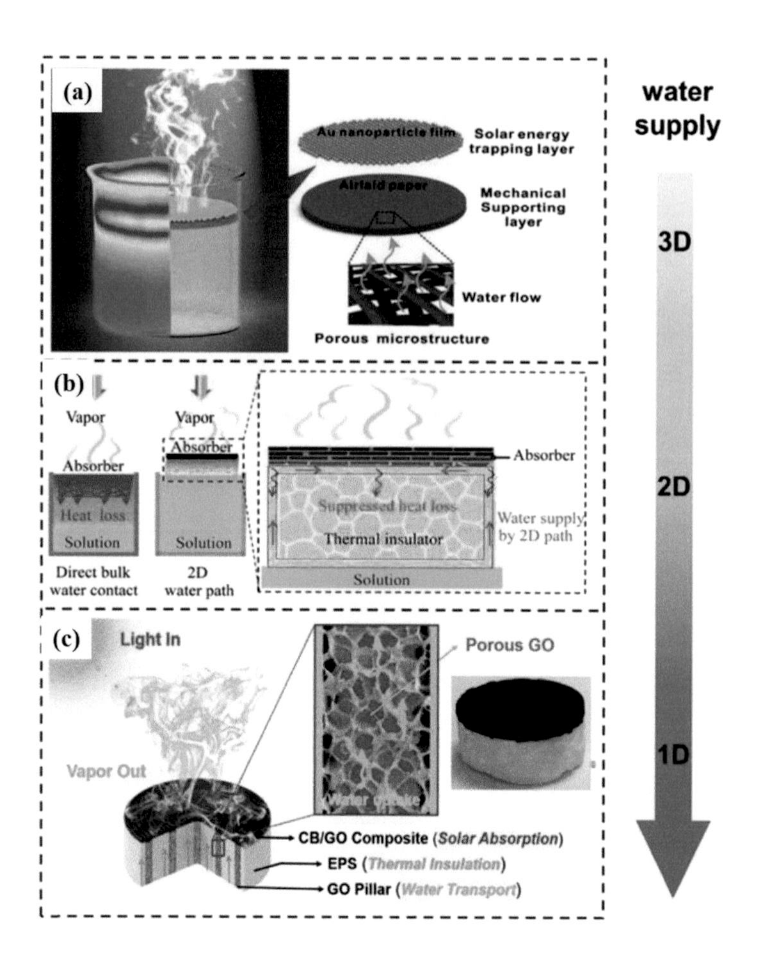

Figure 11.24 Various water supply strategies of interfacial water evaporation. Reproduced with permission from [1157]. Copyright 2013, with permission from Royal Society of Chemistry. (a) Structure schematic illustration of the paper-based gold nanoparticle film. Reproduced with permission from [1160]. Copyright 2015 Wiley-VCH GmbH, Weinheim. (b) Schematics of solar evaporators with direct water contact and 2D water supply. Reproduced with permission from [1161]. Copyright 2016 the Authors. Published by PNAS. (c) Principle illustration of the expedited water transport owing to the capillary effect of 1D porous GO pillars. Reprinted from [1162]. Copyright 2017, with permission from Elsevier.

water supply structures. For example, in [1160], they proposed a bioinspired and paper-based system composed of an airlaid paper substrate and a Au nanoparticle film as shown in Fig. 11.24(a). In this case, the water below can be transferred upward to the heating platform via capillary forces, for the purpose of continuous water supply. Besides, the airlaid substrate has a low conductivity, contributing to reducing heat loss to some extent. To further reduce the unnecessary heat loss, Zhu's group also designed a confined 2D water path using a folded graphene oxide film, as shown in Fig. 11.24(b), which serves as a solar absorber, vapor channel, and thermal insulator with small heat loss and high absorption over the solar spectrum. To this end, the solar-thermal conversion efficiency can be enhanced from 30% to 80% under 1 sun illumination. Similarly, a 1D water supply path has also been proposed as shown in Fig. 11.24(c), in which porous graphene oxide pillars were considered as the 1D water path. Such designs reduce the contact areas between the upper solar absorber and bottom unevaporated waters, leading to the reduction of heat loss. The distributions of 1D pillars can be random, which will accelerate the speed of the water supply.

11.4.2 Solar-Steam Generation: Materials, Structures, and Devices

As introduced in Section 11.4.1, the mentioned features of designs are of equal importance for high efficiency, stable, and sustainable solar-steam generation. In the following, we would like to give a brief introduction of the recent development in materials, structures, and advanced devices for improving the performance of solar absorbers, water supply paths, and thermal insulators.

Materials Selection

As for photon-thermal applications, perfect solar absorbers play a key role to ensure high-efficiency solar-to-thermal conversion. Over the past few years, many efforts have been dedicated to designing high-efficiency and low-cost materials for solar-steam generation, including carbon-based materials [1162–1165], metallic nanomaterials, semiconductors, and organic polymers. In the following, some typical materials are briefly introduced.

Carbon-based materials have been widely used owing to their ultrabroadband perfect absorption performance. Then, carbon-based solar absorbers, such as graphite, carbon sponge, graphene oxide, carbon nanotubes, or carbonization materials, can ensure high absorption efficiency to harvest solar energy. Graphene materials are good candidates in solar-thermal conversion since such materials have relatively low molar specific heat, high Debye temperature, tunable thermal conductivity, and high light-absorption ability. But their inferior hydrophobic features introduce the challenge in interfacial solar-steam generation. Ito et al. [1165] developed multifunctional 3D porous N-doped graphene (35 μm) as illustrated in Fig. 11.25(a). The pore size in graphene can be controlled by changing the temperature during chemical vapor deposition. The 3D porous nanostructures play essential roles in both water supply for capillary effects and heat locations. In their experiment, the porous graphene can realize the solar-thermal conversion at a high efficiency of 80%. Later, by employing the scalable antifreeze-assisted freezing technique, Zhang et al. [1163] fabricated a vertically aligned graphene sheet membrane that possesses high light-absorption capacity, water transport channels, and outstanding stability. In this case, the conversion efficiency can achieve 86.5% under 1 sun illumination. Besides, it is worth noting that, some natural organism also provides inspiring ideas. For

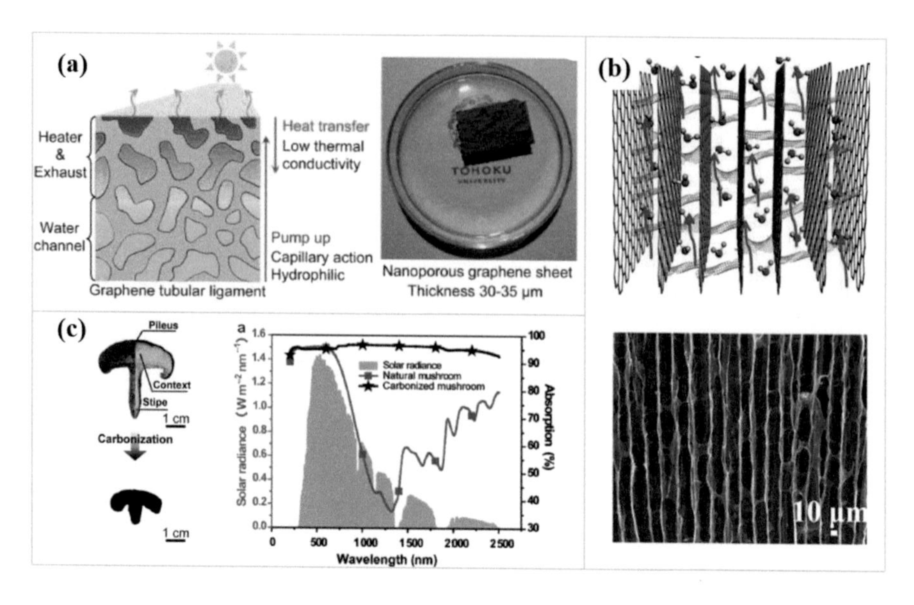

Figure 11.25 Carbon-based solar-thermal absorber. (a) Left: Schematic illustration of heat localization system to convert sunlight into steam using a piece of porous N-doped graphene as the steam generator. The thin porous graphene plays versatile roles in harvesting solar illumination as thermal energy. Right: Optical image of a piece of porous N-doped graphene sheet. Reproduced with permission from [1160]. Copyright 2015 Wiley-VCH GmbH, Weinheim. (b) VA-GSM for solar-steam generation. Reprinted with permission from [1163]. Copyright 2017 American Chemical Society. (c) The physical picture of a shiitake mushroom and a mushroom after carbonization. Solar spectral irradiance (AM 1.5 G) (gray, LHS axis) and absorption (black, RHS axis) of natural and carbonized shiitake mushrooms. Reproduced with permission from [1164]. Copyright 2017 Wiley-VCH GmbH, Weinheim.

example, Xu et al. [1164] first found that mushrooms can be efficient solar-steam-generation devices as shown in Fig. 11.25(c). The authors compared the solar-thermal conversion performance between natural and carbonized mushrooms under 1 sun. The corresponding absorption spectra are given in Fig. 11.25(c) (right), in which broadband perfect absorption over the whole solar spectrum is visible for carbonized mushrooms. Then, the conversion efficiency can reach 62% and 78%, respectively. Even though the efficiency is expected to be further improved, such methods provide inspiration for further bio-inspired solar-thermal devices.

Another widely-used material for solar-steam generation is the metallic absorber. Especially, metallic nanoparticles support localized plasmon resonances near the plasmon frequency, leading to enhanced energy localization. When such on-resonant nanoparticles are randomly dispersed in the water, which can further increase photon absorption due to multiple scattering [1166] as demonstrated in Fig. 11.26(a). Some seminal works were completed by Prof. Zhu's

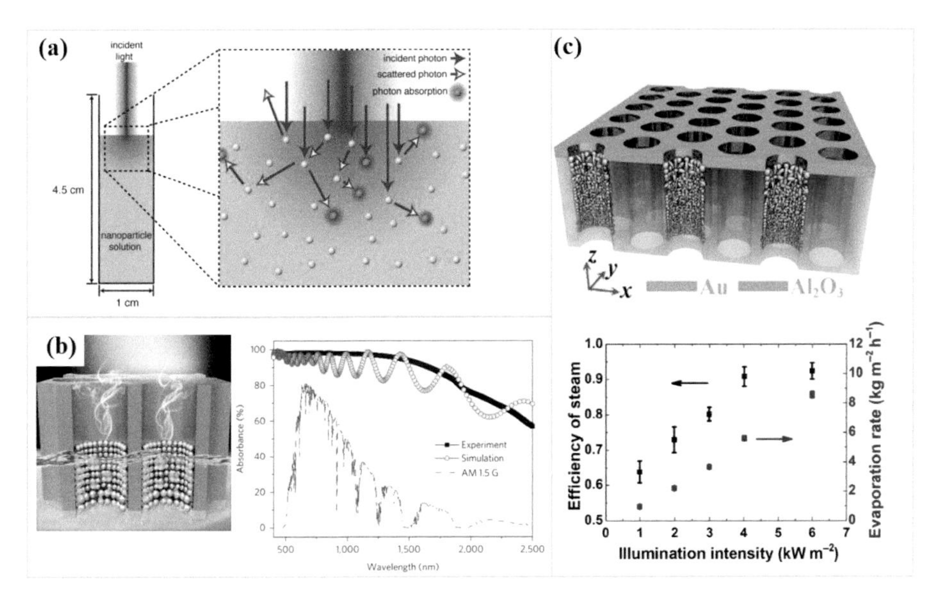

Figure 11.26 Metallic-nanostructures solar-thermal absorber. (a) Schematic illustrating characteristic experiment (left) where a dense solution of nanoparticles contained in a cuvette is illuminated with 808 nm laser light; multiparticle optical interactions in such nanofluids (right) where photons are scattered and/or absorbed. Reprinted with permission from [1166]. Copyright 2014 American Chemical Society. (b) Left: Al NP-based plasmonic structure. Right: experimental and simulated absorption of aluminum-based plasmonic absorbers. Reproduced with permission from [1008]. Copyright 2016, Springer Nature. (c) (Top) 3D schematic of self-assembled plasmonic absorbers. (Bottom) Solar-steam efficiency (black, LHS axis) and evaporation rate (blue, RHS axis) with plasmonic absorbers as a function of illumination intensity on the absorber surface. Reprinted from [1009].

group [1008, 1009]. For example, Zhou et al. [1009] demonstrated a plasmon-enhanced solar desalination device by using self-assembly of Al nanoparticles into a 3D porous membrane in Fig. 11.26(b) (left). Such nanostructures can float on the water and harvest nearly 96% solar energy (right), ensuring 90% desalination efficiency. Similarly, the authors also employed such designs to generate solar steam. Instead, Au nanoparticles are embedded into Al_2O_3 nanoholes structures as shown in Fig. 11.26(c) (top). The measured steam-generation efficiency is shown in the bottom inset, in which nearly 90% solar-thermal conversion efficiency can be available.

In addition, metal oxides like WO_x [1167, 1168], FeO_x [1169] and TiO_x [1170], support intervalence charge transfer, localized surface plasmon resonances of free electron and small polaron absorption, leading better absorption performance covering visible to infrared regions and making them good candidates in solar-thermal applications. In this regard, one of the most popular materials is Tungsten oxide (WO_x) due to the efficient light-to-heat conversion, strong solar spectrum absorption, and surface hydrophobicity. For example, Chala et al. [1167]

proposed reduced $WO_{2.72}$ nanoparticles (absorbers) incorporated into polylactic acid (matrix). The membrane with 7 wt% $WO_{2.27}$ nanoparticles can achieve 81.39% water evaporation efficiency. The TiO_2 nanotubes combined with Ni nanoparticles were designed by Chen et al. [1170]. The combined localized plasmon resonances help to realize 96.83% solar absorption and the reported overall solar-thermal efficiency is about 78.9%. It is noted that, high power density is usually needed when using metallic nanoparticles (including metal oxides), which may be not cost-ineffective.

More recently, a new class of material, namely MXene, has gained much attentions in thermal areas, also including thermal camouflage and solar-steam generation. This type of material is a new family of multifunctional 2D material with the chemical formula of $M_{n+1}X_nT_x$, in which M is an early transition metal, X is carbon or nitrogen and T represents the surface functional groups. It shows high internal light-to-heat conversion efficiency (100%), and thus can be good candidates for steam-generation devices. Fig. 11.27 shows two typical examples. Li et al. [1171] designed hierarchical MXene nanocoatings ($Ti_3C_2T_x$

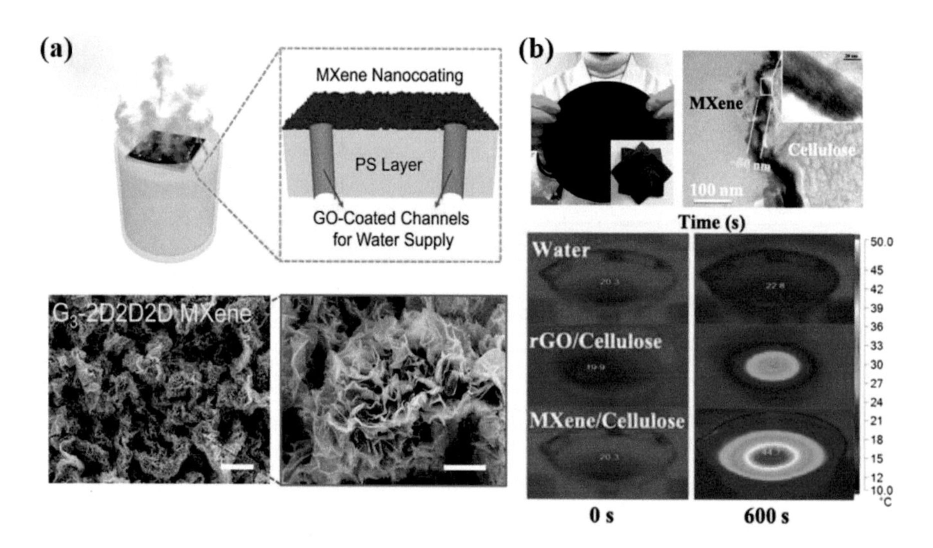

Figure 11.27 MXene-based solar-thermal absorber. (a) (Top) SEM image (left, scale bar: 50 μm) and enlarged image (right, scale bar: 10 μm) of hierarchical G3-2D2D2D MXene structures. (Bottom) Schematic illustration of the solar-steam-generation device with the bioinspired MXene nanocoating for high solar-thermal conversion. Reproduced with permission from [1171]. Copyright 2019, Wiley-VCH Verlag GmbH & Co. KGaA Weinheim. (b) (Top left) Digital image of MXene/cellulose membrane (diameter 15 cm and thickness 0.2 mm). The inserted image of a small flower folded with the MXene/cellulose membrane exhibits good flexibility. (Top right) TEM cross-section image of the MXene/cellulose membrane. The inserted magnified image shows the layer structure of MXene sheets. (Bottom) IR thermal images of bulk water, MXene/cellulose, and rGO/cellulose membrane surface. Reprinted with permission from [1172]. Copyright 2019 American Chemical Society.

MXene) as shown in Fig. 11.27(a) (bottom), which can ensure over 93% broadband absorption. In this case, superior solar-steam generation can also be ensured (1.33 kg m^{-2} h^{-1}) even using a very small portion of materials (0.32 mg cm^{-2}). Notably, the mechanically deformed MXene possesses stretchable and wearable features, which can be reversibly stretched and heated at a high temperature. Similarly, Zha et al. [1172] fabricated fibrous photothermal structures composed of a MXene-coated cellulose membrane, which exhibit ~94% light-absorption efficiency and the evaporation rate can be up to 1.44 kg m^{-2} g^{-1} under one solar illumination. Besides, the MXene/cellulose also has high antibacterial efficiency (99%), showing great potential in long-term, biofouling, and high-efficiency wastewater treatments.

Structures Design

The key feature of structures for solar-thermal conversion should guarantee high absorption, superior thermal insulation, and continuous water supply. Therefore, solar-steam-generation structures normally contain two functional layers, including optical absorbers and thermal insulators. Some composite structures have been discussed above like MXene/cellulose membrane. For example, Min et al. [1177] synthesized a membrane composed of carbon black film as the top absorber layer and a porous poly(vinylidene difluoride) film embedded with pyroelectric $BaTiO_3$ nanoparticles as the bottom layer. The composite membrane reached a fast average rate of temperature change of 3 °C s^{-1} and the instantaneous rate of 13 °C s^{-1}. Similarly, Tao et al. [1178] proposed an integrated structural system consisting of graphite power(GD) and a semipermeable collodion membrane (SCM). The 8 mg-GP/SCM ensures the 56.8% efficiency of water evaporation under 1.5 sun illumination.

Another popular structure for solar-steam generation is the porous configuration, which has been mentioned in the above discussion about solar-steam materials. In this regard, hydrogels and aerogels have excellent solar-thermal conversion performance owing to their porosity micronanostructures and good thermal insulation. For example, Geng et al. [1179] demonstrated the negative temperature response of poly(N-iso-propylacrylamide) hydrogel (PN) anchored onto a super hydrophilic melamine foam skeleton, coated with a layer of PNIPAm-modified graphene filter membrane. The properly designed structures exhibit a collection of 4.2 kg m^{-2} h^{-1} and an ionic rejection of $>$99% under one sun illustration. Besides, aerogels are also a good porous nanostructure. Fu's group has employed modified graphene aerogel as solar absorbers combined with 3D porous networks formed by randomly reduced graphene oxide sheets and hydrophilic surface. Under one sun, the evaporation efficiency can be 76.9%.

On the other hand, some bio-inspired structures have also been proposed in recent years as exampled in Fig. 11.28. Figure 11.28(a) illustrates a tree-like structure, in which the hydrophilic air-laid paper working as the root of a tree, the expanded polyethylene foam as the soil (thermal insulator) and black carbonized wood with distributed channels as the solar absorber. Such an interesting

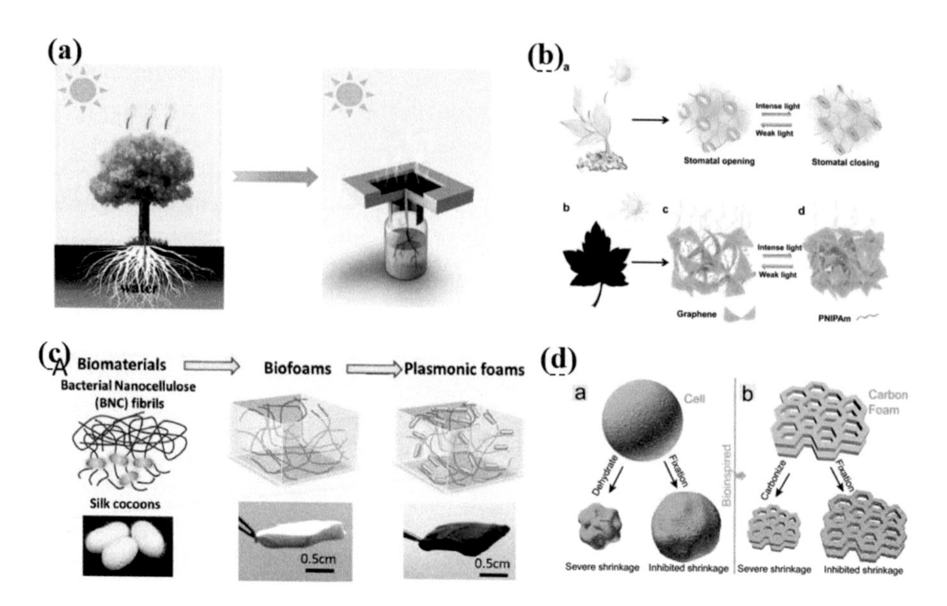

Figure 11.28 Bioinspired solar-thermal structure. (a) Transpiration of trees and MTS under solar illumination for solar-steam generation. Reprinted from [1173], Copyright 2018, with permission from Elsevier. (b) (Top) Schematic illustration of the stomatal opening/closing of leaves under intense and weak sunlight irradiation, respectively. (Bottom) A leaf-like mG/PNIPAm membrane. Under solar irradiation of different intensities, the water transport channels can be autonomously turned on and off by the opening and closing of microstructures. Reproduced with permission from [1174]. Copyright 2018 Wiley-VCH GmbH, Weinheim. (c) Schematic illustration showing the fabrication of plasmonic foams (two photographs in the bottom right show the aerogel before and after dense loading of AuNRs). Reprinted with permission from [1175]. Copyright 2016 American Chemical Society. (d) Left: Schematic diagrams showing shape transformations of biological cells after dehydration with and without fixation treatment. Right: Schematic showing shape transformations of 3D foam after the carbonization process with and without fixation treatment. Reproduced from [1176]. Copyright 2013, with permission from Royal Society of Chemistry.

design can achieve 91.3% photon-thermal conversion efficiency under only one sun. Zhang et al. [1174] proposed an intelligent solar-steam evaporation structure imitating the process of stomatal opening and closing of leaves as shown in Fig. 11.28(b). It contains a microstructured graphene/poly(N-isopropylacrylamide) membrane. The water transfer channels can be autonomously turned on and off by the opening and closing of microstructures depending on the intensity of light. The reported water evaporation rate is 1.66 kg m^{-2} h^{-1} under weak sunlight and 1.24 kg m^{-2} h^{-1} under intense sunlight. Combined with porous biomaterial aerogels, Tian et al. [1175] proposed a feasible method to create plamonically active three-dimensional biofoams (in Fig. 11.28(c)), showing superior solar-steam-generation performance compared with bare aerogel and plasmonic paper. Zhang et al. [1176] have employed a novel "nano-fixation" method, inspired by biological

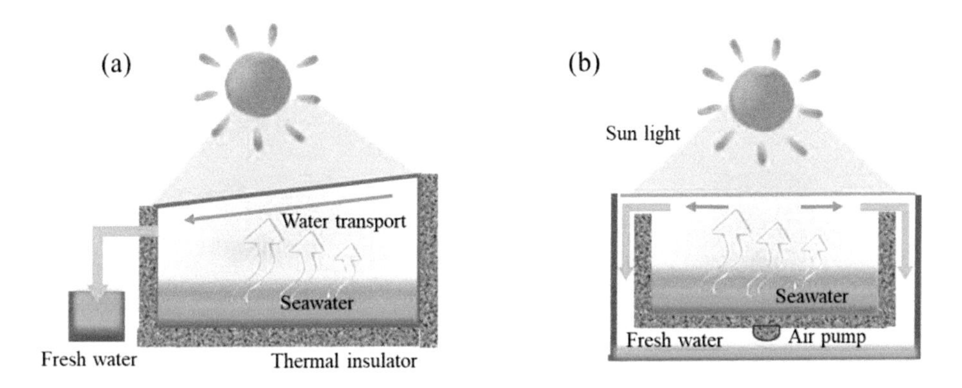

Figure 11.29 Two different water production types of solar-steam generation devices: (a) Passive water transport with gravity structure design and (b) Active water transport with additional power consumption. Reprinted from [1180], Copyright 2020, with permission from Elsevier.

fixation to design solar-thermal nanostructures, which involves in situ crystallization of MoS_2 nanocrystal on a melamine-driven carbon skeleton as illustrated in Fig. 11.28(d). Thanks to the novel properties of skeleton-fixed foam, such designs show a high water evaporation rate (1.458 kg m^{-2} h^{-1}) and high conversion efficiency (90.4%) under one sun.

Solar-Steam-Generation Devices

In the aforementioned sections, several types of materials and structures have been discussed for enhancing the performance of solar-steam generation. While, in practice, large-scale devices are urgently needed and one should consider how to collect water evaporation. Generally, the designed solar-steam-generation device has a single basin, which consists of a monoclinic cover, cone-shaped, quasi-semi-spherical, house model, and pyramid prototype. As introduced in [1180], there are two methods for evaporation collection: one is passive water transport and the other is active, as shown in Fig. 11.29. The first strategy depends on the gravity of the monoclinic cover, and the other relies on additional power supplements. Figure 11.30 shows some typical devices. In Fig. 11.30(a), Wang et al. [1181] designed a device in which the solar absorber floats on the seawater and the glass window slopes as the condensation cover to collect clean water. The whole device is wrapped with a thermal insulator to reduce heat loss. The experimental setup is shown in the bottom inset. Similarly, Higgins et al. [1182] also propose a similar device as shown in Fig. 11.30(b), where the impure water and purified water are separated by a wall. Alternatively, Wu et al. [1183] created a biomimetic 3D structure as evaporators. For example, the brine will be added into the container through the inlet, and steam water evaporates on the designed 3D evaporator surface under the sum illustration. In the following, the vapor will condense on the upper surface and sidewall, which can be collected

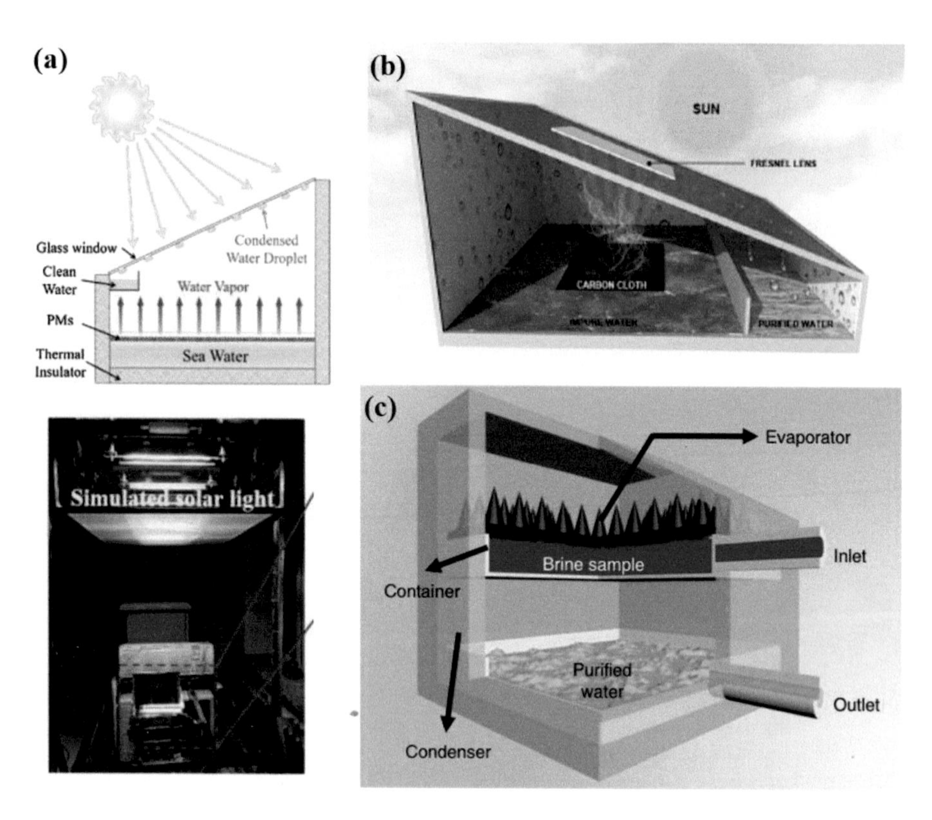

Figure 11.30 Solar-steam-generation devices. (a) (Top) Schematic illustration of a single basin solar still. (Bottom) The experiment was set up for seawater desalination. Reprinted from [1181], Copyright 2017, with permission from Elsevier. (b) The schematic diagram of the prototype used for water purification. Reprinted from [1182], Copyright 2018, with permission from Elsevier. (c) Scheme of the batch purification prototype which simulates the practical solar water purification apparatus. The brine sample is introduced into the container through the inlet. Water evaporates on the 3D evaporator surface under illumination, and then condenses on the upper surface and sidewall of the condenser, which is finally collected by the outlet. Reproduced from [1183]. Copyright 2020, Nature Publishing Group.

by the outlet. The reported water evaporation rate is about 2.63 kg m^{-2} h^{-1}, and the energy conversion efficiency can be up to 96% under one sun condition.

11.4.3 Prospects and Challenges

Over the last decades, solar-steam technology has witnessed great development in several aspects as discussed above. But, there are some nonnegligible problems and challenges. Firstly, there is still no clear criterion to evaluate an individual device, since the proposed designs usually work under different conditions. It is hard to compare their performance and determine which one is superior under a reasonable rule. Besides, most of the works in this area were studied on a lab scale. To meet the requirements of real large-scale situations, several practical problems should be carefully considered, like the cost of the fabrication and the actual working efficiency. In addition, the service life, the stability of the whole system, and the applicability in different working scenarios should be assessed in a scientific evaluation system.

11.5 Applications of Near-Field Radiative Heat Transfer

11.5.1 Near-Field Thermophotovoltaics

Although the far-field TPV has been extensively investigated in the last two decades, the upper limit of heat flow density of macro-scale TPV devices is limited by the temperature of the heat source, which greatly limits the output power of the far-field TPV system. So far, the research on far-field TPVs has not solved the problem.

However, when the gap distance between the emitter and PV cell is in the same magnitude as the wavelength of thermal radiation, the radiative heat transfer between two adjacent objects will directly exceed the limitation of Planck's Blackbody radiation law, which results from the near field effect of the evanescent wave. This type of heat transfer enhancement can bring an increase in the amount of energy transfer inside the TPV system, which will lead to a direct increment of the output power. The concept of near-field thermal photovoltaic provides the possibility to solve the limitation of presented TPV systems.

Early in 2001, the concept of near-field thermal photovoltaic was confirmed by the experiment conducted by DiMatteo et al. [689]. In this experiment, the output current of TPV system with submicron gaps is observed to be 5 times higher than the system with gaps in the far-field. Then in 2007, a near-field TPV system with a continuously variable gap was realized for the first time by Hanamura et al. [1185, 1186]. Their experiment showed that the jump in output current appeared when the gap distance was reduced to less than 2 μm.

In recent years, with the further investigation of near-field radiation, the researchers of near-field TPVs have also made great progress. In 2018, Fiorino et al. [385] developed and fabricated functional NFTPV devices consisting of a microfabricated system and a custom-built nanopositioner. About 40-fold enhancement in the power output at gaps of 60 nm relative to the far field can be realized by this system. Then in 2019, Inoue et al. [1184] designed a one-chip near-field TPV device consisting of a thin-film Si emitter and InGaAs PV cell (Fig. 11.31). A 10-fold enhancement of the photocurrent in the PV cell compared to a far-field device and a large temperature dif-

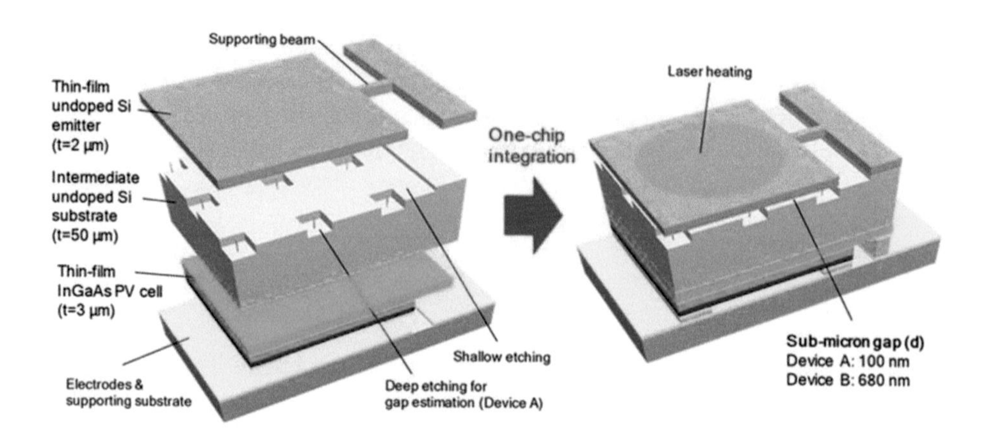

Figure 11.31 Schematic of the proposed one-chip near-field TPV device. Reprinted with permission from [1184]. Copyright 2019 American Chemical Society.

ference (>700 K) between the emitter and PV cell can be achieved by this system.

The fast advances of fundamental understanding in both the theoretical and experimental studies on NFRHT lead to various promising applications. This section provides a survey of recent advances in identifying and demonstrating potential applications in energy conversion, thermal management and thermal circuits, near-field imaging, local heating and cooling, and nanomanufacturing.

11.5.2 Energy Conversion

The pronounced potential in enhancing energy conversion has been one of the main forces driving the recent advances in near-field thermal radiation. TPV devices usually consist of a thermal emitter and a receiver typically made of low-bandgap photovoltaic cells, can convert the thermal energy or heat directly into electricity. When the vacuum gap distance between the TPV emitter and the cell is much smaller than the thermal wavelength, this device called near-field TPV [385, 1184, 1187]. The technology of near-field TPV is advantageous because NFRHT can be increased to exceed the far-field blackbody limit predicted by Planck's law.

11.5.3 Near-Field Thermal Management and Thermal Circuits

One promising application with large modulation of NFRHT is to design vacuum thermal rectifiers or thermal diodes, which could transfer the heat flow in a preferential direction as an analog to electrical diodes. The vacuum thermal rectifier is desired to have a larger radiative heat flux between two materials than that with reverse-biased temperature difference [1188–1190].

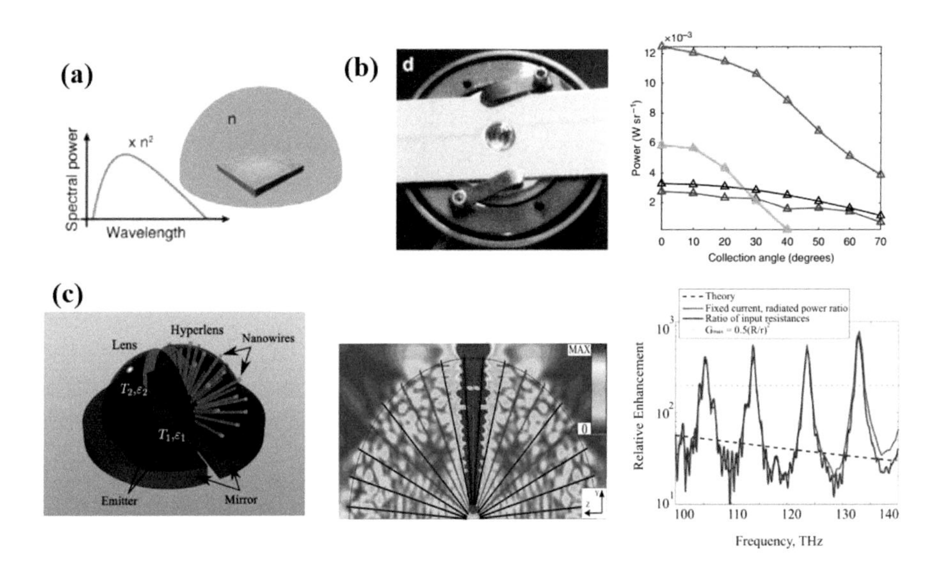

Figure 11.32 Thermal extraction behavior using nanostructures. (a) Schematic of thermal extraction, where the background index is larger than air. Reprinted from [232]. (b) Left: Directional radiation control. Right: Experimentally measured emission power from the carbon dot as a function of angle [1191]. (c) Two hemispherical structures offer super-Planckian thermal radiation from a finite emitter (left). Electric-field amplitude distribution in the H-plane produced by a dipole located in between the internal ends of perfect nanowires of radius (25 nm) forming the hyperlens (middle). Relative enhancement of radiation of a TE dipole due to the presence of a hyperlens formed by perfectly conducting wires calculated (1) directly via the radiated power spectrum (green curve) and (2) through the input resistance of a short wire dipole (red curve). Reprinted from [1192], Copyright 2015, with permission from Elsevier.

11.5.4 Thermal Extraction

As discussed in the last section, near-field extraction of thermal emission can achieve super-Planck's radiation in several orders of magnitude compared with conventional blackbody radiation. Then, another question arises, is it possible to realize super-Planck radiation in the far field?

We would like to overview Planck's law again:

$$P(\omega,\hat{n}) = e(\omega,\hat{n})P_0 \leq P_0 = A \cdot \frac{\omega^2}{4\pi^2 c^2} \frac{\hbar\omega}{e^{\hbar\omega/K_B T} - 1}. \tag{11.5.1}$$

Apparently, the P_0 determines the upper limit of thermal radiation in the far filed for macro objects. While, in a micro or nanoscale, something will be different and interesting. Note that the area A in Eq. (11.5.1) should be considered as the absorption cross section, instead of the geometrical cross section of the emitters. As for a macro object surrounded by a vacuum, its absorption cross section cannot exceed the corresponding geometrical cross section, thus the overall emission power is always smaller than blackbody emission obeying Planck's law. While, when considering placing an emitter in a dielectric medium with a refractive index n as depicted in Fig. 11.32(a), the effective light speed will change to c/n. As a result, the total emission power will be improved by n^2, showing thermal extraction performance.

On the basis of the aforementioned mechanism, Yu et al. proposed a thermal extraction scheme by placing a thermal emitter in ZnSe dome ($n_{ZnSe}=2.4$) as shown in Fig. 11.32(b). The experimental results are also given and verified that the measured thermal emission exceeds the blackbody emission (black line) over all angles in Fig. 11.32(b). Similarly, Simovski et al. further investigated this behavior by considering hyperbolic metamaterials composed of nanowire arrays as shown in Fig. 11.32(c). The role of hyperbolic metamaterials like hyperlens as shown in the middle figures in Fig. 11.32(c) is to convert the emitter's near-fields into propagating waves, thus making the emitter's spectral radiation go beyond the blackbody limit.

11.6 Summary

This chapter has introduced several important applications in terms of metamaterials-based thermal energy harvesting, conversion, and high-efficiency thermal management technologies. The employments of metamaterials-based thermal emitters/absorbers make great contributions to improving the working efficiency, enriching the functions, and renovating several advanced technologies. We anticipate that thermal radiation engineering will continue to improve technologies where radiative heat transfer can play a significant role.

We imagine that there are still several opportunities for thermal metamaterials in various areas. Firstly, metamaterial thermal devices can also be utilized to enhance the efficiency of selective lighting devices. For a general concept, in comparison with conventional thermal lighting sources, a desirable light source should be wavelength (and angle) selective, to avoid unnecessary heat loss emitting from infrared waveband. For example, a tungsten filament can be heated up to 2000K, which leads to the majority of thermal energy working in NIR regions rather than visible region. Thereby, one of the feasible approaches is to suppress infrared emissivity by enclosing wavelength-selective filters around thermal sources [256, 1193, 1194]. Further improvement can be considered to directly design metamaterials-based wavelength-selective thermal sources with shaped spectral and angular responses. Besides, there are enormous opportunities for designing thermal devices combined with nonreciprocal emission/absorption properties as introduced in Chapter 10.2.3. In nonreciprocal thermal devices, the intrinsic loss induced by the equality of emissivity and absorptivity can be reduced, which will further improve the whole working efficiency of several thermal devices such as solar photovoltaic cells, radiative cooling, and TPVs. While the experimental realization largely nonreciprocal thermal radiation under an extremely small external stimulus or without any stimulus is still a big challenge.

The related areas empower various opportunities in both theory and practical application.

On the other hand, the currently employed fabrication methods rely on costly advanced technologies, like electron beam lithography, molecular Beam epitaxy, etc. Those costly methods are also time-consuming, and the effective areas are generally constricted to an extremely small area at a micro-meter scale. Therefore, more efforts should be made towards cost-efficiency and large-area manufacturing, which will be more attractive in practice.

References

[1] Schwinger J, DeRaad Jr LL, Milton K, Tsai Wy. Classical electrodynamics. Westview Press; 1998.

[2] Einstein A. Über einen die Erzeugung und Verwandlung des Lichtes betreffenden heuristischen Gesichtspunkt. Annalen der Physik. 1905;322(6):132–148.

[3] French AP, Taylor EF. An introduction to quantum physics. Routledge; 2018.

[4] Holman JP. Heat transfer. McGraw Hill Higher Education; 2010.

[5] Incropera FP, Dewitt DP, Bergman TL, Lavine AS, India W. Principles of heat and mass transfer, ISV. John Wiley & Sons; 2003.

[6] Zhang ZM. Nano/microscale heat transfer. Springer Nature; 2020.

[7] Howell JR, Mengüç MP, Daun K, Siegel R. Thermal radiation heat transfer. CRC Press; 2020.

[8] Matsumi Y, Kawasaki M. Photolysis of atmospheric ozone in the ultraviolet region. Chemical Reviews. 2003;103(12):4767–4782.

[9] Marin O, Buckius R. A simplified wide band model of the cumulative distribution function for carbon dioxide. International Journal of Heat and Mass Transfer. 1998;41(23):3881–3897.

[10] Marin O, Buckius R. A simplified wide band model of the cumulative distribution function for water vapor. International Journal of Heat and Mass Transfer. 1998;41(19):2877–2892.

[11] Dunkle R. Geometric mean beam lengths for radiant heat-transfer calculations. ASME Journal of Heat and Mass Transfer. 1864;86(1):75–80.

[12] Eckert E, Pfender E. Heat and mass transfer in porous media with phase change. In: International Heat Transfer Conference Digital Library. Begel House Inc.; 1978. pp. 1–12.

[13] Hottel HC. Radiant heat transmission. WH McAdams Heat Transmission; 1954.

[14] Johnson FS. The solar constant. Journal of Atmospheric Sciences. 1954;11(6):431–439.

[15] Bergman TL, Bergman TL, Incropera FP, Dewitt DP, Lavine AS. Fundamentals of heat and mass transfer. John Wiley & Sons; 2011.

[16] Ossipov P. The angular coefficient method for calculating the stationary molecular gas flow for arbitrary reflection law. Vacuum. 1997;48(5):409–412.

[17] Howell JR. A catalog of radiation heat transfer configuration factors. www.thermalradiation.net/intro.html. 2010.

[18] Hamilton Jr DC. Radiant interchange configuration factors. Purdue University; 1949.

[19] Martinek J, Weimer AW. Evaluation of finite volume solutions for radiative heat transfer in a closed cavity solar receiver for high temperature solar thermal processes. International Journal of Heat and Mass Transfer. 2013;58(1–2): 585–596.
[20] Oppenheim A. Radiation analysis by the network method. Transactions of the American Society of Mechanical Engineers. 1956;78(4):725–735.
[21] Chai JC, Lee HS, Patankar SV. Finite volume method for radiation heat transfer. Journal of Thermophysics and Heat Transfer. 1994;8(3):419–425.
[22] Sarkar A, Mahapatra SK. Role of surface radiation on the functionality of thermoelectric cooler with heat sink. Applied Thermal Engineering. 2014;69 (1–2):39–45.
[23] Smith G. Radiation efficiency of electrically small multiturn loop antennas. IEEE Transactions on Antennas and Propagation. 1972;20(5):656–657.
[24] Halama H. Effects of radiation on surface resistance of superconducting niobium cavity. Applied Physics Letters. 1971;19(4):90–91.
[25] Luan ZJ, Zhang GM, Tian MC, Fan MX. Flow resistance and heat transfer characteristics of a new-type plate heat exchanger. Journal of Hydrodynamics. 2008;20(4):524–529.
[26] Gori V, Marincioni V, Biddulph P, Elwell CA. Inferring the thermal resistance and effective thermal mass distribution of a wall from in situ measurements to characterise heat transfer at both the interior and exterior surfaces. Energy and Buildings. 2017;135:398–409.
[27] Halbritter J. On surface resistance of superconductors. Zeitschrift für Physik. 1974;266(3):209–217.
[28] Jang C, Kim J, Song TH. Combined heat transfer of radiation and conduction in stacked radiation shields for vacuum insulation panels. Energy and Buildings. 2011;43(12):3343–3352.
[29] Wang Q, Li J, Yang H, Su K, Hu M, Pei G. Performance analysis on a high-temperature solar evacuated receiver with an inner radiation shield. Energy. 2017;139:447–458.
[30] Rinker G, Solomon L, Qiu S. Optimal placement of radiation shields in the displacer of a Stirling engine. Applied Thermal Engineering. 2018;144:65–70.
[31] Wang Q, Yang H, Zhong S, Huang Y, Hu M, Cao J, et al. Comprehensive experimental testing and analysis on parabolic trough solar receiver integrated with radiation shield. Applied Energy. 2020;268:115004.
[32] Rayleigh L. On the electromagnetic theory of light. The London, Edinburgh, and Dublin Philosophical Magazine and Journal of Science. 1881;12(73):81–101.
[33] Lorenz L. Videnskab Selskab Skrifter. Kongelige Danske Videnskabernes Selskab; 1890.
[34] Mie G. Beiträge zur Optik trüber Medien, speziell kolloidaler Metallösungen. Annalen der physik. 1908;330(3):377–445.
[35] Debye P. Der Lichtdruck auf Kugeln von beliebigem Material. Annalen der Physik. 1909;30(1):157–136.
[36] Kerker M. The scattering of light and other electromagnetic radiation: Physical chemistry: A series of monographs. vol. 16. Academic Press; 2013.
[37] Deirmendjian D. Electromagnetic scattering on spherical polydispersions. RAND Corporation; 1969.

[38] Bohren CF, Huffman DR. Absorption and scattering of light by small particles. John Wiley & Sons; 2008.
[39] Bi L, Yang P. High-frequency extinction efficiencies of spheroids: Rigorous T-matrix solutions and semi-empirical approximations. Optics Express. 2014;22(9):10270–10293.
[40] Haiducek J. Experimental Validation Techniques for the HELEEOS Off-Axis Laser Propagation Model. Theses and Dissertations. 2010.
[41] Kumar S, Mitra K. Microscale aspects of thermal radiation transport and laser applications. vol. 33 of Advances in Heat Transfer. Elsevier; 1999. pp. 187–294.
[42] Hartung LC, Mitcheltree RA, Gnoffo PA. Stagnation point nonequilibrium radiative heating and the influence of energy exchange models. Journal of Thermophysics and Heat Transfer. 1992;6(3):412–418.
[43] Pomraning GC. The equations of radiation hydrodynamics. Courier Corporation; 2005.
[44] Abdallah PB, Le Dez V. Thermal emission of a semi-transparent slab with variable spatial refractive index. Journal of Quantitative Spectroscopy and Radiative Transfer. 2000;67(3):185–198.
[45] Abdallah PB, Le Dez V. Temperature field inside an absorbing–emitting semi-transparent slab at radiative equilibrium with variable spatial refractive index. Journal of Quantitative Spectroscopy and Radiative Transfer. 2000;65(4): 595–608.
[46] Wu CY, Hou MF. Integral equation solutions based on exact ray paths for radiative transfer in a participating medium with formulated refractive index. International Journal of Heat and Mass Transfer. 2012;55(23–24):6600–6608.
[47] Zhao J, Tan J, Liu L. On the derivation of vector radiative transfer equation for polarized radiative transport in graded index media. Journal of Quantitative Spectroscopy and Radiative Transfer. 2012;113(3):239–250.
[48] Modest MF. Fundamentals of thermal radiation. In: Radiative heat transfer; 2003. Academic Press. pp. 1–29.
[49] Viskanta R, Mengüç MP. Radiation heat transfer in combustion systems. Progress in Energy and Combustion Science. 1987;13(2):97–160.
[50] Ruan LM, Tan HP, Yan YY. A Monte Carlo (MC) method applied to the medium with nongray absorbing-emitting-anisotropic scattering particles and gray approximation. Numerical Heat Transfer; Part A: Applications. 2002;42(3):253–268.
[51] Wang A, Modest MF. Spectral Monte Carlo models for nongray radiation analyses in inhomogeneous participating media. International Journal of Heat and Mass Transfer. 2007;50(19):3877–3889.
[52] Guihua W, Huaichun Z, Qiang C, Zhichao W. Equation-solving DRESOR method for radiative transfer in an absorbing-emitting and isotropically scattering slab with diffuse boundaries. Journal of Heat Transfer. 2012;134(12):122702.
[53] MacRobert TM. Spherical harmonics. vol. 98. 3rd ed. Pergamon Press; 1967.
[54] Derby JJ, Brandon S, Salinger AG. The diffusion and P1 approximations for modeling buoyant flow of an optically thick fluid. International Journal of Heat and Mass Transfer. 1998;41(11):1405–1415.
[55] Mark J. The spherical harmonics method, Part I. Atomic Energy Report No MT. 1944;92.

[56] Truelove JS. Discrete-ordinate solutions of the radiation transport equation. Journal of Heat Transfer. 1987;109(4):1048–1051.

[57] Fiveland WA. Selection of discrete ordinate quadrature sets for anisotropic scattering. Fundamentals of Radiation Heat Transfer. American Society of Mechanical Engineers, Heat Transfer Division. 1991. Vol. 160, pp. 89–96.

[58] Fiveland WA. Discrete ordinate methods for radiative heat transfer in isotropically and anisotropically scattering media. Journal of Heat Transfer. 1987;109(3):809–812.

[59] Da Graça Carvalho M, Farias T, Fontes P. Multidimensional modeling of radiative heat transfer in scattering media. Journal of Heat Transfer. 1993;115(2):486–489.

[60] Kuo DC, Morales JC, Ball KS. Combined natural convection and volumetric radiation in a horizontal annulus: Spectral and finite volume predictions. Journal of Heat Transfer. 1999;121(3):610–615.

[61] Farmer JT, Howell JR. Comparison of Monte Carlo strategies for radiative transfer in participating media. In: Advances in heat transfer. vol. 31; 1998. pp. 333–429.

[62] Walters DV, Buckius RO. Monte Carlo methods for radiative heat transfer in scattering media. Annual Review of Heat Transfer. 2013;5(5):131–176.

[63] Modest MF. The Monte Carlo method applied to gases with spectral line structure. Numerical Heat Transfer, Part B: Fundamentals. 1992;22(3):273–284.

[64] Bevilacqua F, Piguet D, Marquet P, Gross JD, Tromberg BJ, Depeursinge C. In vivo local determination of tissue optical properties: Applications to human brain. Applied Optics. 1999;38(22):4939.

[65] Palmer GM, Ramanujam N. Monte Carlo-based inverse model for calculating tissue optical properties. Part I: Theory and validation on synthetic phantoms. Applied Optics. 2006;45(5):1062–1071.

[66] Hayakawa CK, Spanier J, Bevilacqua F, Dunn AK, You JS, Tromberg BJ, et al. Perturbation Monte Carlo methods to solve inverse photon migration problems in heterogeneous tissues. Optics Letters. 2001;26(17):1335.

[67] Alerstam E, Andersson-Engels S, Svensson T. White Monte Carlo for time-resolved photon migration. Journal of Biomedical Optics. 2008;13(4):041304.

[68] Alerstam E, Andersson-Engels S, Svensson T. Improved accuracy in time-resolved diffuse reflectance spectroscopy. Optics Express. 2008;16(14):10440.

[69] Svensson T, Alerstam E, Einarsdóttír M, Svanberg K, Andersson-Engels S. Towards accurate in vivo spectroscopy of the human prostate. Journal of Biophotonics. 2008;1(3):200–203.

[70] Wang L, Jacques SL, Zheng L. MCML-Monte Carlo modeling of light transport in multi-layered tissues. Computer Methods and Programs in Biomedicine. 1995;47(2):131–146.

[71] Zołek NS, Liebert A, Maniewski R. Optimization of the Monte Carlo code for modeling of photon migration in tissue. Computer Methods and Programs in Biomedicine. 2006;84(1):50–57.

[72] Bianchi S, Ferrara A, Giovanardi C. Monte Carlo simulations of dusty spiral galaxies: Extinction and polarization properties. The Astrophysical Journal. 1996;465:127.

[73] Kattawar GW, Adams CN. Stokes vector calculations of the submarine light field in an atmosphere-ocean with scattering according to a Rayleigh phase matrix: Effect of interface refractive index on radiance and polarization. Limnology and Oceanography. 1989;34(8):1453–1472.

[74] Wang X, Wang LV. Propagation of polarized light in birefringent turbid media: A Monte Carlo study. Journal of Biomedical Optics. 2002;7(3):279.

[75] Ambirajan A, Look DC. A backward Monte Carlo study of the multiple scattering of a polarized laser beam. Journal of Quantitative Spectroscopy and Radiative Transfer. 1997;58(2):171–192.

[76] Martinez AS, Maynard R. Polarization statistics in multiple scattering of light: A Monte Carlo approach; 1993. pp. 99–114.

[77] Bartel S, Hielscher AH. Monte Carlo simulations of the diffuse backscattering Mueller matrix for highly scattering media. Applied Optics. 2000;39(10):1580.

[78] Cameron BD, Raković MJ, Mehrübeoğlu M, Kattawar GW, Rastegar S, Wang LV, et al. Measurement and calculation of the two-dimensional backscattering Mueller matrix of a turbid medium: Errata. Optics Letters. 1998;23(20):1630.

[79] Côté D, Vitkin IA. Robust concentration determination of optically active molecules in turbid media with validated three-dimensional polarization sensitive Monte Carlo calculations. Optics Express. 2005;13(1):148.

[80] Raković MJ, Kattawar GW, Mehrűbeoğlu M, Cameron BD, Wang LV, Rastegar S, et al. Light backscattering polarization patterns from turbid media: Theory and experiment. Applied Optics. 1999;38(15):3399.

[81] Tynes HH, Kattawar GW, Zege EP, Katsev IL, Prikhach AS, Chaikovskaya LI. Monte Carlo and multicomponent approximation methods for vector radiative transfer by use of effective Mueller matrix calculations. Applied Optics. 2001;40(3):400.

[82] Kaplan B, Ledanois G, Drévillon B. Mueller matrix of dense polystyrene latex sphere suspensions: Measurements and Monte Carlo simulation. Applied Optics. 2001;40(16):2769.

[83] Lux I, Koblinger L. Monte Carlo particle transport methods: Neutron and photon calculations; 2018.

[84] Xu M. Electric field Monte Carlo simulation of polarized light propagation in turbid media. Optics Express. 2004;12(26):6530.

[85] Cherkaoui M, Dufresne JL, Fournier R, Grandpeix JY, Lahellec A. Monte Carlo simulation of radiation in gases with a narrow-band model and a net-exchange formulation. Journal of Heat Transfer. 1996;118(2):401–407.

[86] Cherkaoui M, Dufresne JL, Fournier R, Grandpeix JY, Lahellec A. Radiative net exchange formulation within one-dimensional gas enclosures with reflective surfaces; 1998.

[87] Tessé L, Dupoirieux F, Zamuner B, Taine J. Radiative transfer in real gases using reciprocal and forward Monte Carlo methods and a correlated-k approach. International Journal of Heat and Mass Transfer. 2002;45(13):2797–2814.

[88] Dupoirieux F, Tessé L, Avila S, Taine J. An optimized reciprocity Monte Carlo method for the calculation of radiative transfer in media of various optical thicknesses. International Journal of Heat and Mass Transfer. 2006;49(7–8):1310–1319.

[89] Tessé L, Dupoirieux F, Taine J. Monte Carlo modeling of radiative transfer in a turbulent sooty flame. International Journal of Heat and Mass Transfer. 2004;47(3):555–572.

[90] Sun HF, Sun FX, Xia XL. Bidirectionally weighted Monte Carlo method for radiation transfer in the participating media. Numerical Heat Transfer, Part B: Fundamentals. 2017;71(2):202–215.

[91] Soucasse L, Rivière P, Soufiani A. Monte Carlo methods for radiative transfer in quasi-isothermal participating media. Journal of Quantitative Spectroscopy and Radiative Transfer. 2013;128:34–42.

[92] Niederreiter H. Random Number generation and Quasi-Monte Carlo methods; 1992.

[93] Sobol IM. Uniformly distributed sequences with an additional uniform property. USSR Computational Mathematics and Mathematical Physics. 1976;16(5): 236–242.

[94] Halton JH. On the efficiency of certain quasi-random sequences of points in evaluating multi-dimensional integrals. Numerische Mathematik. 1960;2(1): 84–90.

[95] Wang F, Liu D, fa Cen K, hua Yan J, xing Huang Q, Chi Y. Efficient inverse radiation analysis of temperature distribution in participating medium based on backward Monte Carlo method. Journal of Quantitative Spectroscopy and Radiative Transfer. 2008;109(12–13):2171–2181.

[96] Jeans JH. The equations of radiative transfer of energy. Monthly Notices of the Royal Astronomical Society. 1917;78(1):28–36.

[97] Gelbard EM. Simplified spherical harmonics equations and their use in shielding problems. Technical Report WAPD-T-1182. 1961.

[98] Bayazitoğlu Y, Higenyi J. Higher-order differential equations of radiative transfer: P3 approximation. AIAA Journal. 1979;17(4):424–431.

[99] Ravishankar M, Mazumder S, Sankar M. Application of the modified differential approximation for radiative transfer to arbitrary geometry. Journal of Quantitative Spectroscopy and Radiative Transfer. 2010;111(14):2052–2069.

[100] Wu CY, Ou NR. Transient two-dimensional radiative and conductive heat transfer in a scattering medium. International Journal of Heat and Mass Transfer. 1994;37(17):2675–2686.

[101] Pal G, Modest MF. Advanced differential approximation formulation of the PN method for radiative transfer. Journal of Heat Transfer. 2015;137(7);072701.

[102] Gerardin J, Seiler N, Ruyer P, Trovalet L, Boulet P. P1 approximation, MDA and IDA for the simulation of radiative transfer in a 3D geometry for an absorbing scattering medium. Journal of Quantitative Spectroscopy and Radiative Transfer. 2012;113(2):140–149.

[103] Chandrasekhar S. Radiative transfer. Dover Publications Inc.; 1960.

[104] Lee CE. The discrete Sn approximation to transport theory. Technical Information Series Report LA. 1962;2595.

[105] Charest MRJ, Groth CPT, Gülder ÖL. Solution of the equation of radiative transfer using a Newton-Krylov approach and adaptive mesh refinement. Journal of Computational Physics. 2012;231(8):3023–3040.

[106] Coelho PJ. Modified discrete ordinates and finite volume methods. In: Thermopedia. Begel House Inc.; 2012.

[107] Zhou HC, Cheng Q, Huang ZF, He C. The influence of anisotropic scattering on the radiative intensity in a gray, plane-parallel medium calculated by the DRESOR method. Journal of Quantitative Spectroscopy and Radiative Transfer. 2007;104(1):99–115.

[108] Thurgood C, Pollard A, Rubini P. Development of TN quadrature sets and heart solution method for calculating radiative heat transfer. International Symposium on Steel Reheat Furnace Technology, Hamilton; 1990.

[109] Asllanaj F, Fumeron S. Modified finite volume method applied to radiative transfer in 2D complex geometries and graded index media. Journal of Quantitative Spectroscopy and Radiative Transfer. 2010;111(2):274–279.

[110] Chai JC, Lee HS, Patankar SV. Finite volume method for radiation heat transfer. Advances in Numerical Heat Transfer. 2000;2:109–141.

[111] Coelho PJ. Advances in the discrete ordinates and finite volume methods for the solution of radiative heat transfer problems in participating media; 2014.

[112] Jeandel G, Boulet P, Morlot G. Radiative transfer through a medium of silica fibres oriented in parallel planes. International Journal of Heat and Mass Transfer. 1993;36(2):531–536.

[113] Argento C, Bouvard D. A ray tracing method for evaluating the radiative heat transfer in porous media. International Journal of Heat and Mass Transfer. 1996;39(15):3175–3180.

[114] Siegel R, Spuckler CM. Approximate solution methods for spectral radiative transfer in high refractive index layers. International Journal of Heat and Mass Transfer. 1994;37(SUPPL. 1):403–413.

[115] Spuckler CM, Siegel R. Two-flux and diffusion methods for radiative transfer in composite layers. Journal of Heat Transfer. 1996;118(1):218–222.

[116] Tremante A, Malpica F. Analysis of the temperature profile of ceramic composite materials exposed to combined conduction–radiation between concentric cylinders. Journal of Engineering for Gas Turbines and Power. 1998 04;120(2): 271–275.

[117] Dembele S, Wen JX, Sacadura JF. Analysis of the two-flux model for predicting water spray transmittance in fire protection application. Journal of Heat Transfer. 2000;122(1):183–186.

[118] Chu CM. Numerical solution of problems in multiple scattering of electromagnetic radiation. Journal of Physical Chemistry. 1955;59(9):855–863.

[119] Chin JH, Churchill SW. Anisotropic, multiply scattered radiation from an arbitrary, cylindrical source in an infinite slab. Journal of Heat Transfer. 1965;87(2):167–172.

[120] Daniel KJ, Laurendeau NM, Incropera FP. Prediction of radiation absorption and scattering in turbid water bodies. Journal of Heat Transfer. 1979;101(1): 63–67.

[121] Sasse C, Koenigsdorff R, Frank S. Evaluation of an improved hybrid six-flux/zone model for radiative transfer in rectangular enclosures. International Journal of Heat and Mass Transfer. 1995;38(18):3423–3431.

[122] Keramida EP, Liakos HH, Founti MA, Boudouvis AG, Markatos NC. Radiative heat transfer in natural gas-fired furnaces. International Journal of Heat and Mass Transfer. 2000;43(10):1801–1809.

[123] Cumber PS. Improvements to the discrete transfer method of calculating radiative heat transfer. International Journal of Heat and Mass Transfer. 1995;38(12):2251–2258.

[124] Cumber PS. Application of adaptive quadrature to fire radiation modeling. Journal of Heat Transfer. 1999;121(1):203–205.

[125] Coelho PJ, Carvalho MG. A conservative formulation of the discrete transfer method. Journal of Heat Transfer. 1997;119(1):118–128.

[126] Versteeg HK, Henson JC, Malalasekera W. Approximation errors in the heat flux integral of the discrete transfer method, part 1: Transparent media. Numerical Heat Transfer, Part B: Fundamentals. 1999;36(4):387–407.

[127] Versteeg HK, Henson JC, Malalasekera W. Approximation errors in the heat flux integral of the discrete transfer method, part 2: Participating media. Numerical Heat Transfer, Part B: Fundamentals. 1999;36(4):409–432.

[128] Malalasekera WMG, James EH. Radiative heat transfer calculations in three-dimensional complex geometries. Journal of Heat Transfer. 1996;118(1):228.

[129] Henson JC, Malalasekera WMG. Comparison of the discrete transfer and monte carlo methods for radiative heat transfer in three-dimensional nonhomogeneous scattering media. Numerical Heat Transfer; Part A: Applications. 1997;32(1):19–36.

[130] Bressloff NW, Moss JB, Rubini PA. CFD prediction of coupled radiation heat transfer and soot production in turbulent flames. Symposium (International) on Combustion. 1996;26(2):2379–2386.

[131] Tan ZM, Hsu PF, Wu SH, Wu CY. Modified YIX method and pseudoadaptive angular quadrature for ray effects mitigation. Journal of Thermophysics and Heat Transfer. 2000;14(3):289–296.

[132] Zhou HC, Chen DL, Cheng Q. A new way to calculate radiative intensity and solve radiative transfer equation through using the Monte Carlo method. Journal of Quantitative Spectroscopy and Radiative Transfer. 2004;83(3–4):459–481.

[133] Cheng Q, Zhou HC. The DRESOR method for a collimated irradiation on an isotropically scattering layer. Journal of Heat Transfer. 2007;129(5):634–645.

[134] Cheng Q, Zhang X, Huang Z, Wang Z, Zhou H. The DRESOR method for radiative heat transfer in semitransparent graded index cylindrical medium. Journal of Quantitative Spectroscopy and Radiative Transfer. 2014;143:16–24.

[135] Zheng S, Qi C, Huang Z, Zhou H. Non-gray radiation study of gas and soot mixtures in one-dimensional planar layer by DRESOR. Journal of Quantitative Spectroscopy and Radiative Transfer. 2018;217:425–431.

[136] Alpaydin E. Machine learning: The new AI. MIT Press; 2016.

[137] Voyant C, Notton G, Kalogirou S, Nivet ML, Paoli C, Motte F, et al. Machine learning methods for solar radiation forecasting: A review. Renewable Energy. 2017;105:569–582.

[138] Gurney K. An introduction to neural networks. CRC Press; 2018.

[139] Deo RC, Ghorbani MA, Samadianfard S, Maraseni T, Bilgili M, Biazar M. Multi-layer perceptron hybrid model integrated with the firefly optimizer algorithm for windspeed prediction of target site using a limited set of neighboring reference station data. Renewable Energy. 2018;116:309–323.

[140] Ren T, Modest MF, Fateev A, Sutton G, Zhao W, Rusu F. Machine learning applied to retrieval of temperature and concentration distributions from infrared emission measurements. Applied Energy. 2019;252:113448.

[141] Mockus J. Bayesian approach to global optimization: Theory and applications. vol. 37. Springer Science & Business Media; 2012.

[142] Sutton G, Fateev A, Rodríguez-Conejo MA, Meléndez J, Guarnizo G. Validation of emission spectroscopy gas temperature measurements using a standard flame traceable to the International Temperature Scale of 1990 (ITS-90). International Journal of Thermophysics. 2019;40(11):1–36.

[143] Nicodemus FE. Reflectance nomenclature and directional reflectance and emissivity. Applied Optics. 1970;9 6:1474–1475.

[144] Kale BM, Broome BG. In SITU Bidirectional Reflectance Distribution Function (BRDF) measurement facility. In: Photonics West – Lasers and Applications in Science and Engineering; 1979.

[145] Bartell FO, Dereniak EL, Wolfe WL. The theory and measurement of Bidirectional Reflectance Distribution Function (BRDF) and Bidirectional Transmittance Distribution Function (BTDF). In: Other Conferences; 1981.

[146] Lee WW, Scherr LM, Barsh MK. Stray light analysis and suppression in small angle BRDF/BTDF measurement. In: Optics & Photonics; 1987.

[147] Zaworski JR, Welty JR, Drost MK. Measurement and use of bi-directional reflectance. International Journal of Heat and Mass Transfer. 1996;39:1149–1156.

[148] Johnson JR, Grundy WM, Shepard MK. Visible/near-infrared spectrogoniometric observations and modeling of dust-coated rocks. Icarus. 2004;171:546–556.

[149] Li H, Foo SC, Torrance KE, Westin SH. Automated three-axis gonioreflectometer for computer graphics applications. In: SPIE Optics + Photonics; 2005.

[150] Yang P, Zhang ZM. Bidirectional reflection of semitransparent polytetrafluoroethylene (PTFE) sheets on a silver film. International Journal of Heat and Mass Transfer. 2020;148:118992.

[151] Xie Y, Tan J, Jing L, Zhang W, Lai Q. Investigating directional reflection characteristics of anisotropic machined surfaces using a self-designed scatterometer. Applied Optics. 2019;58(29):7970–7980.

[152] Jeong SY, Chen C, Ranjan D, Loutzenhiser PG, Zhang ZM. Measurements of scattering and absorption properties of submillimeter bauxite and silica particles. Journal of Quantitative Spectroscopy and Radiative Transfer. 2021;276:107923.

[153] Ferraro JR, Basile LJ. Fourier transform infrared spectroscopy: Applications to chemical systems; 1978.

[154] Quintás G, Lendl B, Garrigues S, de la Guardia M. Univariate method for background correction in liquid chromatography-Fourier transform infrared spectrometry. Journal of Chromatography A. 2008;1190(1–2):102–109.

[155] Ylmén R, Jäglid U. Carbonation of Portland cement studied by diffuse reflection Fourier transform infrared spectroscopy. International Journal of Concrete Structures and Materials. 2013;7:119–125.

[156] Maier SA. Plasmonics: fundamentals and applications. Vol. 1. Springer; 2007.

[157] Michelson AA, Morley EW. On the relative motion of the Earth and the luminiferous ether. American Journal of Science. 1887;34:333–345.

[158] Griffiths PR. Fourier transform infrared spectrometry. Science. 1983;222 4621:297–302.

[159] Lochbaum A, Fedoryshyn Y, Dorodnyy A, Koch U, Hafner C, Leuthold J. On-chip narrowband thermal emitter for Mid-IR optical gas sensing. ACS Photonics. 2017;4:1371–1380.

[160] Song X, Dong W, Yuan Z, Lu X, Li Z, Duanmu Q. Investigation of the linearity of the NIM FTIR infrared spectral emissivity measurement facility by means of flux superposition method. Infrared Physics & Technology. 2020;109:103416.

[161] Pedrotti FL, Pedrotti LS, Pedrotti LM. Introduction to optics; 2017.

[162] Fujiwara H. Spectroscopic ellipsometry: Principles and applications; 2007.

[163] Schöche S, Hofmann T, Korlacki R, Tiwald TE, Schubert M. Infrared dielectric anisotropy and phonon modes of rutile TiO_2. Journal of Applied Physics. 2013;113:164102.

[164] Hajduk B, Bednarski H, Trzebicka B. Temperature-dependent spectroscopic ellipsometry of thin polymer films. The Journal of Physical Chemistry B. 2020;124:3229–3251.

[165] He J, Jiang W, Zhu X, Zhang R, Wang J, Zhu M, et al. Optical properties of thickness-controlled PtSe2 thin films studied via spectroscopic ellipsometry. Physical Chemistry Chemical Physics : PCCP. 2020;22:26383–26389.

[166] Richter S, Rebarz M, Herrfurth O, Espinoza S, Schmidt-Grund R, Andreasson J. Broadband femtosecond spectroscopic ellipsometry. The Review of Scientific Instruments. 2021;92 3:033104.

[167] Karlovets E, Gordon IE, Rothman LS, Hashemi R, Hargreaves RJ, Toon GC, et al. The update of the line positions and intensities in the line list of carbon dioxide for the HITRAN2020 spectroscopic database. Journal of Quantitative Spectroscopy and Radiative Transfer. 2021;276;107896.

[168] Rothman LS, Gordon IE, Barber RJ, Dothe H, Gamache RR, Goldman A, et al. HITEMP, the high-temperature molecular spectroscopic database. Journal of Quantitative Spectroscopy & Radiative Transfer. 2010;111:2139–2150.

[169] Tashkun SA, Perevalov VI. CDSD-4000: High-resolution, high-temperature carbon dioxide spectroscopic databank. Journal of Quantitative Spectroscopy & Radiative Transfer. 2011;112:1403–1410.

[170] Beier K, Lindermeir E. Comparison of line-by-line and molecular band IR modeling of high altitude missile plume. Journal of Quantitative Spectroscopy & Radiative Transfer. 2007;105:111–127.

[171] Goody RM. A statistical model for water-vapour absorption. Quarterly Journal of the Royal Meteorological Society. 1952;78:638–640.

[172] Malkmus W. Random Lorentz band model with exponential-tailed S-1 line-intensity distribution function. Journal of the Optical Society of America. 1967;57:323–329.

[173] Giedt WH, Tien CL. Experimental determination of infrared absorption of high-temperature gases; 1965.

[174] Modest MF. Radiative heat transfer; 1993.

[175] Penner SS, Landshoff RKM. Quantitative molecular spectroscopy and gas emissivities; 1959.

[176] Goldstein RB. Measurements of infrared absorption by water vapor at temperatures to 1000°K. Journal of Quantitative Spectroscopy & Radiative Transfer. 1964;4:343–352.

[177] Eckert ERG, Goldstein RJ. Measurements in heat transfer; 1976.

[178] Hottel HC, Mangelsdorf HG. Heat transmission by radiation from non-luminous gases II. Experimental study of carbon dioxide and water vapor. Transactions of the American Institute of Chemical Engineers. 1935;31:517–549.

[179] Huaichun Z, Zhifang H, Jianjun G, Yaoping L, Jun Z, Ping C, et al. On-line optimization of coal-fired boiler operation in power plants for smart power generation. Distributed Energy Resources. 2019;4(3):1–7.

[180] Bin-shuai Z. Study on online measurement of temperature field of the furnace. Refrigeration air conditioning & electric power machinery; 2010.

[181] Zhou H, Lou X, Deng Y. Measurement method of three-dimensional combustion temperature distribution in utility furnaces based on image processing radiative. Proceedings-Chinese Society of Electrical Engineering. 1997;17:1–4.

[182] Zhou HC, Lou C, Cheng Q, Wei Jiang Z, He J, Huang B, et al. Experimental investigations on visualization of three-dimensional temperature distributions in a large-scale pulverized-coal-fired boiler furnace; 2005.

[183] Luo Z, Zhou HC. A combustion-monitoring system with 3-D temperature reconstruction based on flame-image processing technique. IEEE Transactions on Instrumentation and Measurement. 2007;56:1877–1882.

[184] Huajian W, Zhi-feng H, Dundun W, Zixue L, Yipeng S, Qingyan F, et al. Measurements on flame temperature and its 3D distribution in a 660 MWe arch-fired coal combustion furnace by visible image processing and verification by using an infrared pyrometer. Measurement Science and Technology. 2009;20:114006.

[185] Ni M, Zhang H, Wang F, Xie Z, Huang Q, Yan J, et al. Study on the detection of three-dimensional soot temperature and volume fraction fields of a laminar flame by multispectral imaging system. Applied Thermal Engineering. 2016;96: 421–431.

[186] Achal S, McFee JE, Ivanco T, Anger CD. A thermal infrared hyperspectral imager (tasi) for buried landmine detection. In: SPIE Defense + Commercial Sensing; 2007.

[187] Wu K, Feng Y, Yu G, Liu L, Li J, Xiong Y, et al. Development of an imaging gas correlation spectrometry based mid-infrared camera for two-dimensional mapping of CO in vehicle exhausts. Optics Express. 2018;26(7):8239–8251.

[188] Shepanski JF, Sandor-Leahy S. The NGST long wave hyperspectral imaging spectrometer: Sensor hardware and data processing. In: SPIE Defense + Commercial Sensing; 2006.

[189] Hall JL, Boucher RH, Buckland KN, Gutierrez DJ, Hackwell JA, Johnson BR, et al. MAGI: A new high-performance airborne thermal-infrared imaging spectrometer for earth science applications. IEEE Transactions on Geoscience and Remote Sensing. 2015;53:5447–5457.

[190] Yuan L, He Z, Lv G, Wang Y, Li C, Xie J, et al. Optical design, laboratory test, and calibration of airborne long wave infrared imaging spectrometer. Optics Express. 2017;25 19:22440–22454.

[191] Sun JG, Erdman SV, Connolly L. Measurement of delamination size and depth in ceramic matrix composites using pulsed thermal imaging; 2008.

[192] Sun J. Method for thermal tomography of thermal effusivity from pulsed thermal imaging. Google Patents; 2008.

[193] Sun JG. Thermal conductivity measurement for thermal barrier coatings based on one- and two-sided thermal imaging methods; 2010.

[194] Ringermacher HI, Archacki Jr RJ, Veronesi WA. Nondestructive testing: Transient depth thermography. Google Patents; 1998.

[195] Hanrieder N, Wilbert S, Mancera-Guevara D, Buck R, Giuliano S, Pitz-Paal R. Atmospheric extinction in solar tower plants–A review. Solar Energy. 2017;152:193–207.

[196] Gueymard CA. Parameterized transmittance model for direct beam and circumsolar spectral irradiance. Solar Energy. 2001;71(5):325–346.

[197] Griggs D, Jones D, Ouldridge M, Sparks W. Instruments and Observing Methods. Report No. 41. The first WMO Intercomparison of Visibility Measurements. World Meteorological Organization; 1989.

[198] Koschmieder H. Theorie der horizontalen Sichtweite. Beitrage zur Physik der freien Atmosphare. 1924:33–53.

[199] Vittitoe CN, Biggs F. Terrestrial propagation loss. Denver: Presented at the American Section of the International Solar Energy Society. 1978.

[200] Wendelin TJ. SolTRACE: A new optical modeling tool for concentrating solar optics. Solar Energy. 2003:253–260.

[201] Leary PL, Hankins JD. User's guide for MIRVAL: A computer code for comparing designs of heliostat-receiver optics for central receiver solar power plants; 1979.

[202] Hottel HC. A simple model for estimating the transmittance of direct solar radiation through clear atmospheres. Solar Energy. 1976;18:129–134.

[203] Kistler BL. A user's manual for DELSOL3: A computer code for calculating the optical performance and optimal system design for solar thermal central receiver plants; 1986.

[204] Blair N, Dobos AP, Freeman J, Neises T, Wagner M, Ferguson T, et al. System advisor model, SAM 2014.1. 14: General description. National Renewable Energy Lab.(NREL), Golden, CO (United States); 2014.

[205] Vittitoe CN, Biggs F. User's guide to HELIOS: A computer program for modeling the optical behavior of reflecting solar concentrators. Part III. Appendices concerning HELIOS-code details. Sandia National Lab.(SNL-NM), Albuquerque, NM (United States); 1981.

[206] Pitman C, Vant-Hull L. Atmospheric transmittance model for a solar beam propagating between a heliostat and a receiver. Sandia National Lab.(SNL-NM), Albuquerque, NM (United States); Houston Univ ...; 1984.

[207] Ballestrín J, Monterreal R, Carra M, Fernández-Reche J, Polo J, Enrique R, et al. Solar extinction measurement system based on digital cameras. Application to solar tower plants. Renewable Energy. 2018;125:648–654.

[208] Mishchenko MI, Tishkovets VP, Travis LD, Cairns B, Dlugach JM, Liu L, et al. Electromagnetic scattering by a morphologically complex object: Fundamental concepts and common misconceptions. Journal of Quantitative Spectroscopy and Radiative Transfer. 2011;112(4):671–692.

[209] Tokunaga T, Arai M, Kobayashi K, Hayami W, Suehara S, Shiga T, et al. First-principles calculations of phonon transport across a vacuum gap. Physical Review B. 2022 Jan;105:045410.

[210] Forn-Díaz P, Lamata L, Rico E, Kono J, Solano E. Ultrastrong coupling regimes of light–matter interaction. Review of Modern Physics. 2019 Jun;91:025005.

[211] Khandekar C, Yang L, Rodriguez AW, Jacob Z. Quantum nonlinear mixing of thermal photons to surpass the blackbody limit. Optics Express. 2020 Jan;28(2):2045–2059.

[212] Longtin J, Tien C. Microscale radiation phenomena. Microscale Energy Transport. 1997:119–147.

[213] Costantini D, Lefebvre A, Coutrot AL, Moldovan-Doyen I, Hugonin JP, Boutami S, et al. Plasmonic metasurface for directional and frequency-selective thermal emission. Physical Review Applied. 2015 Jul;4:014023.

[214] Wang BX, Liu MQ, Huang TC, Zhao CY. Micro/nanostructures for far-field thermal emission control: An overview. ES Energy & Environment. 2019 Dec;6:18–38.

[215] Sobhan CB, Peterson GP. Microscale and nanoscale heat transfer: Fundamentals and engineering applications. CRC Press; 2008.

[216] Wolf E. Introduction to the theory of coherence and polarization of light. Cambridge University Press; 2007.

[217] Tien CL, Chen G. Challenges in microscale conductive and radiative heat transfer. Journal of Heat Transfer. 1994 11;116(4):799–807.

[218] Mehta C. Coherence-time and effective bandwidth of blackbody radiation. Il Nuovo Cimento (1955-1965). 1963;28(2):401–408.

[219] Mandel L, Wolf E. Coherence properties of optical fields. Reviews of Modern Physics. 1965 Apr;37:231–287.

[220] Blomstedt K, Friberg AT, Setälä T. Chapter Five – Classical Coherence of Blackbody Radiation. vol. 62 of Progress in Optics. Elsevier; 2017. pp. 293–346.

[221] Klein LJ, Hamann HF, Au YY, Ingvarsson S. Coherence properties of infrared thermal emission from heated metallic nanowires. Applied Physics Letters. 2008;92(21):213102.

[222] Chen G, Tien CL. Partial coherence theory of thin film radiative properties. Journal of Heat Transfer. 1992 08;114(3):636–643.

[223] Anderson CF, Bayazitoglu Y. Radiative properties of films using partial coherence theory. Journal of Thermophysics and Heat Transfer. 1996;10(1):26–32.

[224] Chen G. Nanoscale energy transport and conversion: A parallel treatment of electrons, molecules, phonons, and photons. Oxford University Press; 2005.

[225] Novotny L, Hecht B. Principles of nano-optics. Cambridge University Press; 2012.

[226] Reuter GEH, Sondheimer EH, Wilson AH. The theory of the anomalous skin effect in metals. Proceedings of the Royal Society of London Series A Mathematical and Physical Sciences. 1948;195(1042):336–364.

[227] Duncan AB, Peterson GP. Review of microscale heat transfer. Applied Mechanics Reviews. 1994 09;47(9):397–428.

[228] Xuan Y. An overview of micro/nanoscaled thermal radiation and its applications. Photonics and Nanostructures – Fundamentals and Applications. 2014;12(2):93–113.

[229] Cahill DG, Ford WK, Goodson KE, Mahan GD, Majumdar A, Maris HJ, et al. Nanoscale thermal transport. Journal of Applied Physics. 2003;93(2):793–818.

[230] Cahill DG, Braun PV, Chen G, Clarke DR, Fan S, Goodson KE, et al. Nanoscale thermal transport. II. 2003–2012. Applied Physics Reviews. 2014;1(1):011305.

[231] Song B, Fiorino A, Meyhofer E, Reddy P. Near-field radiative thermal transport: From theory to experiment. AIP Advances. 2015;5(5):053503.

[232] Li W, Fan S. Nanophotonic control of thermal radiation for energy applications. Optics Express. 2018 Jun;26(12):15995–16021.

[233] Cuevas JC, García-Vidal FJ. Radiative heat transfer. ACS Photonics. 2018;5(10):3896–3915.

[234] Baranov DG, Xiao Y, Nechepurenko IA, Krasnok A, Alù A, Kats MA. Nanophotonic engineering of far-field thermal emitters. Nature Materials. 2019 Sep;18(9):920–930.

[235] Rao AM, Eklund PC, Lehman GW, Face DW, Doll GL, Dresselhaus G, et al. Far-infrared optical properties of superconducting $Bi_2Sr_2CaCu_2O_x$ films. Physical Review B. 1990 Jul;42:193–201.

[236] Phelan PE, Flik MI, Tien CL. Radiative properties of superconducting Y-Ba-Cu-O thin films. Journal of Heat Transfer. 1991 05;113(2):487–493.

[237] Choi BI, Zhang ZM, Flik MI, Siegrist T. Radiative properties of Y-Ba-Cu-0 films with variable oxygen content. Journal of Heat Transfer. 1992 11;114(4):958–964.

[238] Zhang ZM, Choi BI, Le TA, Flik MI, Siegal MP, Phillips JM. Infrared refractive index of thin $YBa_2Cu_30_7$ superconducting films. Journal of Heat Transfer. 1992 08;114(3):644–652.

[239] Basov DN, Liang R, Bonn DA, Hardy WN, Dabrowski B, Quijada M, et al. In-Plane anisotropy of the penetration depth in $YBa_2Cu_3O_{7-x}$ and $YBa_2Cu_4O_8$ superconductors. Physical Review Letters. 1995 Jan;74:598–601.

[240] Bednorz JG, Müller KA. Possible high T_c superconductivity in the Ba-La-Cu-O system. Zeitschrift für Physik B Condensed Matter. 1986;64:189–193.

[241] Chu CW, Hor PH, Meng RL, Gao L, Huang ZJ. Superconductivity at 52.5 K in the Lanthanum-Barium-Copper-Oxide system. Science. 1987;235(4788): 567–569.

[242] Wu MK, Ashburn JR, Torng CJ, Hor PH, Meng RL, Gao L, et al. Superconductivity at 93 K in a new mixed-phase Y-Ba-Cu-O compound system at ambient pressure. Physical Review Letters. 1987 Mar;58:908–910.

[243] Zhao ZX, Chen LQ, Yang QS, Huang YZ, Chen GH, Tang RM, et al. 2. In: Superconductivity above liquid nitrogen temperature in Ba-Y-Cu oxides;. pp. 7–11.

[244] Timusk T, Bonn DA, Greedan JE, Stager CV, Garrett JD, O'Reilly AH, et al. Infrared properties of YBa2Cu3O7-. Physica C: Superconductivity. 1988; 153-155:1744–1747.

[245] Plakida NM. High-temperature superconductivity: Experiment and theory. Springer Berlin, Heidelberg; 1995.

[246] Timusk T, Tanner DB. Infrared properties of high-Tc superconductors. Physical Properties of High Temperature Superconductors. 1989;1:339.

[247] Bonn DA, Greedan JE, Stager CV, Timusk T, Doss MG, Herr SL, et al. Far-Infrared conductivity of the high-T_c superconductor $YBa_2Cu_3O_7$. Physical Review Letters. 1987 May;58:2249–2250.

[248] Herr SL, Kamarás K, Porter CD, Doss MG, Tanner DB, Bonn DA, et al. Optical properties of $La_{1.85}Sr_{0.15}CuO_4$: Evidence for strong electron–phonon and electron–electron interactions. Physical Review B. 1987 Jul;36:733–735.

[249] Timusk T, Herr SL, Kamarás K, Porter CD, Tanner DB, Bonn DA, et al. Infrared studies of ab-plane oriented oxide superconductors. Physical Review B. 1988 Oct;38:6683–6688.

[250] Timusk T, Statt B. The pseudogap in high-temperature superconductors: An experimental survey. Reports on Progress in Physics. 1999 Jan;62(1):61–122.

[251] Basov DN, Timusk T. Electrodynamics of high-T_c superconductors. Review of Modern Physics. 2005 Aug;77:721–779.

[252] Basov DN, Timusk T. Chapter 202 Infrared properties of high-Tc superconductors: An experimental overview. In: High-Temperature Superconductors – II. vol. 31 of Handbook on the Physics and Chemistry of Rare Earths. Elsevier; 2001. pp. 437–507.

[253] Basov DN, Averitt RD, van der Marel D, Dressel M, Haule K. Electrodynamics of correlated electron materials. Review of Modern Physics. 2011 Jun;83: 471–541.

[254] Vedeneev SI. Pseudogap problem in high-temperature superconductors. Physics-Uspekhi. 2021 Dec;64(9):890–922.

[255] Yang J, Du W, Su Y, Fu Y, Gong S, He S, et al. Observing of the super-Planckian near-field thermal radiation between graphene sheets. Nature Communications. 2018 Oct;9(1):4033.

[256] Shiue RJ, Gao Y, Tan C, Peng C, Zheng J, Efetov DK, et al. Thermal radiation control from hot graphene electrons coupled to a photonic crystal nanocavity. Nature Communications. 2019;10(1):1–7.

[257] Watanabe K, Taniguchi T, Niiyama T, Miya K, Taniguchi M. Far-ultraviolet plane-emission handheld device based on hexagonal boron nitride. Nature Photonics. 2009;3(10):591.

[258] Song JCW, Gabor NM. Electron quantum metamaterials in van der Waals heterostructures. Nature Nanotechnology. 2018;13:986–993.

[259] Qi XL, Zhang SC. Topological insulators and superconductors. Review of Modern Physics. 2011 Oct;83:1057–1110.

[260] Xie M, Zhang S, Cai B, Gu Y, Liu X, Kan E, et al. Van der Waals bilayer antimonene: A promising thermophotovoltaic cell material with 31% energy conversion efficiency. Nano Energy. 2017;38:561–568.

[261] Majumdar A, Carrejo JP, Lai J. Thermal imaging using the atomic force microscope. Applied Physics Letters. 1993;62(20):2501–2503.

[262] Marschall J, Majumdar A. Charge and energy transport by tunneling thermoelectric effect. Journal of Applied Physics. 1993;74(6):4000–4005.

[263] Cui L, Miao R, Jiang C, Meyhofer E, Reddy P. Perspective: Thermal and thermoelectric transport in molecular junctions. The Journal of Chemical Physics. 2017;146(9):092201.

[264] Mader H. Microstructuring in semiconductor technology. Thin Solid Films. 1989;175:1–16.

[265] Reichelt K, Jiang X. The preparation of thin films by physical vapour deposition methods. Thin Solid Films. 1990;191(1):91–126.

[266] Randhawa H. Review of plasma-assisted deposition processes. Thin Solid Films. 1991;196(2):329–349.

[267] Flik MI, Choi BI, Anderson AC, Westerheim AC. Thermal analysis and control for sputtering deposition of high-T_c superconducting films. Journal of Heat Transfer. 1992 02;114(1):255–263.

[268] Vandenabeele P, Maex K. Influence of temperature and backside roughness on the emissivity of Si wafers during rapid thermal processing. Journal of Applied Physics. 1992;72(12):5867–5875.

[269] Yen A, Anderson EH, Ghanbari RA, Schattenburg ML, Smith HI. Achromatic holographic configuration for 100-nm-period lithography. Applied Optics. 1992 Aug;31(22):4540–4545.

[270] Ursu I, Mihailescu IN, Nistor LC, Teodorescu VS, Prokhorov AM, Konov VI, et al. Periodic structures on the surface of fused silica under multipulse 10.6-μm laser irradiation. Applied Optics. 1985 Nov;24(22):3736–3739.

[271] Pettit GH, Sauerbrey RA. Pulsed ultraviolet laser ablation. Applied Physics A. 1993;56:51–63.

[272] Grigoropoulos CP, Buckholz RH, Domoto GA. The role of reflectivity change in optically induced recrystallization of thin silicon films. Journal of Applied Physics. 1986;59(2):454–458.

[273] Qiu TQ, Tien CL. Heat transfer mechanisms during short-pulse laser heating of metals. Journal of Heat Transfer. 1993 Nov;115(4):835–841.

[274] Brorson SD, Fujimoto JG, Ippen EP. Femtosecond electronic heat-transport dynamics in thin gold films. Physical Review Letters. 1987 Oct;59:1962–1965.

[275] Heltzel A, Battula A, Howell JR, Chen S. Nanostructuring borosilicate glass with near-field enhanced energy using a femtosecond laser pulse. Journal of Heat Transfer. 2006 May;129(1):53–59.

[276] Wong PY, Hess CK, Miaoulis IN. Thermal radiation modeling in multilayer thin film structures. International Journal of Heat and Mass Transfer. 1992;35(12):3313–3321.

[277] Wong PY, Hess CK, Miaoulis IN. Coherent thermal radiation effects on temperature-dependent emissivity of thin-film structures on optically thick substrates. Optical Engineering. 1995;34(6):1776–1781.

[278] Ray AK. RTP temperature control requirement for submicron device fabrication. In: Kwong DL, Mueller HG, editors. Rapid Thermal and Laser Processing. vol. 1804. International Society for Optics and Photonics. SPIE; 1993. pp. 2–12.

[279] Sorrell FY, Fordham MJ, Ozturk MC, Wortman JJ. Temperature uniformity in RTP furnaces. IEEE Transactions on Electron Devices. 1992;39(1):75–80.

[280] Vandenabeele P, Maex K, De Keersmaecker R. Impact of patterned layers on temperature non-uniformity during rapid thermal processing for VLSI applications. MRS Proceedings. 1989;146:149.

[281] Ko¨ylu¨ UO, Faeth GM. Radiative properties of flame-generated soot. Journal of Heat Transfer. 1993 May;115(2):409–417.

[282] Ku JC, Shim KH. Optical diagnostics and radiative properties of simulated soot agglomerates. Journal of Heat Transfer. 1991 Nov;113(4):953–958.

[283] Howell JR. Thermal radiation in participating media: The past, the present, and some possible futures. Journal of Heat Transfer. 1988 Nov;110(4b):1220–1229.

[284] Tien CL. Thermal radiation in packed and fluidized Beds. Journal of Heat Transfer. 1988 Nov;110(4b):1230–1242.
[285] Born M, Wolf E. Principles of optics: Electromagnetic theory of propagation, interference and diffraction of light. Elsevier; 2013.
[286] Min-Dianey KAA, Zhang HC, M'Bouana NLP, Su CS, Xia XL. Modeling of spectral energy density as thermal radiation characteristic on the basis of porous silicon photonic crystals. Computational Materials Science. 2017;136:306–314.
[287] García PD, Sapienza R, Blanco A, Lopez C. Photonic glass: A novel random material for light. Advanced Materials. 2007;19(18):2597–2602.
[288] Rojas-Ochoa LF, Mendez-Alcaraz JM, Sáenz JJ, Schurtenberger P, Scheffold F. Photonic properties of strongly correlated colloidal liquids. Physical Review Letters. 2004 Aug;93:073903.
[289] Kulkarni A, Wang Z, Nakamura T, Sampath S, Goland A, Herman H, et al. Comprehensive microstructural characterization and predictive property modeling of plasma-sprayed zirconia coatings. Acta Materialia. 2003;51(9):2457–2475.
[290] Lu TJ, Stone HA, Ashby MF. Heat transfer in open-cell metal foams. Acta Materialia. 1998;46(10):3619–3635.
[291] Jiang F, Liu H, Li Y, Kuang Y, Xu X, Chen C, et al. Lightweight, mesoporous, and highly absorptive all-nanofiber aerogel for efficient solar steam generation. ACS Applied Materials & Interfaces. 2018;10(1):1104–1112.
[292] Sun YP, Lou C, Zhou HC. Estimating soot volume fraction and temperature in flames using stochastic particle swarm optimization algorithm. International Journal of Heat and Mass Transfer. 2011;54(1):217–224.
[293] Wang F, Tan J, Ma L, Shuai Y, Tan H, Leng Y. Thermal performance analysis of porous medium solar receiver with quartz window to minimize heat flux gradient. Solar Energy. 2014;108:348–359.
[294] Yang G, Zhao CY, Wang BX. Experimental study on radiative properties of air plasma sprayed thermal barrier coatings. International Journal of Heat and Mass Transfer. 2013;66:695–698.
[295] Zhai Y, Ma Y, David SN, Zhao D, Lou R, Tan G, et al. Scalable-manufactured randomized glass-polymer hybrid metamaterial for daytime radiative cooling. Science. 2017;355(6329):1062–1066.
[296] Lagendijk A, Van Tiggelen BA. Resonant multiple scattering of light. Physics Reports. 1996;270(3):143–215.
[297] van Rossum MCW, Nieuwenhuizen TM. Multiple scattering of classical waves: Microscopy, mesoscopy, and diffusion. Reviews of Modern Physics. 1999;71: 313–371.
[298] Tsang L, Kong JA. Scattering of electromagnetic waves: Advanced topics. John Wiley & Sons; 2004.
[299] Sheng P. Introduction to wave scattering, localization and mesoscopic phenomena. Springer Science & Business Media; 2006.
[300] Mishchenko MI, Rosenbush V, Kiselev N, Lupishko D, Tishkovets V, Kaydash V, et al. Polarimetric remote sensing of solar system objects. arXiv preprint arXiv:10101171. 2010.
[301] Mishchenko MI, Travis LD, Lacis AA. Multiple scattering of light by particles: Radiative transfer and coherent backscattering. Cambridge University Press; 2006.

[302] García PD, Sapienza R, Bertolotti J, Martín MD, Blanco A, Altube A, et al. Resonant light transport through Mie modes in photonic glasses. Physical Review A. 2008 Aug;78:023823.

[303] Wang BX, Zhao CY. Modeling radiative properties of air plasma sprayed thermal barrier coatings in the dependent scattering regime. International Journal of Heat and Mass Transfer. 2015;89:920–928.

[304] Rezvani Naraghi R, Sukhov S, Sáenz JJ, Dogariu A. Near-field effects in mesoscopic light transport. Physical Review Letters. 2015;115:203903.

[305] Wang BX, Zhao CY. Effect of dependent scattering on light absorption in highly scattering random media. International Journal of Heat and Mass Transfer. 2018;125:1069–1078.

[306] Tien CL, Drolen B. Thermal radiation in part1culate media with dependent and independent scattering. Annual Review of Heat Transfer. 1987;1(1):1–32.

[307] Kumar S, Tien C. Dependent absorption and extinction of radiation by small particles. Journal of Heat Transfer. 1990;112(1):178–185.

[308] Lee SC. Dependent scattering by parallel fibers: Effects of multiple scattering and wave interference. Journal of Thermophysics and Heat Transfer. 1992;6(4):589–595.

[309] Ivezić Z, Mengüç MP. An investigation of dependent/independent scattering regimes using a discrete dipole approximation. International Journal of Heat and Mass Transfer. 1996;39(4):811–822.

[310] Durant S, Calvo-Perez O, Vukadinovic N, Greffet JJ. Light scattering by a random distribution of particles embedded in absorbing media: Full-wave Monte Carlo solutions of the extinction coefficient. Journal of the Optical Society of America A. 2007 Sep;24(9):2953–2962.

[311] Nguyen VD, Faber DJ, van der Pol E, van Leeuwen TG, Kalkman J. Dependent and multiple scattering in transmission and backscattering optical coherence tomography. Optics Express. 2013 Dec;21(24):29145–29156.

[312] Ma L, Tan J, Zhao J, Wang F, Wang C. Multiple and dependent scattering by densely packed discrete spheres: Comparison of radiative transfer and Maxwell theory. Journal of Quantitative Spectroscopy and Radiative Transfer. 2017;187:255–266.

[313] Wang BX, Zhao CY. Achieving a strongly negative scattering asymmetry factor in random media composed of dual-dipolar particles. Physical Review A. 2018 Feb;97:023836.

[314] Wang BX, Zhao CY. Analysis of dependent scattering mechanism in hard-sphere Yukawa random media. Journal of Applied Physics. 2018;123(22):223101.

[315] Tinsley S, Bowman A, Phil D. Rutile type titanium pigments. Journal of the Oil and Colour Chemist Association. 1949;32(348):233–270.

[316] Stieg Jr F. The effect of extenders on the hiding power of titanium pigments. Official Digest Federation of Paint and Varnish Production Clubs. 1959;31(408):52–64.

[317] Hulst HC. Light scattering by small particles. Courier Corporation; 1957.

[318] Churchill SW, Clark GC, Sliepcevich CM. Light-scattering by very dense monodispersions of latex particles. Discussions of the Faraday Society. 1960;30:192–199.

[319] Harding RH, Golding B, Morgen RA. Optics of light-scattering films. Study of effects of pigment size and concentration. Journal of the Optical Society of America. 1960 May;50(5):446–455.

[320] Blevin WR, Brown WJ. Effect of particle separation on the reflectance of semi-infinite diffusers. Journal of the Optical Society America. 1961 Feb;51(2): 129–134.

[321] Rozenberg G. Optical characteristics of thick weakly absorbing scattering layers. In: Doklady Akademii Nauk: Archive. 1962;145:775–777.

[322] Foldy LL. The multiple scattering of waves. I. General theory of isotropic scattering by randomly distributed scatterers. Physical Review. 1945 Feb;67:107–119.

[323] Lax M. Multiple scattering of waves. Reviews of Modern Physics. 1951 Oct;23:287–310.

[324] Lax M. Multiple scattering of waves. II. the effective field in dense systems. Physical Review. 1952;85(4):621–629.

[325] Twersky V. Multiple scattering of radiation by an arbitrary configuration of parallel cylinders. The Journal of the Acoustical Society of America. 1952;24(1): 42–46.

[326] Hottel HC, Sarofim AF, Vasalos IA, Dalzell WH. Multiple scatter: Comparison of theory with experiment. Journal of Heat Transfer. 1970 May;92(2):285–291.

[327] Hottel HC, Sarofim AF, Dalzeil WH, Vasalos IA. Optical properties of coatings. Effect of pigment concentration. AIAA Journal. 1971 Oct;9(10):1895–1898.

[328] Brewster MQ, Tien CL. Radiative transfer in packed fluidized beds: Dependent versus independent scattering. Journal of Heat Transfer. 1982 Nov;104(4): 573–579.

[329] Yamada Y, Cartigny JD, Tien CL. Radiative transfer with dependent scattering by particles: Part 2—Experimental investigation dependent. Journal of Heat Transfer. 1986;108(August 1986).

[330] Cartigny J, Yamada Y, Tien C. Radiative transfer with dependent scattering by particles: Part 1—Theoretical investigation. Journal of Heat Transfer. 1986;108(3):608–613.

[331] Drolen BL, Tien CL. Independent and dependent scattering in packed-sphere systems. Journal of Thermophysics and Heat Transfer. 1987;1(1):63–68.

[332] Singh BP, Kaviany M. Modelling radiative heat transfer in packed beds. International Journal of Heat and Mass Transfer. 1992;35(6):1397–1405.

[333] Cornelius CM, Dowling JP. Modification of Planck blackbody radiation by photonic band-gap structures. Physical Review A. 1999 Jun;59:4736–4746.

[334] Lin SY, Fleming JG, Chow E, Bur J, Choi KK, Goldberg A. Enhancement and suppression of thermal emission by a three-dimensional photonic crystal. Physical Review B. 2000 Jul;62:R2243–R2246.

[335] Greffet JJ, Carminati R, Joulain K, Mulet JP, Mainguy S, Chen Y. Coherent emission of light by thermal sources. Nature. 2002;416(6876):61–64.

[336] Lee B, Fu C, Zhang Z. Coherent thermal emission from one-dimensional photonic crystals. Applied Physics Letters. 2005;87(7):071904.

[337] Liu X, Tyler T, Starr T, Starr AF, Jokerst NM, Padilla WJ. Taming the blackbody with infrared metamaterials as selective thermal emitters. Physical Review Letters. 2011;107(4):045901.

[338] Liu B, Gong W, Yu B, Li P, Shen S. Perfect thermal emission by nanoscale transmission line resonators. Nano Letters. 2017;17(2):666–672.

[339] Pralle MU, Moelders N, McNeal MP, Puscasu I, Greenwald AC, Daly JT, et al. Photonic crystal enhanced narrow-band infrared emitters. Applied Physics Letters. 2002;81(25):4685–4687.

[340] Celanovic I, Perreault D, Kassakian J. Resonant-cavity enhanced thermal emission. Physical Review B. 2005 Aug;72:075127.

[341] Dahan N, Niv A, Biener G, Gorodetski Y, Kleiner V, Hasman E. Enhanced coherency of thermal emission: Beyond the limitation imposed by delocalized surface waves. Physical Review B. 2007 Jul;76:045427.

[342] Rephaeli E, Fan S. Absorber and emitter for solar thermo-photovoltaic systems to achieve efficiency exceeding the Shockley-Queisser limit. Optics Express. 2009 Aug;17(17):15145–15159.

[343] Maruyama S, Kashiwa T, Yugami H, Esashi M. Thermal radiation from two-dimensionally confined modes in microcavities. Applied Physics Letters. 2001;79(9):1393–1395.

[344] De Zoysa M, Asano T, Mochizuki K, Oskooi A, Inoue T, Noda S. Conversion of broadband to narrowband thermal emission through energy recycling. Nature Photonics. 2012 Aug;6(8):535–539.

[345] Inoue T, De Zoysa M, Asano T, Noda S. Single-peak narrow-bandwidth mid-infrared thermal emitters based on quantum wells and photonic crystals. Applied Physics Letters. 2013;102(19):191110.

[346] Guo Y, Fan S. Narrowband thermal emission from a uniform tungsten surface critically coupled with a photonic crystal guided resonance. Optics Express. 2016 Dec;24(26):29896–29907.

[347] Inoue T, Zoysa MD, Asano T, Noda S. High-Q mid-infrared thermal emitters operating with high power-utilization efficiency. Optics Express. 2016 Jun;24(13):15101–15109.

[348] Dyachenko PN, Molesky S, Petrov AY, Störmer M, Krekeler T, Lang S, et al. Controlling thermal emission with refractory epsilon-near-zero metamaterials via topological transitions. Nature Communications. 2016;7(1):1–8.

[349] Molesky S, Dewalt CJ, Jacob Z. High temperature epsilon-near-zero and epsilon-near-pole metamaterial emitters for thermophotovoltaics. Optics Express. 2013;21(101):A96–A110.

[350] Hesketh PJ, Zemel JN, Gebhart B. Organ pipe radiant modes of periodic micromachined silicon surfaces. Nature. 1986;324:549–551.

[351] Hesketh PJ, Zemel JN, Gebhart B. Polarized spectral emittance from periodic micromachined surfaces. I. Doped silicon: The normal direction. Physical Review B. 1988;37(18):10795.

[352] Hesketh PJ, Zemel JN, Gebhart B. Polarized spectral emittance from periodic micromachined surfaces. II. Doped silicon: Angular variation. Physical Review B. 1988 Jun;37:10803–10813.

[353] Cui Y, Fung KH, Xu J, Ma H, Jin Y, He S, et al. Ultrabroadband light absorption by a sawtooth anisotropic metamaterial slab. Nano Letters. 2012;12(3):1443–1447.

[354] Wang H, Wang L. Perfect selective metamaterial solar absorbers. Optics Express. 2013 Nov;21(S6):A1078–A1093.

[355] Liu Y, Qiu J, Zhao J, Liu L. General design method of ultra-broadband perfect absorbers based on magnetic polaritons. Optics Express. 2017 Oct;25(20): A980–A989.

[356] Zhao B, Zhang ZM. Perfect absorption with trapezoidal gratings made of natural hyperbolic materials. Nanoscale and Microscale Thermophysical Engineering. 2017;21(3):123–133.

[357] Aydin K, Ferry VE, Briggs RM, Atwater HA. Broadband polarization-independent resonant light absorption using ultrathin plasmonic super absorbers. Nature Communications. 2011 Nov;2(1):517.

[358] Cheng CW, Abbas MN, Chiu CW, Lai KT, Shih MH, Chang YC. Wide-angle polarization independent infrared broadband absorbers based on metallic multi-sized disk arrays. Optics Express. 2012 Apr;20(9):10376–10381.

[359] Zhou M, Yi S, Luk TS, Gan Q, Fan S, Yu Z. Analog of superradiant emission in thermal emitters. Physical Review B. 2015 Jul;92:024302.

[360] Yeng YX, Ghebrebrhan M, Bermel P, Chan WR, Joannopoulos JD, Soljačić M, et al. Enabling high-temperature nanophotonics for energy applications. Proceedings of the National Academy of Sciences. 2012;109(7):2280–2285.

[361] Guo Y, Cortes CL, Molesky S, Jacob Z. Broadband super-Planckian thermal emission from hyperbolic metamaterials. Applied Physics Letters. 2012;101(13):131106.

[362] Schuller JA, Taubner T, Brongersma ML. Optical antenna thermal emitters. Nature Photonics. 2009;3(11):658–661.

[363] Ingvarsson S, Klein LJ, Au YY, Lacey JA, Hamann HF. Enhanced thermal emission from individual antenna-like nanoheaters. Optics Express. 2007 Sep;15(18):11249–11254.

[364] Chan DLC, Soljačić M, Joannopoulos JD. Thermal emission and design in one-dimensional periodic metallic photonic crystal slabs. Physical Review E. 2006 Jul;74:016609.

[365] Miyazaki HT, Ikeda K, Kasaya T, Yamamoto K, Inoue Y, Fujimura K, et al. Thermal emission of two-color polarized infrared waves from integrated plasmon cavities. Applied Physics Letters. 2008;92(14):141114.

[366] Lee JCW, Chan CT. Circularly polarized thermal radiation from layer-by-layer photonic crystal structures. Applied Physics Letters. 2007;90(5):051912.

[367] Wu C, Arju N, Kelp G, Fan JA, Dominguez J, Gonzales E, et al. Spectrally selective chiral silicon metasurfaces based on infrared Fano resonances. Nature Communications. 2014;5:3892.

[368] Shitrit N, Yulevich I, Maguid E, Ozeri D, Veksler D, Kleiner V, et al. Spin-optical metamaterial route to spin-controlled photonics. Science. 2013;340(6133): 724–726.

[369] Han SE, Norris DJ. Beaming thermal emission from hot metallic bull's eyes. Optics Express. 2010 Mar;18(5):4829–4837.

[370] Park JH, Han SE, Nagpal P, Norris DJ. Observation of thermal beaming from tungsten and molybdenum bull's eyes. ACS Photonics. 2016;3(3):494–500.

[371] Laroche M, Arnold C, Marquier F, Carminati R, Greffet JJ, Collin S, et al. Highly directional radiation generated by a tungsten thermal source. Optics Letters. 2005 Oct;30(19):2623–2625.

[372] Laroche M, Carminati R, Greffet JJ. Coherent thermal antenna using a photonic crystal slab. Physical Review Letters. 2006;96(12):123903.

[373] Shen Y, Ye D, Celanovic I, Johnson SG, Joannopoulos JD, Soljačić M. Optical Broadband Angular Selectivity. Science. 2014;343(6178):1499–1501.

[374] Kosten ED, Atwater JH, Parsons J, Polman A, Atwater HA. Highly efficient GaAs solar cells by limiting light emission angle. Light: Science & Applications. 2013 Jan;2(1):e45–e45.

[375] Chalabi H, Alù A, Brongersma ML. Focused thermal emission from a nanostructured SiC surface. Physical Review B. 2016 Sep;94:094307.

[376] Yu N, Capasso F. Flat optics with designer metasurfaces. Nature Materials. 2014;13(2):139–150.

[377] Ozawa T, Price HM, Amo A, Goldman N, Hafezi M, Lu L, et al. Topological photonics. Review of Modern Physics. 2019 Mar;91:015006.

[378] El-Ganainy R, Makris KG, Khajavikhan M, Musslimani ZH, Rotter S, Christodoulides DN. Non-Hermitian physics and PT symmetry. Nature Physics. 2018;14(1):11.

[379] Chang DE, Douglas JS, González-Tudela A, Hung CL, Kimble HJ. Colloquium: Quantum matter built from nanoscopic lattices of atoms and photons. Review of Modern Physics. 2018 Aug;90:031002.

[380] Silveirinha MG. Proof of the bulk-edge correspondence through a link between topological photonics and fluctuation-electrodynamics. Physical Review X. 2019 Feb;9:011037.

[381] Doiron CF, Naik GV. Non-Hermitian selective thermal emitters using metal-semiconductor hybrid resonators. Advanced Materials. 2019;31(44):1904154.

[382] Ridolfo A, Savasta S, Hartmann MJ. Nonclassical radiation from thermal cavities in the ultrastrong coupling regime. Physical Review Letter. 2013 Apr;110:163601.

[383] Song B, Ganjeh Y, Sadat S, Thompson D, Fiorino A, Fernández-Hurtado V, et al. Enhancement of near-field radiative heat transfer using polar dielectric thin films. Nature Nanotechnology. 2015;10(3):253–258.

[384] Kim K, Song B, Fernández-Hurtado V, Lee W, Jeong W, Cui L, et al. Radiative heat transfer in the extreme near field. Nature. 2015 Dec;528(7582):387–391.

[385] Fiorino A, Zhu L, Thompson D, Mittapally R, Reddy P, Meyhofer E. Nanogap near-field thermophotovoltaics. Nature Nanotechnology. 2018;13(9):806–811.

[386] Fiorino A, Thompson D, Zhu L, Song B, Reddy P, Meyhofer E. Giant enhancement in radiative heat transfer in sub-30 nm gaps of plane parallel surfaces. Nano Letters. 2018;18(6):3711–3715.

[387] Thompson D, Zhu LX, Mittapally R, Sadat S, Xing Z, McArdle P, et al. Hundred-fold enhancement in far-field radiative heat transfer over the blackbody limit. Nature. 2018;561(7722):216–+.

[388] DeSutter J, Tang L, Francoeur M. A near-field radiative heat transfer device. Nature Nanotechnology. 2019;14(8):751–755.

[389] Tang L, DeSutter J, Francoeur M. Near-field radiative heat transfer between dissimilar materials mediated by coupled surface phonon- and plasmon-polaritons. 2020;7(5):1304–1311.

[390] Rytov S. Theory of electrical fluctuation and thermal radiation. Academy of Science of USSR, Moscow; 1953.

[391] Rytov SM, Kravtsov YA, Tatarskii VI. Principles of statistical radiophysics. 3. Elements of random fields; 1989.

[392] Joulain K, Mulet JP, Marquier F, Carminati R, Greffet JJ. Surface electromagnetic waves thermally excited: Radiative heat transfer, coherence properties and Casimir forces revisited in the near field. Surface Science Reports. 2005;57(3): 59–112.

[393] Polder D, Van Hove M. Theory of radiative heat transfer between closely spaced bodies. Physical Review B. 1971;4(10):3303–3314.

[394] Narayanaswamy A, Chen G. Thermal near-field radiative transfer between two spheres. Physical Review B. 2008 Feb;77:075125.

[395] Otey C, Fan S. Numerically exact calculation of electromagnetic heat transfer between a dielectric sphere and plate. Physical Review B. 2011 Dec;84:245431.

[396] Wen SB. Direct numerical simulation of near field thermal radiation based on Wiener Chaos expansion of thermal fluctuating current. Journal of Heat Transfer. 2010;132(7):072704.

[397] Rodriguez AW, Ilic O, Bermel P, Celanovic I, Joannopoulos JD, Soljačić M, et al. Frequency-selective near-field radiative heat transfer between photonic crystal slabs: A computational approach for arbitrary geometries and materials. Physical Review Letters. 2011 Sep;107:114302.

[398] Rodriguez AW, Reid MTH, Johnson SG. Fluctuating-surface-current formulation of radiative heat transfer for arbitrary geometries. Physical Review B. 2012 Dec;86:220302.

[399] Reid MTH, Johnson SG. Efficient computation of power, force, and torque in BEM scattering calculations. IEEE Transactions on Antennas and Propagation. 2015;63(8):3588–3598.

[400] Edalatpour S, Francoeur M. The Thermal Discrete Dipole Approximation (T-DDA) for near-field radiative heat transfer simulations in three-dimensional arbitrary geometries. Journal of Quantitative Spectroscopy and Radiative Transfer. 2014;133:364–373.

[401] Chapuis PO, Volz S, Henkel C, Joulain K, Greffet JJ. Effects of spatial dispersion in near-field radiative heat transfer between two parallel metallic surfaces. Physical Review B. 2008 Jan;77:035431.

[402] Singer F, Ezzahri Y, Joulain K. Near field radiative heat transfer between two nonlocal dielectrics. Journal of Quantitative Spectroscopy and Radiative Transfer. 2015;154:55–62.

[403] Xiong S, Yang K, Kosevich YA, Chalopin Y, D'Agosta R, Cortona P, et al. Classical to quantum transition of heat transfer between two silica clusters. Physical Review Letters. 2014 Mar;112:114301.

[404] Klöckner JC, Siebler R, Cuevas JC, Pauly F. Thermal conductance and thermoelectric figure of merit of C_{60}-based single-molecule junctions: Electrons, phonons, and photons. Physical Review B. 2017 Jun;95:245404.

[405] Chiloyan V, Garg J, Esfarjani K, Chen G. Transition from near-field thermal radiation to phonon heat conduction at sub-nanometre gaps. Nature Communications. 2015 Apr;6(1):6755.

[406] Cravalho E, Domoto G, Tien C. Measurements of thermal radiation of solids at liquid-helium temperatures. In: 3rd Thermophysics Conference; 1968. p. 774.

[407] Hargreaves CM. Anomalous radiative transfer between closely-spaced bodies. Physics Letters A. 1969;30(9):491–492.

[408] Domoto GA, Boehm RF, Tien CL. Experimental investigation of radiative transfer between metallic surfaces at cryogenic temperatures. Journal of Heat Transfer. 1970 Aug;92(3):412–416.

[409] Narayanaswamy A, Shen S, Chen G. Near-field radiative heat transfer between a sphere and a substrate. Physical Review B. 2008 Sep;78:115303.

[410] Shen S, Narayanaswamy A, Chen G. Surface phonon polaritons mediated energy transfer between nanoscale gaps. Nano Letters. 2009;9(8):2909–2913.

[411] Rousseau E, Siria A, Jourdan G, Volz S, Comin F, Chevrier J, et al. Radiative heat transfer at the nanoscale. Nature Photonics. 2009 Sep;3(9):514–517.

[412] Hu L, Narayanaswamy A, Chen X, Chen G. Near-field thermal radiation between two closely spaced glass plates exceeding Planck's blackbody radiation law. Applied Physics Letters. 2008;92(13):133106.

[413] Ottens RS, Quetschke V, Wise S, Alemi AA, Lundock R, Mueller G, et al. Near-field radiative heat transfer between macroscopic planar surfaces. Physical Review Letters. 2011 Jun;107:014301.

[414] Bernardi MP, Milovich D, Francoeur M. Radiative heat transfer exceeding the blackbody limit between macroscale planar surfaces separated by a nanosize vacuum gap. Nature Communications. 2016 Sep;7(1):12900.

[415] Ghashami M, Geng H, Kim T, Iacopino N, Cho SK, Park K. Precision measurement of phonon-polaritonic near-field energy transfer between macroscale planar structures under large thermal gradients. Physical Review Letters. 2018 Apr;120:175901.

[416] St-Gelais R, Guha B, Zhu LX, Fan SH, Lipson M. Demonstration of strong near-field radiative heat transfer between integrated nanostructures. Nano Letters. 2014;14(12):6971–6975.

[417] St-Gelais R, Zhu LX, Fan SH, Lipson M. Near-field radiative heat transfer between parallel structures in the deep subwavelength regime. Nature Nanotechnology. 2016;11(6):515–+.

[418] Song B, Thompson D, Fiorino A, Ganjeh Y, Reddy P, Meyhofer E. Radiative Heat Conductances between Dielectric and Metallic Parallel Plates with Nanoscale Gaps. Nature Nanotechnology. 2016;11(6):509–+.

[419] Kittel A, Müller-Hirsch W, Parisi J, Biehs SA, Reddig D, Holthaus M. Near-field heat transfer in a scanning thermal microscope. Physical Review Letters. 2005 Nov;95:224301.

[420] Worbes L, Hellmann D, Kittel A. Enhanced near-field heat flow of a monolayer dielectric Island. Physical Review Letters. 2013 Mar;110:134302.

[421] Cui L, Jeong W, Fernández-Hurtado V, Feist J, García-Vidal FJ, Cuevas JC, et al. Study of radiative heat transfer in Ångström- and nanometre-sized gaps. Nature Communications. 2017 Feb;8(1):14479.

[422] Kloppstech K, Könne N, Biehs SA, Rodriguez AW, Worbes L, Hellmann D, et al. Giant heat transfer in the crossover regime between conduction and radiation. Nature Communications. 2017 Feb;8(1):14475.

[423] Cui L, Jeong W, Hur S, Matt M, Klöckner JC, Pauly F, et al. Quantized thermal transport in single-atom junctions. Science. 2017;355(6330):1192–1195.

[424] Mosso N, Drechsler U, Menges F, Nirmalraj P, Karg S, Riel H, et al. Heat transport through atomic contacts. Nature Nanotechnology. 2017 May;12(5):430–433.

[425] Jackson JD. Classical electrodynamics. 3rd ed. John Wiley & Sons; 1998.

[426] Greiner W. Classical electrodynamics. Springer; 1998.

[427] Landau LD, Bell J, Kearsley M, Pitaevskii L, Lifshitz E, Sykes J. Electrodynamics of continuous media. vol. 8. Elsevier; 2013.

[428] Chen G, Borca-Tasciuc D, Yang R. Nanoscale heat transfer. Encyclopedia of Nanoscience and Nanotechnology. 2004;7(1):429–459.

[429] Wolf E, Nieto-Vesperinas M. Analyticity of the angular spectrum amplitude of scattered fields and some of its consequences. Journal of the Optical Society of America A. 1985 Jun;2(6):886–890.

[430] Callaway J. Quantum theory of the solid state. Academic Press; 1991.

[431] Fox M. Optical properties of solids. vol. 3. Oxford University Press; 2010.

[432] Johnson PB, Christy RW. Optical constants of the noble metals. Physical Review B. 1972 Dec;6:4370–4379.

[433] Ordal MA, Bell RJ, Alexander RW, Long LL, Querry MR. Optical properties of fourteen metals in the infrared and far infrared: Al, Co, Cu, Au, Fe, Pb, Mo, Ni, Pd, Pt, Ag, Ti, V, and W. Applied Optics. 1985 Dec;24(24):4493–4499.

[434] Kerker M. The scattering of light and other electromagnetic radiation. Academic Press; 1969.

[435] Ishimaru A. Wave propagation and scattering in random media. vol. 2. Academic Press; 1978.

[436] Leung Tsang JAK, Ding KH. Scattering of electromagnetic waves, theories and applications. vol. 1. Wiley; 2000.

[437] Stratton J. Electromagnetic theory. McGraw-Hill; 1941.

[438] Tribelsky MI, Luk'yanchuk BS. Anomalous light scattering by small particles. Physical Review Letters. 2006 Dec;97:263902.

[439] Kuznetsov AI, Miroshnichenko AE, Brongersma ML, Kivshar YS, Luk'yanchuk B. Optically resonant dielectric nanostructures. Science. 2016; 354(6314):aag2472.

[440] Abramowitz M, Stegun IA. Handbook of mathematical functions: With formulas, graphs, and mathematical tables. vol. 55. Courier Corporation; 1964.

[441] Bohren CF, Koh G. Forward-scattering corrected extinction by nonspherical particles. Applied Optics. 1985 Apr;24(7):1023–1029.

[442] Mishchenko MI, Hovenier JW, Travis LD. Light scattering by nonspherical particles: Theory, measurements, and applications. Academic Press; 1999.

[443] Yeh C. Perturbation approach to the diffraction of electromagnetic waves by arbitrarily shaped dielectric obstacles. Physical Review. 1964 Aug;135: A1193–A1201.

[444] Yeh C. Scattering by liquid-coated prolate spheroids. The Journal of the Acoustical Society of America. 1969;46(3B):797–801.

[445] Waterman PC. Matrix formulation of electromagnetic scattering. Proceedings of the IEEE. 1965 Aug;53(8):805–812.

[446] Peterson B, Ström S. T-matrix formulation of electromagnetic scattering from multilayered scatterers. Physical Review D. 1974 Oct;10:2670–2684.

[447] Mackowski DW, Mishchenko MI. Calculation of the T matrix and the scattering matrix for ensembles of spheres. Journal of the Optical Society of America A. 1996 Nov;13(11):2266–2278.
[448] Mackowski DW, Mishchenko MI. Direct simulation of extinction in a slab of spherical particles. Journal of Quantitative Spectroscopy and Radiative Transfer. 2013;123:103–112.
[449] Yurkin MA, Hoekstra AG. The discrete dipole approximation: An overview and recent developments. Journal of Quantitative Spectroscopy and Radiative Transfer. 2007;106(1):558–589.
[450] Purcell EM, Pennypacker CR. Scattering and absorption of light by nonspherical dielectric grains. The Astrophysical Journal. 1973 Dec;186:705–714.
[451] Draine BT. The discrete-dipole approximation and its application to interstellar graphite grains. The Astrophysical Journal. 1988 Oct;333:848.
[452] Draine BT, Goodman J. Beyond Clausius-Mossotti: Wave propagation on a polarizable point lattice and the discrete dipole approximation. The Astrophysical Journal. 1993 Mar;405:685.
[453] Draine BT, Flatau PJ. Discrete-dipole approximation for scattering calculations. Journal of the Optical Society of America A. 1994 Apr;11(4):1491–1499.
[454] Markel VA. Extinction, scattering and absorption of electromagnetic waves in the coupled-dipole approximation. Journal of Quantitative Spectroscopy and Radiative Transfer. 2019;236:106611.
[455] Yurkin MA, Hoekstra AG. The discrete-dipole-approximation code ADDA: Capabilities and known limitations. Journal of Quantitative Spectroscopy and Radiative Transfer. 2011;112(13):2234–2247.
[456] Mahan GD. Many-particle physics. Springer Science & Business Media; 2013.
[457] Doicu A, Mishchenko MI. Overview of methods for deriving the radiative transfer theory from the Maxwell equations. I: Approach based on the far-field Foldy equations. Journal of Quantitative Spectroscopy and Radiative Transfer. 2018;220:123–139.
[458] Doicu A, Mishchenko MI. An overview of methods for deriving the radiative transfer theory from the Maxwell equations. II: Approach based on the Dyson and Bethe–Salpeter equations. Journal of Quantitative Spectroscopy and Radiative Transfer. 2019;224:25–36.
[459] Doicu A, Mishchenko MI. Electromagnetic scattering by discrete random media. I: The dispersion equation and the configuration-averaged exciting field. Journal of Quantitative Spectroscopy and Radiative Transfer. 2019;230:282–303.
[460] Doicu A, Mishchenko MI. Electromagnetic scattering by discrete random media. II: The coherent field. Journal of Quantitative Spectroscopy and Radiative Transfer. 2019;230:86–105.
[461] Doicu A, Mishchenko MI. Electromagnetic scattering by discrete random media. III: The vector radiative transfer equation. Journal of Quantitative Spectroscopy and Radiative Transfer. 2019;236:106564.
[462] Doicu A, Mishchenko MI. Electromagnetic scattering by discrete random media. IV: Coherent backscattering. Journal of Quantitative Spectroscopy and Radiative Transfer. 2019;236:106565.
[463] Čapeta D, Radić J, Szameit A, Segev M, Buljan H. Anderson localization of partially incoherent light. Physical Review A. 2011 Jul;84:011801.

[464] Auger JC, Stout B. Dependent light scattering in white paint films: Clarification and application of the theoretical concepts. Journal of Coatings Technology and Research. 2012 May;9(3):287–295.

[465] Labeyrie G, Vaujour E, Müller CA, Delande D, Miniatura C, Wilkowski D, et al. Slow diffusion of light in a cold atomic cloud. Physical Review Letters. 2003 Nov;91:223904.

[466] Stephen MJ, Cwilich G. Intensity correlation functions and fluctuations in light scattered from a random medium. Physical Review Letters. 1987 Jul;59:285–287.

[467] Stephen MJ. Temporal fluctuations in wave propagation in random media. Physical Review B. 1988 Jan;37:1–5.

[468] Mishchenko MI, Dlugach JM, Yurkin MA, Bi L, Cairns B, Liu L, et al. First-principles modeling of electromagnetic scattering by discrete and discretely heterogeneous random media. Physics Reports. 2016;632:1–75.

[469] Etemad S, Thompson R, Andrejco MJ. Weak localization of photons: Universal fluctuations and ensemble averaging. Physical Review Letters. 1986 Aug;57: 575–578.

[470] Feng S, Kane C, Lee PA, Stone AD. Correlations and fluctuations of coherent wave transmission through disordered media. Physical Review Letters. 1988 Aug;61:834–837.

[471] Kaiser R. Quantum multiple scattering. Journal of Modern Optics. 2009; 56(18–19):2082–2088.

[472] Smolka S, Huck A, Andersen UL, Lagendijk A, Lodahl P. Observation of spatial quantum correlations induced by multiple scattering of nonclassical light. Physical Review Letters. 2009 May;102:193901.

[473] Ott JR, Mortensen NA, Lodahl P. Quantum interference and entanglement induced by multiple scattering of light. Physical Review Letters. 2010 Aug;105:090501.

[474] Alberucci A, Jisha CP, Bolis S, Beeckman J, Nolte S. Interplay between multiple scattering and optical nonlinearity in liquid crystals. Optics Letters. 2018 Aug;43(15):3461–3464.

[475] Angelani L, Conti C, Ruocco G, Zamponi F. Glassy behavior of light. Physical Review Letters. 2006 Feb;96:065702.

[476] Conti C, Angelani L, Ruocco G. Light diffusion and localization in three-dimensional nonlinear disordered media. Physical Review A. 2007 Mar;75:033812.

[477] Lee S. Dependent scattering of an obliquely incident plane wave by a collection of parallel cylinders. Journal of Applied Physics. 1990;68(10):4952–4957.

[478] Lee SC. Scattering by closely-spaced radially-stratified parallel cylinders. Journal of Quantitative Spectroscopy and Radiative Transfer. 1992;48(2):119–130.

[479] Lee SC. Dependent vs independent scattering in fibrous composites containing parallel fibers. Journal of Thermophysics and Heat Transfer. 1994;8(4):641–646.

[480] Lee SC. A general solution for scattering by infinite cylinders in the presence of an interface. Journal of Quantitative Spectroscopy and Radiative Transfer. 2019;235:140–161.

[481] Tsang L, Kong JA, Ding KH, Ao CO. Scattering of electromagnetic waves: Numerical simulations. vol. 25. John Wiley & Sons; 2004.

[482] Frisch U. Wave propagation in random media. Probabilistic methods in applied mathematics. 1968:75–198.

[483] Ishimaru A. Theory and application of wave propagation and scattering in random media. Proceedings of the IEEE. 1977 July;65(7):1030–1061.

[484] Barabanenkov YN, Kravtsov YA, Rytov SM, Tamarskiĭ VI. Status of the theory of propagation of waves in a randomly inhomogeneous medium. Soviet Physics Uspekhi. 1971 May;13(5):551–575.

[485] Barabanenkov YN, Zurk LM, Barabanenkov MY. Poynting's theorem and electromagnetic wave multiple scattering in dense media near resonance: Modified radiative transfer equation. Journal of Electromagnetic Waves and Applications. 1995;9(11–12):1393–1420.

[486] Tsang L, Kong JA. Multiple scattering of electromagnetic waves by random distributions of discrete scatterers with coherent potential and quantum mechanical formalism. Journal of Applied Physics. 1980;51(7):3465–3485.

[487] Tsang L, Kong JA, Shin RT. Theory of microwave remote sensing. John Wiley & Sons; 1985.

[488] Wang BX, Zhao CY. The dependent scattering effect on radiative properties of micro/nanoscale discrete disordered media. Annual Review of Heat Transfer. 2020;23:231–353.

[489] Tsang L. Van de Hulst essay: Multiple scattering of waves by discrete scatterers and rough surfaces. Journal of Quantitative Spectroscopy and Radiative Transfer. 2019;224:566–587.

[490] Page JH, Sheng P, Schriemer HP, Jones I, Jing X, Weitz DA. Group velocity in strongly scattering media. Science. 1996;271(5249):634–637.

[491] Barrera RG, Reyes-Coronado A, García-Valenzuela A. Nonlocal nature of the electrodynamic response of colloidal systems. Physical Review B. 2007 May;75:184202.

[492] Gower AL, Abrahams ID, Parnell WJ. A proof that multiple waves propagate in ensemble-averaged particulate materials. Proceedings of the Royal Society A: Mathematical, Physical and Engineering Sciences. 2019;475(2229):20190344.

[493] de Vries P, van Coevorden DV, Lagendijk A. Point scatterers for classical waves. Reviews of Modern Physics. 1998 Apr;70:447–466.

[494] Cherroret N, Delande D, van Tiggelen BA. Induced dipole-dipole interactions in light diffusion from point dipoles. Physical Review A. 2016;94:012702.

[495] Barabanenkov YN. Multiple scattering of waves by ensembles of particles and the theory of radiation transport. Soviet Physics Uspekhi. 1975;18(9):673.

[496] Kugo T, Mitchard MG. The chiral Ward-Takahashi identity in the ladder approximation. Physics Letters B. 1992;282(1):162–170.

[497] Varadan VV, Ma Y, Varadan VK. Propagator model including multipole fields for discrete random media. Journal of Optical Society America A. 1985 Dec;2(12):2195–2201.

[498] Lamb W, Wood DM, Ashcroft NW. Long-wavelength electromagnetic propagation in heterogeneous media. Physical Review B. 1980 Mar;21:2248–2266.

[499] Tsang L, Kong JA. Effective propagation constants for coherent electromagnetic wave propagation in media embedded with dielectric scatters. Journal of Applied Physics. 1982;53(11):7162–7173.

[500] Xu YL, Gustafson BAS. A generalized multiparticle Mie-solution: Further experimental verification. Journal of Quantitative Spectroscopy and Radiative Transfer. 2001;70(4):395–419.

[501] Mackowski DW, Mishchenko MI. A multiple sphere T-matrix Fortran code for use on parallel computer clusters. Journal of Quantitative Spectroscopy and Radiative Transfer. 2011;112(13):2182–2192.

[502] Pitman KM, Kolokolova L, Verbiscer AJ, Mackowski DW, Joseph ECS. Coherent backscattering effect in spectra of icy satellites and its modeling using multi-sphere T-matrix (MSTM) code for layers of particles. Planetary and Space Science. 2017;149:23–31.

[503] Stout B, Auger JC, Lafait J. A transfer matrix approach to local field calculations in multiple-scattering problems. Journal of Modern Optics. 2002;49(13):2129–2152.

[504] Wang Y, Chew WC. A recursive T-matrix approach for the solution of electromagnetic scattering by many spheres. IEEE Transactions on Antennas and Propagation. 1993 Dec;41(12):1633–1639.

[505] Egel A, Pattelli L, Mazzamuto G, Wiersma DS, Lemmer U. CELES: CUDA-accelerated simulation of electromagnetic scattering by large ensembles of spheres. Journal of Quantitative Spectroscopy and Radiative Transfer. 2017;199:103–110.

[506] Varadan VK, Bringi VN, Varadan VV. Coherent electromagnetic wave propagation through randomly distributed dielectric scatterers. Physical Review D. 1979 Apr;19:2480–2489.

[507] Bertrand M, Devilez A, Hugonin JP, Lalanne P, Vynck K. Global polarizability matrix method for efficient modeling of light scattering by dense ensembles of non-spherical particles in stratified media. Journal of the Optical Society of America A. 2020 Jan;37(1):70–83.

[508] Doicu A, Wriedt T. Near-field computation using the null-field method. Journal of Quantitative Spectroscopy and Radiative Transfer. 2010;111(3):466–473.

[509] Forestiere C, Iadarola G, Negro LD, Miano G. Near-field calculation based on the T-matrix method with discrete sources. Journal of Quantitative Spectroscopy and Radiative Transfer. 2011;112(14):2384–2394.

[510] Egel A, Theobald D, Donie Y, Lemmer U, Gomard G. Light scattering by oblate particles near planar interfaces: On the validity of the T-matrix approach. Optics Express. 2016 Oct;24(22):25154–25168.

[511] Egel A, Eremin Y, Wriedt T, Theobald D, Lemmer U, Gomard G. Extending the applicability of the T-matrix method to light scattering by flat particles on a substrate via truncation of sommerfeld integrals. Journal of Quantitative Spectroscopy and Radiative Transfer. 2017;202:279–285.

[512] Theobald D, Egel A, Gomard G, Lemmer U. Plane-wave coupling formalism for T-matrix simulations of light scattering by nonspherical particles. Physical Review A. 2017 Sep;96:033822.

[513] Gimbutas Z, Greengard L. Fast multi-particle scattering: A hybrid solver for the Maxwell equations in microstructured materials. Journal of Computational Physics. 2013;232(1):22–32.

[514] Lai J, Kobayashi M, Greengard L. A fast solver for multi-particle scattering in a layered medium. Optics Express. 2014 Aug;22(17):20481–20499.

[515] Blankrot B, Heitzinger C. Efficient computational design and optimization of dielectric metamaterial structures. IEEE Journal on Multiscale and Multiphysics Computational Techniques. 2019;4:234–244.

[516] Martin PA. On connections between boundary integral equations and T-matrix methods. Engineering Analysis with Boundary Elements. 2003;27(7):771–777.

[517] Gumerov NA, Duraiswami R. A scalar potential formulation and translation theory for the time-harmonic Maxwell equations. Journal of Computational Physics. 2007;225(1):206–236.

[518] Greengard L, Rokhlin V. A fast algorithm for particle simulations. Journal of Computational Physics. 1987;73(2):325–348.

[519] Engheta N, Murphy WD, Rokhlin V, Vassiliou MS. The fast multipole method (FMM) for electromagnetic scattering problems. IEEE Transactions on Antennas and Propagation. 1992 June;40(6):634–641.

[520] Cheng H, Crutchfield WY, Gimbutas Z, Greengard LF, Ethridge JF, Huang J, et al. A wideband fast multipole method for the Helmholtz equation in three dimensions. Journal of Computational Physics. 2006;216(1):300–325.

[521] Markkanen J, Yuffa AJ. Fast superposition T-matrix solution for clusters with arbitrarily-shaped constituent particles. Journal of Quantitative Spectroscopy and Radiative Transfer. 2017;189:181–188.

[522] Maier SA, Brongersma ML, Kik PG, Atwater HA. Observation of near-field coupling in metal nanoparticle chains using far-field polarization spectroscopy. Phys Rev B. 2002 May;65:193408.

[523] Gunnarsson L, Rindzevicius T, Prikulis J, Kasemo B, Käll M, Zou S, et al. Confined plasmons in nanofabricated single silver particle pairs: Experimental observations of strong interparticle interactions. The Journal of Physical Chemistry B. 2005;109(3):1079–1087.

[524] Auguié B, Barnes WL. Collective resonances in gold nanoparticle arrays. Physical Review Letters. 2008 Sep;101:143902.

[525] Auguié B, Barnes WL. Diffractive coupling in gold nanoparticle arrays and the effect of disorder. Optics Letters. 2009 Feb;34(4):401–403.

[526] Taylor RW, Esteban R, Mahajan S, Coulston R, Scherman OA, Aizpurua J, et al. Simple composite dipole model for the optical modes of strongly-coupled plasmonic nanoparticle aggregates. The Journal of Physical Chemistry C. 2012;116(47):25044–25051.

[527] Kravets VG, Schedin F, Pisano G, Thackray B, Thomas PA, Grigorenko AN. Nanoparticle arrays: From magnetic response to coupled plasmon resonances. Physical Review B. 2014 Sep;90:125445.

[528] Kupriyanov DV, Sokolov IM, Havey MD. Mesoscopic coherence in light scattering from cold, optically dense and disordered atomic systems. Physics Reports. 2017;671:1–60.

[529] Guerin W, Araújo MO, Kaiser R. Subradiance in a large cloud of cold atoms. Physical Review Letters. 2016 Feb;116:083601.

[530] Chandrasekhar S. Radiative transfer. Clarendon Press; 1950.

[531] Collett E, Foley JT, Wolf E. On an investigation of Tatarskii into the relationship between coherence theory and the theory of radiative transfer∗. Journal of the Optical Society of America. 1977 Apr;67(4):465–467.

[532] Fante RL. Relationship between radiative-transport theory and Maxwell's equations in dielectric media. Journal of the Optical Society of America. 1981 Apr;71(4):460–468.

[533] van Tiggelen B, Stark H. Nematic liquid crystals as a new challenge for radiative transfer. Reviews of Modern Physics. 2000 Oct;72:1017–1039.

[534] Cazé A, Schotland JC. Diagrammatic and asymptotic approaches to the origins of radiative transport theory: Tutorial. Journal of the Optical Society of America A. 2015 Aug;32(8):1475–1484.

[535] Vynck K, Pierrat R, Carminati R. Multiple scattering of polarized light in disordered media exhibiting short-range structural correlations. Physical Review A. 2016 Sep;94:033851.

[536] van der Mark MB, van Albada MP, Lagendijk A. Light scattering in strongly scattering media: Multiple scattering and weak localization. Physical Review B. 1988 Mar;37:3575–3592.

[537] Akkermans E, Wolf PE, Maynard R. Coherent backscattering of light by disordered media: Analysis of the peak line shape. Physical Review Letters. 1986 Apr;56:1471–1474.

[538] Akkermans E, Wolf P, Maynard R, Maret G. Theoretical study of the coherent backscattering of light by disordered media. J de Phys (France). 1988;49(1): 77–98.

[539] Mishchenko MI, Dlugach JM. Can weak localization of photons explain the opposition effect of Saturn's rings? Monthly Notices of the Royal Astronomical Society. 1992 01;254(1):15P–18P.

[540] Mandt CE, Kuga Y, Tsang L, Ishimaru A. Microwave propagation and scattering in a dense distribution of non-tenuous spheres: Experiment and theory. Waves in Random Media. 1992;2(3):225–234.

[541] Tsang L, Chang TC. Dense media radiative transfer theory based on quasicrystalline approximation with applications to passive microwave remote sensing of snow. Radio Science. 2000;35(3):731–749.

[542] Liang D, Xu X, Tsang L, Andreadis KM, Josberger EG. The effects of layers in dry snow on its passive microwave emissions using dense media radiative transfer theory based on the quasicrystalline approximation (QCA/DMRT). IEEE Transactions on Geoscience and Remote Sensing. 2008 Nov;46(11):3663–3671.

[543] Picard G, Brucker L, Roy A, Dupont F, Fily M, Royer A, et al. Simulation of the microwave emission of multi-layered snowpacks using the Dense Media Radiative transfer theory: The DMRT-ML model. Geoscientific Model Development. 2013;6(4):1061–1078.

[544] van Albada MP, van Tiggelen BA, Lagendijk A, Tip A. Speed of propagation of classical waves in strongly scattering media. Physical Review Letters. 1991 Jun;66:3132–3135.

[545] Störzer M, Gross P, Aegerter CM, Maret G. Observation of the critical regime near Anderson localization of light. Physical Review Letters. 2006 Feb;96:063904.

[546] Aubry GJ, Schertel L, Chen M, Weyer H, Aegerter CM, Polarz S, et al. Resonant transport and near-field effects in photonic glasses. Physical Review A. 2017 Oct;96:043871.

[547] van Tiggelen BA, Lagendijk A, Wiersma DS. Reflection and transmission of waves near the localization threshold. Physical Review Letters. 2000 May;84:4333–4336.

[548] Tian CS, Cheung SK, Zhang ZQ. Local diffusion theory for localized waves in open media. Physical Review Letters. 2010 Dec;105:263905.

[549] Yamilov AG, Sarma R, Redding B, Payne B, Noh H, Cao H. Position-dependent diffusion of light in disordered waveguides. Physics Review Letters. 2014 Jan;112:023904.

[550] Hu H, Strybulevych A, Page JH, Skipetrov SE, van Tiggelen BA. Localization of ultrasound in a three-dimensional elastic network. Nature Physics. 2008;4(12):945–948.

[551] Zhang ZQ, Chabanov AA, Cheung SK, Wong CH, Genack AZ. Dynamics of localized waves: Pulsed microwave transmissions in quasi-one-dimensional media. Physical Review B. 2009 Apr;79:144203.

[552] Haberko J, Froufe-Pérez LS, Scheffold F. Transition from light diffusion to localization in three-dimensional hyperuniform dielectric networks near the band edge. arXiv e-prints. 2018 Dec:arXiv:1812.02095.

[553] Akkermans E, Montambaux G. Mesoscopic physics of electrons and photons. Cambridge University Press; 2007.

[554] Reufer M, Rojas-Ochoa LF, Eiden S, Sáenz JJ, Scheffold F. Transport of light in amorphous photonic materials. Applied Physics Letters. 2007;91(17):171904.

[555] Xiao M, Hu Z, Wang Z, Li Y, Tormo AD, Le Thomas N, et al. Bioinspired bright noniridescent photonic melanin supraballs. Science Advances. 2017;3(9):e1701151.

[556] Liew SF, Forster J, Noh H, Schreck CF, Saranathan V, Lu X, et al. Short-range order and near-field effects on optical scattering and structural coloration. Optics Express. 2011;19(9):8208–8217.

[557] Skipetrov SE, Sokolov IM. Absence of Anderson localization of light in a random ensemble of point scatterers. Physical Review Letters. 2014;112:023905.

[558] M EJ, E SS. Longitudinal optical fields in light scattering from dielectric spheres and Anderson localization of light. Annalen der Physik. 2017;529(8):1700039.

[559] Schirmacher W, Abaie B, Mafi A, Ruocco G, Leonetti M. What is the right theory for Anderson localization of light? An experimental test. Physical Review Letters. 2018 Feb;120:067401.

[560] Silies M, Mascheck M, Leipold D, Kollmann H, Schmidt S, Sartor J, et al. Near-field-assisted localization: Effect of size and filling factor of randomly distributed zinc oxide nanoneedles on multiple scattering and localization of light. Applied Physics B. 2016;122(7):181.

[561] Pierrat R, Carminati R. Spontaneous decay rate of a dipole emitter in a strongly scattering disordered environment. Physical Review A. 2010;81:063802.

[562] Tishkovets VP, Jockers K. Multiple scattering of light by densely packed random media of spherical particles: Dense media vector radiative transfer equation. Journal of Quantitative Spectroscopy and Radiative Transfer. 2006;101(1): 54–72.

[563] Petrova EV, Tishkovets VP, Jockers K. Modeling of opposition effects with ensembles of clusters: Interplay of various scattering mechanisms. Icarus. 2007;188(1):233–245.

[564] Tishkovets VP. Light scattering by closely packed clusters: Shielding of particles by each other in the near field. Journal of Quantitative Spectroscopy and Radiative Transfer. 2008;109(16):2665–2672.

[565] Tishkovets VP, Petrova EV, Mishchenko MI. Scattering of electromagnetic waves by ensembles of particles and discrete random media. Journal of Quantitative Spectroscopy and Radiative Transfer. 2011;112(13):2095–2127.

[566] Tishkovets VP, Petrova EV. Light scattering by densely packed systems of particles: Near-field effects. Springer Berlin Heidelberg; 2013. pp. 3–36.

[567] Shen Z, Dogariu A. Meaning of phase in subwavelength elastic scattering. Optica. 2019 Apr;6(4):455–459.

[568] van Tiggelen BA, Lagendijk A, Tip A. Multiple-scattering effects for the propagation of light in 3D slabs. Journal of Physics: Condensed Matter. 1990;2(37):7653.

[569] van Tiggelen BA, Lagendijk A. Resonantly induced dipole-dipole interactions in the diffusion of scalar waves. Physical Review B. 1994 Dec;50:16729–16732.

[570] Fraden S, Maret G. Multiple light scattering from concentrated, interacting suspensions. Physical Review Letters. 1990 Jul;65:512–515.

[571] Froufe-Pérez LS, Engel M, Sáenz JJ, Scheffold F. Band gap formation and Anderson localization in disordered photonic materials with structural correlations. Proceedings of the National Academy of Sciences. 2017;114(36):9570–9574.

[572] Liu MQ, Zhao CY, Wang BX, Fang X. Role of short-range order in manipulating light absorption in disordered media. Journal of the Optical Society of America B. 2018 Mar;35(3):504–513.

[573] Wertheim MS. Exact solution of the Percus-Yevick integral equation for hard spheres. Physical Review Letters. 1963 Apr;10:321–323.

[574] Mishchenko MI, Goldstein DH, Chowdhary J, Lompado A. Radiative transfer theory verified by controlled laboratory experiments. Optics Letters. 2013 Sep;38(18):3522–3525.

[575] Mishchenko MI. Asymmetry parameters of the phase function for densely packed scattering grains. Journal of Quantitative Spectroscopy and Radiative Transfer. 1994;52(1):95–110.

[576] Leseur O, Pierrat R, Carminati R. High-density hyperuniform materials can be transparent. Optica. 2016 Jul;3(7):763–767.

[577] Froufe-Pérez LS, Engel M, Damasceno PF, Muller N, Haberko J, Glotzer SC, et al. Role of short-range order and hyperuniformity in the formation of band gaps in disordered photonic materials. Physical Review Letters. 2016;117:053902.

[578] Baxter RJ. Percus–Yevick equation for hard spheres with surface adhesion. The Journal of the Chemical Physics. 1968;49(6):2770–2774.

[579] Frenkel D. Playing tricks with designer "Atoms." Science. 2002;296(5565):65–66.

[580] Bressel L, Hass R, Reich O. Particle sizing in highly turbid dispersions by Photon Density Wave spectroscopy. Journal of Quantitative Spectroscopy and Radiative Transfer. 2013;126:122–129.

[581] Sudiarta IW, Chylek P. Mie-scattering formalism for spherical particles embedded in an absorbing medium. Journal of the Optical Society of America A. 2001 Jun;18(6):1275–1278.

[582] Fu Q, Sun W. Mie theory for light scattering by a spherical particle in an absorbing medium. Applied Optics. 2001 Mar;40(9):1354–1361.

[583] Yang P, Gao BC, Wiscombe WJ, Mishchenko MI, Platnick SE, Huang HL, et al. Inherent and apparent scattering properties of coated or uncoated spheres embedded in an absorbing host medium. Applied Optics. 2002 May;41(15): 2740–2759.

[584] Videen G, Sun W. Yet another look at light scattering from particles in absorbing media. Applied Optics. 2003 Nov;42(33):6724–6727.

[585] Yin J, Pilon L. Efficiency factors and radiation characteristics of spherical scatterers in an absorbing medium. Journal of the Optical Society of America A. 2006 Nov;23(11):2784–2796.

[586] Aernouts B, Watté R, Beers RV, Delport F, Merchiers M, Block JD, et al. Flexible tool for simulating the bulk optical properties of polydisperse spherical particles in an absorbing host: Experimental validation. Optics Express. 2014 Aug;22(17):20223–20238.

[587] Mishchenko MI, Videen G, Yang P. Extinction by a homogeneous spherical particle in an absorbing medium. Optics Letters. 2017 Dec;42(23):4873–4876.

[588] Mishchenko MI, Yang P. Far-field Lorenz–Mie scattering in an absorbing host medium: Theoretical formalism and FORTRAN program. Journal of Quantitative Spectroscopy and Radiative Transfer. 2018;205:241–252.

[589] Mishchenko MI, Yurkin MA, Cairns B. Scattering of a damped inhomogeneous plane wave by a particle in a weakly absorbing medium. OSA Continuum. 2019 Aug;2(8):2362–2368.

[590] Lee JY, Peumans P. The origin of enhanced optical absorption in solar cells with metal nanoparticles embedded in the active layer. Optics Express. 2010 May;18(10):10078–10087.

[591] Nagel JR, Scarpulla MA. Enhanced absorption in optically thin solar cells by scattering from embedded dielectric nanoparticles. Optics Express. 2010 Jun;18(S2):A139–A146.

[592] Chen L, Zhou Y, Li Y, Hong M. Microsphere enhanced optical imaging and patterning: From physics to applications. Applied Physics Reviews. 2019;6(2):021304.

[593] Mishchenko MI. Multiple scattering by particles embedded in an absorbing medium. 1. Foldy–Lax equations, order-of-scattering expansion, and coherent field. Optics Express. 2008 Feb;16(3):2288–2301.

[594] Mishchenko MI. Multiple scattering by particles embedded in an absorbing medium. 2. Radiative transfer equation. Journal of Quantitative Spectroscopy and Radiative Transfer. 2008;109(14):2386–2390.

[595] Durant S, Calvo-Perez O, Vukadinovic N, Greffet JJ. Light scattering by a random distribution of particles embedded in absorbing media: Diagrammatic expansion of the extinction coefficient. Journal of the Optical Society of America A. 2007 Sep;24(9):2943–2952.

[596] Dick VP, Ivanov AP. Extinction of light in dispersive media with high particle concentrations : Applicability limits of the interference approximation. Journal of the Optical Society of America A. 1999;16(5):1034–1039.

[597] Xiao M, Hu Z, Gartner TE, Yang X, Li W, Jayaraman A, et al. Experimental and theoretical evidence for molecular forces driving surface segregation in photonic colloidal assemblies. Science Advances. 2019;5(9):eaax1254.

[598] Twersky V. Acoustic bulk parameters in distributions of pair-correlated scatterers. The Journal of the Acoustical Society of America. 1978;64(6):1710–1719.

[599] Holthoff H, Borkovec M, Schurtenberger P. Determination of light-scattering form factors of latex particle dimers with simultaneous static and dynamic light scattering in an aggregating suspension. Physical Review E. 1997 Dec;56: 6945–6953.

[600] Conley GM, Burresi M, Pratesi F, Vynck K, Wiersma DS. Light transport and localization in two-dimensional correlated disorder. Physical Review Letters. 2014 Apr;112:143901.

[601] Lagendijk A, Nienhuis B, van Tiggelen BA, de Vries P. Microscopic approach to the Lorentz cavity in dielectrics. Physical Review Letters. 1997;79:657–660.

[602] Mallet P, Guérin CA, Sentenac A. Maxwell-Garnett mixing rule in the presence of multiple scattering: Derivation and accuracy. Physical Review B. 2005 Jul;72:014205.

[603] Grimes CA, Grimes DM. Permeability and permittivity spectra of granular materials. Physical Review B. 1991 May;43:10780–10788.

[604] Chaumet PC, Rahmani A. Coupled-dipole method for magnetic and negative-refraction materials. Journal of Quantitative Spectroscopy and Radiative Transfer. 2009;110(1):22–29.

[605] Ruppin R. Evaluation of extended Maxwell-Garnett theories. Optics Communications. 2000;182(4):273–279.

[606] Wheeler MS, Aitchison JS, Chen JIL, Ozin GA, Mojahedi M. Infrared magnetic response in a random silicon carbide micropowder. Physical Review B. 2009 Feb;79:073103.

[607] Wang BX, Zhao CY. Light propagation in two-dimensional cold atomic clouds with positional correlations. In: Gong Q, Guo GC, Ham BS, editors. Quantum and nonlinear optics VI. vol. 11195. International Society for Optics and Photonics. SPIE; 2019. pp. 27–36.

[608] Karal FC, Keller JB. Elastic, electromagnetic, and other waves in a random medium. Journal of Mathematical Physics. 1964;5(4):537–547.

[609] Keller JB. Stochastic equations and wave propagation in random media. Stochastic Processes in Mathematical Physics and Engineering. 1964. Springer. Vol. 16, p. 145.

[610] Keller JB, Karal FC. Effective dielectric constant, permeability, and conductivity of a random medium and the velocity and attenuation coefficient of coherent waves. Journal of Mathematical Physics. 1966;7(4):661–670.

[611] Hespel L, Mainguy S, Greffet JJ. Theoretical and experimental investigation of the extinction in a dense distribution of particles: Nonlocal effects. Journal of the Optical Society of America A. 2001 Dec;18(12):3072–3076.

[612] Ishimaru A, Kuga Y. Attenuation constant of a coherent field in a dense distribution of particles. Journal of the Optical Society of America. 1982 Oct;72(10): 1317–1320.

[613] Derode A, Mamou V, Tourin A. Influence of correlations between scatterers on the attenuation of the coherent wave in a random medium. Physical Review E. 2006 Sep;74:036606.

[614] Chanal H, Segaud JP, Borderies P, Saillard M. Homogenization and scattering from heterogeneous media based on finite-difference-time-domain Monte Carlo computations. Journal of the Optical Society of America A. 2006 Feb;23(2): 370–381.

[615] Bringi V, Varadan VV, Varadan VK. Coherent wave attenuation by a random distribution of particles. Radio Science. 1982;17(05):946–952.

[616] Ma Y, Varadan VV, Varadan VK. Scattered intensity of a wave propagating in a discrete random medium. Applied Optics. 1988;27(12):2469–2477.

[617] West R, Gibbs D, Tsang L, Fung AK. Comparison of optical scattering experiments and the quasi-crystalline approximation for dense media. Journal of the Optical Society of America. 1994 Jun;11(6):1854–1858.

[618] Nashashibi A, Sarabandi K. Experimental characterization of the effective propagation constant of dense random media. IEEE Transactions on Antennas and Propagation. 1999;47(9):1454–1462.

[619] Prasher R. Thermal radiation in dense nano- and microparticulate media. Journal of Applied Physics. 2007;102(7):074316.

[620] Vander Meulen F, Feuillard G, Matar OB, Levassort F, Lethiecq M. Theoretical and experimental study of the influence of the particle size distribution on acoustic wave properties of strongly inhomogeneous media. Journal of the Acoustical Society of America. 2001;110(5 Pt 1):2301–2307.

[621] Gyorffy BL. Electronic states in liquid metals: A generalization of the coherent-potential approximation for a system with short-range order. Physical Review B. 1970 Apr;1:3290–3299.

[622] Kwong CC, Wilkowski D, Delande D, Pierrat R. Coherent light propagation through cold atomic clouds beyond the independent scattering approximation. Physical Review A. 2019 Apr;99:043806.

[623] Wang BX, Zhao CY. Role of near-field interaction on light transport in disordered media. arXiv preprint arXiv:180709953. 2018.

[624] Soven P. Coherent-potential model of substitutional disordered alloys. Physical Review. 1967 Apr;156:809–813.

[625] Roth LM. Tight-binding model of electronic states in a liquid metal. Physical Review Letters. 1972 Jun;28:1570–1573.

[626] Slovick BA, Yu ZG, Krishnamurthy S. Generalized effective-medium theory for metamaterials. Physical Review B. 2014 Apr;89:155118.

[627] Slovick BA. Negative refractive index induced by percolation in disordered metamaterials. Physical Review B. 2017 Mar;95:094202.

[628] Huang TC, Wang BX, Zhao CY. Negative refraction in metamaterials based on dielectric spherical particles. Journal of Quantitative Spectroscopy and Radiative Transfer. 2018;214:82–93.

[629] Soukoulis CM, Datta S, Economou EN. Propagation of classical waves in random media. Physical Review B. 1994 Feb;49:3800–3810.

[630] Busch K, Soukoulis CM. Transport properties of random media: A new effective medium theory. Physical Review Letters. 1995 Nov;75:3442–3445.

[631] Busch K, Soukoulis CM. Transport properties of random media: An energy-density CPA approach. Physical Review B. 1996 Jul;54:893–899.

[632] Schertel L, Wimmer I, Besirske P, Aegerter CM, Maret G, Polarz S, et al. Tunable high-index photonic glasses. Physics Review Materials. 2019 Jan;3:015203.

[633] Schertel L, Siedentop L, Meijer JM, Keim P, Aegerter CM, Aubry GJ, et al. The structural colors of photonic glasses. Advanced Optical Materials. 2019;7(15):1900442.

[634] Peng XT, Dinsmore AD. Light propagation in strongly scattering, random colloidal films: The role of the packing geometry. Physical Review Letters. 2007;99:143902.

[635] Bekshaev AY, Bliokh KY, Nori F. Mie scattering and optical forces from evanescent fields: A complex-angle approach. Optics Express. 2013 Mar;21(6): 7082–7095.

[636] Baudouin Q, Guerin W, Kaiser R. Cold and hot atomic vapors: A testbed for astrophysics? In: Annual Review of Cold Atoms and Molecules. World Scientific; 2014. pp. 251–311.

[637] Sokolov IM, Guerin W. Comparison of three approaches to light scattering by dilute cold atomic ensembles. Journal of the Optical Society of America B. 2019 Aug;36(8):2030–2037.

[638] Guerin W, Rouabah M, Kaiser R. Light interacting with atomic ensembles: Collective, cooperative and mesoscopic effects. Journal of Modern Optics. 2017;64(9):895–907.

[639] Lagendijk A, Tiggelen BV, Wiersma DS. Fifty years of Anderson localization. Physics Today. 2009;62(8):24–29.

[640] Dogariu A, Carminati R. Electromagnetic field correlations in three-dimensional speckles. Physics Reports. 2015;559:1–29.

[641] Berkovits R, Feng S. Correlations in coherent multiple scattering. Physics Reports. 1994;238(3):135–172.

[642] Mishchenko MI. Maxwell's equations, radiative transfer, and coherent backscattering: A general perspective. Journal of Quantitative Spectroscopy and Radiative Transfer. 2006;101(3):540–555.

[643] Fiebig S, Aegerter CM, Bührer W, Störzer M, Akkermans E, Montambaux G, et al. Conservation of energy in coherent backscattering of light. EPL (Europhysics Letters). 2008 Feb;81(6):64004.

[644] Wolf PE, Maret G. Weak localization and coherent backscattering of photons in disordered media. Physical Review Letters. 1985 Dec;55:2696–2699.

[645] Etemad S, Thompson R, Andrejco MJ, John S, MacKintosh FC. Weak localization of photons: Termination of coherent random walks by absorption and confined geometry. Physical Review Letters. 1987 Sep;59:1420–1423.

[646] Daozhong Z, Wei H, Youlong Z, Zhaolin L, Bingying C, Guozhen Y. Experimental verification of light localization for disordered multilayers in the visible-infrared spectrum. Physical Review B. 1994 Oct;50:9810–9814.

[647] Sheinfux HH, Lumer Y, Ankonina G, Genack AZ, Bartal G, Segev M. Observation of Anderson localization in disordered nanophotonic structures. Science. 2017;356(6341):953–956.

[648] Shi WB, Liu LZ, Peng R, Xu DH, Zhang K, Jing H, et al. Strong localization of surface plasmon polaritons with engineered disorder. Nano Letters. 2018;18(3):1896–1902.

[649] Caselli N, Intonti F, La China F, Biccari F, Riboli F, Gerardino A, et al. Near-field speckle imaging of light localization in disordered photonic systems. Applied Physics Letters. 2017;110(8):081102.

[650] Segev M, Silberberg Y, Christodoulides DN. Anderson localization of light. Nature Photonics. 2013;7(February):197–204.

[651] Schwartz T, Bartal G, Fishman S, Segev M. Transport and Anderson localization in disordered two-dimensional photonic lattices. Nature (London). 2007;446(7131):52–55.

[652] Wiersma DS, Bartolini P, Lagendijk A, Righini R. Localization of light in a disordered medium. Nature (London). 1997;390(6661):671–673.

[653] Sperling T, Schertel L, Ackermann M, Aubry GJ, Aegerter CM, Maret G. Can 3D light localization be reached in 'white paint'? New Journal of Physics. 2016 Jan;18(1):013039.

[654] Aegerter CM, Störzer M, Maret G. Experimental determination of critical exponents in Anderson localisation of light. Europhysics Letters (EPL). 2006 Aug;75(4):562–568.

[655] Aegerter CM, Störzer M, Fiebig S, Bührer W, Maret G. Observation of Anderson localization of light in three dimensions. Journal of the Optical Society of America A. 2007 Oct;24(10):A23–A27.

[656] Skipetrov SE, Page JH. Red light for Anderson localization. New Journal of Physics. 2016 Jan;18(2):021001.

[657] Anderson PW. The question of classical localization A theory of white paint? Philosophical Magazine B. 1985;52(3):505–509.

[658] Scheffold F, Lenke R, Tweer R, Maret G. Localization or classical diffusion of light? Nature. 1999;398(6724):206–207.

[659] Wiersma DS, Rivas JG, Bartolini P, Lagendijk A, Righini R. Localization or classical diffusion of light? Nature. 1999;398(6724):206–207.

[660] Sperling T, Bührer W, Aegerter CM, Maret G. Direct determination of the transition to localization of light in three dimensions. Nature Photonics. 2013;7(1):48–52.

[661] Scheffold F, Wiersma D. Inelastic scattering puts in question recent claims of Anderson localization of light. Nature Photonics. 2013;7(12):934–934.

[662] John S. Strong localization of photons in certain disordered dielectric superlattices. Physical Review Letters. 1987 Jun;58:2486–2489.

[663] Jeon SY, Kwon H, Hur K. Intrinsic photonic wave localization in a three-dimensional icosahedral quasicrystal. Nature Physics. 2017;13(4):363–368.

[664] Bromberg Y, Cao H. Generating Non-Rayleigh speckles with tailored intensity statistics. Physical Review Letters. 2014 May;112:213904.

[665] Shapiro B. Large intensity fluctuations for wave propagation in random media. Physical Review Letters. 1986 Oct;57:2168–2171.

[666] Genack AZ, Garcia N, Polkosnik W. Long-range intensity correlation in random media. Physical Review Letters. 1990 Oct;65:2129–2132.

[667] Emiliani V, Intonti F, Cazayous M, Wiersma DS, Colocci M, Aliev F, et al. Near-field short range correlation in optical waves transmitted through random media. Physical Review Letters. 2003 Jun;90:250801.

[668] Carminati R. Subwavelength spatial correlations in near-field speckle patterns. Physical Review A. 2010 May;81:053804.

[669] Boas DA, Dunn AK. Laser speckle contrast imaging in biomedical optics. Journal of Biomedical Optics. 2010;15(1):011109.

[670] Briers D, Duncan DD, Hirst ER, Kirkpatrick SJ, Larsson M, Steenbergen W, et al. Laser speckle contrast imaging: Theoretical and practical limitations. Journal of Biomedical Optics. 2013;18(6):066018.

[671] Heeman W, Steenbergen W, van Dam GM, Boerma EC. Clinical applications of laser speckle contrast imaging: A review. Journal of Biomedical Optics. 2019;24(8):080901.

[672] Borycki D, Kholiqov O, Srinivasan VJ. Interferometric near-infrared spectroscopy directly quantifies optical field dynamics in turbid media. Optica. 2016 Dec;3(12):1471–1476.

[673] Cheng X, Tamborini D, Carp SA, Shatrovoy O, Zimmerman B, Tyulmankov D, et al. Time domain diffuse correlation spectroscopy: Modeling the effects of laser coherence length and instrument response function. Optics Letters. 2018 Jun;43(12):2756–2759.

[674] Wiersma DS. Disordered photonics. Nature Photonics. 2013;7(3):188–196.

[675] Katz O, Heidmann P, Fink M, Gigan S. Non-invasive single-shot imaging through scattering layers and around corners via speckle correlations. Nature Photonics. 2014 Aug;8:784 EP –.

[676] Salhov O, Weinberg G, Katz O. Depth-resolved speckle-correlations imaging through scattering layers via coherence gating. Optic Letters. 2018 Nov;43(22):5528–5531.

[677] Stern G, Katz O. Noninvasive focusing through scattering layers using speckle correlations. Optics Letters. 2019 Jan;44(1):143–146.

[678] McCabe DJ, Tajalli A, Austin DR, Bondareff P, Walmsley IA, Gigan S, et al. Spatio-temporal focusing of an ultrafast pulse through a multiply scattering medium. Nature Communications. 2011;2(1):447.

[679] Aulbach J, Gjonaj B, Johnson PM, Mosk AP, Lagendijk A. Control of light transmission through opaque scattering media in space and time. Physical Review Letters. 2011 Mar;106:103901.

[680] Riboli F, Caselli N, Vignolini S, Intonti F, Vynck K, Barthelemy P, et al. Engineering of light confinement in strongly scattering disordered media. Nature Materials. 2014 May;13:720 EP –.

[681] Bruck R, Vynck K, Lalanne P, Mills B, Thomson DJ, Mashanovich GZ, et al. All-optical spatial light modulator for reconfigurable silicon photonic circuits. Optica. 2016 Apr;3(4):396–402.

[682] Rotter S, Gigan S. Light fields in complex media: Mesoscopic scattering meets wave control. Review of Modern Physics. 2017;89:015005.

[683] Geffrin JM, García-Cámara B, Gómez-Medina R, Albella P, Froufe-Pérez LS, Eyraud C, et al. Magnetic and electric coherence in forward- and back-scattered electromagnetic waves by a single dielectric subwavelength sphere. Nature Communications. 2012 Nov;3(1):1171.

[684] Carminati R, Greffet JJ. Near-field effects in spatial coherence of thermal sources. Physical Review Letters. 1999 Feb;82:1660–1663.

[685] Henkel C, Joulain K, Carminati R, Greffet JJ. Spatial coherence of thermal near fields. Optics Communications. 2000;186(1):57–67.

[686] Joulain K. Radiative transfer on short length scales. In: Microscale and nanoscale heat transfer. Springer; 2007. pp. 107–131.

[687] Arnold C, Marquier F, Garin M, Pardo F, Collin S, Bardou N, et al. Coherent thermal infrared emission by two-dimensional silicon carbide gratings. Physical Review B. 2012;86(3):035316.

[688] Whale MD, Cravalho EG. Modeling and performance of microscale thermophotovoltaic energy conversion devices. IEEE Transactions on Energy Conversion. 2002;17(1):130–142.

[689] DiMatteo RS, Greiff P, Finberg SL, Young-Waithe KA, Choy H, Masaki MM, et al. Enhanced photogeneration of carriers in a semiconductor via coupling across a nonisothermal nanoscale vacuum gap. Applied Physics Letters. 2001;79(12):1894–1896.

[690] Laroche M, Carminati R, Greffet JJ. Near-field thermophotovoltaic energy conversion. Journal of Applied Physics. 2006;100(6):063704.

[691] Narayanaswamy A, Chen G. Thermal emission control with one-dimensional metallodielectric photonic crystals. Physical Review B. 2004;70(12):125101.

[692] Narayanaswamy A, Chen G. Thermal radiation in 1D photonic crystals. Journal of Quantitative Spectroscopy and Radiative Transfer. 2005;93(1–3):175–183.

[693] Francoeur M, Mengüç MP, Vaillon R. Near-field radiative heat transfer enhancement via surface phonon polaritons coupling in thin films. Applied Physics Letters. 2008;93(4):043109.

[694] Ben-Abdallah P, Biehs SA. Near-field thermal transistor. Physical Review Letters. 2014 Jan;112:044301.

[695] Messina R, Ben-Abdallah P. Many-body near-field radiative heat pumping. Physical Review B. 2020 Apr;101:165435.

[696] Yang Y, Basu S, Wang L. Radiation-based near-field thermal rectification with phase transition materials. Applied Physics Letters. 2013;103(16):163101.

[697] De Wilde Y, Formanek F, Carminati R, Gralak B, Lemoine PA, Joulain K, et al. Thermal radiation scanning tunnelling microscopy. Nature. 2006 Dec;444(7120):740–743.

[698] Jones AC, Raschke MB. Thermal infrared near-field spectroscopy. Nano Letters. 2012;12(3):1475–1481.

[699] O'Callahan BT, Raschke MB. Laser heating of scanning probe tips for thermal near-field spectroscopy and imaging. APL Photonics. 2017;2(2):021301.

[700] Scully MO, Zubairy MS. Quantum optics. Cambridge University Press; 1997.

[701] Mandel L, Wolf E. Optical coherence and quantum optics. Cambridge University Press; 1995.

[702] Tai CT. Dyadic Green functions in electromagnetic theory. IEEE; 1994.

[703] Francoeur M, Pinar Mengüç M, Vaillon R. Solution of near-field thermal radiation in one-dimensional layered media using dyadic Green's functions and the scattering matrix method. Journal of Quantitative Spectroscopy and Radiative Transfer. 2009;110(18):2002–2018.

[704] Joulain K, Carminati R, Mulet JP, Greffet JJ. Definition and measurement of the local density of electromagnetic states close to an interface. Physical Review B. 2003 Dec;68:245405.

[705] Pendry JB. Radiative exchange of heat between nanostructures. Journal of Physics: Condensed Matter. 1999 Aug;11(35):6621–6633.

[706] Mulet JP, Joulain K, Carminati R, Greffet JJ. Enhanced radiative heat transfer at naometirc distances. Microscale Thermophysical Engineering. 2002;6(3): 209–222.

[707] Yeh P. Optical waves in layered media. John Wiley & Sons; 2005.

[708] Palik ED. Handbook of optical constants of solids. Academic Press; 1985.

[709] Fu CJ, Zhang ZM. Nanoscale radiation heat transfer for silicon at different doping levels. International Journal of Heat and Mass Transfer. 2006;49(9): 1703–1718.

[710] Boehm RF, Tien CL. Small spacing analysis of radiative transfer between parallel metallic surfaces. Journal of Heat Transfer. 1970 Aug;92(3):405–411.

[711] Mulet JP, Joulain K, Carminati R, Greffet JJ. Nanoscale radiative heat transfer between a small particle and a plane surface. Applied Physics Letters. 2001;78(19):2931–2933.

[712] Lee BJ, Zhang ZM. Lateral shifts in near-field thermal radiation with surface phonon polaritons. Nanoscale and Microscale Thermophysical Engineering. 2008;12(3):238–250.

[713] Biehs SA. Thermal heat radiation, near-field energy density and near-field radiative heat transfer of coated materials. The European Physical Journal B. 2007;58(4):423–431.

[714] Volokitin AI, Persson BNJ. Radiative heat transfer between nanostructures. Physical Review B. 2001 Apr;63:205404.

[715] Domingues G, Volz S, Joulain K, Greffet JJ. Heat transfer between two nanoparticles through near field interaction. Physical Review Letters. 2005 Mar;94:085901.

[716] Biehs SA, Huth O, Rüting F. Near-field radiative heat transfer for structured surfaces. Physical Review B. 2008 Aug;78:085414.

[717] Zhang Y, Yi HL, Tan HP. Near-field radiative heat transfer between black phosphorus sheets via anisotropic surface plasmon polaritons. ACS Photonics. 2018;5(9):3739–3747.

[718] Zhang WB, Zhao CY, Wang BX. Enhancing near-field heat transfer between composite structures through strongly coupled surface modes. Physical Review B. 2019;100(7):075425.

[719] Biehs SA, Messina R, Venkataram PS, Rodriguez AW, Cuevas JC, Ben-Abdallah P. Near-field radiative heat transfer in many-body systems. Review of Modern Physics. 2021 Jun;93:025009.

[720] Song J, Cheng Q, Zhang B, Lu L, Zhou X, Luo Z, et al. Many-body near-field radiative heat transfer: Methods, functionalities and applications. Reports on Progress in Physics. 2021 Mar;84(3):036501.

[721] Rodriguez AW, Reid MTH, Johnson SG. Fluctuating-surface-current formulation of radiative heat transfer: Theory and applications. Physical Review B. 2013 Aug;88:054305.

[722] Didari A, Mengüç MP. Analysis of near-field radiation transfer within nanogaps using FDTD method. Journal of Quantitative Spectroscopy and Radiative Transfer. 2014;146:214–226.

[723] Basu S, Lee BJ, Zhang ZM. Infrared radiative properties of heavily doped silicon at room temperature. Journal of Heat Transfer. 2009 Nov;132(2):023301.

[724] Jarzembski A, Tokunaga T, Crossley J, Yun J, Shaskey C, Murdick RA, et al. Role of acoustic phonon transport in near- to asperity-contact heat transfer. arXiv; 2019.

[725] Yee K. Numerical solution of initial boundary value problems involving Maxwell's equations in isotropic media. IEEE Transactions on Antennas and Propagation. 1966;14(3):302–307.

[726] Taflove A, Brodwin ME. Numerical solution of steady-state electromagnetic scattering problems using the time-dependent Maxwell's equations. IEEE Transactions on Microwave Theory and Techniques. 1975;23(8):623–630.

[727] Taflove A. Application of the finite-difference time-domain method to sinusoidal steady-state electromagnetic-penetration problems. IEEE Transactions on Electromagnetic Compatibility. 1980;(3):191–202.

[728] Umashankar K, Taflove A. A novel method to analyze electromagnetic scattering of complex objects. IEEE Transactions on Electromagnetic Compatibility. 1982;(4):397–405.

[729] Taflove A, Umashankar K. A hybrid moment method/finite-difference time-domain approach to electromagnetic coupling and aperture penetration into complex geometries. IEEE Transactions on Antennas and Propagation. 1982;30(4):617–627.

[730] Sullivan DM. Electromagnetic simulation using the FDTD method. John Wiley & Sons; 2013.

[731] Taflove A, Hagness SC. Computational electromagnetics: The finite-difference time-domain method. Artech House, Inc.; 2005.

[732] Sadiku MN. Numerical techniques in electromagnetics. CRC Press; 2000.

[733] Smith GD. Numerical solution of partial differential equations: Finite difference methods. Oxford University Press; 1985.

[734] Thomas J. Numerical partial differential equations: Finite difference methods. vol. 22. Springer Science & Business Media; 1998.

[735] Lavrinenko AV, Lægsgaard J, Gregersen N, Schmidt F, Søndergaard T. Numerical methods in photonics. CRC Press; 2018.

[736] Engquist B, Majda A. Absorbing boundary conditions for numerical simulation of waves. Proceedings of the National Academy of Sciences. 1977;74(5): 1765–1766.

[737] Berenger JP. A perfectly matched layer for the absorption of electromagnetic waves. Journal of Computational Physics. 1994;114(2):185–200.

[738] Gedney SD. An anisotropic perfectly matched layer-absorbing medium for the truncation of FDTD lattices. IEEE Transactions on Antennas and Propagation. 1996;44(12):1630–1639.

[739] Petropoulos PG, Zhao L, Cangellaris AC. A reflectionless sponge layer absorbing boundary condition for the solution of Maxwell's equations with high-order staggered finite difference schemes. Journal of Computational Physics. 1998;139(1):184–208.

[740] Lavrinenko A, Borel PI, Frandsen LH, Thorhauge M, Harpøth A, Kristensen M, et al. Comprehensive FDTD modelling of photonic crystal waveguide components. Optics Express. 2004 Jan;12(2):234–248.
[741] Shyroki DM, Lavrinenko AV. Perfectly matched layer method in the finite-difference time-domain and frequency-domain calculations. physica status solidi (b). 2007;244(10):3506–3514.
[742] Kittel C. Introduction to solid state physics. John Wiley & Sons; 2005.
[743] Kashiwa T, Fukai I. A treatment by the FD-TD method of the dispersive characteristics associated with electronic polarization. Microwave and Optical Technology Letters. 1990;3(6):203–205.
[744] Joseph RM, Hagness SC, Taflove A. Direct time integration of Maxwell's equations in linear dispersive media with absorption for scattering and propagation of femtosecond electromagnetic pulses. Optics Letters. 1991 Sep;16(18):1412–1414.
[745] Inan US, Marshall RA. Numerical electromagnetics: The FDTD method. Cambridge University Press; 2011.
[746] Moharam MG, Grann EB, Pommet DA, Gaylord TK. Formulation for stable and efficient implementation of the rigorous coupled-wave analysis of binary gratings. Journal of the Optical Society of America A. 1995 May;12(5): 1068–1076.
[747] Li L. New formulation of the Fourier modal method for crossed surface-relief gratings. Journal of the Optical Society of America A. 1997 Oct;14(10): 2758–2767.
[748] Bienstman P, Baets R. Optical modelling of photonic crystals and VCSELs using eigenmode expansion and perfectly matched layers. Optical and Quantum Electronics. 2001;33(4):327–341.
[749] Gregersen N, Reitzenstein S, Kistner C, Strauss M, Schneider C, Höfling S, et al. Numerical and experimental study of the Q factor of high-Q micropillar cavities. IEEE Journal of Quantum Electronics. 2010;46(10):1470–1483.
[750] Hugonin JP, Lalanne P. Perfectly matched layers as nonlinear coordinate transforms: a generalized formalization. Journal of the Optical Society of America A. 2005 Sep;22(9):1844–1849.
[751] Bigourdan F, Hugonin JP, Lalanne P. Aperiodic-Fourier modal method for analysis of body-of-revolution photonic structures. Journal of the Optical Society of America A. 2014 Jun;31(6):1303–1311.
[752] Photon Design Ltd, United Kingdom, www.photond.com/index.htm
[753] Peter Bienstman, CAMFR: An efficient eigenmode expansion tool, 2001, Photonics Research Group, http://photonics.intec.ugent.be/research/topics .asp?ID=17
[754] Hugonin JP, Lalanne P. RETICOLO software for grating analysis. arXiv; 2021.
[755] Sagan H. Boundary and eigenvalue problems in mathematical physics. Courier Corporation; 1989.
[756] Li L. Formulation and comparison of two recursive matrix algorithms for modeling layered diffraction gratings. Journal of the Optical Society of America A. 1996 May;13(5):1024–1035.
[757] Kim H, Park J, Lee B. Fourier Modal method and its applications in computational nanophotonics. CRC Press; 2012.

[758] Courant R, Hilbert D. Methods of mathematical physics-Vol. 1; Vol. 2. Interscience Publication; 1953.

[759] Li L. Use of Fourier series in the analysis of discontinuous periodic structures. Journal of the Optical Society of America A. 1996 Sep;13(9):1870–1876.

[760] Bonod N, Popov E, Nevière M. Differential theory of diffraction by finite cylindrical objects. Journal of the Optical Society of America A. 2005 Mar;22(3): 481–490.

[761] Basu S. Near-field radiative heat transfer across nanometer vacuum gaps: Fundamentals and applications. William Andrew; 2016.

[762] Otey CR, Zhu L, Sandhu S, Fan S. Fluctuational electrodynamics calculations of near-field heat transfer in non-planar geometries: A brief overview. Journal of Quantitative Spectroscopy and Radiative Transfer. 2014;132:3–11.

[763] Bimonte G. Scattering approach to Casimir forces and radiative heat transfer for nanostructured surfaces out of thermal equilibrium. Physical Review A. 2009 Oct;80:042102.

[764] Didari A, Pinar Mengüç M. Near-field thermal emission between corrugated surfaces separated by nano-gaps. Journal of Quantitative Spectroscopy and Radiative Transfer. 2015;158:43–51.

[765] Didari A, Menguc MP. Computational near-field radiative transfer and nf-rt-fdtd algorithm. Annual Review of Heat Transfer. 2020;23.

[766] McCauley AP, Reid MTH, Krüger M, Johnson SG. Modeling near-field radiative heat transfer from sharp objects using a general three-dimensional numerical scattering technique. Physical Review B. 2012 Apr;85:165104.

[767] Reid MTH, White J, Johnson SG. Fluctuating surface currents: An algorithm for efficient prediction of Casimir interactions among arbitrary materials in arbitrary geometries. Physical Review A. 2013 Aug;88:022514.

[768] Polimeridis AG, Reid MTH, Jin W, Johnson SG, White JK, Rodriguez AW. Fluctuating volume-current formulation of electromagnetic fluctuations in inhomogeneous media: Incandescence and luminescence in arbitrary geometries. Physical Review B. 2015 Oct;92:134202.

[769] Liu XL, Zhang ZM. Graphene-assisted near-field radiative heat transfer between corrugated polar materials. Applied Physics Letters. 2014;104(25):251911.

[770] Guérout R, Lussange J, Rosa FSS, Hugonin JP, Dalvit DAR, Greffet JJ, et al. Enhanced radiative heat transfer between nanostructured gold plates. Physical Review B. 2012 May;85:180301.

[771] Lussange J, Guérout R, Rosa FSS, Greffet JJ, Lambrecht A, Reynaud S. Radiative heat transfer between two dielectric nanogratings in the scattering approach. Physical Review B. 2012 Aug;86:085432.

[772] Liu X, Zhao B, Zhang ZM. Enhanced near-field thermal radiation and reduced Casimir stiction between doped-Si gratings. Physical Review A. 2015 Jun;91:062510.

[773] Hu Y, Li H, Zhang Y, Zhu Y, Yang Y. Enhanced near-field radiation in both TE and TM waves through excitation of Mie resonance. Physical Review B. 2020 Sep;102:125434.

[774] Edalatpour S, Čuma M, Trueax T, Backman R, Francoeur M. Convergence analysis of the thermal discrete dipole approximation. Physical Review E. 2015 Jun;91:063307.

[775] Edalatpour S, Francoeur M. Near-field radiative heat transfer between arbitrarily shaped objects and a surface. Physical Review B. 2016 Jul;94:045406.

[776] Edalatpour S, Hatamipour V, Francoeur M. Spectral redshift of the thermal near field scattered by a probe. Physical Review B. 2019 Apr;99:165401.

[777] Badieirostami M, Adibi A, Zhou HM, Chow SN. Model for efficient simulation of spatially incoherent light using the Wiener chaos expansion method. Optics Lett. 2007 Nov;32(21):3188–3190.

[778] Hou TY, Luo W, Rozovskii B, Zhou HM. Wiener Chaos expansions and numerical solutions of randomly forced equations of fluid mechanics. Journal of Computational Physics. 2006;216(2):687–706.

[779] Margengo EA, Rappaport CM, Miller EL. Optimum PML ABC conductivity profile in FDFD. IEEE Transactions on Magnetics. 1999;35(3):1506–1509.

[780] Xu F, Zhang Y, Hong W, Wu K, Cui TJ. Finite-difference frequency-domain algorithm for modeling guided-wave properties of substrate integrated waveguide. IEEE Transactions on Microwave Theory and Techniques. 2003;51(11): 2221–2227.

[781] Li Z, Li J, Liu X, Salihoglu H, Shen S. Wiener chaos expansion method for thermal radiation from inhomogeneous structures. Physical Review B. 2021 Nov;104:195426.

[782] Edalatpour S. Near-field thermal emission by periodic arrays. Physical Review E. 2019 Jun;99:063308.

[783] Abraham Ekeroth RM, García-Martín A, Cuevas JC. Thermal discrete dipole approximation for the description of thermal emission and radiative heat transfer of magneto-optical systems. Physical Review B. 2017 Jun;95:235428.

[784] Fernández-Hurtado V, Fernández-Domínguez AI, Feist J, García-Vidal FJ, Cuevas JC. Super-Planckian far-field radiative heat transfer. Physical Review B. 2018 Jan;97:045408.

[785] Hillenbrand R, Taubner T, Keilmann F. Phonon-enhanced light–matter interaction at the nanometre scale. Nature. 2002;418(6894):159.

[786] Fei Z, Andreev GO, Bao W, Zhang LM, McLeod AS, Wang C, et al. Infrared nanoscopy of dirac plasmons at the graphene–SiO_2 interface. Nano Letters. 2011;11(11):4701–4705.

[787] Peragut F, Cerutti L, Baranov A, Hugonin JP, Taliercio T, Wilde YD, et al. Hyperbolic metamaterials and surface plasmon polaritons. Optica. 2017 Nov;4(11):1409–1415.

[788] Muller EA, Pollard B, Raschke MB. Infrared chemical nano-imaging: Accessing structure, coupling, and dynamics on molecular length scales. The Journal of Physical Chemistry Letters. 2015;6(7):1275–1284.

[789] Rao VJ, Matthiesen M, Goetz KP, Huck C, Yim C, Siris R, et al. AFM-IR and IR-SNOM for the characterization of small molecule organic semiconductors. The Journal of Physical Chemistry C. 2020;124(9):5331–5344.

[790] Nabetani Y, Yamasaki M, Miura A, Tamai N. Fluorescence dynamics and morphology of electroluminescent polymer in small domains by time-resolved SNOM. Thin Solid Films. 2001;393(1):329–333.

[791] POHL DW. Scanning Near-field Optical Microscopy (SNOM). vol. 12 of Advances in Optical and Electron Microscopy. Elsevier; 1991. pp. 243–312.

[792] Heinzelmann H, Pohl D. Scanning near-field optical microscopy. Applied Physics A. 1994;59(2):89–101.

[793] Girard C, Dereux A. Near-field optics theories. Reports on Progress in Physics. 1996 May;59(5):657–699.

[794] Dunn RC. Near-field scanning optical microscopy. Chemical Reviews. 1999;99(10):2891–2928.

[795] Hecht B, Sick B, Wild UP, Deckert V, Zenobi R, Martin OJF, et al. Scanning near-field optical microscopy with aperture probes: Fundamentals and applications. The Journal of Chemical Physics. 2000;112(18):7761–7774.

[796] Barbara PF, Adams DM, O'Connor DB. Characterization of organic thin film materials with near-field scanning optical microscopy (NSOM). Annual Review of Materials Science. 1999;29(1):433–469.

[797] Rotenberg N, Kuipers L. Mapping nanoscale light fields. Nature Photonics. 2014;8(12):919–926.

[798] Goodman JW. Introduction to Fourier optics. McGraw Hill; 1996.

[799] Wilson T, Sheppard C. Theory and practice of scanning optical microscopy. vol. 180. Academic Press; 1984.

[800] Pawley J. Handbook of biological confocal microscopy. vol. 236. Springer Science & Business Media; 2006.

[801] Synge EH. XXXVIII. A suggested method for extending microscopic resolution into the ultra-microscopic region. The London, Edinburgh, and Dublin Philosophical Magazine and Journal of Science. 1928;6(35):356–362.

[802] Synge EH. III. A microscopic method. The London, Edinburgh, and Dublin Philosophical Magazine and Journal of Science. 1931;11(68):65–80.

[803] Binnig G, Rohrer H. Rastertunnelmikroskopie. Helvetica Physica Acta. 1982;55:726.

[804] Pohl DW, Denk W, Lanz M. Optical stethoscopy: Image recording with resolution $\lambda/20$. Applied Physics Letters. 1984;44(7):651–653.

[805] Dürig U, Pohl DW, Rohner F. Near-field optical-scanning microscopy. Journal of Applied Physics. 1986;59(10):3318–3327.

[806] Lewis A, Isaacson M, Harootunian A, Muray A. Development of a 500 Å spatial resolution light microscope: I. light is efficiently transmitted through $\lambda/16$ diameter apertures. Ultramicroscopy. 1984;13(3):227–231.

[807] Harootunian A, Betzig E, Isaacson M, Lewis A. Super-resolution fluorescence near-field scanning optical microscopy. Applied Physics Letters. 1986;49(11):674–676.

[808] Betzig E, Isaacson M, Lewis A. Collection mode near-field scanning optical microscopy. Applied Physics Letters. 1987;51(25):2088–2090.

[809] Fischer UC, Zingsheim HP. Submicroscopic contact imaging with visible light by energy transfer. Applied Physics Letters. 1982;40(3):195–197.

[810] Fischer UC. Optical characteristics of 0.1 μm circular apertures in a metal film as light sources for scanning ultramicroscopy. Journal of Vacuum Science & Technology B: Microelectronics Processing and Phenomena. 1985;3(1):386–390.

[811] Fischer UC. Submicrometer aperture in a thin metal film as a probe of its microenvironment through enhanced light scattering and fluorescence. Journal of the Optical Society of America B. 1986 Oct;3(10):1239–1244.

[812] Hecht B, Heinzelmann H, Pohl DW. Combined aperture SNOM/PSTM: Best of both worlds? Ultramicroscopy. 1995;57(2):228–234.

[813] Wessel J. Surface-enhanced optical microscopy. Journal of the Optical Society of America B. 1985 Sep;2(9):1538–1541.

[814] Fischer UC, Pohl DW. Observation of single-particle plasmons by near-field optical microscopy. Physical Review Letters. 1989 Jan;62:458–461.

[815] Zenhausern F, Martin Y, Wickramasinghe HK. Scanning interferometric apertureless microscopy: Optical imaging at 10 Angstrom resolution. Science. 1995;269(5227):1083–1085.

[816] Knoll B, Keilmann F. Near-field probing of vibrational absorption for chemical microscopy. Nature. 1999;399(6732):134–137.

[817] Keilmann F, Hillenbrand R. Near-field microscopy by elastic light scattering from a tip. Philosophical Transactions of the Royal Society of London Series A: Mathematical, Physical and Engineering Sciences. 2004;362(1817):787–805.

[818] Anger P, Bharadwaj P, Novotny L. Enhancement and quenching of single-molecule fluorescence. Physical Review Letters. 2006 Mar;96:113002.

[819] Kühn S, Håkanson U, Rogobete L, Sandoghdar V. Enhancement of single-molecule fluorescence using a gold nanoparticle as an optical nanoantenna. Physical Review Letters. 2006 Jul;97:017402.

[820] Novotny L, Stranick SJ. Near-field optical microscopy and spectroscopy with pointed probes. Annual Review of Physical Chemistry. 2006;57(1):303–331.

[821] Chen X, Hu D, Mescall R, You G, Basov DN, Dai Q, et al. Modern scattering-type scanning near-field optical microscopy for advanced material research. Advanced Materials. 2019;31(24):1804774.

[822] Adams W, Sadatgol M, Güney D. Review of near-field optics and superlenses for sub-diffraction-limited nano-imaging. AIP Advances. 2016;6(10):100701.

[823] Courjon D, Sarayeddine K, Spajer M. Scanning tunneling optical microscopy. Optics Communications. 1989;71(1):23–28.

[824] Reddick RC, Warmack RJ, Ferrell TL. New form of scanning optical microscopy. Physical Review B. 1989 Jan;39:767–770.

[825] Marti O, Bielefeldt H, Hecht B, Herminghaus S, Leiderer P, Mlynek J. Near-field optical measurement of the surface plasmon field. Optics Communications. 1993;96(4):225–228.

[826] Krenn JR, Dereux A, Weeber JC, Bourillot E, Lacroute Y, Goudonnet JP, et al. Squeezing the optical near-field zone by plasmon coupling of metallic nanoparticles. Physical Review Letters. 1999 Mar;82:2590–2593.

[827] Wieghold S, Nienhaus L. Probing semiconductor properties with optical scanning tunneling microscopy. Joule. 2020;4(3):524–538.

[828] Schrader M, Hell SW, van der Voort HTM. Potential of confocal microscopes to resolve in the 50–100 nm range. Applied Physics Letters. 1996;69(24):3644–3646.

[829] Toledo-Crow R, Yang PC, Chen Y, Vaez-Iravani M. Near-field differential scanning optical microscope with atomic force regulation. Applied Physics Letters. 1992;60(24):2957–2959.

[830] Betzig E, Finn PL, Weiner JS. Combined shear force and near-field scanning optical microscopy. Applied Physics Letters. 1992;60(20):2484–2486.

[831] Karrai K, Grober RD. Piezoelectric tip–sample distance control for near field optical microscopes. Applied Physics Letters. 1995;66(14):1842–1844.

[832] Ruiter AGT, Veerman JA, van der Werf KO, van Hulst NF. Dynamic behavior of tuning fork shear-force feedback. Applied Physics Letters. 1997;71(1):28–30.

[833] Pfeffer M, Lambelet P, Marquis-Weible F. Shear-force detection based on an external cavity laser interferometer for a compact scanning near field optical microscope. Review of Scientific Instruments. 1997;68(12):4478–4482.

[834] Huser T, Novotny L, Lacoste T, Eckert R, Heinzelmann H. Observation and analysis of near-field optical diffraction. Journal of the Optical Society of America A. 1999 Jan;16(1):141–148.

[835] Hecht B, Bielefeldt H, Pohl DW, Novotny L, Heinzelmann H. Influence of detection conditions on near-field optical imaging. Journal of Applied Physics. 1998;84(11):5873–5882.

[836] Münster S, Werner S, Mihalcea C, Scholz W, Oesterschulze E. Novel micromachined cantilever sensors for scanning near-field optical microscopy. Journal of Microscopy. 1997;186(1):17–22.

[837] Noell W, Abraham M, Mayr K, Ruf A, Barenz J, Hollricher O, et al. Micromachined aperture probe tip for multifunctional scanning probe microscopy. Applied Physics Letters. 1997;70(10):1236–1238.

[838] Nanonics Imaging Ltd. NSOM SNOM Probes, www.nanonics.co.il/products/nsom-snom-probes

[839] Stöckle R, Fokas C, Deckert V, Zenobi R, Sick B, Hecht B, et al. High-quality near-field optical probes by tube etching. Applied Physics Letters. 1999;75(2):160–162.

[840] Burgos P, Lu Z, Ianoul A, Hnatovsky C, Viriot ML, Johnston LJ, et al. Near-field scanning optical microscopy probes: A comparison of pulled and double-etched bent NSOM probes for fluorescence imaging of biological samples. Journal of Microscopy. 2003;211(1):37–47.

[841] Betzig E, Trautman JK, Harris TD, Weiner JS, Kostelak RL. Breaking the diffraction barrier: Optical microscopy on a nanometric scale. Science. 1991;251(5000):1468–1470.

[842] Valaskovic GA, Holton M, Morrison GH. Parameter control, characterization, and optimization in the fabrication of optical fiber near-field probes. Applied Optics. 1995 Mar;34(7):1215–1228.

[843] Hoffmann P, Dutoit B, Salathé RP. Comparison of mechanically drawn and protection layer chemically etched optical fiber tips. Ultramicroscopy. 1995;61(1):165–170.

[844] Zeisel D, Nettesheim S, Dutoit B, Zenobi R. Pulsed laser-induced desorption and optical imaging on a nanometer scale with scanning near-field microscopy using chemically etched fiber tips. Applied Physics Letters. 1996;68(18):2491–2492.

[845] Yatsui T, Kourogi M, Ohtsu M. Increasing throughput of a near-field optical fiber probe over 1000 times by the use of a triple-tapered structure. Applied Physics Letters. 1998;73(15):2090–2092.

[846] Lambelet P, Sayah A, Pfeffer M, Philipona C, Marquis-Weible F. Chemically etched fiber tips for near-field optical microscopy: A process for smoother tips. Appl Opt. 1998 Nov;37(31):7289–7292.

[847] Shi J, Qin XR. Formation of glass fiber tips for scanning near-field optical microscopy by sealed- and open-tube etching. Review of Scientific Instruments. 2005;76(1):013702.

[848] Patanè S, Cefalì E, Arena A, Gucciardi PG, Allegrini M. Wide angle near-field optical probes by reverse tube etching. Ultramicroscopy. 2006;106(6): 475–479.

[849] Yang J, Zhang J, Li Z, Gong Q. Fabrication of high-quality SNOM probes by pre-treating the fibres before chemical etching. Journal of Microscopy. 2007; 228(1):40–44.

[850] Hollars CW, Dunn RC. Evaluation of thermal evaporation conditions used in coating aluminum on near-field fiber-optic probes. Review of Scientific Instruments. 1998;69(4):1747–1752.

[851] Pilevar S, Edinger K, Atia W, Smolyaninov I, Davis C. Focused ion-beam fabrication of fiber probes with well-defined apertures for use in near-field scanning optical microscopy. Applied Physics Letters. 1998;72(24):3133–3135.

[852] Veerman JA, Otter AM, Kuipers L, van Hulst NF. High definition aperture probes for near-field optical microscopy fabricated by focused ion beam milling. Applied Physics Letters. 1998;72(24):3115–3117.

[853] Richards D. Near-field microscopy: Throwing light on the nanoworld. Philosophical Transactions of the Royal Society of London Series A: Mathematical, Physical and Engineering Sciences. 2003;361(1813):2843–2857.

[854] Greffet JJ, Carminati R. Image formation in near-field optics. Progress in Surface Science. 1997;56(3):133–237.

[855] Wu XY, Lin S, Tan QF, Wang J. A novel phase-sensitive scanning near-field optical microscope. Chinese Physics B. 2015 Mar;24(5):054204.

[856] Nesci A, Dändliker R, Herzig HP. Quantitative amplitude and phase measurement by use of a heterodyne scanning near-field optical microscope. Optics Letters. 2001 Feb;26(4):208–210.

[857] Burresi M, Engelen RJP, Opheij A, van Oosten D, Mori D, Baba T, et al. Observation of polarization singularities at the nanoscale. Physical Review Letters. 2009 Jan;102:033902.

[858] Rotenberg N, le Feber B, Visser TD, Kuipers L. Tracking nanoscale electric and magnetic singularities through three-dimensional space. Optica. 2015 Jun;2(6):540–546.

[859] Inouye Y, Kawata S. Near-field scanning optical microscope with a metallic probe tip. Optics Letters. 1994 Feb;19(3):159–161.

[860] Lahrech A, Bachelot R, Gleyzes P, Boccara AC. Infrared-reflection-mode near-field microscopy using an apertureless probe with a resolution of $\lambda/600$. Optics Letters. 1996 Sep;21(17):1315–1317.

[861] Bechtel HA, Muller EA, Olmon RL, Martin MC, Raschke MB. Ultrabroadband infrared nanospectroscopic imaging. Proceedings of the National Academy of Sciences. 2014;111(20):7191–7196.

[862] Brehm M, Taubner T, Hillenbrand R, Keilmann F. Infrared spectroscopic mapping of single nanoparticles and viruses at nanoscale resolution. Nano Letters. 2006;6(7):1307–1310.

[863] Huth F, Chuvilin A, Schnell M, Amenabar I, Krutokhvostov R, Lopatin S, et al. Resonant antenna probes for tip-enhanced infrared near-field microscopy. Nano Letters. 2013;13(3):1065–1072.

[864] Martin OJF, Girard C. Controlling and tuning strong optical field gradients at a local probe microscope tip apex. Applied Physics Letters. 1997;70(6):705–707.

[865] Neacsu CC, Dreyer J, Behr N, Raschke MB. Scanning-probe Raman spectroscopy with single-molecule sensitivity. Physical Review B. 2006 May;73: 193406.

[866] Cvitkovic A, Ocelic N, Hillenbrand R. Analytical model for quantitative prediction of material contrasts in scattering-type near-field optical microscopy. Optics Express. 2007 Jul;15(14):8550–8565.

[867] Renger J, Grafström S, Eng LM, Deckert V. Evanescent wave scattering and local electric field enhancement at ellipsoidal silver particles in the vicinity of a glass surface. Journal of the Optical Society of America A. 2004 Jul;21(7): 1362–1367.

[868] Zhang LM, Andreev GO, Fei Z, McLeod AS, Dominguez G, Thiemens M, et al. Near-field spectroscopy of silicon dioxide thin films. Physical Review B. 2012 Feb;85:075419.

[869] Cvitkovic A, Ocelic N, Aizpurua J, Guckenberger R, Hillenbrand R. Infrared imaging of single nanoparticles via strong field enhancement in a scanning nanogap. Physics Review Letters. 2006 Aug;97:060801.

[870] Brehm M, Schliesser A, Čajko F, Tsukerman I, Keilmann F. Antenna-mediated back-scattering efficiency in infrared near-field microscopy. Optics Express. 2008 Jul;16(15):11203–11215.

[871] Novotny L, Bian RX, Xie XS. Theory of nanometric optical tweezers. Physics Review Letters. 1997 Jul;79:645–648.

[872] Zayats AV. Electromagnetic field enhancement in the context of apertureless near-field microscopy. Optics Communications. 1999;161(1):156–162.

[873] Martin YC, Hamann HF, Wickramasinghe HK. Strength of the electric field in apertureless near-field optical microscopy. Journal of Applied Physics. 2001;89(10):5774–5778.

[874] Lu F, Jin M, Belkin MA. Tip-enhanced infrared nanospectroscopy via molecular expansion force detection. Nature Photonics. 2014 Apr;8(4):307–312.

[875] Madrazo A, Carminati R, Nieto-Vesperinas M, Greffet JJ. Polarization effects in the optical interaction between a nanoparticle and a corrugated surface: Implications for apertureless near-field microscopy. Journal of the Optical Society of America A. 1998 Jan;15(1):109–119.

[876] Hillenbrand R, Keilmann F. Optical oscillation modes of plasmon particles observed in direct space by phase-contrast near-field microscopy. Applied Physics B. 2001;73(3):239–243.

[877] McLeod AS, Kelly P, Goldflam MD, Gainsforth Z, Westphal AJ, Dominguez G, et al. Model for quantitative tip-enhanced spectroscopy and the extraction of nanoscale-resolved optical constants. Physical Review B. 2014 Aug;90:085136.

[878] Huber AJ, Kazantsev D, Keilmann F, Wittborn J, Hillenbrand R. Simultaneous IR material recognition and conductivity mapping by nanoscale near-field microscopy. Advanced Materials. 2007;19(17):2209–2212.

[879] Osad'ko IS. The near-field microscope as a tool for studying nanoparticles. Physics-Uspekhi. 2010 Jan;53(1):77–81.

[880] Taubner T, Keilmann F, Hillenbrand R. Nanoscale-resolved subsurface imaging by scattering-type near-field optical microscopy. Optics Express. 2005 Oct; 13(22):8893–8899.

[881] Hillenbrand R, Keilmann F. Complex optical constants on a subwavelength scale. Physical Review Letters. 2000 Oct;85:3029–3032.

[882] Huber A, Ocelic N, Taubner T, Hillenbrand R. Nanoscale resolved infrared probing of crystal structure and of plasmon-phonon coupling. Nano Letters. 2006;6(4):774–778.

[883] Taubner T, Hillenbrand R, Keilmann F. Performance of visible and mid-infrared scattering-type near-field optical microscopes. Journal of Microscopy. 2003;210(3):311–314.

[884] Vaez-Iravani M, Toledo-Crow R. Phase contrast and amplitude pseudoheterodyne interference near field scanning optical microscopy. Applied Physics Letters. 1993;62(10):1044–1046.

[885] Ignatovich FV, Novotny L. Real-time and background-free detection of nanoscale particles. Physical Review Letters. 2006 Jan;96:013901.

[886] Gomez L, Bachelot R, Bouhelier A, Wiederrecht GP, hui Chang S, Gray SK, et al. Apertureless scanning near-field optical microscopy: A comparison between homodyne and heterodyne approaches. Journal of the Optical Society of America B. 2006 May;23(5):823–833.

[887] Zhong Q, Inniss D, Kjoller K, Elings VB. Fractured polymer/silica fiber surface studied by tapping mode atomic force microscopy. Surface Science. 1993;290(1):L688–L692.

[888] Labardi M, Patanè S, Allegrini M. Artifact-free near-field optical imaging by apertureless microscopy. Applied Physics Letters. 2000;77(5):621–623.

[889] Maghelli N, Labardi M, Patanè S, Irrera F, Allegrini M. Optical near-field harmonic demodulation in apertureless microscopy. Journal of Microscopy. 2001;202(1):84–93.

[890] Steinle T, Neubrech F, Steinmann A, Yin X, Giessen H. Mid-infrared Fourier-transform spectroscopy with a high-brilliance tunable laser source: Investigating sample areas down to 5 μm diameter. Optics Express. 2015 May;23(9): 11105–11113.

[891] Steinle T, Mörz F, Steinmann A, Giessen H. Ultra-stable high average power femtosecond laser system tunable from 1.33 to 20 μm. Optics Letters. 2016 Nov;41(21):4863–4866.

[892] Mörz F, Semenyshyn R, Steinle T, Neubrech F, Zschieschang U, Klauk H, et al. Nearly diffraction limited FTIR mapping using an ultrastable broadband femtosecond laser tunable from 1.33 to 8 μm. Optics Express. 2017 Dec;25(26):32355–32363.

[893] Paulite M, Fakhraai Z, Li ITS, Gunari N, Tanur AE, Walker GC. Imaging secondary structure of individual amyloid fibrils of a β2-microglobulin fragment using near-field infrared spectroscopy. Journal of the American Chemical Society. 2011;133(19):7376–7383.

[894] Amarie S, Zaslansky P, Kajihara Y, Griesshaber E, Schmahl WW, Keilmann F. Nano-FTIR chemical mapping of minerals in biological materials. Beilstein Journal of Nanotechnology. 2012;3:312–323.

[895] Khatib O, Bechtel HA, Martin MC, Raschke MB, Carr GL. Far infrared synchrotron near-field nanoimaging and nanospectroscopy. ACS Photonics. 2018;5(7):2773–2779.

[896] Huber AJ, Wittborn J, Hillenbrand R. Infrared spectroscopic near-field mapping of single nanotransistors. Nanotechnology. 2010 May;21(23):235702.

[897] Dominguez G, Mcleod AS, Gainsforth Z, Kelly P, Bechtel HA, Keilmann F, et al. Nanoscale infrared spectroscopy as a non-destructive probe of extraterrestrial samples. Nature Communications. 2014 Dec;5(1):5445.

[898] Dai S, Ma Q, Liu MK, Andersen T, Fei Z, Goldflam MD, et al. Graphene on hexagonal boron nitride as a tunable hyperbolic metamaterial. Nature Nanotechnology. 2015 Aug;10(8):682–686.

[899] Stiegler JM, Abate Y, Cvitkovic A, Romanyuk YE, Huber AJ, Leone SR, et al. Nanoscale infrared absorption spectroscopy of individual nanoparticles enabled by scattering-type near-field microscopy. ACS Nano. 2011;5(8):6494–6499.

[900] Jacob R, Winnerl S, Schneider H, Helm M, Wenzel MT, von Ribbeck HG, et al. Quantitative determination of the charge carrier concentration of ion implanted silicon by IR-near-field spectroscopy. Optics Express. 2010 Dec;18(25): 26206–26213.

[901] Mattis Hoffmann J, Hauer B, Taubner T. Antenna-enhanced infrared near-field nanospectroscopy of a polymer. Applied Physics Letters. 2012;101(19):193105.

[902] Hermann P, Hoehl A, Patoka P, Huth F, Rühl E, Ulm G. Near-field imaging and nano-Fourier-transform infrared spectroscopy using broadband synchrotron radiation. Optics Express. 2013 Feb;21(3):2913–2919.

[903] Hermann P, Kästner B, Hoehl A, Kashcheyevs V, Patoka P, Ulrich G, et al. Enhancing the sensitivity of nano-FTIR spectroscopy. Optics Express. 2017 Jul;25(14):16574–16588.

[904] Pollard B, Maia FCB, Raschke MB, Freitas RO. Infrared vibrational nanospectroscopy by self-referenced interferometry. Nano Letters. 2016;16(1):55–61.

[905] Amenabar I, Poly S, Nuansing W, Hubrich EH, Govyadinov AA, Huth F, et al. Structural analysis and mapping of individual protein complexes by infrared nanospectroscopy. Nature Communications. 2013;4:2890.

[906] Patoka P, Ulrich G, Nguyen AE, Bartels L, Dowben PA, Turkowski V, et al. Nanoscale plasmonic phenomena in CVD-grown MoS2 monolayer revealed by ultra-broadband synchrotron radiation based nano-FTIR spectroscopy and near-field microscopy. Optics Express. 2016 Jan;24(2):1154–1164.

[907] Dai S, Fei Z, Ma Q, Rodin AS, Wagner M, McLeod AS, et al. Tunable phonon polaritons in atomically thin van der Waals crystals of boron nitride. Science. 2014;343(6175):1125–1129.

[908] Fei Z, Rodin AS, Andreev GO, Bao W, McLeod AS, Wagner M, et al. Gate-tuning of graphene plasmons revealed by infrared nano-imaging. Nature. 2012 Jul;487(7405):82–85.

[909] Fei Z, Rodin AS, Gannett W, Dai S, Regan W, Wagner M, et al. Electronic and plasmonic phenomena at graphene grain boundaries. Nature Nanotechnology. 2013 Nov;8(11):821–825.

[910] Hu G, Ma W, Hu D, Wu J, Zheng C, Liu K, et al. Real-space nanoimaging of hyperbolic shear polaritons in a monoclinic crystal. Nature Nanotechnology. 2022 Dec.

[911] O'Callahan BT, Park KD, Novikova IV, Jian T, Chen CL, Muller EA, et al. In liquid infrared scattering scanning near-field optical microscopy for chemical and biological nanoimaging. Nano Letters. 2020;20(6):4497–4504.

[912] Wu CY, Wolf WJ, Levartovsky Y, Bechtel HA, Martin MC, Toste FD, et al. High-spatial-resolution mapping of catalytic reactions on single particles. Nature. 2017 Jan;541(7638):511–515.

[913] Liu M, Sternbach AJ, Wagner M, Slusar TV, Kong T, Bud'ko SL, et al. Phase transition in bulk single crystals and thin films of VO_2 by nanoscale infrared spectroscopy and imaging. Physical Review B. 2015 Jun;91:245155.

[914] Huang TC, Wang BX, Zhang WB, Zhao CY. Ultracompact energy transfer in anapole-based metachains. Nano Letters. 2021;21(14):6102–6110.

[915] Ni GX, McLeod AS, Sun Z, Wang L, Xiong L, Post KW, et al. Fundamental limits to graphene plasmonics. Nature. 2018 May;557(7706):530–533.

[916] Basov DN, Fogler MM, de Abajo FJG. Polaritons in van der Waals materials. Science. 2016;354(6309):aag1992.

[917] McLeod AS, van Heumen E, Ramirez JG, Wang S, Saerbeck T, Guenon S, et al. Nanotextured phase coexistence in the correlated insulator V_2O_3. Nature Physics. 2017 Jan;13(1):80–86.

[918] Atkin JM, Berweger S, Jones AC, Raschke MB. Nano-optical imaging and spectroscopy of order, phases, and domains in complex solids. Advances in Physics. 2012;61(6):745–842.

[919] Woessner A, Lundeberg MB, Gao Y, Principi A, Alonso-González P, Carrega M, et al. Highly confined low-loss plasmons in graphene–boron nitride heterostructures. Nature Materials. 2015 Apr;14(4):421–425.

[920] Ni GX, Wang L, Goldflam MD, Wagner M, Fei Z, McLeod AS, et al. Ultrafast optical switching of infrared plasmon polaritons in high-mobility graphene. Nature Photonics. 2016 Apr;10(4):244–247.

[921] Alonso-González P, Nikitin AY, Golmar F, Centeno A, Pesquera A, Vélez S, et al. Controlling graphene plasmons with resonant metal antennas and spatial conductivity patterns. Science. 2014;344(6190):1369–1373.

[922] Wang L, Meric I, Huang PY, Gao Q, Gao Y, Tran H, et al. One-dimensional electrical contact to a two-dimensional material. Science. 2013;342(6158): 614–617.

[923] Kaelberer T, Fedotov VA, Papasimakis N, Tsai DP, Zheludev NI. Toroidal dipolar response in a metamaterial. Science. 2010;330(6010):1510–1512.

[924] Miroshnichenko AE, Evlyukhin AB, Yu YF, Bakker RM, Chipouline A, Kuznetsov AI, et al. Nonradiating anapole modes in dielectric nanoparticles. Nature Communications. 2015 Aug;6(1):8069.

[925] Grinblat G, Li Y, Nielsen MP, Oulton RF, Maier SA. Efficient third harmonic generation and nonlinear subwavelength imaging at a higher-order anapole mode in a single germanium nanodisk. ACS Nano. 2017;11(1):953–960.

[926] Xu L, Rahmani M, Zangeneh Kamali K, Lamprianidis A, Ghirardini L, Sautter J, et al. Boosting third-harmonic generation by a mirror-enhanced anapole resonator. Light: Science & Applications. 2018 Jul;7(1):44.

[927] Zhang T, Che Y, Chen K, Xu J, Xu Y, Wen T, et al. Anapole mediated giant photothermal nonlinearity in nanostructured silicon. Nature Communications. 2020 Jun;11(1):3027.

[928] Ospanova AK, Stenishchev IV, Basharin AA. Anapole mode sustaining silicon metamaterials in visible spectral range. Laser & Photonics Reviews. 2018;12(7):1800005.

[929] Basharin AA, Chuguevsky V, Volsky N, Kafesaki M, Economou EN. Extremely high Q-factor metamaterials due to anapole excitation. Physical Review B. 2017 Jan;95:035104.

[930] Totero Gongora JS, Miroshnichenko AE, Kivshar YS, Fratalocchi A. Anapole nanolasers for mode-locking and ultrafast pulse generation. Nature Communications. 2017 May;8(1):15535.

[931] Mazzone V, Totero Gongora JS, Fratalocchi A. Near-field coupling and mode competition in multiple anapole systems. Applied Sciences. 2017;7(6):542.

[932] Zenin VA, Evlyukhin AB, Novikov SM, Yang Y, Malureanu R, Lavrinenko AV, et al. Direct amplitude-phase near-field observation of higher-order anapole states. Nano Letters. 2017;17(11):7152–7159.

[933] Weng Q, Panchal V, Lin KT, Sun L, Kajihara Y, Tzalenchuk A, et al. Comparison of active and passive methods for the infrared scanning near-field microscopy. Applied Physics Letters. 2019;114(15):153101.

[934] Keilmann F, Hillenbrand R. Near-field nanoscopy by elastic light scattering from a tip. In: Nano-optics and near-field optical microscopy. Artech House; 2009. pp. 235–265.

[935] Huth F, Schnell M, Wittborn J, Ocelić N, Hillenbrand R. Infrared-spectroscopic nanoimaging with a thermal source. Nature Materials. 2011;10 5:352–356.

[936] O'Callahan BT, Lewis WE, Möbius S, Stanley JC, Muller EA, Raschke MB. Broadband infrared vibrational nano-spectroscopy using thermal blackbody radiation. Optics Express. 2015 Dec;23(25):32063–32074.

[937] Babuty A, Joulain K, Chapuis PO, Greffet JJ, De Wilde Y. Blackbody spectrum revisited in the near field. Physics Review Letters. 2013 Apr;110:146103.

[938] Jarzembski A, Shaskey C, Park K. Tip-based vibrational spectroscopy for nanoscale analysis of emerging energy materials. Frontiers in Energy. 2018;12(1):43–71.

[939] O'Callahan BT, Lewis WE, Jones AC, Raschke MB. Spectral frustration and spatial coherence in thermal near-field spectroscopy. Physics Review B. 2014 Jun;89:245446.

[940] Herz F, An Z, Komiyama S, Biehs SA. Revisiting the dipole model for a thermal infrared near-field spectroscope. Physical Review Applied. 2018 Oct;10:044051.

[941] Lin KT, Komiyama S, Kim S, Kawamura Ki, Kajihara Y. A high signal-to-noise ratio passive near-field microscope equipped with a helium-free cryostat. Review of Scientific Instruments. 2017;88(1):013706.

[942] Weng Q, Yang L, An Z, Chen P, Tzalenchuk A, Lu W, et al. Quasiadiabatic electron transport in room temperature nanoelectronic devices induced by hot-phonon bottleneck. Nature Communications. 2021 Aug;12(1):4752.

[943] Kajihara Y, Kosaka K, Komiyama S. A sensitive near-field microscope for thermal radiation. Review of Scientific Instruments. 2010;81(3):033706.

[944] Weng Q, Komiyama S, Yang L, An Z, Chen P, Biehs SA, et al. Imaging of nonlocal hot-electron energy dissipation via shot noise. Science. 2018;360(6390): 775–778.

[945] Weng Q, Lin KT, Yoshida K, Nema H, Komiyama S, Kim S, et al. Near-field radiative nanothermal imaging of nonuniform Joule heating in narrow metal wires. Nano Letters. 2018;18(7):4220–4225.

[946] Komiyama S. Perspective: Nanoscopy of charge kinetics via terahertz fluctuation. Journal of Applied Physics. 2019;125(1):010901.

[947] Sakuma R, Lin KT, Kim S, Kimura F, Kajihara Y. Passive near-field imaging via grating-based spectroscopy. Review of Scientific Instruments. 2022;93(1):013704.

[948] Xu JB, Läuger K, Möller R, Dransfeld K, Wilson IH. Heat transfer between two metallic surfaces at small distances. Journal of Applied Physics. 1994;76(11):7209–7216.

[949] Xu JB, Läuger K, Dransfeld K, Wilson IH. Thermal sensors for investigation of heat transfer in scanning probe microscopy. Review of Scientific Instruments. 1994;65(7):2262–2266.

[950] Müller-Hirsch W, Kraft A, Hirsch MT, Parisi J, Kittel A. Heat transfer in ultrahigh vacuum scanning thermal microscopy. Journal of Vacuum Science & Technology A. 1999;17(4):1205–1210.

[951] Lang S, Sharma G, Molesky S, Kränzien PU, Jalas T, Jacob Z, et al. Dynamic measurement of near-field radiative heat transfer. Scientific Reports. 2017 Oct;7(1):13916.

[952] Shi J, Li P, Liu B, Shen S. Tuning near field radiation by doped silicon. Applied Physics Letters. 2013;102(18):183114.

[953] Shi J, Liu B, Li P, Ng LY, Shen S. Near-field energy extraction with hyperbolic metamaterials. Nano Letters. 2015;15(2):1217–1221.

[954] van Zwol PJ, Thiele S, Berger C, de Heer WA, Chevrier J. Nanoscale radiative heat flow due to surface plasmons in graphene and doped silicon. Physical Review Letters. 2012 Dec;109:264301.

[955] van Zwol PJ, Ranno L, Chevrier J. Tuning near field radiative heat flux through surface excitations with a metal insulator transition. Physical Review Letters. 2012 Jun;108:234301.

[956] Kralik T, Hanzelka P, Zobac M, Musilova V, Fort T, Horak M. Strong near-field enhancement of radiative heat transfer between metallic surfaces. Physical Review Letters. 2012 Nov;109:224302.

[957] Ijiro T, Yamada N. Near-field radiative heat transfer between two parallel SiO2 plates with and without microcavities. Applied Physics Letters. 2015;106(2):023103.

[958] Ito K, Miura A, Iizuka H, Toshiyoshi H. Parallel-plate submicron gap formed by micromachined low-density pillars for near-field radiative heat transfer. Applied Physics Letters. 2015;106(8):083504.

[959] Lim M, Lee SS, Lee BJ. Near-field thermal radiation between doped silicon plates at nanoscale gaps. Physical Review B. 2015 May;91:195136.

[960] Watjen JI, Zhao B, Zhang ZM. Near-field radiative heat transfer between doped-Si parallel plates separated by a spacing down to 200 nm. Applied Physics Letters. 2016;109(20):203112.

[961] Ying X, Sabbaghi P, Sluder N, Wang L. Super-Planckian radiative heat transfer between macroscale surfaces with vacuum gaps down to 190 nm directly created by SU-8 posts and characterized by capacitance method. ACS Photonics. 2020;7(1):190–196.

[962] Feng C, Tang Z, Yu J, Sun C. A MEMS device capable of measuring near-field thermal radiation between membranes. Sensors. 2013;13(2):1998–2010.

[963] Ghashami M, Jarzembski A, Lim M, Lee BJ, Park K. Experimental exploration of near-field radiative heat transfer. Annual Review of Heat Transfer. 2020;23:13–58.

[964] Lim M, Song J, Lee SS, Lee BJ. Tailoring near-field thermal radiation between metallo-dielectric multilayers using coupled surface plasmon polaritons. Nature Communications. 2018 Oct;9(1):4302.

[965] Basu S, Zhang ZM. Ultrasmall penetration depth in nanoscale thermal radiation. Applied Physics Letters. 2009;95(13):133104.

[966] Fong KY, Li HK, Zhao R, Yang S, Wang Y, Zhang X. Phonon heat transfer across a vacuum through quantum fluctuations. Nature. 2019 Dec;576(7786):243–247.

[967] Jarzembski A, Tokunaga T, Crossley J, Yun J, Shaskey C, Murdick RA, et al. Force-induced acoustic phonon transport across single-digit nanometre vacuum gaps. arXiv; 2019.

[968] Le Gall J, Olivier M, Greffet JJ. Experimental and theoretical study of reflection and coherent thermal emissionby a SiC grating supporting a surface-phonon polariton. Physical Review B. 1997;55(15):10105.

[969] Liu J, Guler U, Lagutchev A, Kildishev A, Malis O, Boltasseva A, et al. Quasi-coherent thermal emitter based on refractory plasmonic materials. Optical Materials Express. 2015;5(12):2721–2728.

[970] Chen YB, Zhang Z. Design of tungsten complex gratings for thermophotovoltaic radiators. Optics Communications. 2007;269(2):411–417.

[971] He X, Jie J, Yang J, Han Y, Zhang S. Asymmetric dielectric grating on metallic film enabled dual-and narrow-band absorbers. Optics Express. 2020;28(4): 4594–4602.

[972] Heinzel A, Boerner V, Gombert A, Bläsi B, Wittwer V, Luther J. Radiation filters and emitters for the NIR based on periodically structured metal surfaces. Journal of Modern Optics. 2000;47(13):2399–2419.

[973] Kohiyama A, Shimizu M, Iguchi F, Yugami H. Narrowband thermal radiation from closed-end microcavities. Journal of Applied Physics. 2015;118(13):133102.

[974] Zhao B, Zhao J, Zhang Z. Enhancement of near-infrared absorption in graphene with metal gratings. Applied Physics Letters. 2014;105(3):031905.

[975] Joannopoulos JD, Johnson SG, Winn JN, Meade RD. Molding the flow of light. Princeton University Press; 2008.

[976] Lin SY, Fleming J, El-Kady I. Three-dimensional photonic-crystal emission through thermal excitation. Optics Letters. 2003;28(20):1909–1911.

[977] Chan DL, Soljačić M, Joannopoulos J. Thermal emission and design in 2D-periodic metallic photonic crystal slabs. Optics Express. 2006;14(19):8785–8796.

[978] Niu X, Qi D, Wang X, Cheng Y, Chen F, Li B, et al. Improved broadband spectral selectivity of absorbers/emitters for solar thermophotovoltaics based on 2D photonic crystal heterostructures. JOSA A. 2018;35(11):1832–1838.

[979] Fleming J, Lin S, El-Kady I, Biswas R, Ho K. All-metallic three-dimensional photonic crystals with a large infrared bandgap. Nature. 2002;417(6884):52–55.

[980] Lin SY, Moreno J, Fleming J. Three-dimensional photonic-crystal emitter for thermal photovoltaic power generation. Applied Physics Letters. 2003; 83(2):380–382.

[981] Han SE, Stein A, Norris DJ. Tailoring self-assembled metallic photonic crystals for modified thermal emission. Physical Review Letters. 2007;99(5):053906.

[982] Arpin KA, Losego MD, Cloud AN, et al. Three-dimensional self-assembled photonic crystals with high temperature stability for thermal emission modification[J]. Nature Communications, 2013;4(1):2630.

[983] Liu N, Mesch M, Weiss T, Hentschel M, Giessen H. Infrared perfect absorber and its application as plasmonic sensor. Nano Letters. 2010;10(7):2342–2348.

[984] Wang L, Zhang Z. Wavelength-selective and diffuse emitter enhanced by magnetic polaritons for thermophotovoltaics. Applied Physics Letters. 2012;100(6):063902.

[985] Chen YK, Wang BX, Zhao CY. Dual-band spatially-distinguishable metasurface thermal emitter for filterless mid-infrared gas sensing. International Journal of Thermal Sciences. 2023;185:108069.

[986] Gong Y, Wang Z, Li K, Uggalla L, Huang J, Copner N, et al. Highly efficient and broadband mid-infrared metamaterial thermal emitter for optical gas sensing. Optics Letters. 2017;42(21):4537–4540.

[987] Aigner A, Dawes JM, Maier SA, Ren H. Nanophotonics shines light on hyperbolic metamaterials. Light, Science & Applications. 2022;11.

[988] Huo P, Zhang S, Liang Y, Lu Y, Xu T. Hyperbolic metamaterials and metasurfaces: Fundamentals and applications. Advanced Optical Materials. 2019;7(14):1801616.

[989] Poddubny A, Iorsh I, Belov P, Kivshar Y. Hyperbolic metamaterials. Nature Photonics. 2013;7(12):948–957.

[990] Campione S, Marquier F, Hugonin JP, Ellis AR, Klem JF, Sinclair MB, et al. Directional and monochromatic thermal emitter from epsilon-near-zero conditions in semiconductor hyperbolic metamaterials. Scientific Reports. 2016;6(1):1–9.

[991] Zhao B, Zhang ZM. Perfect mid-infrared absorption by hybrid phonon-plasmon polaritons in hBN/metal-grating anisotropic structures. International Journal of Heat and Mass Transfer. 2017;106:1025–1034.

[992] Kan YH, Zhao CY, Zhang ZM. Compact mid-infrared broadband absorber based on hBN/metal metasurface. International Journal of Thermal Sciences. 2018;130:192–199.

[993] Hendrickson JR, Vangala S, Dass C, Gibson R, Goldsmith J, Leedy K, et al. Coupling of epsilon-near-zero mode to gap plasmon mode for flat-top wideband perfect light absorption. ACS Photonics. 2018;5(3):776–781.

[994] Niu X, Hu X, Xu Y, Yang H, Gong Q. Ultrafast all-optical polarization switching based on composite metasurfaces with gratings and an Epsilon-Near-Zero film. Advanced Photonics Research. 2021;2(4):2000167.

[995] Feng S, Halterman K. Coherent perfect absorption in epsilon-near-zero metamaterials. Physical Review B. 2012;86(16):165103.

[996] Liao YL, Zhao Y, Zhang X, Chen Z. An ultra-narrowband absorber with a compound dielectric grating and metal substrate. Optics Communications. 2017;385:172–176.

[997] Shamkhi HK, Sayanskiy A, Valero AC, Kupriianov AS, Kapitanova P, Kivshar YS, et al. Transparency and perfect absorption of all-dielectric resonant metasurfaces governed by the transverse Kerker effect. Physical Review Materials. 2019;3(8):085201.

[998] Liu MQ, Zhao CY. Ultranarrow and wavelength-scalable thermal emitters driven by high-order antiferromagnetic resonances in dielectric nanogratings. ACS Applied Materials & Interfaces. 2021.

[999] Sun S, He Q, Hao J, Xiao S, Zhou L. Electromagnetic metasurfaces: Physics and applications. Advances in Optics and Photonics. 2019;11(2):380–479.

[1000] Wang R, Dal Negro L. Engineering non-radiative anapole modes for broadband absorption enhancement of light. Optics Express. 2016;24(17):19048–19062.

[1001] Tian J, Li Q, Belov PA, Sinha RK, Qian W, Qiu M. High-Q all-dielectric metasurface: Super and suppressed optical absorption. ACS Photonics. 2020;7(6):1436–1443.

[1002] Yang CY, Yang JH, Yang ZY, Zhou ZX, Sun MG, Babicheva VE, et al. Nonradiating silicon nanoantenna metasurfaces as narrowband absorbers. Acs Photonics. 2018;5(7):2596–2601.

[1003] Fang X, Lou MH, Bao H, Zhao CY. Thin films with disordered nanohole patterns for solar radiation absorbers. Journal of Quantitative Spectroscopy and Radiative Transfer. 2015;158:145–153.

[1004] Mao P, Liu C, Song F, Han M, Maier SA, Zhang S. Manipulating disordered plasmonic systems by external cavity with transition from broadband absorption to reconfigurable reflection. Nature Communications. 2020;11(1):1–7.

[1005] Chen X, Gong H, Dai S, Zhao D, Yang Y, Li Q, et al. Near-infrared broadband absorber with film-coupled multilayer nanorods. Optics Letters. 2013;38(13):2247–2249.

[1006] Sakurai A, Yada K, Simomura T, Ju S, Kashiwagi M, Okada H, et al. Ultranarrow-band wavelength-selective thermal emission with aperiodic multilayered metamaterials designed by Bayesian optimization. ACS Central Science. 2019;5(2):319–326.

[1007] Vynck K, Burresi M, Riboli F, Wiersma DS. Photon management in two-dimensional disordered media. Nature Materials. 2012;11(12):1017–1022.

[1008] Zhou L, Tan Y, Wang J, Xu W, Yuan Y, Cai W, et al. 3D self-assembly of aluminium nanoparticles for plasmon-enhanced solar desalination. Nature Photonics. 2016;10(6):393–398.

[1009] Zhou L, Tan Y, Ji D, Zhu B, Zhang P, Xu J, et al. Self-assembly of highly efficient, broadband plasmonic absorbers for solar steam generation. Science Advances. 2016;2(4):e1501227.

[1010] Zhang WB, Wang BX, Zhao CY. Selective thermophotovoltaic emitter with aperiodic multilayer structures designed by machine learning. ACS Applied Energy Materials. 2021;4(2):2004–2013.

[1011] Long L, Taylor S, Wang L. Enhanced infrared emission by thermally switching the excitation of magnetic polariton with scalable microstructured VO_2 metasurfaces. ACS Photonics. 2020;7(8):2219–2227.

[1012] Qu Y, Li Q, Cai L, Pan M, Ghosh P, Du K, et al. Thermal camouflage based on the phase-changing material GST. Light: Science & Applications. 2018;7(1): 1–10.

[1013] Park J, Kang JH, Liu X, Maddox SJ, Tang K, McIntyre PC, et al. Dynamic thermal emission control with InAs-based plasmonic metasurfaces. Science Advances. 2018;4(12):eaat3163.

[1014] Zylbersztejn A, Mott NF. Metal-insulator transition in vanadium dioxide. Physical Review B. 1975;11(11):4383.

[1015] Tang K, Wang X, Dong K, Li Y, Li J, Sun B, et al. A thermal radiation modulation platform by emissivity engineering with graded metal-insulator transition. Advanced Materials. 2020;32(36):1907071.

[1016] Qu Y, Cai L, Luo H, Lu J, Qiu M, Li Q. Tunable dual-band thermal emitter consisting of single-sized phase-changing GST nanodisks. Optics Express. 2018;26(4):4279–4287.

[1017] Luo F, Fan Y, Peng G, Xu S, Yang Y, Yuan K, et al. Graphene thermal emitter with enhanced joule heating and localized light emission in air. ACS Photonics. 2019;6(8):2117–2125.

[1018] Xiao Y, Charipar NA, Salman J, Piqué A, Kats MA. Nanosecond mid-infrared pulse generation via modulated thermal emissivity. Light: Science & Applications. 2019;8(1):1–8.

[1019] Coppens ZJ, Valentine JG. Spatial and temporal modulation of thermal emission. Advanced Materials. 2017;29(39):1701275.

[1020] Inampudi S, Mosallaei H. Tunable wideband-directive thermal emission from SiC surface using bundled graphene sheets. Physical Review B. 2017 Sep;96:125407.

[1021] Liu MQ, Zhao CY, Bao H. Transverse Kerker scattering governed by two nondegenerate electric dipoles and its application in arbitrary beam steering. Journal of Quantitative Spectroscopy and Radiative Transfer. 2021;262:107514.

[1022] Liu W, Kivshar YS. Generalized Kerker effects in nanophotonics and metaoptics. Optics Express. 2018;26(10):13085–13105.

[1023] Liu MQ, Zhao CY, Wang BX. Polarization management based on dipolar interferences and lattice couplings. Optics Express. 2018;26(6):7235–7252.

[1024] Wang Z, Clark JK, Ho YL, Volz S, Daiguji H, Delaunay JJ. Ultranarrow and wavelength-tunable thermal emission in a hybrid metal–optical tamm state structure. ACS Photonics. 2020;7(6):1569–1576.

[1025] Bossard JA, Lin L, Yun S, Liu L, Werner DH, Mayer TS. Near-ideal optical metamaterial absorbers with super-octave bandwidth. ACS Nano. 2014;8(2):1517–1524.

[1026] Dahan N, Niv A, Biener G, Kleiner V, Hasman E. Space-variant polarization manipulation of a thermal emission by a SiO_2 subwavelength grating supporting surface phonon-polaritons. Applied Physics Letters. 2005;86(19):191102.

[1027] Dyakov SA, Semenenko VA, Gippius NA, Tikhodeev SG. Magnetic field free circularly polarized thermal emission from a chiral metasurface. Physical Review B. 2018 Dec;98:235416.

[1028] Li W, Coppens ZJ, Besteiro LV, Wang W, Govorov AO, Valentine J. Circularly polarized light detection with hot electrons in chiral plasmonic metamaterials. Nature Communications. 2015 Sep;6(1):8379.

[1029] Narayanaswamy A, Chen G. Surface modes for near field thermophotovoltaics. Applied Physics Letters. 2003;82(20):3544–3546.

[1030] Volokitin AI, Persson BNJ. Near-field radiative heat transfer between closely spaced graphene and amorphous SiO_2. Physical Review B. 2011;83(24):241407.

[1031] Ilic O, Jablan M, Joannopoulos JD, Celanovic I, Buljan H, Soljačić M. Near-field thermal radiation transfer controlled by plasmons in graphene. Physical Review B. 2012;85(15):155422.

[1032] Messina R, Hugonin JP, Greffet JJ, Marquier F, De Wilde Y, Belarouci A, et al. Tuning the electromagnetic local density of states in graphene-covered systems via strong coupling with graphene plasmons. Physical Review B. 2013;87(8):085421.

[1033] Zhang WB, Wang BX, Zhao CY. Active control and enhancement of near- field heat transfer between dissimilar materials by strong coupling effects. International Journal of Heat and Mass Transfer. 2022;188:122588.

[1034] Svetovoy VB, van Zwol PJ, Chevrier J. Plasmon enhanced near-field radiative heat transfer for graphene covered dielectrics. Physical Review B. 2012;85(15):155418.

[1035] Messina R, Ben-Abdallah P, Guizal B, Antezza M. Graphene-based amplification and tuning of near-field radiative heat transfer between dissimilar polar materials. Physical Review B. 2017;96(4):045402.

[1036] Liu XL, Zhang ZM. Giant enhancement of nanoscale thermal radiation based on hyperbolic graphene plasmons. Applied Physics Letters. 2015;107(14):143114.

[1037] Liu XL, Zhang RZ, Zhang ZM. Near-field radiative heat transfer with doped-silicon nanostructured metamaterials. International Journal of Heat and Mass Transfer. 2014;73:389–398.

[1038] Liu XL, Wang LP, Zhang ZMM. Near-field thermal radiation: Recent progress and outlook. Nanoscale and Microscale Thermophysical Engineering. 2015;19(2):98–126.

[1039] Ikeda T, Ito K, Iizuka H. Tunable quasi-monochromatic near-field radiative heat transfer in s and p polarizations by a hyperbolic metamaterial layer. Journal of Applied Physics. 2017;121(1):013106.

[1040] Zhang WB, Wang BX, Xu JM, Zhao CY. High-quality quasi-monochromatic near-field radiative heat transfer designed by adaptive hybrid Bayesian optimization. Science China Technological Sciences. 2022;(1674–7321).

[1041] Jin W, Molesky S, Lin Z, Rodriguez AW. Material scaling and frequency-selective enhancement of near-field radiative heat transfer for lossy metals in two dimensions via inverse design. Physical Review B. 2019;99(4):041403.

[1042] García-Esteban JJ, Bravo-Abad J, Cuevas JC. Deep learning for the modeling and inverse design of radiative heat transfer. Physical Review Applied. 2021;16(6):064006.

[1043] Wen S, Dang C, Liu X. A machine learning strategy for modeling and optimal design of near-field radiative heat transfer. Applied Physics Letters. 2022;121(7):071101.

[1044] Messina R, Antezza M, Ben-Abdallah P. Three-body amplification of photon heat tunneling. Physical Review Letters. 2012;109(24):244302.

[1045] Kan YH, Zhao CY, Zhang ZM. Near-field radiative heat transfer in three-body systems with periodic structures. Physical Review B. 2019;99(3):035433.

[1046] Simchi H. Graphene-based three-body amplification of photon heat tunneling. Journal of Applied Physics. 2017;121(9):094301.

[1047] Song J, Lu L, Cheng Q, Luo Z. Three-body heat transfer between anisotropic magneto-dielectric hyperbolic metamaterials. Journal of Heat Transfer. 2018;140(8).

[1048] Latella I, Pérez-Madrid A, Rubi JM, Biehs SA, Ben-Abdallah P. Heat engine driven by photon tunneling in many-body systems. Physical Review Applied. 2015;4(1):011001.

[1049] Messina R, Antezza M. Scattering-matrix approach to Casimir-Lifshitz force and heat transfer out of thermal equilibrium between arbitrary bodies. Physical Review A. 2011 Oct;84:042102.

[1050] Chen J, Wang BX, Zhao CY. Near-field heat transport between nanoparticles inside a cavity configuration. International Journal of Heat and Mass Transfer. 2022;196:123213.

[1051] Han S. Theory of thermal emission from periodic structures. Physical Review B. 2009;80(15):155108.

[1052] Zhu L, Fan S. Near-complete violation of detailed balance in thermal radiation. Physical Review B. 2014;90(22):220301.

[1053] Zhao B, Shi Y, Wang J, Zhao Z, Zhao N, Fan S. Near-complete violation of Kirchhoff's law of thermal radiation with a 0.3 T magnetic field. Optics Letters. 2019;44(17):4203–4206.

[1054] Zhao B, Guo C, Garcia CA, Narang P, Fan S. Axion-field-enabled nonreciprocal thermal radiation in Weyl semimetals. Nano Letters. 2020;20(3):1923–1927.

[1055] Tsurimaki Y, Qian X, Pajovic S, Han F, Li M, Chen G. Large nonreciprocal absorption and emission of radiation in type-I Weyl semimetals with time reversal symmetry breaking. Physical Review B. 2020;101(16):165426.

[1056] Shayegan KJ, Zhao B, Kim Y, Fan S, Atwater HA. Nonreciprocal infrared absorption via resonant magneto-optical coupling to InAs. Science Advances. 2022;8(18):eabm4308.

[1057] Liu MQ, Xia S, Wan WJ, Qin J, H L, Zhao CY, et al. Broadband mid-infrared non-reciprocal absorption using magnetized gradient epsilon-near-zero thin films. Nature Materials 2023;22(10):1196–1202.

[1058] Xu SY, Belopolski I, Alidoust N, Neupane M, Bian G, Zhang C, et al. Discovery of a Weyl fermion semimetal and topological Fermi arcs. Science. 2015;349(6248):613–617.

[1059] Wang Q, Xu Y, Lou R, Liu Z, Li M, Huang Y, et al. Large intrinsic anomalous Hall effect in half-metallic ferromagnet $Co_3Sn_2S_2$ with magnetic Weyl fermions. Nature Communications. 2018;9(1):3681.

[1060] Ghimire NJ, Botana A, Jiang J, Zhang J, Chen YS, Mitchell J. Large anomalous Hall effect in the chiral-lattice antiferromagnet $CoNb_3S_6$. Nature Communications. 2018;9(1):3280.

[1061] Hofmann J, Sarma SD. Surface plasmon polaritons in topological Weyl semimetals. Physical Review B. 2016;93(24):241402.

[1062] Sushkov AB, Hofmann JB, Jenkins GS, Ishikawa J, Nakatsuji S, Sarma SD, et al. Optical evidence for a Weyl semimetal state in pyrochlore $Eu_2Ir_2O_7$. Physical Review B. 2015;92(24):241108.

[1063] Pajovic S, Tsurimaki Y, Qian X, Chen G. Intrinsic nonreciprocal reflection and violation of Kirchhoff's law of radiation in planar type-I magnetic Weyl semimetal surfaces. Physical Review B. 2020;102(16):165417.

[1064] Burkov A. Anomalous Hall effect in Weyl metals. Physical Review Letters. 2014;113(18):187202.

[1065] Raman AP, Abou Anoma M, Zhu L, Rephaeli E, Fan S. Passive radiative cooling below ambient air temperature under direct sunlight. Nature. 2014;515(7528):540–544.

[1066] Rephaeli E, Raman A, Fan S. Ultrabroadband photonic structures to achieve high-performance daytime radiative cooling. Nano Letters. 2013;13(4): 1457–1461.

[1067] Chae D, Kim M, Jung PH, Son S, Seo J, Liu Y, et al. Spectrally selective inorganic-based multilayer emitter for daytime radiative cooling. ACS Applied Materials & Interfaces. 2020;12(7):8073–8081.

[1068] Kim M, Seo J, Yoon S, Lee H, Lee J, Lee BJ. Optimization and performance analysis of a multilayer structure for daytime radiative cooling. Journal of Quantitative Spectroscopy and Radiative Transfer. 2021;260:107475.

[1069] Bao H, Yan C, Wang BX, Fang X, Zhao CY, Ruan XL. Double-layer nanoparticle-based coatings for efficient terrestrial radiative cooling. Solar Energy Materials and Solar Cells. 2017;168:78–84.

[1070] Mandal J, Fu Y, Overvig AC, Jia M, Sun K, Shi NN, et al. Hierarchically porous polymer coatings for highly efficient passive daytime radiative cooling. Science. 2018;362(6412):315–319.

[1071] Li T, Zhai Y, He S, Gan W, Wei Z, Heidarinejad M, et al. A radiative cooling structural material. Science. 2019;364(6442):760–763.

[1072] Zeng S, Pian S, Su M, Wang Z, Wu M, Liu X, et al. Hierarchical-morphology metafabric for scalable passive daytime radiative cooling. Science. 2021;373(6555):692–696.

[1073] Yang M, Zou W, Guo J, Qian Z, Luo H, Yang S, et al. Bioinspired "skin" with cooperative thermo-optical effect for daytime radiative cooling. ACS Applied Materials & Interfaces. 2020;12(22):25286–25293.

[1074] Jeon S, Son S, Lee SY, Chae D, Bae JH, Lee H, et al. Multifunctional daytime radiative cooling devices with simultaneous light-emitting and radiative cooling functional layers. ACS Applied Materials & Interfaces. 2020;12(49): 54763–54772.

[1075] Wang T, Wu Y, Shi L, Hu X, Chen M, Wu L. A structural polymer for highly efficient all-day passive radiative cooling. Nature Communications. 2021;12(1): 1–11.

[1076] Li X, Peoples J, Huang Z, Zhao Z, Qiu J, Ruan X. Full daytime sub-ambient radiative cooling in commercial-like paints with high figure of merit. Cell Reports Physical Science. 2020;1(10):100221.

[1077] Li W, Shi Y, Chen K, Zhu L, Fan S. A comprehensive photonic approach for solar cell cooling. ACS Photonics. 2017;4(4):774–782.

[1078] Wu D, Liu C, Xu Z, Liu Y, Yu Z, Yu L, et al. The design of ultra-broadband selective near-perfect absorber based on photonic structures to achieve near-ideal daytime radiative cooling. Materials & Design. 2018;139:104–111.

[1079] Kong A, Cai B, Shi P, Yuan Xc. Ultra-broadband all-dielectric metamaterial thermal emitter for passive radiative cooling. Optics Express. 2019;27(21):30102–30115.

[1080] Banik U, Agrawal A, Meddeb H, Sergeev O, Reininghaus N, Gotz-Kohler M, et al. Efficient thin polymer coating as a selective thermal emitter for passive daytime radiative cooling. ACS Applied Materials & Interfaces. 2021;13(20):24130–24137.

[1081] Chen Z, Zhu L, Raman A, Fan S. Radiative cooling to deep sub-freezing temperatures through a 24-h day–night cycle. Nature Communications. 2016;7(1):1–5.

[1082] Mira ZF, Heo SY, Lee GJ, Song YM, et al. Multilayer selective passive daytime radiative cooler optimization utilizing memetic algorithm. Journal of Quantitative Spectroscopy and Radiative Transfer. 2021;272:107774.

[1083] Hossain MM, Jia B, Gu M. A metamaterial emitter for highly efficient radiative cooling. Advanced Optical Materials. 2015;3(8):1047–1051.

[1084] Liu X, Tian Y, Chen F, Ghanekar A, Antezza M, Zheng Y. Continuously variable emission for mechanical deformation induced radiative cooling. Communications Materials. 2020;1(1):1–7.

[1085] Jeon S, Shin J. Directional radiation for optimal radiative cooling. Optics Express. 2021;29(6):8376–8386.

[1086] Chamoli SK, Li W, Guo C, ElKabbash M. Angularly selective thermal emitters for deep subfreezing daytime radiative cooling. Nanophotonics. 2022;11(16):3709–3717.

[1087] Fan S, Li W. Photonics and thermodynamics concepts in radiative cooling. Nature Photonics. 2022;16(3):182–190.

[1088] Chow T. Performance analysis of photovoltaic-thermal collector by explicit dynamic model. Solar Energy. 2003;75(2):143–152.

[1089] Kou Jl, Jurado Z, Chen Z, Fan S, Minnich AJ. Daytime radiative cooling using near-black infrared emitters. ACS Photonics. 2017;4(3):626–630.

[1090] Wang W, Fernandez N, Katipamula S, Alvine K. Performance assessment of a photonic radiative cooling system for office buildings. Renewable Energy. 2018;118:265–277.

[1091] Goldstein EA, Raman AP, Fan S. Sub-ambient non-evaporative fluid cooling with the sky. Nature Energy. 2017;2(9):1–7.

[1092] Smith G, Gentle A. Radiative cooling: Energy savings from the sky. Nature Energy. 2017;2(9):1–2.

[1093] Lu Y, Chen Z, Ai L, Zhang X, Zhang J, Li J, et al. A universal route to realize radiative cooling and light management in photovoltaic modules. Solar RRL. 2017;1(10):1700084.

[1094] Wang Z, Kortge D, Zhu J, Zhou Z, Torsina H, Lee C, et al. Lightweight, passive radiative cooling to enhance concentrating photovoltaics. Joule. 2020; 4(12):2702–2717.

[1095] Cho JW, Park SJ, Park SJ, Kim YB, Kim KY, Bae D, et al. Scalable on-chip radiative coolers for concentrated solar energy devices. ACS Photonics. 2020;7(10):2748–2755.

[1096] Zhou M, Song H, Xu X, Shahsafi A, Xia Z, Ma Z, et al. Accelerating vapor condensation with daytime radiative cooling. In: New concepts in solar and thermal radiation conversion II. vol. 11121. International Society for Optics and Photonics; 2019. p. 1112107.

[1097] Li W, Dong M, Fan L, John JJ, Chen Z, Fan S. Nighttime radiative cooling for water harvesting from solar panels. ACS Photonics. 2020;8(1):269–275.

[1098] Cai L, Song AY, Wu P, Hsu PC, Peng Y, Chen J, et al. Warming up human body by nanoporous metallized polyethylene textile. Nature Communications. 2017;8(1):1–8.

[1099] Hsu PC, Liu C, Song AY, Zhang Z, Peng Y, Xie J, et al. A dual-mode textile for human body radiative heating and cooling. Science Advances. 2017; 3(11):e1700895.

[1100] Kang MH, Lee GJ, Lee JH, Kim MS, Yan Z, Jeong JW, et al. Outdoor-useable, wireless/battery-free patch-type tissue oximeter with radiative cooling. Advanced Science. 2021:2004885.

[1101] Raman AP, Li W, Fan S. Generating light from darkness. Joule. 2019; 3(11):2679–2686.

[1102] Moghimi MJ, Lin G, Jiang H. Broadband and ultrathin infrared stealth sheets. Advanced Engineering Materials. 2018;20(11):1800038.

[1103] Li L, Shi M, Liu X, Jin X, Cao Y, Yang Y, et al. Ultrathin titanium carbide (MXene) films for high-temperature thermal camouflage. Advanced Functional Materials. 2021:2101381.

[1104] Wang QW, Zhang HB, Liu J, Zhao S, Xie X, Liu L, et al. Multifunctional and water-resistant MXene-decorated polyester textiles with outstanding electromagnetic interference shielding and Joule heating performances. Advanced Functional Materials. 2019;29(7):1806819.

[1105] Yan J, Ren CE, Maleski K, Hatter CB, Anasori B, Urbankowski P, et al. Flexible MXene/graphene films for ultrafast supercapacitors with outstanding volumetric capacitance. Advanced Functional Materials. 2017;27(30):1701264.

[1106] Kim T, Bae JY, Lee N, Cho HH. Hierarchical metamaterials for multispectral camouflage of infrared and microwaves. Advanced Functional Materials. 2019;29(10):1807319.

[1107] Zhu H, Li Q, Tao C, Hong Y, Xu Z, Shen W, et al. Multispectral camouflage for infrared, visible, lasers and microwave with radiative cooling. Nature Communications. 2021;12(1):1–8.

[1108] Wu C, Yu H, Li H, Zhang X, Takeuchi I, Li M. Low-loss integrated photonic switch using subwavelength patterned phase change material. ACS Photonics. 2018;6(1):87–92.

[1109] Zhang H, Zhou L, Lu L, Xu J, Wang N, Hu H, et al. Miniature multilevel optical memristive switch using phase change material. ACS Photonics. 2019;6(9): 2205–2212.

[1110] Carrillo SGC, Nash GR, Hayat H, Cryan MJ, Klemm M, Bhaskaran H, et al. Design of practicable phase-change metadevices for near-infrared absorber and modulator applications. Optics Express. 2016;24(12):13563–13573.

[1111] Ríos C, Stegmaier M, Hosseini P, Wang D, Scherer T, Wright CD, et al. Integrated all-photonic non-volatile multi-level memory. Nature Photonics. 2015; 9(11):725–732.

[1112] Jian J, Wang X, Li L, Fan M, Zhang W, Huang J, et al. Continuous tuning of phase transition temperature in VO_2 thin films on c-cut sapphire substrates via strain variation. ACS Applied Materials & Interfaces. 2017;9(6):5319–5327.

[1113] Salihoglu O, Uzlu HB, Yakar O, Aas S, Balci O, Kakenov N, et al. Graphene-based adaptive thermal camouflage. Nano Letters. 2018;18(7):4541–4548.

[1114] Hong S, Shin S, Chen R. An adaptive and wearable thermal camouflage device. Advanced Functional Materials. 2020;30(11):1909788.

[1115] Inoue T, De Zoysa M, Asano T, Noda S. Realization of dynamic thermal emission control. Nature Materials. 2014;13(10):928–931.

[1116] Li M, Liu D, Cheng H, Peng L, Zu M. Manipulating metals for adaptive thermal camouflage. Science Advances. 2020;6(22):eaba3494.

[1117] Duan X, Kamin S, Liu N. Dynamic plasmonic colour display. Nature Communications. 2017;8(1):1–9.

[1118] Kolm H. Solar-battery power source. Solar-Battery Power Source Quarterly Progress Report Solid State Research, Group 35 (Lexington, MA: MIT Lincoln Laboratory, 1956).

[1119] Burger T, Sempere C, Roy-Layinde B, Lenert A. Present efficiencies and future opportunities in thermophotovoltaics. Joule. 2020;4(8):1660–1680.

[1120] Licht A, Pfiester N, DeMeo D, Chivers J, Vandervelde TE. A review of advances in thermophotovoltaics for power generation and waste heat harvesting. MRS Advances. 2019;4(41–42):2271–2282.

[1121] Yang W, Chou S, Li J. Microthermophotovoltaic power generator with high power density. Applied Thermal Engineering. 2009;29(14–15):3144–3148.

[1122] Chou S, Yang W, Chua K, Li J, Zhang K. Development of micro power generators–a review. Applied Energy. 2011;88(1):1–16.

[1123] Schock A, Mukunda M, Or C, Kumar V, Summers G. Design, analysis, and optimization of a radioisotope thermophotovoltaic (RTPV) generator, and its applicability to an illustrative space mission. Acta Astronautica. 1995;37:21–57.

[1124] Wang X, Chan W, Stelmakh V, Celanovic I, Fisher P. Toward high performance radioisotope thermophotovoltaic systems using spectral control. Nuclear Instruments and Methods in Physics Research Section A: Accelerators, Spectrometers, Detectors and Associated Equipment. 2016;838:28–32.

[1125] Teofilo V, Choong P, Chang J, Tseng YL, Ermer S. Thermophotovoltaic energy conversion for space. The Journal of Physical Chemistry C. 2008;112(21):7841–7845.

[1126] Thekdi A, Nimbalkar SU. Industrial waste heat recovery-potential applications, available technologies and crosscutting R&D opportunities. Oak Ridge National Lab.(ORNL), Oak Ridge, TN (United States); 2015.

[1127] Datas A, Algora C. Development and experimental evaluation of a complete solar thermophotovoltaic system. Progress in Photovoltaics: Research and Applications. 2013;21(5):1025–1039.

[1128] Ungaro C, Gray SK, Gupta MC. Solar thermophotovoltaic system using nanostructures. Optics Express. 2015;23(19):A1149–A1156.

[1129] Good BS, Chubb DL, Lowe RA. Comparison of selective emitter and filter thermophotovoltaic systems. In: AIP Conference Proceedings. vol. 358. American Institute of Physics; 1996. pp. 16–34.

[1130] Fan D, Burger T, McSherry S, Lee B, Lenert A, Forrest SR. Near-perfect photon utilization in an air-bridge thermophotovoltaic cell. Nature. 2020;586(7828): 237–241.

[1131] Bitnar B, Durisch W, Mayor JC, Sigg H, Tschudi H. Characterisation of rare earth selective emitters for thermophotovoltaic applications. Solar Energy Materials and Solar Cells. 2002;73(3):221–234.

[1132] Fraas L, Samaras J, Avery J, Minkin L. Antireflection coated refractory metal matched emitters for use with GaSb thermophotovoltaic generators. In: Conference Record of the Twenty-Eighth IEEE Photovoltaic Specialists Conference-2000 (Cat. No. 00CH37036). IEEE; 2000. pp. 1020–1023.

[1133] Fraas L, Avery J, Huang H. Thermophotovoltaic furnace–generator for the home using low bandgap GaSb cells. Semiconductor Science and Technology. 2003;18(5):S247.

[1134] Heinzel A, Boerner V, Gombert A, Wittwer V, Luther J. Microstructured tungsten surfaces as selective emitters. In: AIP Conference Proceedings. vol. 460. American Institute of Physics; 1999. pp. 191–196.

[1135] Sai H, Yugami H, Akiyama Y, Kanamori Y, Hane K. Surface microstructured selective emitters for TPV systems. In: Conference Record of the Twenty-Eighth IEEE Photovoltaic Specialists Conference-2000 (Cat. No. 00CH37036). IEEE; 2000. pp. 1016–1019.

[1136] Sai H, Yugami H, Akiyama Y, Kanamori Y, Hane K. Spectral control of thermal emission by periodic microstructured surfaces in the near-infrared region. JOSA A. 2001;18(7):1471–1476.

[1137] Sai H, Kanamori Y, Yugami H. High-temperature resistive surface grating for spectral control of thermal radiation. Applied Physics Letters. 2003;82(11): 1685–1687.

[1138] Sai H, Yugami H. Thermophotovoltaic generation with selective radiators based on tungsten surface gratings. Applied Physics Letters. 2004;85(16):3399–3401.

[1139] Celanovic I, Jovanovic N, Kassakian J. Two-dimensional tungsten photonic crystals as selective thermal emitters. Applied Physics Letters. 2008;92(19):193101.

[1140] Chen YB, Tan KH. The profile optimization of periodic nano-structures for wavelength-selective thermophotovoltaic emitters. International Journal of Heat and Mass Transfer. 2010;53(23–24):5542–5551.

[1141] Araghchini M, Yeng Y, Jovanovic N, Bermel P, Kolodziejski L, Soljacic M, et al. Fabrication of two-dimensional tungsten photonic crystals for high-temperature applications. Journal of Vacuum Science & Technology B, Nanotechnology and Microelectronics: Materials, Processing, Measurement, and Phenomena. 2011;29(6):061402.

[1142] Gee JM, Moreno JB, Lin SY, Fleming JG. Selective emitters using photonic crystals for thermophotovoltaic energy conversion. In: Conference Record of the Twenty-Ninth IEEE Photovoltaic Specialists Conference, 2002. IEEE; 2002. pp. 896–899.

[1143] Nagpal P, Han SE, Stein A, Norris DJ. Efficient low-temperature thermophotovoltaic emitters from metallic photonic crystals. Nano Letters. 2008;8(10): 3238–3243.

[1144] Huang TC, Wang BX, Zhang WB, Zhao CY. A novel selective thermophotovoltaic emitter based on multipole resonances. International Journal of Heat and Mass Transfer. 2022;182:122039.

[1145] Rinnerbauer V, Ndao S, Xiang Yeng Y, Senkevich JJ, Jensen KF, Joannopoulos JD, et al. Large-area fabrication of high aspect ratio tantalum photonic crystals for high-temperature selective emitters. Journal of Vacuum Science & Technology B, Nanotechnology and Microelectronics: Materials, Processing, Measurement, and Phenomena. 2013;31(1):011802.

[1146] Rinnerbauer V, Yeng YX, Chan WR, Senkevich JJ, Joannopoulos JD, Soljačić M, et al. High-temperature stability and selective thermal emission of polycrystalline tantalum photonic crystals. Optics Express. 2013;21(9):11482–11491.

[1147] Fraas L, Ballantyne R, Hui S, Ye SZ, Gregory S, Keyes J, et al. Commercial GaSb cell and circuit development for the midnight sun® TPV stove. In: AIP Conference Proceedings. vol. 460. American Institute of Physics; 1999. pp. 480–487.

[1148] Fraas L, Minkin L. TPV history from 1990 to present & future trends. In: AIP Conference Proceedings. vol. 890. American Institute of Physics; 2007. pp. 17–23.

[1149] Nakagawa N, Ohtsubo H, Waku Y, Yugami H. Thermal emission properties of $Al_2O_3/Er_3Al_5O_{12}$ eutectic ceramics. Journal of the European Ceramic Society. 2005;25(8):1285–1291.

[1150] Bitnar B, Durisch W, Holzner R. Thermophotovoltaics on the move to applications. Applied Energy. 2013;105:430–438.

[1151] Durisch W, Grob B, Mayor JC, Panitz JC, Rosselet A. Interfacing a small thermophotovoltaic generator to the grid. In: AIP Conference Proceedings. vol. 460. American Institute of Physics; 1999. pp. 403–416.

[1152] Omair Z, Scranton G, Pazos-Outón LM, Xiao TP, Steiner MA, Ganapati V, et al. Ultraefficient thermophotovoltaic power conversion by band-edge spectral filtering. Proceedings of the National Academy of Sciences. 2019;116(31): 15356–15361.

[1153] Swanson RM. Recent developments in thermophotovoltaic conversion. In: 1980 International Electron Devices Meeting. IEEE; 1980. pp. 186–189.

[1154] Fernández J, Dimroth F, Oliva E, Hermle M, Bett A. Back-surface optimization of germanium TPV cells. In: AIP Conference Proceedings. vol. 890. American Institute of Physics; 2007. pp. 190–197.

[1155] Fraas L, Samaras J, Huang H, Minkin L, Avery J, Daniels W, et al. TPV generators using the radiant tube burner configuration. In: Proceedings of 17th European PV Solar Energy Conference, Munich, Germany. vol. 26; 2001.

[1156] LaPotin A, Schulte KL, Steiner MA, Buznitsky K, Kelsall CC, Friedman DJ, et al. Thermophotovoltaic efficiency of 40%. Nature. 2022;604(7905):287–291.

[1157] Dong X, Gao S, Li S, Zhu T, Huang J, Chen Z, et al. Bioinspired structural and functional designs towards interfacial solar steam generation for clean water production. Materials Chemistry Frontiers. 2021;5(4):1510–1524.

[1158] Neumann O, Urban AS, Day J, Lal S, Nordlander P, Halas NJ. Solar vapor generation enabled by nanoparticles. ACS Nano. 2013;7(1):42–49.

[1159] Ni G, Miljkovic N, Ghasemi H, Huang X, Boriskina SV, Lin CT, et al. Volumetric solar heating of nanofluids for direct vapor generation. Nano Energy. 2015;17:290–301.

[1160] Liu Y, Yu S, Feng R, Bernard A, Liu Y, Zhang Y, et al. A bioinspired, reusable, paper-based system for high-performance large-scale evaporation. Advanced Materials. 2015;27(17):2768–2774.

[1161] Li X, Xu W, Tang M, Zhou L, Zhu B, Zhu S, et al. Graphene oxide-based efficient and scalable solar desalination under one sun with a confined 2D water path. Proceedings of the National Academy of Sciences. 2016;113(49):13953–13958.

[1162] Liu Y, Liu Z, Huang Q, Liang X, Zhou X, Fu H, et al. A high-absorption and self-driven salt-resistant black gold nanoparticle-deposited sponge for highly efficient, salt-free, and long-term durable solar desalination. Journal of Materials Chemistry A. 2019;7(6):2581–2588.

[1163] Zhang P, Li J, Lv L, Zhao Y, Qu L. Vertically aligned graphene sheets membrane for highly efficient solar thermal generation of clean water. ACS Nano. 2017;11(5):5087–5093.

[1164] Xu N, Hu X, Xu W, Li X, Zhou L, Zhu S, et al. Mushrooms as efficient solar steam-generation devices. Advanced Materials. 2017;29(28):1606762.

[1165] Ito Y, Tanabe Y, Han J, Fujita T, Tanigaki K, Chen M. Multifunctional porous graphene for high-efficiency steam generation by heat localization. Advanced Materials. 2015;27(29):4302–4307.

[1166] Hogan NJ, Urban AS, Ayala-Orozco C, Pimpinelli A, Nordlander P, Halas NJ. Nanoparticles heat through light localization. Nano Letters. 2014;14(8): 4640–4645.

[1167] Chala TF, Wu CM, Chou MH, Guo ZL. Melt electrospun reduced tungsten oxide/polylactic acid fiber membranes as a photothermal material for light-driven interfacial water evaporation. ACS Applied Materials & Interfaces. 2018;10(34):28955–28962.

[1168] Sun L, Li Z, Su R, Wang Y, Li Z, Du B, et al. Phase-transition induced conversion into a photothermal material: Quasi-metallic $WO_{2.9}$ nanorods for solar water evaporation and anticancer photothermal therapy. Angewandte Chemie. 2018;130(33):10826–10831.

[1169] Tao P, Shu L, Zhang J, Lee C, Ye Q, Guo H, et al. Silicone oil-based solar-thermal fluids dispersed with PDMS-modified Fe_3O_4@ graphene hybrid nanoparticles. Progress in Natural Science: Materials International. 2018; 28(5):554–562.

[1170] Chen J, Zhou Y, Li R, Wang X, Chen GZ. Highly-dispersed nickel nanoparticles decorated titanium dioxide nanotube array for enhanced solar light absorption. Applied Surface Science. 2019;464:716–724.

[1171] Li K, Chang TH, Li Z, Yang H, Fu F, Li T, et al. Biomimetic MXene textures with enhanced light-to-heat conversion for solar steam generation and wearable thermal management. Advanced Energy Materials. 2019;9(34):1901687.

[1172] Zha XJ, Zhao X, Pu JH, Tang LS, Ke K, Bao RY, et al. Flexible anti-biofouling MXene/cellulose fibrous membrane for sustainable solar-driven water purification. ACS Applied Materials & Interfaces. 2019;11(40):36589–36597.

[1173] Liu PF, Miao L, Deng Z, Zhou J, Su H, Sun L, et al. A mimetic transpiration system for record high conversion efficiency in solar steam generator under one-sun. Materials Today Energy. 2018;8:166–173.

[1174] Zhang P, Liu F, Liao Q, Yao H, Geng H, Cheng H, et al. A microstructured graphene/poly (N-isopropylacrylamide) membrane for intelligent solar water evaporation. Angewandte Chemie International Edition. 2018;57(50): 16343–16347.

[1175] Tian L, Luan J, Liu KK, Jiang Q, Tadepalli S, Gupta MK, et al. Plasmonic biofoam: A versatile optically active material. Nano Letters. 2016;16(1): 609–616.

[1176] Zhang W, Zhu W, Shi S, Hu N, Suo Y, Wang J. Bioinspired foam with large 3D macropores for efficient solar steam generation. Journal of Materials Chemistry A. 2018;6(33):16220–16227.

[1177] Min M, Liu Y, Song C, Zhao D, Wang X, Qiao Y, et al. Photothermally enabled pyro-catalysis of a $BaTiO_3$ nanoparticle composite membrane at the liquid/air interface. ACS Applied Materials & Interfaces. 2018;10(25):21246–21253.

[1178] Tao F, Zhang Y, Wang B, Zhang F, Chang X, Fan R, et al. Graphite powder/semipermeable collodion membrane composite for water evaporation. Solar Energy Materials and Solar Cells. 2018;180:34–45.

[1179] Geng H, Xu Q, Wu M, Ma H, Zhang P, Gao T, et al. Plant leaves inspired sunlight-driven purifier for high-efficiency clean water production. Nature Communications. 2019;10(1):1–10.

[1180] Shi L, Wang X, Hu Y, He Y, Yan Y. Solar-thermal conversion and steam generation: A review. Applied Thermal Engineering. 2020;179:115691.

[1181] Wang X, He Y, Liu X, Cheng G, Zhu J. Solar steam generation through bio-inspired interface heating of broadband-absorbing plasmonic membranes. Applied Energy. 2017;195:414–425.

[1182] Higgins M, Rahmaan AS, Devarapalli RR, Shelke MV, Jha N. Carbon fabric based solar steam generation for waste water treatment. Solar Energy. 2018;159:800–810.

[1183] Wu L, Dong Z, Cai Z, Ganapathy T, Fang NX, Li C, et al. Highly efficient three-dimensional solar evaporator for high salinity desalination by localized crystallization. Nature Communications. 2020;11(1):1–12.

[1184] Inoue T, Koyama T, Kang DD, Ikeda K, Asano T, Noda S. One-chip near-field thermophotovoltaic device integrating a thin-film thermal emitter and photovoltaic cell. Nano Letters. 2019;19(6):3948–3952.

[1185] Hanamura K, Fukai H, Srinivasan E, Asano M, Masuhara T. Photovoltaic generation of electricity using near-field radiation. In: ASME/JSME Thermal Engineering Joint Conference. vol. 38921; 2011. p. T20066.

[1186] Hanamura K, Mori K. Nano-gap tpv generation of electricity through evanescent wave in near-field above emitter surface. In: AIP Conference Proceedings. vol. 890. American Institute of Physics; 2007. pp. 291–296.

[1187] Zhao B, Chen K, Buddhiraju S, Bhatt G, Lipson M, Fan S. High-performance near-field thermophotovoltaics for waste heat recovery. Nano Energy. 2017;41:344–350.

[1188] Wang LP, Zhang ZM. Thermal rectification enabled by near-field radiative heat transfer between intrinsic silicon and a dissimilar material. Nanoscale and Microscale Thermophysical Engineering. 2013;17(4):337–348.

[1189] Otey CR, Lau WT, Fan S. Thermal rectification through vacuum. Physical Review Letters. 2010;104(15):154301.

[1190] Ghanekar A, Ji J, Zheng Y. High-rectification near-field thermal diode using phase change periodic nanostructure. Applied Physics Letters. 2016; 109(12):123106.

[1191] Yu Z, Sergeant NP, Skauli T, Zhang G, Wang H, Fan S. Enhancing far-field thermal emission with thermal extraction. Nature Communications. 2013;4(1):1–7.

[1192] Simovski C, Maslovski S, Nefedov I, Kosulnikov S, Belov P, Tretyakov S. Hyperlens makes thermal emission strongly super-Planckian. Photonics and Nanostructures-Fundamentals and Applications. 2015;13:31–41.

[1193] Ilic O, Bermel P, Chen G, Joannopoulos JD, Celanovic I, Soljačić M. Tailoring high-temperature radiation and the resurrection of the incandescent source. Nature Nanotechnology. 2016;11(4):320–324.

[1194] Leroy A, Wilke K, Soljačić M, Wang EN, Bhatia B, Ilic O. High performance incandescent light bulb using a selective emitter and nanophotonic filters. SPIE; 2017.

Index

Printed in the United States
by Baker & Taylor Publisher Services